MW00838193

# Multibody Mechanics and Visualization

Harry Dankowicz

# Multibody Mechanics and Visualization

With 159 Figures

Harry J. Dankowicz, PhD
Engineering Science and Mechanics,
Virginia Polytechnic Institute and State University,
Blacksburg, Virginia 24061, USA

*Cover illustration:* Image created by the author using MAMBO.

British Library Cataloguing in Publication Data
Dankowicz, Harry
Multibody mechanics and visualization
1. Mambo (Computer file) 2. Mechanics, Applied - Computer simulation 3. Mechanics, Applied - Mathematical models
I. Title
610.1′0015118
ISBN 1852337990

Library of Congress Cataloging-in-Publication Data
A catalog record for this book is available from the Library of Congress



ISBN 1-85233-799-0 Springer London Berlin Heidelberg
Springer is a part of Springer Science+Business Media
springeronline.com


Printed in the United States of America



Typesetting: Camera-ready by author
69/3830-543210 Printed on acid-free paper SPIN 10966105

# Preface

Traditional entry-level mechanics courses serve two fundamentally different objectives. On the one hand, they present a self-contained progression of problem-solving paradigms addressing particular categories of engineering situations without any specific reference to higher-level thinking or the challenges of actual systems (for which the traditional methods typically fall short). They provide a necessary backdrop for the further professional development of an engineering-science or mechanical engineering student but, typically, do not generate much interest in other populations of engineering students, as evidenced, for example, by the lack of required fundamental mechanics courses for computer and electrical engineers in many colleges of engineering.

On the other hand, undergraduate instruction in the subject of classical mechanics constitutes a first attempt at incorporating the mathematics taught in the undergraduate linear-algebra and calculus sequences with real-world applications, developing ideas of physical and mathematical modeling, assessing the relevance of physical phenomena, the appreciation of modeling assumptions, and the formulation of scientific inquiry. These are skills that we expect of all engineering students but that typically are not strongly developed in existing curricula. There is a strong need for courses designed with the goal of bridging the gap between the stated objectives; courses that also attract non-traditional engineering students while ensuring a solid scientific and mathematical training.

To address these shortcomings, in collaboration with colleagues in the Department of Mechanics at the Royal Institute of Technology in Stockholm, Sweden, I recently developed a course that relies on the concept of problem-based learning to allow the student to accumulate theoretical knowledge, develop intuitive insight into, and perfect a practical know-how in the modeling and visualization of complex mechanical systems and their motions. Particular emphasis is placed on a framework that appeals to the educational background, interests, and perspectives of a modern engineering student. The problem-based approach encompasses an understanding of the theoretical concepts, the ability to implement this understanding in concrete applications, and the skill to disseminate the results of one's efforts in oral and written presentations.

The course is unique in its combination of content and form. It is designed to appeal to the interests of computer-savvy students who, in the process of producing attractive computer simulations and animations, acquire significant skills in mathematical and physical modeling of mechanical systems. In particular, the emphasis here is on general skills rather than the ability to solve cooked-up problems. Active-learning strategies and truly cooperative learning constitute an overwhelming part of the course design, the culmination of which is a team project incorporating material from throughout the course and accounting for a majority of the course grade.

## This Text

The instructional objectives for the course discussed above are to prepare the students to:

- Model the kinematics and dynamics of an arbitrary multibody mechanism;
- Formulate a mathematical description of a general motion of the mechanism in terms of sets of descriptive variables and systems of differential equations governing their evolution;
- Implement this description in a computer-graphics application for animating and visualizing a desired or observed motion of the mechanism.

In stark contrast to traditional mechanics courses, the act of analyzing a given set of differential equations to determine and predict the subsequent dynamics is entirely de-emphasized. Indeed, I strongly believe that such analysis should be the subject of a separate, subsequent course coupled with issues of design of mechanical systems for achieving desired behavior and so on. Eliminating such discussions from the present course enables a clarity of presentation, thought, and message, and increases the likelihood that the students firmly establish the mathematical background necessary to proceed with such analysis as compared to traditional courses, where the material is closely interwoven.

The text you have in your hands is the result of several iterations of development of the educational material for this course. Four main pedagogical principles form the foundation for the current edition, namely:

- An inductive approach to learning, whereby general patterns are discerned from observations made in particular instances;
- A need for repetition and review of important concepts and their reinforcement through numerous examples;

- Visual guidance to allow the reader to differentiate between different levels of knowledge;
- Deep incorporation of computer tools, visual representations, and elements of active learning to appeal to a broad spectrum of learning strategies and preferences.

The primary goal in composing this text has been to provide an extensive resource that presents a self-contained and careful exposition of all relevant topics for the sequential reader while containing enough repetition and examples to allow numerous points of entry.

## The MAMBO Toolbox

Parallel to the theoretical presentation, the book contains a track implementing a series of computer-algebra procedures for enabling advanced computations on complex multibody mechanisms. This package – the MAMBO toolbox – bears a general resemblance to a collection of procedures developed by Professor Martin Lesser and Dr Anders Lennartsson in the Department of Mechanics at the Royal Institute of Technology in Stockholm, Sweden, between 1991 and 1999, and named SOPHIA after the Polish-Swedish mathematician Sofja Kowalewskaja (1850–1891). Sufficient changes have been made, however, in all parts of the implementation, to warrant a new name for the software. Nevertheless, I gratefully acknowledge the intellectual heritage from the original package and the efforts of its originators.

This text presents version 1.0 of the MAMBO toolbox. To use the MAMBO toolbox on your computer, download the necessary files from the web site:

www.esm.vt.edu/~danko/Mambo

## MAMBO

The computer-graphics application MAMBO described in this text has been developed with the purpose of allowing the student to visualize the results of their efforts while retaining the need for careful mathematical analysis. In contrast with existing commercially available educational software tools, MAMBO requires detailed input from the user both in order to define the specific geometry of the mechanism as well as the differential equations governing its behavior. With this tool, the student is able to see the implications of decisions made throughout the modeling stage and to check the mathematical analysis.

The following individuals have been involved with the development and coding of MAMBO: Jesper Adolfsson, Kalle Andersson, Arne Nordmark, Gabriel Ortiz, Anders Lennartsson, Petri Piiroinen, Justin Hutchison, and myself. Since the program is continually developing, I have omitted any detailed description of its implementation in this text and instead refer to the MAMBO reference manual.

To use MAMBO on your computer, download the necessary files from the web site:

www.esm.vt.edu/~danko/Mambo

## How This Text is Organized

- Visual cues have been included in the margin to distinguish between different levels of importance of material as illustrated by the following table:

| Symbol | Meaning |
|---|---|
| (hand with pen) | Important terminology |
| 1 | Material which can be skipped upon first reading |
| 2 | Optional material for further study |

- The eleven chapters can be separated into three categories, based on their emphasis on theory, applications, or general introduction and review as illustrated in the following table:

| Applications | Overview | Theory |
|---|---|---|
| | Chapter 1 | |
| Chapter 2 | | |
| | | Chapter 3 |
| Chapter 4 | | |
| | | Chapter 5 |
| Chapter 6 | | |
| | Chapter 7 | |
| Chapter 9 | | Chapter 8 |
| | Chapter 10 | |
| | Chapter 11 | |

- The following tables provide relevant page references for different categories of material:



- Each chapter is concluded with a summary of notation and terminology;
- A collection of animation and modeling projects suitable for semester-long team assignments is included in Appendix C.

# Acknowledgments

The current edition of this text is the result of a sincere effort on my behalf to address the concerns, comments, suggestions, and complaints expressed to me and to my co-instructors since the inception of the course and the first iterations of the course text. Through lengthy and critical reflection upon the issues raised in conversations with students and through surveys and evaluations, I have attempted to formulate a product that better serves the needs of the students without compromising its intellectual integrity or the ultimate purposes of the course. I wish to express my thanks to those anonymous masses of students who have encouraged the further development of this course and this text.

During my years at the Royal Institute of Technology in Stockholm, Sweden, I had many stimulating discussions with various colleagues, notably my friend Professor Martin Lesser, and with a number of graduate students, on the teaching of mechanics. These discussions sharpened my senses and brought out the essence of the topic. My appreciation to these named and unnamed individuals for their patience and willingness to engage in these conversations.

I am further grateful for the help given to me by Dr Petri Piiroinen and Dr Arne Nordmark at the Royal Institute of Technology in proofreading earlier versions of the manuscript. Also, special kudos to Arne Nordmark for his efforts on an earlier version of the MAMBO toolbox.

Finally, with much love to my wife and children, my gratitude for your patience, support, and presence.

Blacksburg, Virginia, USA
November 2003

Harry Dankowicz

# Table of Contents

# Chapter 1

# A First Look

*wherein the reader learns of:*

- *Fundamental ways of representing the motion of rigid bodies;*
- *Describing the configuration of a rigid body in terms of a position and an orientation;*
- *Pure translations and pure rotations and their properties;*
- *Using coordinates to uniquely determine the configuration of a rigid body;*
- *Constraints on the coordinates;*
- *Introducing collections of observers.*

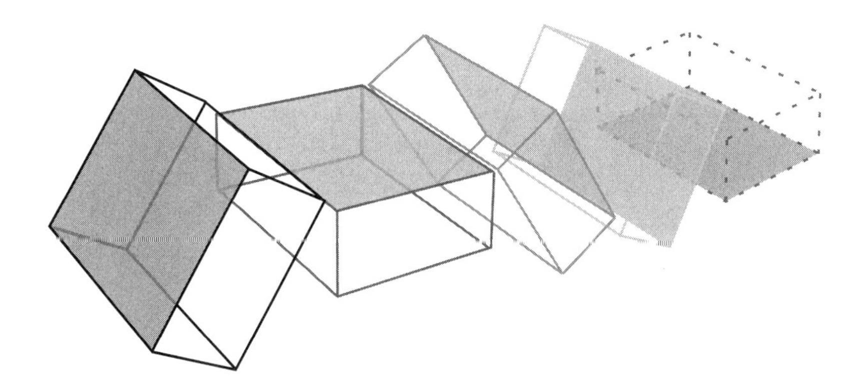

### Practicum

You may read this chapter any number of times and feel quite comfortable with its propositions and arguments. But true understanding is a combination of intuition and experience. This experience comes from engaging in practical, hands-on experimentation with concrete, physical objects.

Just about anything in your immediate environment will probably qualify as a block or a rigid body. Hold the object with both hands and follow along with the discussion in this chapter by moving and rotating the object as suggested. Most certainly, this will enhance your three-dimensional experience of the graphics in this chapter. It will strengthen your geometric intuition. It will be excellent practice for the things to come.

# 1.1 A First Look at Motion

(Ex. 1.1 – Ex. 1.14)

## 1.1.1 Reference Configurations

A rectangular block moves across your visual field. At each moment in time, you describe the block's *configuration* – the spatial arrangement of all its points – by its *position* and *orientation* relative to some *reference position* and *reference orientation*, constituting a *reference configuration*.

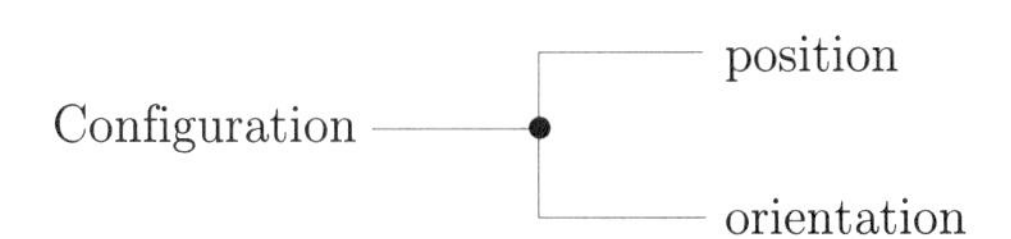


Perhaps you envision the reference configuration as a stationary virtual block whose dimensions agree with those of the actual block.

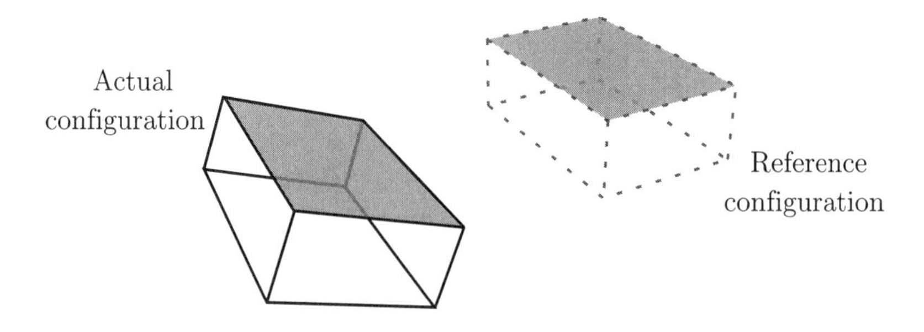


The reference configuration enables you to make meaningful statements about the geometry of space. The locations of points in space are made clear by referring to the reference configuration.

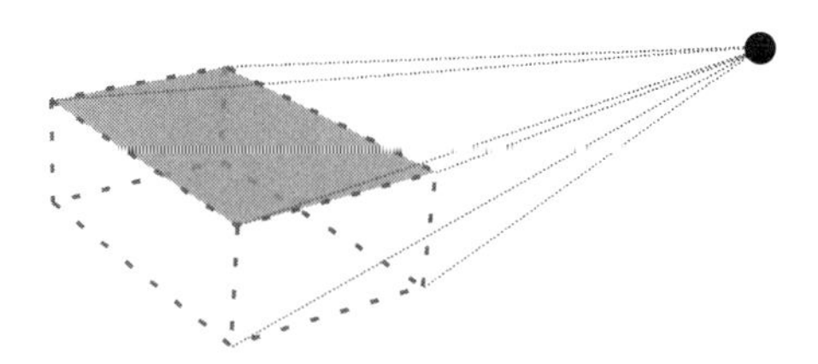

The orientations of straight lines in space are made clear by referring to the reference configuration.

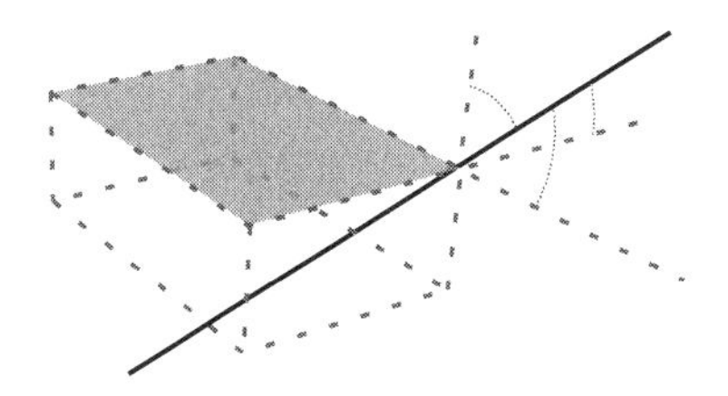

Having established a reference configuration, the notion of the **block's configuration** makes intuitive sense for arbitrary positions and orientations.

The reference configuration also enables you to make meaningful statements about the time-dependence of the geometry of space. For example, a point is said to be *fixed* or *stationary* relative to the reference configuration if its relation to the faces and edges of the virtual block does not change with time.

Having established a reference configuration, it makes intuitive sense to describe the **block's motion** in terms of the time-dependence of its configuration relative to the reference configuration.

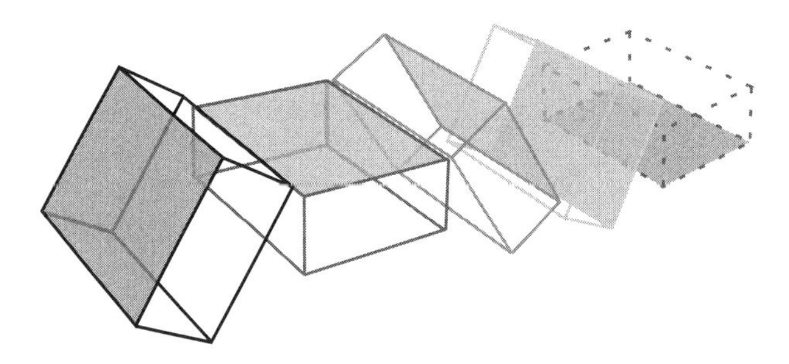

### 1.1.2 Pure Translations and Rotations

As the position and orientation of the block change with time, the block exhibits a motion through space that involves *pure translation*, *pure rotation*, or a combination thereof.

#### Pure translation

A motion of the block that results in a change in the **block's position**, but involves no change in the block's orientation, is called a pure translation. In a pure translation, all points in the block are shifted by equal amounts along parallel paths relative to the reference configuration.

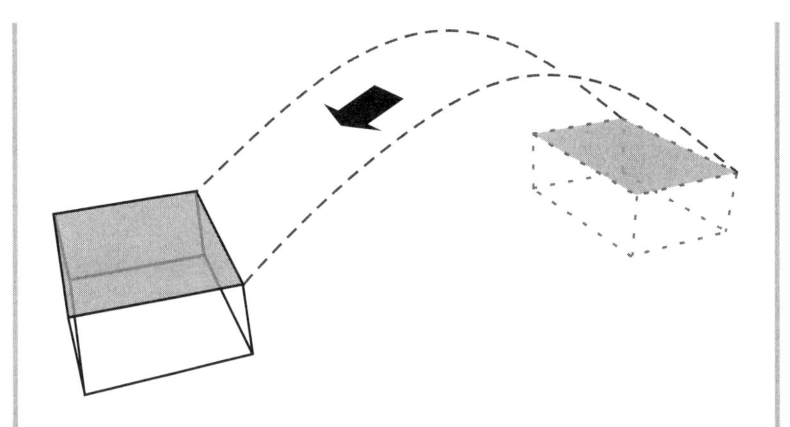

Two configurations of the block are said to be *related through a pure translation* if there exists a pure translation that brings the block from the first to the second configuration. It should follow that two configurations of the block that are related through a pure translation have the same orientation.

Two pure translations that result in the same final configuration when applied to the block in an initial configuration are said to be *equivalent*. Equivalent pure translations result in the same net change of position while involving no change in orientation.

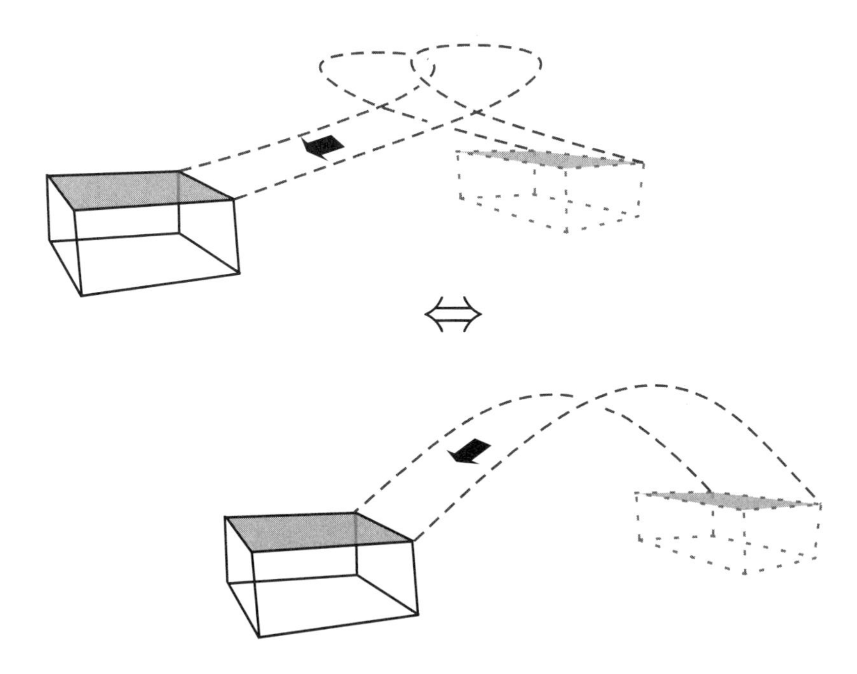

**Illustration 1.1**

Suppose that the initial and final configurations of a block are related through a pure translation. Let $A$ and $B$ be two arbitrary points on the block and denote by $A_{\mathrm{initial}}$, $B_{\mathrm{initial}}$ and $A_{\mathrm{final}}$, $B_{\mathrm{final}}$ the corresponding points in space in the initial and final configurations, respectively.

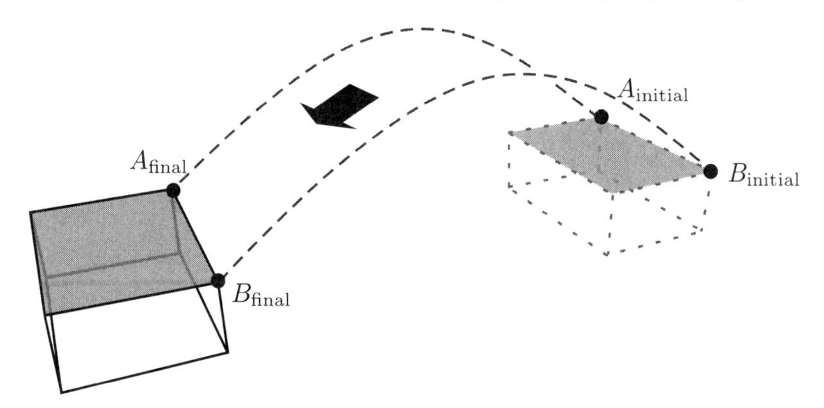


Then, the straight-line segment between $A_{\mathrm{initial}}$ and $A_{\mathrm{final}}$ is parallel to and of equal length as the straight-line segment between $B_{\mathrm{initial}}$ and $B_{\mathrm{final}}$.

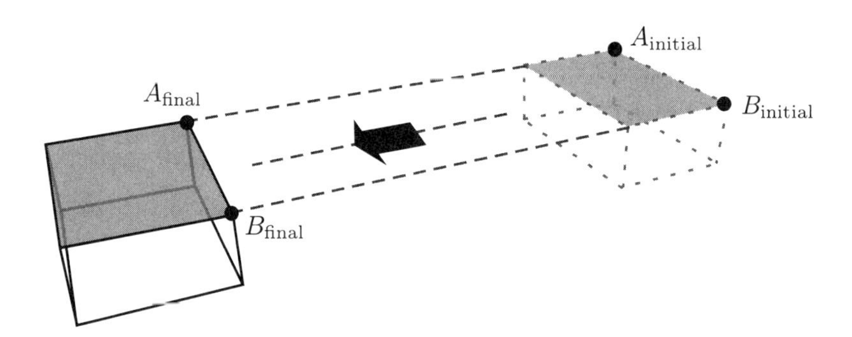


As this observation holds for arbitrary pairs of points, it follows that the initial and final configurations of the block are related through a pure translation that shifts all points in the block by an equal amount along a common fixed direction relative to the reference configuration.

The result of the illustration shows that for every pure translation there is an equivalent pure translation that shifts all points in the block by an equal amount along a common fixed direction relative to the reference configuration.

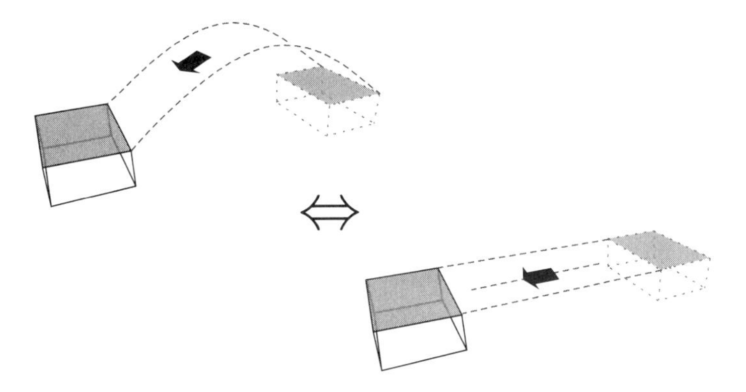

The final configuration that results from the application of any one member of a family of equivalent pure translations to a block in an initial configuration is identical to that which results from the application of any other member of this family. We, therefore, often choose to refer to the whole family collectively by the equivalent pure translation that shifts all points in the block along a common fixed direction from their initial to their final locations in space. Here, we are more concerned with the relative configuration of the initial and final configurations than with the path by which one was brought to the other.

As pure translations preserve the orientation of the block while only affecting its position, it is reasonable to expect that successive compositions of pure translations result in no net change in orientation.

**Illustration 1.2**

Consider the final configuration that results from a shift of the block by one unit of length along a given direction and, subsequently, by two units of length along a different direction as shown in the figure below.

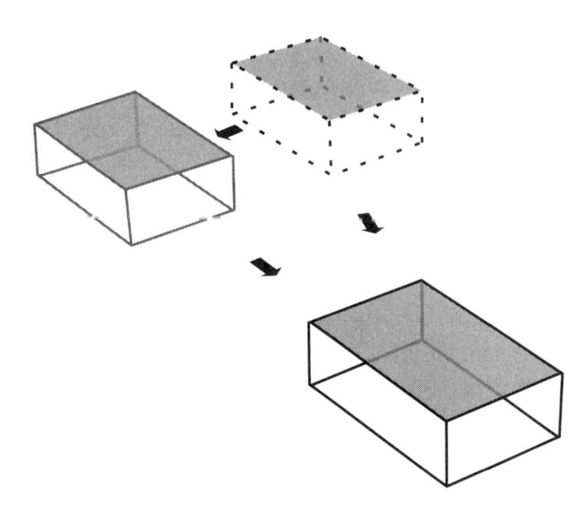

All the points on the block in the final configuration are shifted from their original positions by the same amount and along a common direction. The final configuration is thus related to the initial configuration by a single pure translation.

In general, a pure translation of the block from an initial configuration to some intermediate configuration, followed by a second pure translation to a final configuration, is equivalent to a single pure translation of the block from the initial configuration to the final configuration. This is consistent with the notion that configurations related by pure translations have the same orientation. Indeed, as is suggested by the above observation, the composition of pure translations, each of which preserves the block's orientation, results in no net change in orientation. This supports describing the position of the block relative to the reference configuration in terms of the pure translation that relates the block's configuration to the reference configuration, and vice versa.

**Illustration 1.3**

As shown in the figure below, the final configuration that results from a shift of the block by one unit of length along a given direction and, subsequently, by two units of length along a different direction could also have been achieved by switching the order of the shifts.

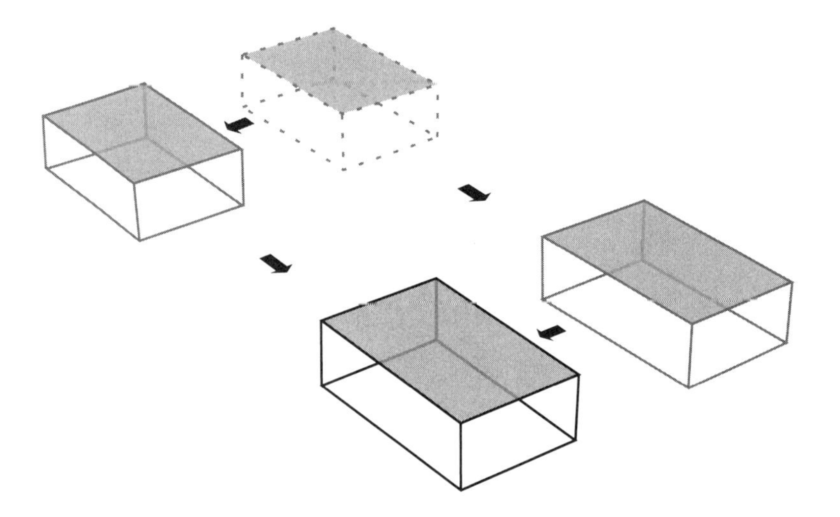

In general, the order in which two pure translations are effected is immaterial to the final configuration of the block. We say that pure translations *commute under composition*.

### Pure rotation

A motion of the block that results in a change in the **block's orientation**, but involves no change in the block's position, is called a pure rotation. In a pure rotation, one point in the block remains fixed relative to the reference configuration.

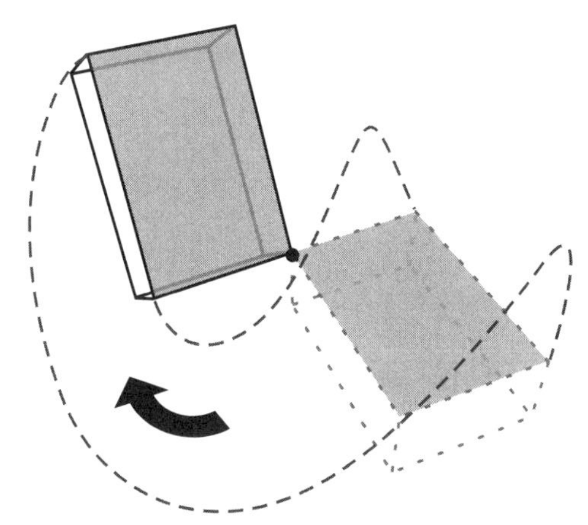

Two configurations of the block are said to be *related through a pure rotation* if there exists a pure rotation that brings the block from the first to the second configuration. It should follow that two configurations of the block that are related through a pure rotation have the same position.

Two pure rotations that result in the same final configuration when applied to the block in an initial configuration are said to be *equivalent*. Equivalent pure rotations result in the same net change of orientation while involving no change in position. The result of Exercises 1.8 and 1.9 shows that for every pure rotation there is an equivalent pure rotation that rotates the block by a given amount about a fixed axis relative to the reference configuration.

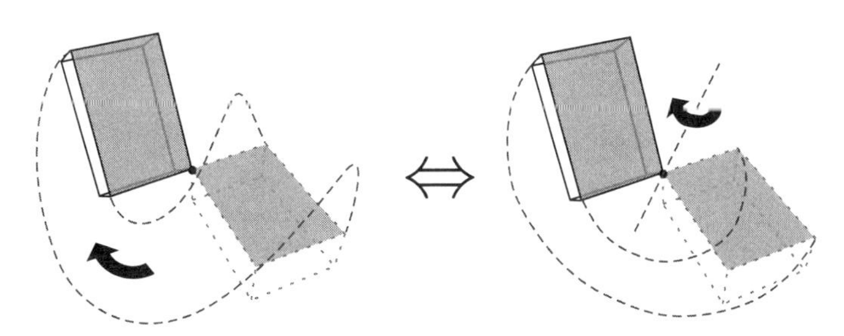

The final configuration that results from the application of any one member of a family of equivalent pure rotations to a block in an initial configuration is identical to that which results from the application of any other member of this family. We, therefore, often choose to refer to the whole family collectively by the equivalent pure rotation that rotates the block about a fixed axis. Here, we are more concerned with the relative configuration of the initial and final configurations than with the path by which one was brought to the other.

As pure rotations preserve the position of the block while only affecting its orientation, it is reasonable to expect that successive compositions of pure rotations result in no net change in position.

**Illustration 1.4**

Consider the final configuration that results from a rotation of the block by a quarter turn about a given direction (keeping one of the corners fixed) and, subsequently, by a quarter turn about the same direction (keeping a different corner fixed) as shown in the figure.

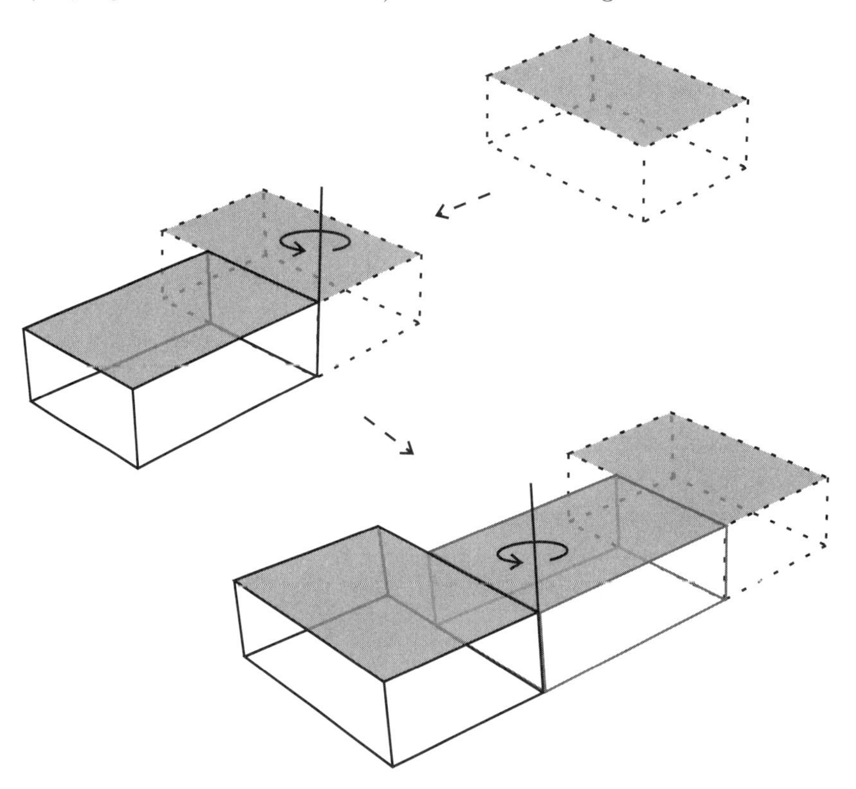

Then, no point of the block in the final configuration coincides with the corresponding point in the initial configuration. The final configuration is thus **not** related to the initial configuration by a single pure rotation.

In contrast to the case of pure translations, a pure rotation of the block from an initial configuration to some intermediate configuration, followed by a second pure rotation to a final configuration, is, in general, **not** equivalent to a single pure rotation of the block from the initial configuration to the final configuration. Although there is no change in the block's position during the two pure rotations, the initial and final configurations do **not** have the same position. It appears that the act of switching the point to be kept fixed by subsequent pure rotations puts the association between pure rotations and unchanging position in jeopardy.

> The final configuration is related to the initial configuration through a pure rotation **if and only if** at least one point in the block in the final configuration coincides with the corresponding point in the initial configuration. This outcome is guaranteed if we require that all pure rotations keep the same point fixed relative to the reference configuration.

With this added condition, a pure rotation of the block from an initial configuration to some intermediate configuration, followed by a second pure rotation to a final configuration, is equivalent to a single pure rotation of the block from the initial configuration to the final configuration. This is consistent with the notion that configurations related by pure rotations have the same position. Indeed, under these conditions, the composition of pure rotations, each of which preserves the block's position, involves no net change in position.

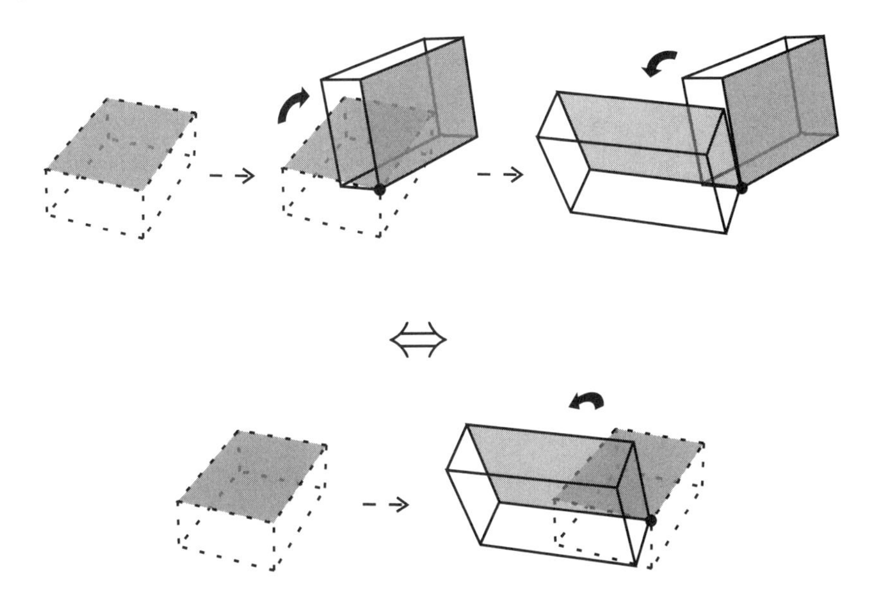

As was the case with pure translations, this supports describing the orientation of the block relative to the reference configuration in terms of the pure rotation that relates the block's configuration to the reference configuration, and vice versa, **provided that the pure rotation always keeps the same point on the block fixed relative to the reference configuration.**

**Illustration 1.5**

As shown in the figure, the final configuration that results from a quarter turn of the block about a given direction (keeping one of the corners fixed) and, subsequently, by a quarter turn about a different direction (keeping the same corner fixed) differs substantially from that achieved by switching thc order of the turns.

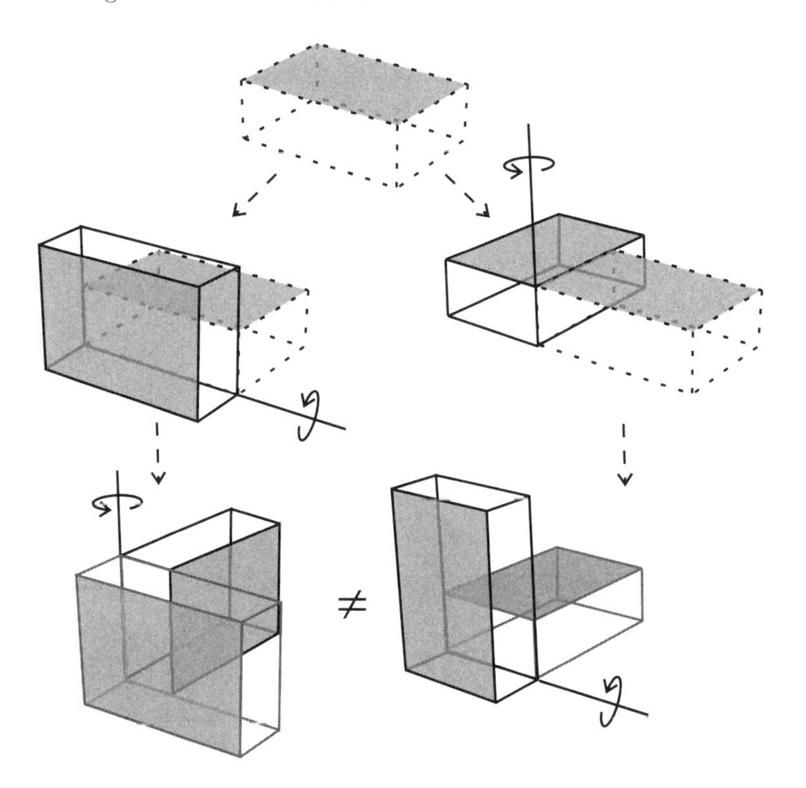


Contrary to the case of pure translations, the order in which two pure rotations are effected is, in general, crucial to the final configuration of the block. We say that pure rotations **do not** commute under composition.

A *non-trivial* pure translation, i.e., one for which the net shift is non-zero, cannot be a pure rotation, since the latter requires one point to be fixed in space. Similarly, a *non-trivial* pure rotation, i.e., one for which the net turning angle is non-zero, cannot be a pure translation, since the latter requires that all points shift by an equal amount along parallel paths. The collections of all pure translations and all pure rotations have only one equivalent element in common, namely the special case of zero net shift and zero net turning angle corresponding to the absence of motion.

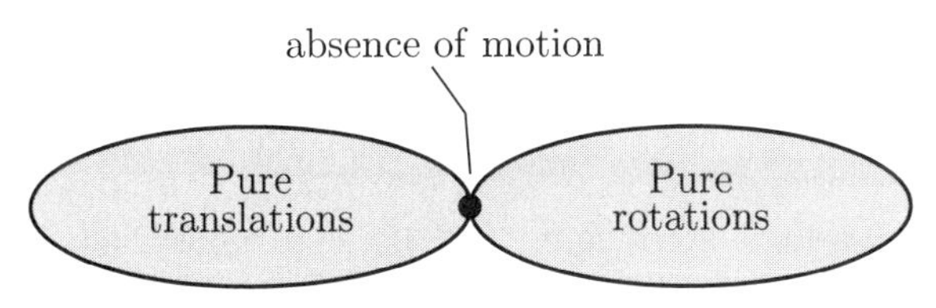


## 1.1.3 Instantaneous Motion

### Pure translations

In a pure translation, all points are shifted by an equal amount along parallel paths relative to the reference configuration. The previous discussion showed that a pure translation is equivalent to a shift by a specific amount along a fixed direction relative to the reference configuration. The amount and direction of the shift associated with a pure translation from the reference configuration to the final configuration **depend** on the choice of reference configuration.

Assume that the reference configuration coincides with the configuration of the block at some instant in time $t$ and that the configuration of the block at time $t + \Delta t$ is related to the reference configuration through a pure translation for all sufficiently small $\Delta t$. For each $\Delta t$, the corresponding pure translation is equivalently described by a direction of translation and a shifting distance. Clearly, the shifting distance goes to zero as $\Delta t$ becomes arbitrarily small.

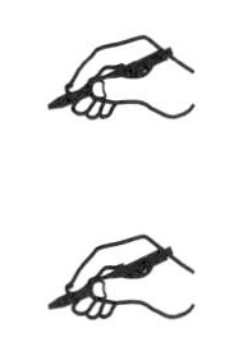

If the direction of translation limits on some specific direction as $\Delta t$ goes to zero, the limiting direction is called the *instantaneous direction of translation* of the block relative to the reference configuration. If the shifting distance divided by $\Delta t$ limits on some specific value as $\Delta t$ goes to zero, the limiting value is called the *linear speed* of the block relative to the reference configuration.

### Pure rotations

In a pure rotation, one point remains fixed relative to the reference configuration. The previous discussion showed that a pure rotation corresponds to a rotation by a specific amount about a fixed axis relative to the reference configuration. The axis of rotation and the amount of rotation associated with a pure rotation from the reference configuration to the final configuration **depend** on the choice of reference configuration.

Assume that the reference configuration coincides with the configuration of the block at some instant in time $t$ and that the configuration of the block at time $t + \Delta t$ is related to the reference configuration through a pure rotation keeping the same point fixed for all sufficiently small $\Delta t$. For each $\Delta t$, the corresponding pure rotation is equivalently described by an axis of rotation and a turning angle. Clearly, the turning angle goes to zero as $\Delta t$ becomes arbitrarily small.

If the axis of rotation limits on some specific axis as $\Delta t$ goes to zero, the limiting axis is called the *instantaneous axis of rotation* of the block relative to the reference configuration. If the turning angle divided by $\Delta t$ limits on some specific value as $\Delta t$ goes to zero, the limiting value is called the *angular speed*[1] of the block relative to the reference configuration.

### 1.1.4 Curved Space

The assertion that a sequence of pure translations is equivalent to a single pure translation is actually not quite as obvious as might appear from the discussion following Illustration 1.2. It certainly agrees with our general impression of the geometry of the space we live in, but it is quite possible to conceive of spaces with different inherent geometries in which the assertion is false.

As an example, consider motions constrained to the surface of a sphere.

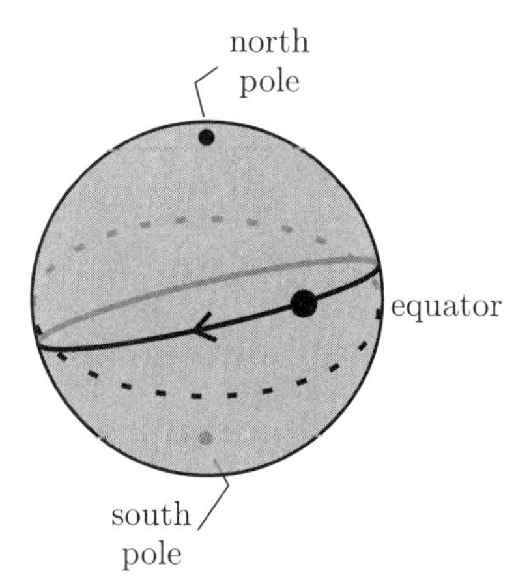


For reference, identify two diametrically opposite points on the sphere's surface as the *north* and *south poles* of the sphere and let the circle located halfway between the north and south poles be called the sphere's

[1] A more common terminology for this quantity is *angular velocity*. The term velocity, however, is typically intended to refer to a quantity that has magnitude as well as direction. The magnitude of a velocity is called the *speed*. The terminology used here is consistent with this usage and agrees with that used for the case of pure translations.

*equator*. A great circle through a point on this surface is characterized by its tangent direction at the point and a positive direction of travel along the circle. We can represent the great circle by an arrow based at the point, tangential to the circle and pointing in the positive direction of travel along the circle. The angle between two great circles intersecting at a point is then defined as the angle between the two corresponding arrows.

Think of the configuration of a "block" in this two-dimensional world as a point on the sphere's surface and a great circle through this point. Consider two configurations of the block in the two-dimensional world and let

$$(A_{\text{reference}}, \Psi_{\text{reference}}) \text{ and } (A_{\text{final}}, \Psi_{\text{final}})$$

represent the corresponding pairs of a point and a great circle, respectively. Denote by $\Psi_{\text{reference}\to\text{final}}$ the great circle through the two points $A_{\text{reference}}$ and $A_{\text{final}}$, such that the positive direction of travel agrees with the shortest path from $A_{\text{reference}}$ to $A_{\text{final}}$. The two configurations are related by a pure translation if the angle

$$\measuredangle\left(\Psi_{\text{reference}\to\text{final}}, \Psi_{\text{reference}}\right)$$

between $\Psi_{\text{reference}\to\text{final}}$ and $\Psi_{\text{reference}}$ equals the angle

$$\measuredangle\left(\Psi_{\text{reference}\to\text{final}}, \Psi_{\text{final}}\right)$$

between $\Psi_{\text{reference}\to\text{final}}$ and $\Psi_{\text{final}}$.

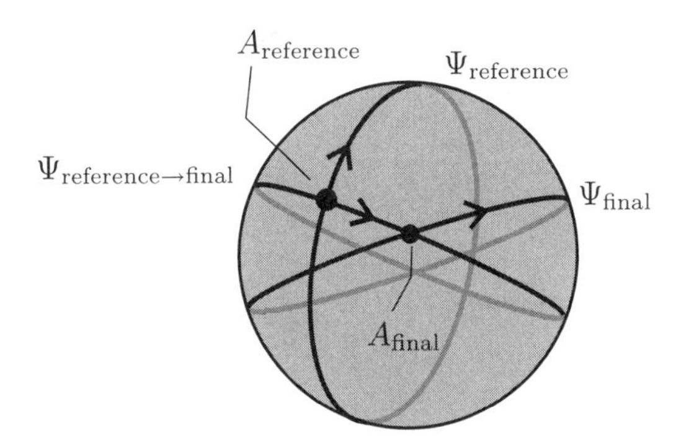


**Illustration 1.6**
Let the initial configuration of the block

$$(A_{\text{initial}}, \Psi_{\text{initial}})$$

be given by a point $A_{\text{initial}}$ on the sphere's equator and the great circle $\Psi_{\text{initial}}$ through $A_{\text{initial}}$ and the north pole, such that the corresponding arrow at $A_{\text{initial}}$ points toward the north pole.

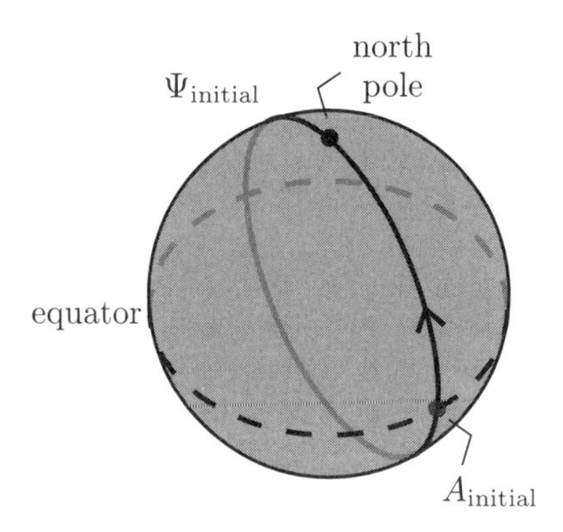


Let the intermediate configuration of the block

$$(A_{\text{intermediate}}, \Psi_{\text{intermediate}})$$

be given by a point $A_{\text{intermediate}}$ on the equator a quarter of the circumference from the initial point; and the great circle $\Psi_{\text{intermediate}}$ through $A_{\text{intermediate}}$ and the north pole, such that the corresponding arrow at $A_{\text{intermediate}}$ points toward the north pole. Then, $\Psi_{\text{initial}\to\text{intermediate}}$ coincides with the equator with direction of travel from $A_{\text{initial}}$ to $A_{\text{intermediate}}$. Since

$$\measuredangle\,(\Psi_{\text{initial}\to\text{intermediate}}, \Psi_{\text{initial}}) = \measuredangle\,(\Psi_{\text{initial}\to\text{intermediate}}, \Psi_{\text{intermediate}}) = 90^\circ,$$

the intermediate configuration is related to the initial configuration by a pure translation.

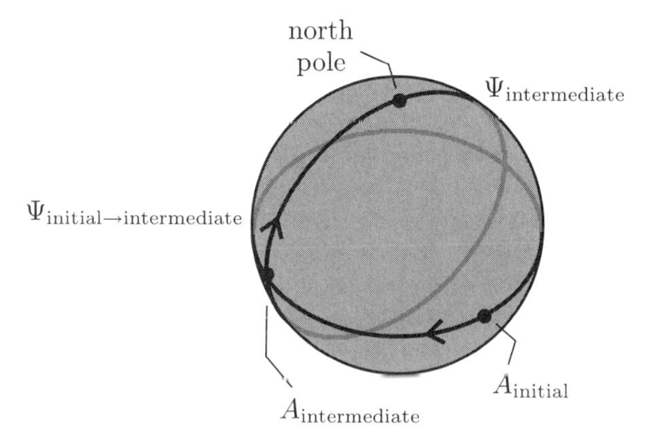


Now, let the final configuration of the block

$$(A_{\text{final}}, \Psi_{\text{final}})$$

be given by the point $A_{\text{final}}$ at the north pole and the great circle $\Psi_{\text{final}} = \Psi_{\text{intermediate}}$, i.e., such that the corresponding arrow at $A_{\text{final}}$ points away from $A_{\text{intermediate}}$. Let $\Psi_{\text{intermediate}\to\text{final}}$ equal $\Psi_{\text{intermediate}}$. Then, since

$$\begin{aligned}&\measuredangle\left(\Psi_{\text{intermediate}\to\text{final}}, \Psi_{\text{intermediate}}\right)\\ &\qquad = \measuredangle\left(\Psi_{\text{intermediate}\to\text{final}}, \Psi_{\text{final}}\right) = 0^\circ,\end{aligned}$$

the final configuration is related to the intermediate configuration by a pure translation.

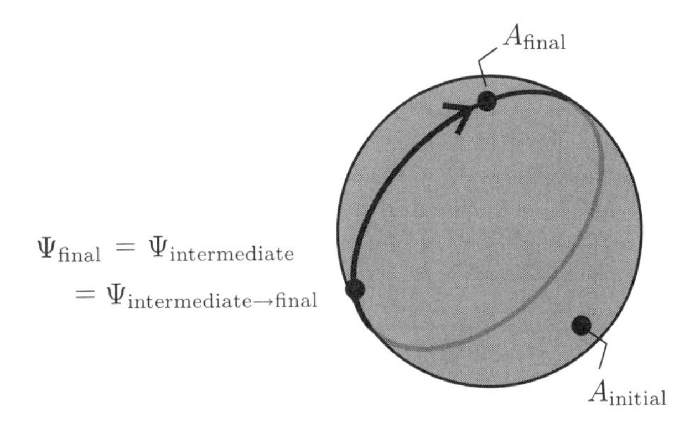


The two pure translations result in the block positioned at the north pole with an orientation given by a great circle $\Psi_{\text{final}} \neq \Psi_{\text{initial}}$. But, since the great circle $\Psi_{\text{initial}\to\text{final}} = \Psi_{\text{initial}}$,

$$0^\circ = \measuredangle\left(\Psi_{\text{initial}\to\text{final}}, \Psi_{\text{initial}}\right) \neq \measuredangle\left(\Psi_{\text{initial}\to\text{final}}, \Psi_{\text{final}}\right) = 90^\circ.$$

The final configuration is therefore **not** related to the initial configuration through a pure translation. Instead, the operation equivalent to the combined effect of the two pure translations is a pure translation followed by a pure rotation, in stark contrast to the claims made in a previous section.

The geometry of the spherical surface is that of a *curved space*. In contrast, in a *flat space*, arbitrary combinations of pure translations are equivalent to a single pure translation. Our everyday experience certainly suggests that our space is flat. But it is possible to show that arbitrary combinations of sufficiently small pure translations in a curved space may be closely approximated by a single pure translation. Indeed, this approximation becomes increasingly accurate as the amount of shift of the pure translations decreases. This observation expresses the fact that a curved space is *locally flat*. In the case of the sphere, a small patch of the sphere's surface centered on some point is closely approximated by the plane tangent to the sphere's surface at that point. It is, thus, quite conceivable that our experience of the flatness of our space is born from

observations only on very small motions relative to the length scales over which curvature plays a role[2].

### 1.1.5 Combinations of Translations and Rotations

Two motions that result in the same final configuration when applied to a block in an initial configuration are said to be *equivalent*. Equivalent motions result in the same net change of position and orientation. For example, the final configuration of the block that results from a pure translation to an intermediate configuration followed by a pure rotation is identical to that obtained by switching the order of the operations, provided that the same point on the block is kept fixed by the pure rotations as shown in the figure.

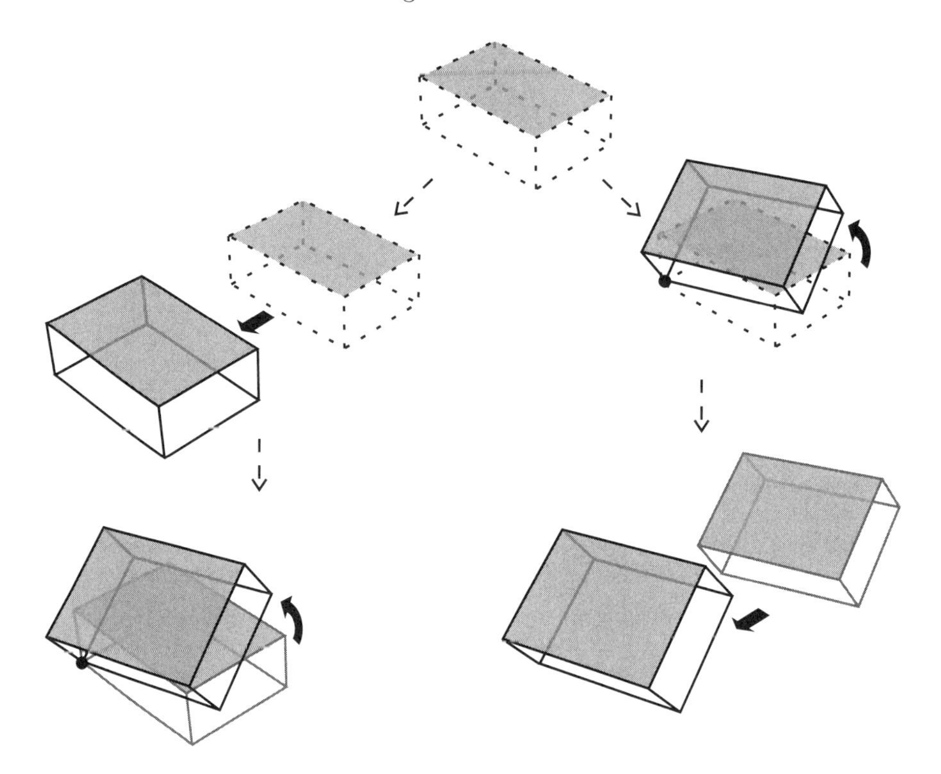

In the first case, all points of the block are shifted by an equal amount along a common direction; and the block is subsequently rotated while keeping one point in the block fixed relative to the reference configuration. In the second case, the block is first rotated while keeping the same point in the block fixed relative to the reference configuration; and all points of

[2]It is more than just conceivable; it is a fact, as suggested by Einstein's General Theory of Relativity.

the block are subsequently shifted by an equal amount along a common direction.

**Illustration 1.7**

Consider a sequence of pure translations and pure rotations, where all the pure rotations keep the same point on the block fixed relative to the reference configuration as suggested in the figure.

By the above observation, the order of pure translations and pure rotations may be switched, so as to collect all translations at the beginning of the sequence and all rotations at the end of the sequence.

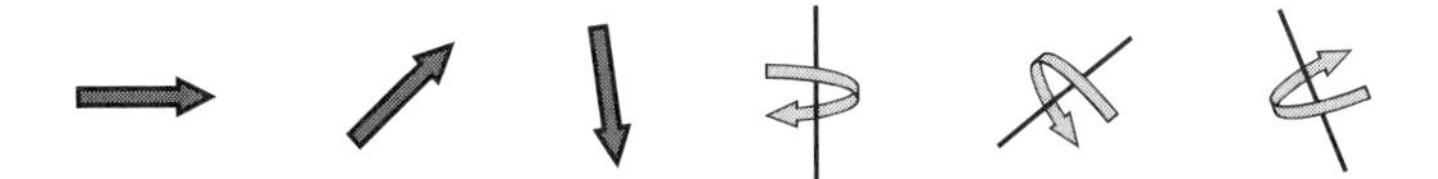

From the discussion of pure translations, we conclude that the pure translations may be combined into a single pure translation. Similarly, since the pure rotations all keep the same point fixed, they, too, may be combined into a single pure rotation.

### 1.1.6 Decompositions of Configurations

An arbitrary configuration of the block can be thought of as the result of a pure translation from the reference configuration to an intermediate configuration followed by a pure rotation. In particular, let the pure translation be such that one point on the block in the intermediate configuration coincides with the corresponding point in the final configuration. This will be the point kept fixed relative to the reference configuration by the subsequent pure rotation.

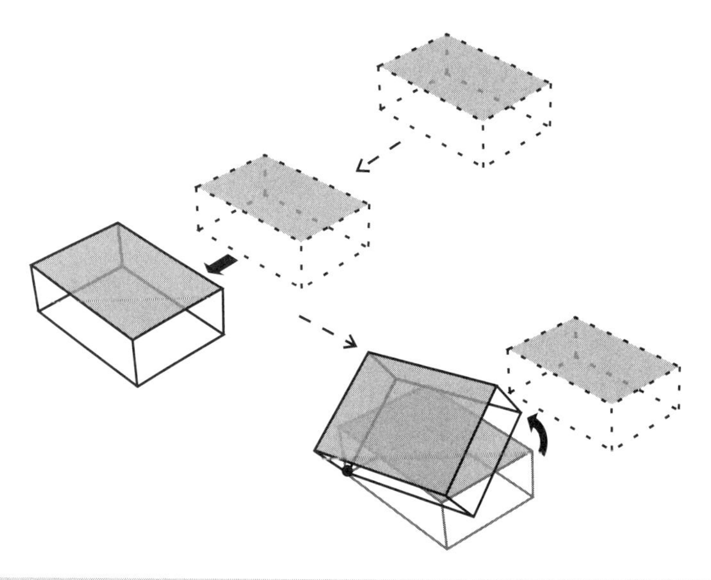

**Illustration 1.8**
To bring the block from the reference configuration to the final configuration, apply a pure translation so that one corner of the block coincides with the corresponding corner in the final configuration, as shown in the left path of the figure. Then, apply a pure rotation to line up the block with the final configuration while keeping this corner fixed.

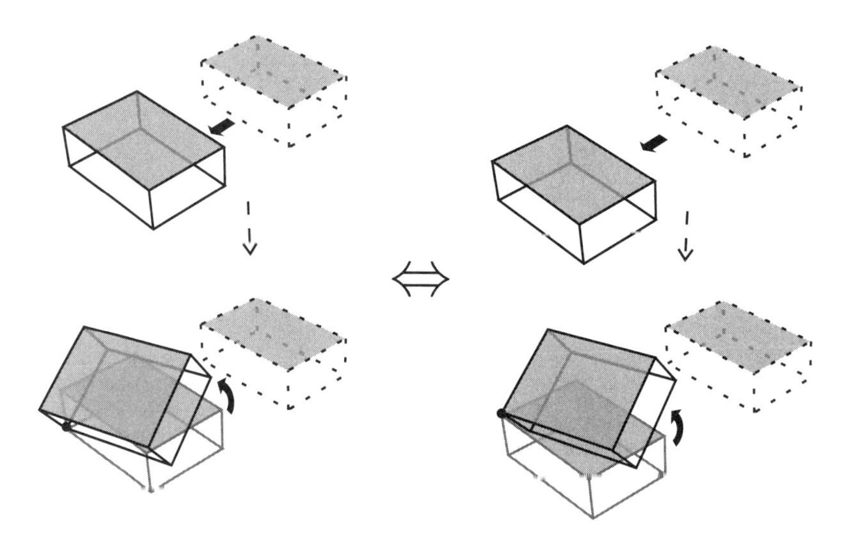

Alternatively, apply a different pure translation to the block in the reference configuration so that a different corner of the block coincides with the corresponding corner in the final configuration as shown in the right path of the figure. Then, apply a different pure rotation to line up the block with the final configuration while keeping this corner fixed.

There is no unique way to decompose an arbitrary configuration of the block relative to the reference configuration into a combination of a pure translation and a pure rotation. That there are multiple (actually, infinitely many) ways of doing this follows from the freedom to choose the point in the intermediate configuration that will coincide with the corresponding point in the final configuration. Once the point about which the pure rotation will take place has been selected, however, both **the pure translation and the pure rotation are uniquely determined** (at least within equivalence).

> The unique pure translation and pure rotation that relate the actual configuration to the reference configuration – given the selection of the point kept fixed by the rotation – provide the clearest description so far of the position and orientation, respectively, of the block relative to the reference configuration. That the order in which the pure translation and the pure rotation are applied is immaterial to reaching the actual configuration implies that the position of the block may be described independently from its orientation, and vice versa.
>
> Position $\leftrightarrow$ Translation
> Orientation $\leftrightarrow$ Rotation

## 1.2 A First Look at Degrees of Freedom

(Ex. 1.15 – Ex. 1.17)

### 1.2.1 Position

**Illustration 1.9**

Assume that one corner of the block in the final configuration is located one unit of length from the corresponding point in the initial configuration along a direction parallel to one of the edges of the block, as shown in the figure on the next page.

The final configuration of the block is then related to the initial configuration by a pure translation shifting all the points of the block by one unit of length along the same direction; and a subsequent pure rotation keeping the corner point fixed.

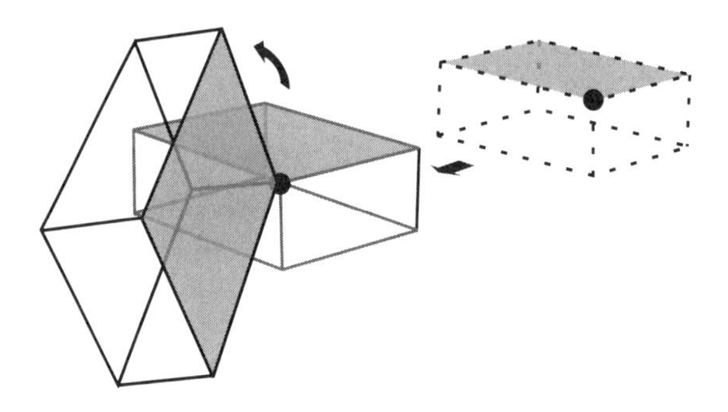

Consider a block in some arbitrary configuration and denote by $A$ the point in the block that has been selected to be kept fixed by the pure rotation in the decomposition discussed in the previous section. Then $A_{\text{reference}}$ and $A_{\text{final}}$ are the points in space that coincide with $A$ when the block is in the reference configuration and final configuration, respectively.

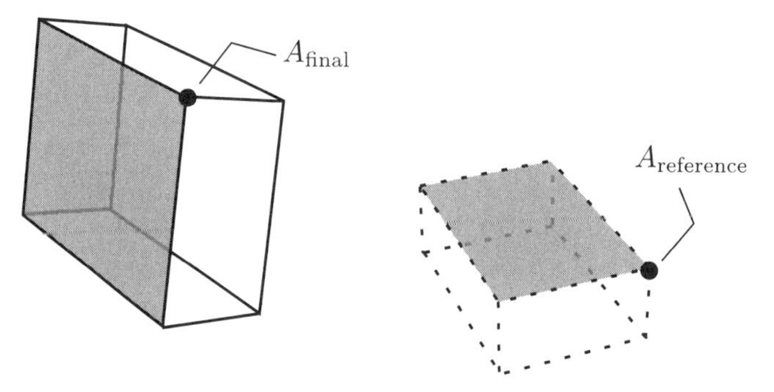


The pure translation that brings the block from the reference configuration to the intermediate configuration is uniquely determined by the location of the point $A_{\text{final}}$ relative to the reference configuration.

### 1.2.2 Orientation

**Illustration 1.10**

Apply a pure rotation to a block, keeping one of the corners fixed as shown in the figure on the next page. Once the location of two of the other corners has been determined, the locations of all other points on the block are known. In contrast, knowing the location of two other points along an axis through the corner kept fixed by the pure rotation **does not** imply that the locations of all other points on the block are known.

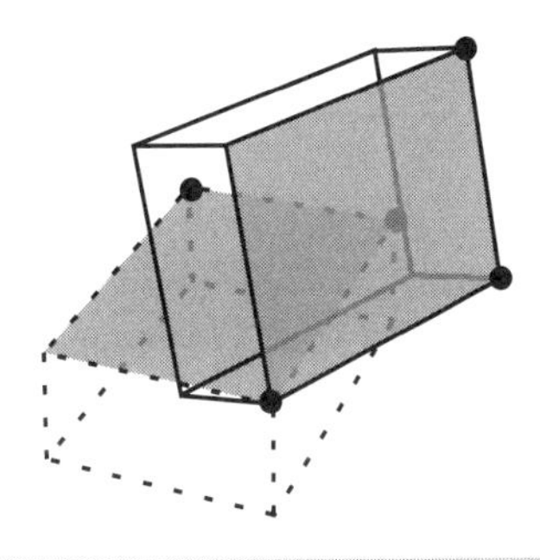

There are infinitely many configurations of the block for which $A$ coincides with $A_{\mathrm{final}}$. These are related to each other through arbitrary pure rotations keeping $A$ fixed.

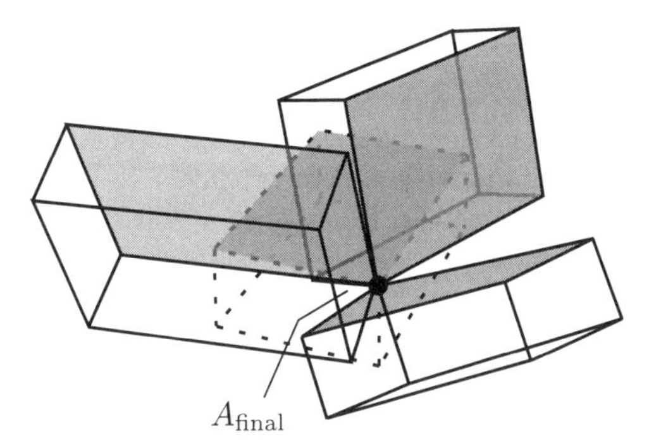


With the introduction of a second point $B$ in the block that is not coincident with $A$, there are still infinitely many configurations of the block for which $A$ and $B$ coincide with $A_{\mathrm{final}}$ and $B_{\mathrm{final}}$, respectively. These are related to each other through arbitrary pure rotations about the straight line through $A_{\mathrm{final}}$ and $B_{\mathrm{final}}$.

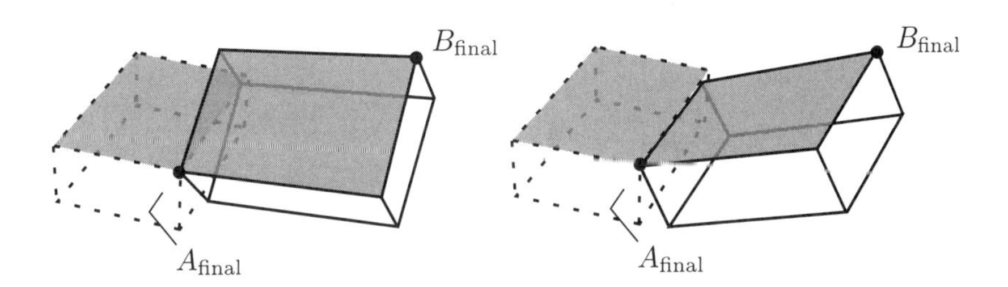


With the introduction of a third point $C$ in the block that does not lie on the line through $A$ and $B$, there is one and only one configuration of the block for which $A$, $B$, and $C$ coincide with $A_{\mathrm{final}}$, $B_{\mathrm{final}}$, and $C_{\mathrm{final}}$, respectively.

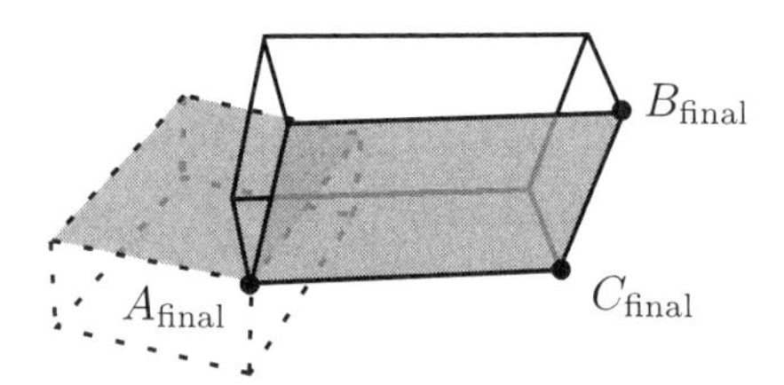


The pure rotation that brings the block from the intermediate configuration to the final configuration is thus entirely determined by the location of the points $A_{\text{final}}$, $B_{\text{final}}$, and $C_{\text{final}}$ relative to the reference configuration.

1

### 1.2.3 Coordinates

Consider a coordinate system with origin at $A_{\text{reference}}$ and axes parallel to the edges of the block in the reference configuration. The coordinates $x_A$, $y_A$, and $z_A$ of the point $A_{\text{final}}$ with respect to this coordinate system quantitatively describe the pure translation that shifts the block from the reference configuration to an intermediate configuration, with the point $A$ coinciding with $A_{\text{final}}$.

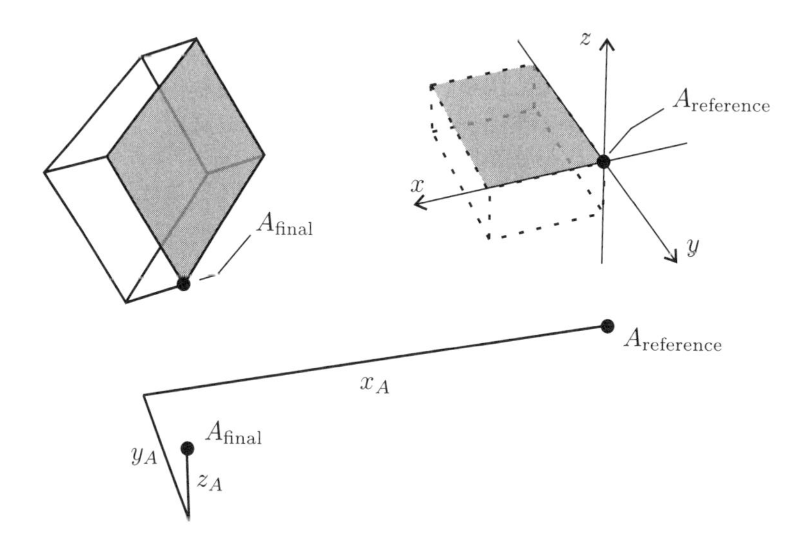


In particular, the pure translation is equivalent to a combination of three pure translations along each of the three coordinate axes by $x_A$, $y_A$, and $z_A$ units of length, respectively.

**Illustration 1.11**

Let the coordinates of one of the corners of the block, with respect to a coordinate system with origin at the corresponding corner in the reference configuration, be 1, −1, and 0 units of length, respectively. The final configuration of the block is related to the reference configuration by a combination of a pure translation shifting all points by one unit of length in the positive direction of the first coordinate axis, a pure translation shifting all points by one unit of length in the negative direction of the second coordinate axis, and a pure rotation keeping the corner point fixed.

Since the distance between $A$ and $B$ must remain unchanged under arbitrary motions, the point $B_{\text{final}}$ is restricted to the surface of a sphere centered on $A_{\text{final}}$. It follows that the location of $B_{\text{final}}$ on this sphere is determined by the values of two independent angles $\theta_1$ and $\theta_2$, e.g., the latitude and longitude of the point $B_{\text{final}}$ relative to some arbitrarily chosen equator and zero meridian.

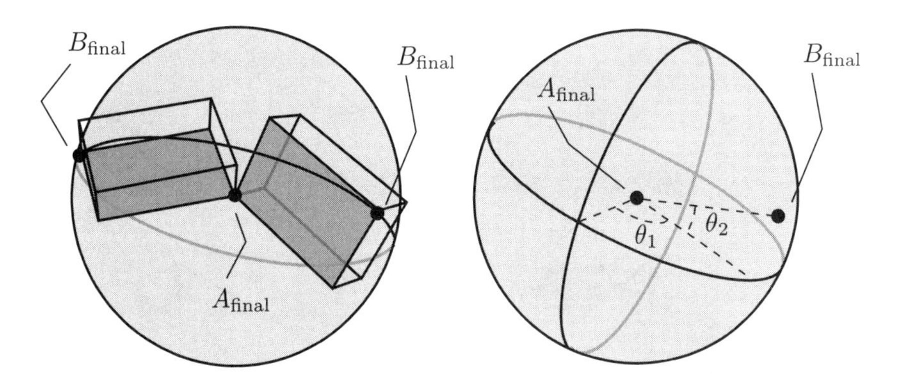


Finally, since the distances between $A$ and $C$ and between $B$ and $C$ must remain unchanged under arbitrary motions, the point $C_{\text{final}}$ is restricted to a circle centered on and perpendicular to the straight line through $A_{\text{final}}$ and $B_{\text{final}}$. It follows that the location of $C_{\text{final}}$ on this circle is determined by the value of a single angle $\theta_3$ relative to some reference position.

The three angles $\theta_1$, $\theta_2$, and $\theta_3$ quantitatively describe the pure rotation that turns the block from the intermediate configuration to the final configuration while keeping the point $A$ fixed.

**Illustration 1.12**

Apply a pure rotation to the block, keeping one corner (denoted by $A$) fixed as shown in the figure.

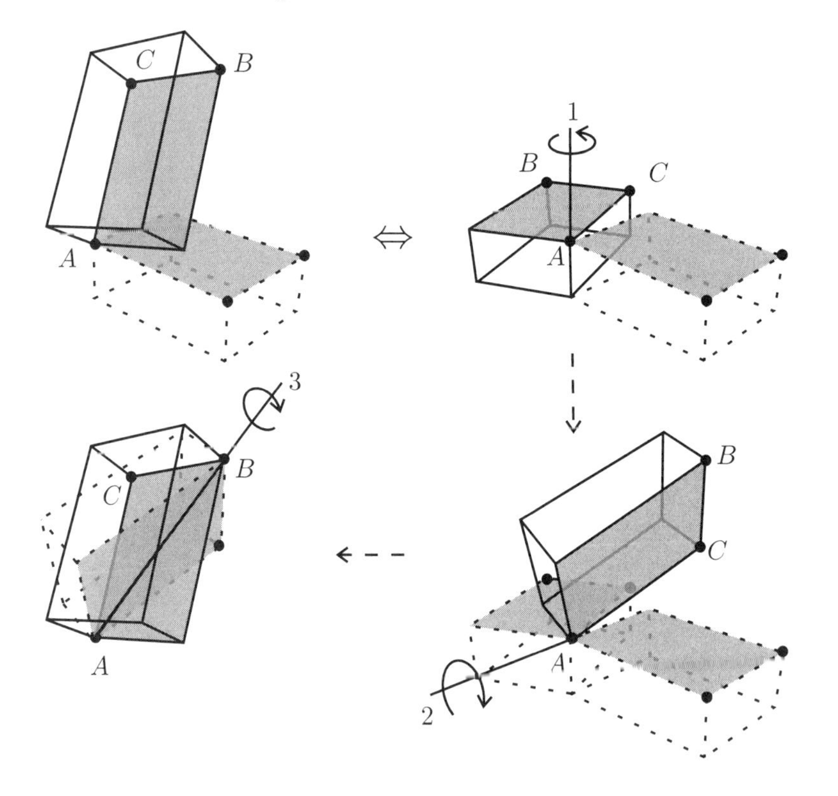


The final configuration of the block is related to the reference configuration by a combination of a pure rotation by an angle of 115° about the axis through $A$ labeled 1, followed by a pure rotation by an angle of 115° about the axis through $A$ labeled 2, followed by a pure rotation by an angle of 75° about the axis labeled 3 through $A$ and the corner denoted by $B$.

Indeed, any arbitrary orientation of the block may be obtained by varying the three angles introduced here.

### 1.2.4 Independence

Every configuration of the block corresponds to some choice of values for the quantities $x_A$, $y_A$, $z_A$, $\theta_1$, $\theta_2$, and $\theta_3$. Similarly, every choice of values for these quantities corresponds to some configuration. By the mutual independence of pure translations and pure rotations, it follows that the values of the angle coordinates are independent of the values of the distance coordinates, and vice versa.

Since our physical space is three-dimensional, it is clear that all three distance coordinates are generally required to describe the block's position relative to the reference configuration. Similarly, all three angle coordinates are generally required to describe the block's orientation relative to the reference configuration. We express these observations by stating that the block has six *geometric degrees of freedom*. We conclude that an arbitrary motion of the block can be translated into specific time histories $x_A(t)$, $y_A(t)$, $z_A(t)$, $\theta_1(t)$, $\theta_2(t)$, and $\theta_3(t)$, and vice versa.

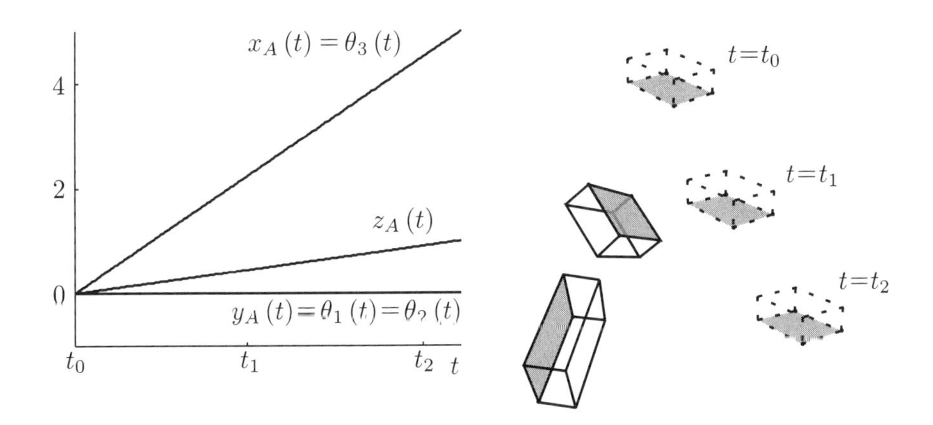


**Illustration 1.13**

The configuration of the block is uniquely determined by the location of the three points $A_{\text{final}}$, $B_{\text{final}}$, and $C_{\text{final}}$ relative to the reference configuration. Thus, every configuration corresponds to some unique choice of values for the coordinates of these three points in the previously introduced coordinate system with origin at $A_{\text{reference}}$. Denote these coordinates by $x_A$, $y_A$, $z_A$, $x_B$, $y_B$, $z_B$, $x_C$, $y_C$, and $z_C$.

In contrast to the set of coordinates $x_A$, $y_A$, $z_A$, $\theta_1$, $\theta_2$, and $\theta_3$, not every choice of values for the collection of nine coordinates corresponds to an actual configuration of the block. The coordinates $x_A$, $y_A$, $z_A$, $x_B$, $y_B$, $z_B$, $x_C$, $y_C$, and $z_C$ are *constrained* to take on values that ensure that the distances between $A_{\text{final}}$ and $B_{\text{final}}$, between $A_{\text{final}}$ and $C_{\text{final}}$, and between $B_{\text{final}}$ and $C_{\text{final}}$ equal those in the reference configuration. Each such constraint is equivalent to an equation that the coordinates must satisfy. For example, if the distance between $A$ and $B$ in the block equals $d_{AB}$, then invoking Pythagoras' theorem yields

$$(x_B - x_A)^2 + (y_B - y_A)^2 + (z_B - z_A)^2 = (d_{AB})^2 .$$

The coordinates $x_A$, $y_A$, $z_A$, $x_B$, $y_B$, $z_B$, $x_C$, $y_C$, and $z_C$ are clearly **not** independent.

## 1.3 A First Look at Rigid Bodies

In the previous sections, the configuration of the block was defined as the position and orientation of the block relative to a reference position and a reference orientation. It was tacitly assumed that no changes could occur to the block other than a change in position and a change in orientation. The three arbitrary points $A$, $B$, and $C$ were useful only under the assumption that the distances between any two of these points remained unchanged under arbitrary motions. From this, we found that six independent quantities describe the configuration of the block at any arbitrary moment during its motion.

Deeper reflection shows the level of idealization employed in this discussion. Clearly, actual physical blocks, even when manufactured to perfection, have shapes that change with time. We say that actual physical bodies are *deformable*, whereas the idealized body whose shape is unchanged during its motion is said to be *rigid*.

When the shape deformations are large, the concepts of position and orientation of the body as a whole are no longer adequate for describing the body's configuration. Instead, it becomes necessary to describe the position of each point of the body relative to its position in the reference configuration. Such bodies require infinitely many quantities to describe their configuration and will not be the subject of this text.

When the shape deformations are small, the rigid-body approximation may prove sufficient for an initial study of the body motion. It is an attractive simplification, given the dramatic reduction in the number of independent quantities necessary for describing the configuration of a rigid body, and given our intuitive understanding of the notions of translation and rotation. All bodies in this text will be assumed to be rigid.

**Illustration 1.14**

Every rigid body can be inscribed within a rectangular block that moves rigidly with the body. It follows that the discussion presented in the previous section applies to an arbitrary rigid body, where reference to the faces and edges of the block refers to the rectangular block within which the body is inscribed.

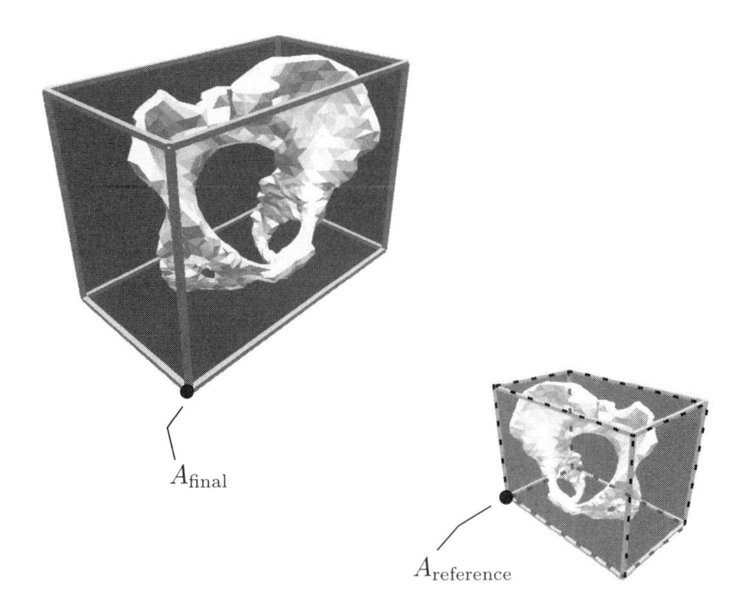


In other words, the configuration of a rigid body can be uniquely determined by the values of six independent quantities. Moreover, arbitrary rigid-body motions relative to a reference configuration can be decomposed into pure translations followed by pure rotations.

## 1.4 A First Look at Observers

(Ex. 1.18)

In the previous sections, we repeatedly emphasized the need for a reference configuration relative to which the current configuration was described. No information was offered, however, about how this reference configuration should be selected. Naturally, the configuration of a rigid body relative to one reference configuration would be quite different from that relative to a different reference configuration.

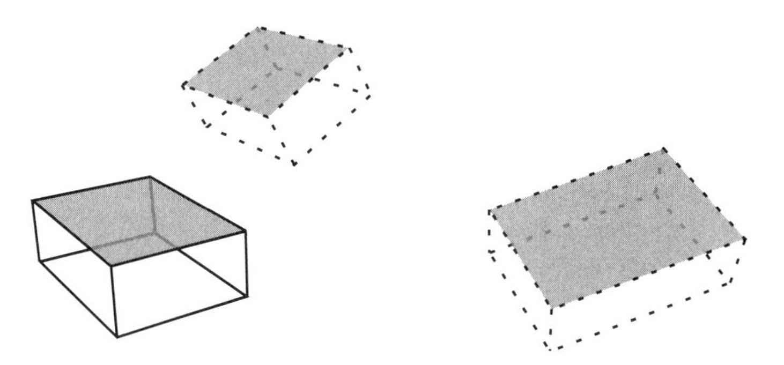

Even if the same three points $A$, $B$, and $C$ on the body were selected, the quantities $x_A$, $y_A$, $z_A$, $\theta_1$, $\theta_2$, and $\theta_3$ would take on different values for the same configuration. In fact, while the body might appear stationary relative to one reference configuration, it might exhibit a complicated tumbling motion relative to another.

The selection of a reference configuration is intimately related to the notion of an *observer*. After all, the initial discussion of rigid-body motion was introduced in the context of a block moving through the reader's visual field.

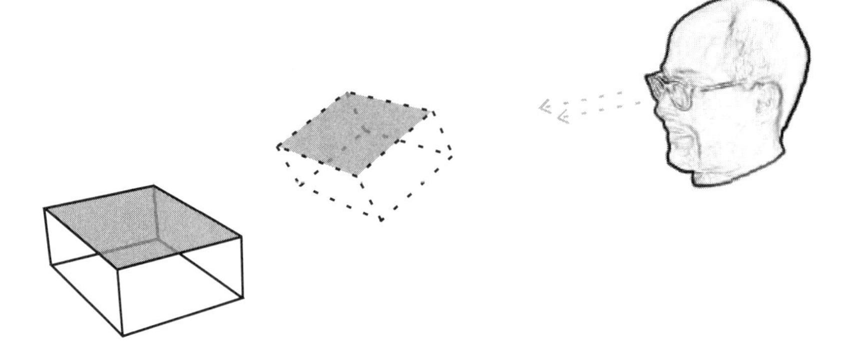

As suggested there, your intuitive reaction to the concepts of a reference position and a reference orientation may have been to imagine a collection of points fixed in your visual field that collectively represent a virtual block of the same dimensions as the actual block. It was not necessary to describe this virtual block any further, since its properties and geometry were self-evident to you. With the help of the reference configuration, you were able to describe the configuration of an arbitrary rigid body. With the help of the reference configuration, you were able to describe the motion of an arbitrary rigid body.

With this image in mind, consider the possibility of an observer observing another observer. This is suggested by the idea that a reference configuration could be used to describe the configuration of the virtual block corresponding to another reference configuration. In this fashion, the configuration of a rigid body relative to one observer could be described as a combination of the configuration of the rigid body relative to some auxiliary observer, and the configuration of the auxiliary observer relative to the original observer.

If the rigid body were stationary relative to the auxiliary observer, any motion relative to the original observer would be described by the motion of the auxiliary observer relative to the original observer.

**Illustration 1.15**

As an example, consider the following breakdown of the configuration of a rectangular block relative to some observer denoted by $\mathcal{W}$. As suggested in previous sections, the configuration of the block relative to $\mathcal{W}$ is given by a pure translation and a subsequent pure rotation about some fixed point $A$ in the block.

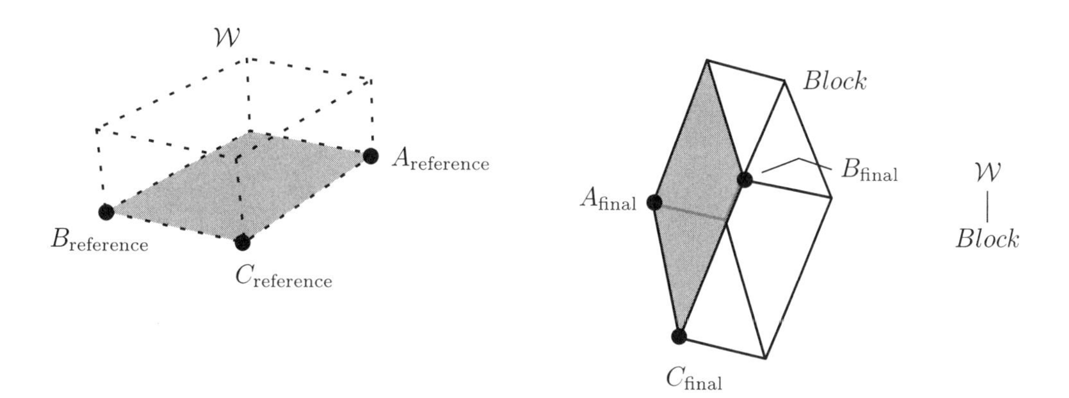


We may introduce an auxiliary observer $\mathcal{A}$, such that the configuration of $\mathcal{A}$ relative to $\mathcal{W}$ is given by the pure translation and such that the configuration of the block relative to $\mathcal{A}$ is given by the pure rotation. Here, the point $A$ on the virtual block representing the $\mathcal{A}$ observer coincides with $A_{\text{final}}$.

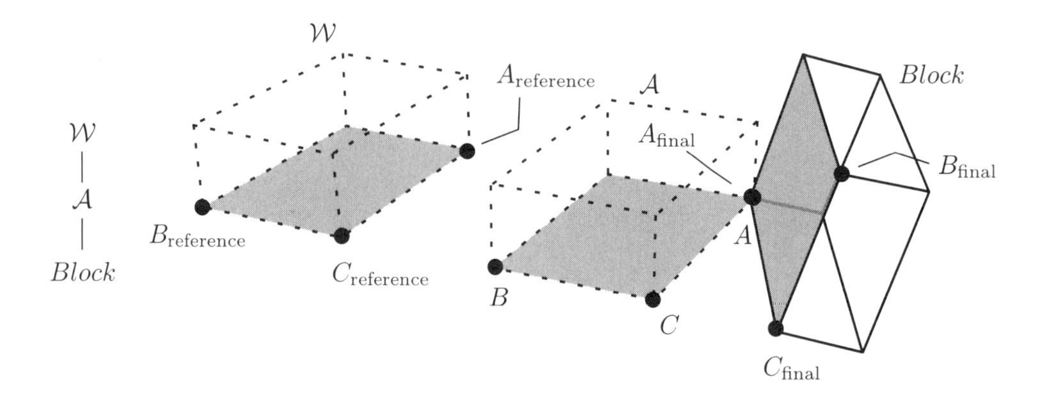


A second auxiliary observer $\mathcal{B}$ may subsequently be introduced, such that the configuration of $\mathcal{B}$ relative to $\mathcal{A}$ is given by some pure rotation keeping $A$ fixed and ensuring that the point $B$ on the virtual block representing the $\mathcal{B}$ observer coincides with $B_{\text{final}}$. The block's configuration relative to $\mathcal{B}$ is then given by the pure rotation about the straight line through $A_{\text{final}}$ and $B_{\text{final}}$ that ensures that the point $C$ coincides with $C_{\text{final}}$.

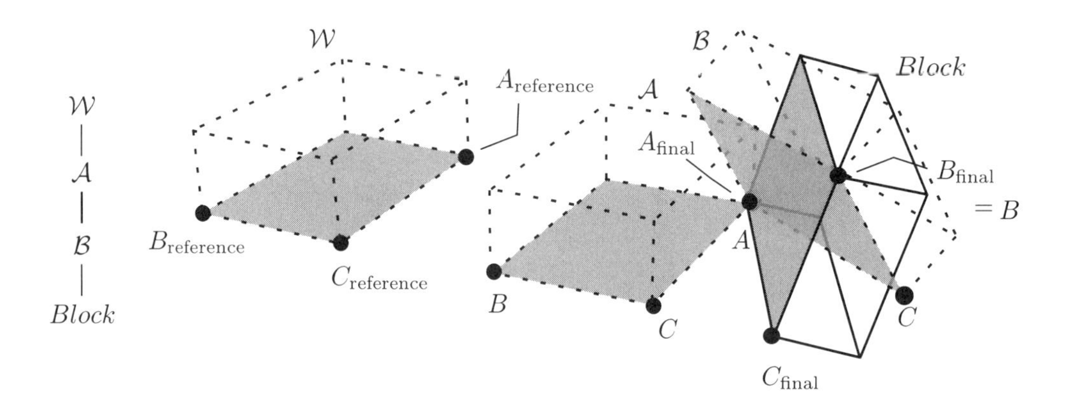


Although the configuration of the block relative to the $\mathcal{W}$ observer may be directly described using a single pure translation and a single pure rotation, the introduction of one or several auxiliary observers serves to simplify the description of each translation and rotation. There is no right answer here, only a question of convenience.

## 1.5 Exercises

**Exercise 1.1** The configuration of a block relative to a reference configuration is given by a pure translation corresponding to a shift of the block by two units of length along a given direction, followed by a pure translation corresponding to a shift of the block by one unit of length along a direction that makes an angle of 42° with the first direction. Then, the configuration of the block relative to the reference configuration is given by a single pure translation corresponding to a shift of $d$ units of length along a direction that makes an angle $\varphi$ with the direction of the first pure translation. Compute $d$ and $\varphi$.

**Solution.** Let $A$ be some arbitrary point on the block and denote by $A_0$, $A_1$, and $A_2$ the corresponding points on the block in the reference, intermediate, and final configurations, respectively. Then, the geometry is described by the figure below.

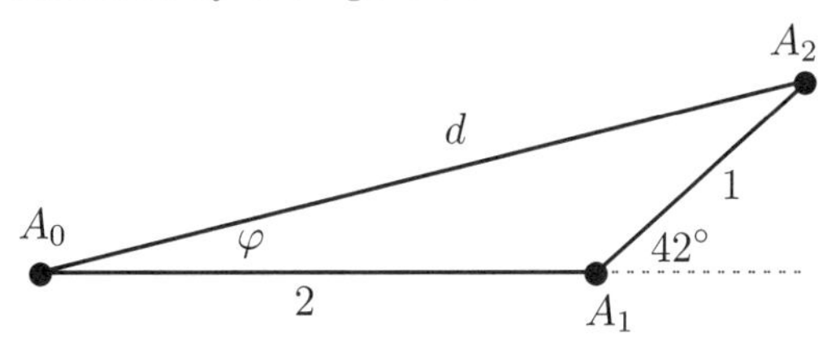


The direction and amount of shift corresponding to the single pure translation relating the block's configuration to the reference configuration is given by the straight-line segment from $A_0$ to $A_2$. By the cosine theorem, the length of this straight-line segment equals

$$d = \sqrt{1^2 + 2^2 - 2 * 1 * 2 * \cos(180° - 42°)} \approx 2.82.$$

Similarly, the angle $\varphi$ between this straight-line segment and the direction of the first pure translation is given by the sine theorem

$$\varphi = \sin^{-1}\left(1 * \frac{\sin(180° - 42°)}{d}\right) \approx 13.7°.$$

**Exercise 1.2** The configuration of a block relative to a reference configuration is given by a pure translation corresponding to a shift of the block by $d_1$ units of length along a given direction followed by a pure translation corresponding to a shift of the block by $d_2$ units of length along a direction that makes an angle $\theta$ with the first direction. Then, the configuration of the block relative to the reference configuration is given by a single pure translation corresponding to a shift of $d$ units of length along a direction that makes an angle $\varphi$ with the direction of the first pure translation. Compute $d$ and $\varphi$ when

a) $d_1 = 2$, $d_2 = 1$, $\theta = 30°$
b) $d_1 = 0.5$, $d_2 = 1.5$, $\theta = 130°$
c) $d_1 = 1$, $d_2 = 1.5$, $\theta = 14°$
d) $d_1 = 2.4$, $d_2 = 0.4$, $\theta = 98°$
e) $d_1 = 2.04$, $d_2 = 4.10$, $\theta = 12.5°$
f) $d_1 = 0.43$, $d_2 = 0.43$, $\theta = 135°$

[Answer: a) $d \approx 2.91$, $\varphi \approx 9.9°$, b) $d \approx 1.24$, $\varphi \approx 112.0°$, c) $d \approx 2.48$, $\varphi \approx 8.4°$, d) $d \approx 2.38$, $\varphi \approx 9.6°$, e) $d \approx 6.11$, $\varphi \approx 8.4°$, f) $d \approx 0.33$, $\varphi \approx 67.5°$]

**Exercise 1.3** Consider applying a pure rotation to a block in its reference configuration by a given amount while keeping all points on one edge of the block fixed, followed by a pure rotation by the same amount but in the opposite direction keeping a different, but parallel, edge fixed relative to the reference configuration, as shown in the figure on the next page.

Show that the final configuration is related to the reference configuration through a pure translation.

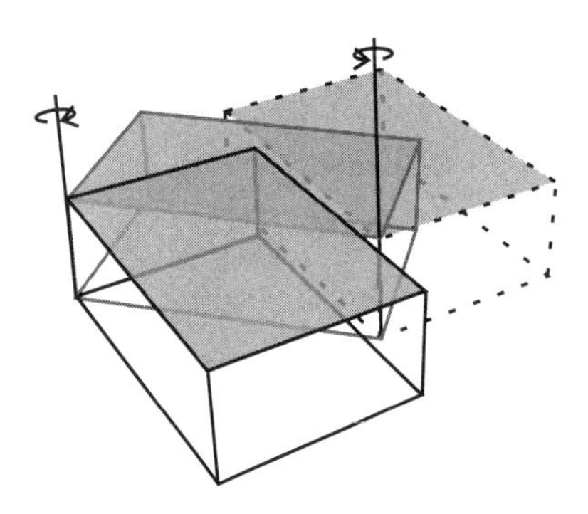

**Solution.** As the paths of all points on the block during the subsequent rotations lie in planes perpendicular to the axes of rotation, it suffices to consider the points on the upper surface of the block.

Specifically, denote by $A$ and $B$ the two points on the upper surface of the block that lie on the two axes of rotation. Let $A_{\text{reference}}$, $B_{\text{reference}}$, $A_{\text{final}}$, and $B_{\text{final}}$ denote the corresponding points in the reference and final configurations, respectively, as shown in the figure.

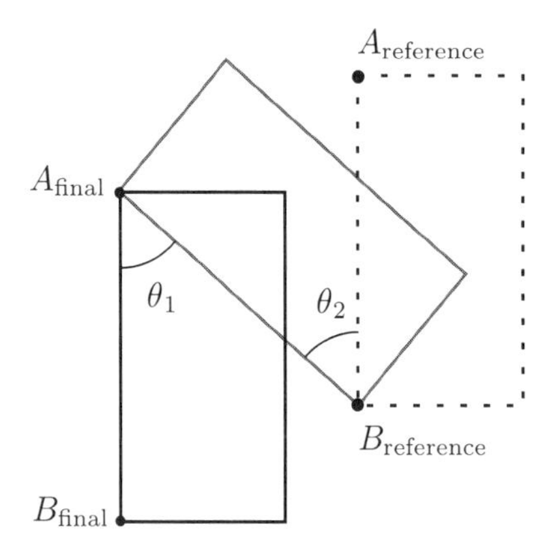


Since the opposing angles $\theta_1$ and $\theta_2$ are equal, it follows that the straight-line segments between $A_{\text{reference}}$ and $B_{\text{reference}}$ and between $A_{\text{final}}$ and $B_{\text{final}}$ are parallel and have the same orientation, and thus that the final configuration is related to the initial configuration through a single pure translation.

**Exercise 1.4** The configuration that results from a pure rotation of a block about a fixed axis through a point $A_1$ by a given amount differs from the configuration that results from a pure rotation of the block about a parallel axis through a point $A_2$ by the same amount.

a) Show that the two configurations are related through a pure translation;

b) Show that the pure translation is along a direction perpendicular to the axes of rotation;

c) Show that the shifting distance is proportional to the perpendicular distance between the two axes of rotation and that the proportionality constant equals $2\sin\frac{\theta}{2}$, where $\theta$ is the turning angle.

[Hint: Consider the figure in Exercise 1.3.]

**Exercise 1.5** Consider applying a pure rotation to a block in its reference configuration corresponding to a half turn about an edge through a given corner on the block, followed by a pure rotation corresponding to a quarter turn about a different edge through the same corner as shown in the figure on the next page.

Show that the final configuration is related to the reference configuration by a single pure rotation about an axis through the corner making an angle of 45° with the first edge and 90° with the second edge.

**Solution.** Since the corner point is kept fixed by both pure rotations, the final configuration is related to the reference configuration by a single pure rotation.

Let $A$ be some arbitrary point on the straight line through the corner making an angle of 45° with the first edge and perpendicular to the second edge as shown in the figure. Denote by $A_0$, $A_1$, and $A_2$ the corresponding points in the reference, intermediate, and final configurations, respectively.

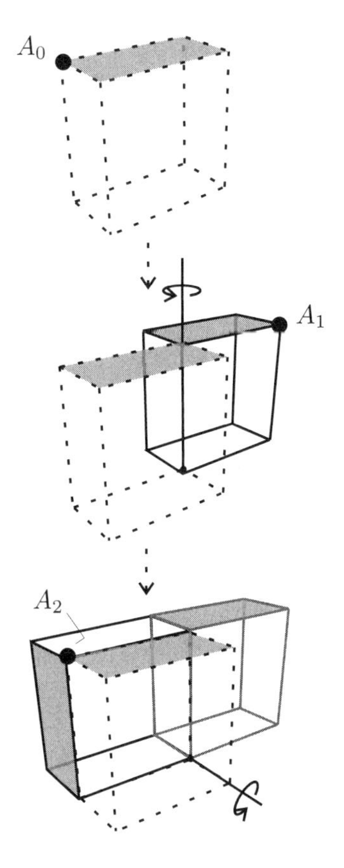


It is clear from the figure that $A_2 = A_0$, i.e., all points on the straight line are kept fixed by the pure rotation from the reference configuration to the final configuration.

**Exercise 1.6** Consider applying a pure rotation to a block in its reference configuration corresponding to a half turn about an edge through a given corner on a block followed by a pure rotation by an angle $\theta$ about a different edge through the same corner. The final configuration is related to the reference configuration by a single pure rotation about an axis through the corner making an angle $\phi$ with the first edge and perpendicular to the second edge. Show that

$$\phi = \frac{|\theta|}{2}.$$

**Exercise 1.7** Consider applying a pure rotation to a block in its reference configuration corresponding to a half turn about some axis through a given corner on a block followed by a pure rotation corresponding to a quarter turn about a different axis through the same corner making an angle 45° with the first axis. The final configuration is related to the reference configuration by a single pure rotation about an axis through the corner making an angle $\phi_1$ with the first axis and $\phi_2$ with the second axis. Show that

$$\phi_1 \approx 35.3^\circ \text{ and } \phi_2 \approx 54.7^\circ.$$

**Exercise 1.8** Show that if two configurations of a block are related through a pure rotation keeping a point $A$ on the block fixed relative to the reference configuration, then the two configurations are related through a pure rotation keeping an entire straight line of points through $A$ fixed relative to the reference configuration.

[Hint: If a block rotates about a straight line through $A$, then all points on the straight line remain fixed relative to the reference configuration. The claim follows if we can show that every pure rotation that keeps the point $A$ fixed is equivalent to a pure rotation of the block about some straight line through $A$. Consider an additional point $B$ on the block that is not coincident with $A$. Let $B_{\text{initial}}$ and $B_{\text{final}}$ be the points in space that coincide with $B$ when the block is in the initial and final configurations, respectively. Then, the claim follows if you show that, for any pure rotation:

1. The points $B_{\text{initial}}$ and $B_{\text{final}}$ lie on the surface of a sphere centered on $A$;
2. There are infinitely many circular arcs on the surface of the sphere that connect $B_{\text{initial}}$ and $B_{\text{final}}$;
3. Each circular arc lies on the intersection of the sphere with a plane that contains the points $B_{\text{initial}}$ and $B_{\text{final}}$;
4. The shortest such circular arc is part of a circle centered on $A$, i.e., the great circle through $B_{\text{initial}}$ and $B_{\text{final}}$;
5. The longest such circular arc is part of a circle centered on the point in space halfway between $B_{\text{initial}}$ and $B_{\text{final}}$;
6. Every such circular arc corresponds to the motion of the point $B$ from $B_{\text{initial}}$ to $B_{\text{final}}$ when the block is rotated about a straight line through $A$ and the center of the corresponding circle;
7. The collection of such straight lines forms a plane that contains $A$ and the point in space halfway between $B_{\text{initial}}$ and $B_{\text{final}}$ and is perpendicular to the line between $B_{\text{initial}}$ and $B_{\text{final}}$;
8. Let $C$ be an additional point on the block that is not colinear with $A$ and $B$. Then, the plane constructed in the previous step intersects the plane constructed by replacing $B$ with $C$ in steps 1. through 7. in a straight line through $A$;
9. The initial and final configurations of the block are related through a pure rotation about the straight line found in the previous step.]

**Exercise 1.9** Use the result of the previous exercise to show that any pure rotation can be equivalently described by specifying an axis about which to turn the block and a turning angle (cf. the equivalent description of a pure translation in terms of a direction along which to shift the block and a shifting distance).

**Exercise 1.10** Show that the order in which two pure rotations are applied to a block is immaterial to the final configuration of the block **if and only if** i) the pure rotations rotate the block about the same axis **or** ii) the pure rotations rotate the block about perpendicular axes by a half turn each.

**Exercise 1.11** Suppose that the actual configuration of a block is related to its reference configuration by a pure translation and a pure rotation, where the axis of rotation is perpendicular to the axis of translation. Show that it is always possible to rigidly embed the block in a larger block, such that the actual configuration of the larger block is related to its reference configuration by a single pure rotation.

[Hint: Consider the solution to Exercise 1.8.]

**Exercise 1.12** Suppose that the actual configuration of a block is related to its reference configuration by a pure translation and a pure rotation, where the axis of rotation is in some arbitrary orientation relative to the axis of translation. Show that it is always possible to rigidly embed the block in a larger block, such that the actual configuration of the larger block is related to its reference configuration by a pure translation and a pure rotation, where the axis of rotation is parallel to the axis of translation. This combination is known as a *screw*.

[Hint: Decompose the pure translation into a component whose axis of translation is parallel to the axis of the pure rotation and a component whose axis of translation is perpendicular to the axis of the pure rotation. Then appeal to the solution to the previous exercise.]

**Exercise 1.13** Consider motions constrained to a plane. A straight line through a point on this plane is characterized by its tangent direction at the point and a positive direction of travel along the line. We can represent the straight line by an arrow based at the point, tangential to the line and pointing in the positive direction of travel along the line. The angle between two straight lines intersecting at a point is then defined as the angle between the two corresponding arrows. Use a construction analogous to that for the spherical surface to define pure translations in this plane and assess whether the plane is a flat space.

**Exercise 1.14** Cut a rectangular strip out of a flat surface and attach the short edges to each other after applying half a turn to one of the edges. Discuss the definition of pure translations for motions on the resulting surface. Consider, in particular, the motion of a two-dimensional block on this surface along the centerline of the strip.

**Exercise 1.15** From a previous exercise, we recall that every pure rotation of the block corresponds to a rotation of the block about some fixed axis. Use this observation to propose an alternative to the angles $\theta_1$, $\theta_2$, and $\theta_3$ for specifying the pure rotation that turns the block from the intermediate configuration to the final configuration while keeping the point $A$ fixed.

[Hint: The pure rotation is uniquely described by specifying an axis through $A$ about which to turn the block and a turning angle. The axis through $A$ is uniquely described by specifying its intersection with a sphere centered on $A$.]

**Exercise 1.16** From a previous exercise, we recall that every configuration of the block is related to its reference configuration by a screw corresponding to a pure rotation and a pure translation by given amounts along a common axis. Use this observation to propose an alternative to the coordinates $x_A$, $y_A$, $z_A$, $\theta_1$, $\theta_2$, and $\theta_3$ for specifying the configuration of the block. Can you propose a set of independent coordinates based on the screw representation?

**Exercise 1.17** Use an actual block to represent the configurations corresponding to the following values for the coordinates:

a) $x_A = 0, y_A = 1, z_A = 1,$ $\theta_1 = 30°, \theta_2 = 0°, \theta_3 = 0°$

b) $x_A = -1, y_A = 1, z_A = 0,$ $\theta_1 = 30°, \theta_2 = 40°, \theta_3 = 0°$

c) $x_A = 0, y_A = 1, z_A = 1,$ $\theta_1 = 0°, \theta_2 = 90°, \theta_3 = 90°$

d) $x_A = 0, y_A = 0, z_A = 0,$ $\theta_1 = 0°, \theta_2 = -45°, \theta_3 = 0°$

e) $x_A = 1, y_A = 0, z_A = 1,$ $\theta_1 = 30°, \theta_2 = 30°, \theta_3 = -90°$

f) $x_A = -1, y_A = 0, z_A = -1,$ $\theta_1 = 180°, \theta_2 = 90°, \theta_3 = 45°$

**Exercise 1.18** The configuration of a block relative to an observer $\mathcal{W}$ corresponds to a pure rotation by a given amount about an axis through a point $O_1$ on the block. Similarly, the configuration of the block relative to an auxiliary observer $\mathcal{B}$ corresponds to a pure rotation about a parallel axis through the block's center $O_2$ by an equal amount. Describe the relative configurations of the observers $\mathcal{W}$ and $\mathcal{B}$ and indicate the relationship graphically.

[Hint: Consider the figure in Exercise 1.3.]

## Summary of notation

The symbols $A$, $B$, $C$, and $O$ were used in this chapter to denote some arbitrary points on a rigid body. The same notation, but with subscripts, e.g., $A_{\text{initial}}$, $A_{\text{intermediate}}$, $A_{\text{reference}}$, or $A_{\text{final}}$, was used to represent points in space that coincided with the corresponding points on the rigid body in the initial, intermediate, reference, or final configurations, respectively.

The symbols $t$ and $\Delta t$ were used in this chapter to represent the time of an event and the difference in time between two events.

The symbol $\Psi$ (*psi*) was used in this chapter to denote a great circle on a sphere. The same notation, but with subscripts, e.g., $\Psi_{\text{initial}}$, $\Psi_{\text{intermediate}}$, $\Psi_{\text{reference}}$, or $\Psi_{\text{final}}$, was used to represent the great circles corresponding to the initial, intermediate, reference, or final configurations, respectively, of a two-dimensional block on the surface of the sphere. Similarly, $\Psi_{\text{reference}\to\text{final}}$, $\Psi_{\text{initial}\to\text{intermediate}}$, and $\Psi_{\text{intermediate}\to\text{final}}$ denoted the great circles through pairs of points on the sphere corresponding to the initial, intermediate, reference, and final configurations, respectively, of the block.

The symbols $x$, $y$, and $z$ were used in this chapter to represent the coordinates of a point on a rigid body with respect to a coordinate system with origin at some point in the reference configuration and axes parallel to the edges of the virtual block corresponding to the reference configuration. The same notation, but with subscripts, such as $x_A$, $y_A$, $z_A$, $x_B$, $y_B$, $z_B$, $x_C$, $y_C$, or $z_C$, was used to represent the coordinates of the points $A_{\text{final}}$, $B_{\text{final}}$, or $C_{\text{final}}$, respectively.

The symbol $d$ was used in this chapter to represent distances between points in space. The same notation, but with subscripts, e.g., $d_{AB}$, was used to represent the distance between points $A$ and $B$ on a rigid body.

The symbols $\theta$ and $\varphi$ (*theta* and *phi*) were used in this chapter to denote an angle. The same notation, but with subscripts, was used to differentiate between the angles that fix the location of the point $B_{\text{final}}$ relative to $A_{\text{final}}$ ($\theta_1$ and $\theta_2$) and the angle that fixes the location of the point $C_{\text{final}}$ relative to $B_{\text{final}}$ and $A_{\text{final}}$ ($\theta_3$).

The symbols $\mathcal{A}$, $\mathcal{B}$, and $\mathcal{W}$ were used in this chapter to denote observers.

## Summary of terminology

The *configuration* of a rigid body is a spatial arrangement of all its points. (Page 3)

At each moment in time, a rigid body's configuration is described by its *position* and *orientation* relative to some *reference position* and *reference orientation*, constituting a *reference configuration.* (Page 3)

A motion of a rigid body that results in a change of its **position**, but involves no change in its orientation, is called a *pure translation.* In a pure translation, all points in the rigid body are shifted by an equal amount along parallel paths relative to the reference configuration. Every pure translation is equivalent to some pure translation that shifts all points in the rigid body by an equal amount along a common fixed direction relative to the reference configuration. (Page 4)

A motion of a rigid body that results in a change of its **orientation**, but involves no change in its position, is called a *pure rotation.* In a pure rotation, one point in the rigid body remains fixed relative to the reference configuration. Every pure rotation is equivalent to some pure rotation that rotates the rigid body by a given amount about an axis fixed relative to the reference configuration. (Page 9)

The configuration of a rigid body relative to a reference configuration can be uniquely *decomposed* into a combination of a single pure translation and a single pure rotation provided that the pure rotation keeps a preselected point in the rigid body fixed relative to the reference configuration. (Page 19)

The configuration of a rigid body is **uniquely determined** by the location of three arbitrary points in the rigid body that do not lie on a single straight line. (Page 22)

In the unique decomposition of the configuration of the rigid body, the pure translation is uniquely determined by the three *coordinates* of the first point in the rigid body with respect to a coordinate system with origin at the corresponding point of the reference configuration. (Page 24)

Similarly, the pure rotation that keeps the first point fixed is uniquely determined by the two angles that describe the location of the second point on the surface of a sphere centered on the first point; and by the single angle that describes the location of the third point along a circle centered on and perpendicular to the line through the first two points. (Page 25)

The rigid body has six *geometric degrees of freedom.* (Page 27)

(Page 30) An *observer* uses a stationary reference configuration to describe the configuration of arbitrary rigid bodies.

(Page 31) The *relative configuration* of two observers is given by the configuration of the reference configuration of one of the observers relative to the reference configuration of the other.

# Chapter 2

# Observers

*wherein the reader learns of:*

- *A notation to represent pure translations, pure rotations, and combinations thereof;*
- *A general methodology for describing complicated arrangements of rigid bodies;*
- *Software tools for multibody analysis and visualization.*

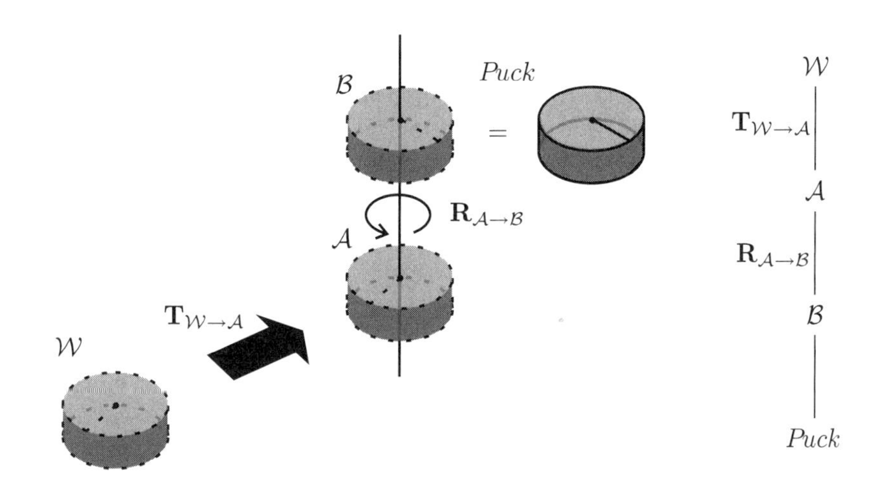

### Practicum

As you read through this chapter, take the opportunity to analyze the geometry of every single object or machine in your immediate environment. Identify its constituent rigid bodies. Introduce a main observer and as many auxiliary observers as you deem necessary, and draw the corresponding tree structures. In particular, consider household machines and tools, such as scissors, tweezers, food processors, can openers, faucets, the flushing mechanism in toilets, window shades, door locks, lawn mowers, weed whackers, electric tooth brushes, eye glasses, tricycles, and so on.

The software package MAMBO and the associated computer-algebra toolbox are excellent tools to illustrate the general notions. But they are much more than that. They offer you the means to create your own reality – one over which you have full control and where any motion is allowed. Code and algorithms become alive with use. So use!

# 2.1 Algebra of Translations and Rotations

(Ex. 2.1 – Ex. 2.7)

We describe the configuration of a rigid body relative to an observer by relating its position and orientation to the reference position and reference orientation of a virtual rigid body that is fixed relative to the observer. Changes in the configuration of the rigid body relative to the observer occur as a result of changes in position and orientation that involve *pure translations*, *pure rotations*, or a combination thereof.

It is **always possible** to describe the *position* and *orientation* of a rigid body relative to some *reference configuration* through a combination of a single pure translation and a single pure rotation. Equivalently, it is **always possible** to describe the *configuration* of a rigid body relative to some *observer* through a combination of a single pure translation and a single pure rotation.

The pure translation and the pure rotation are **unique,** provided that a single point has been selected on the rigid body to act as the point that is kept fixed relative to the reference configuration by any pure rotation. The pure translation and the pure rotation are **independent,** suggesting that we may describe the position of the rigid body independently of describing its orientation, and vice versa.

In the following, we associate with each rigid body and each observer a unique, pre-selected point to be kept fixed by the application of any pure rotation. This enables us to represent the position and orientation of a rigid body relative to an observer or the relative position and orientation of two observers through a unique pure translation and a unique pure rotation.

## 2.1.1 Notation

In a pure translation, all points in a rigid body are shifted by equal amounts along parallel paths. Every pure translation applied to a rigid body is equivalent to a shift of all points in the rigid body by an equal amount along a common fixed direction relative to the reference configuration. We will use the symbol $\mathbf{T}$ to denote the operation corresponding to such a pure translation.

$$\xrightarrow{\mathbf{T}} \mathbf{T}(\;)$$

To *apply* or *perform* the operation $\mathbf{T}$ is to shift all the points on the body by an equal amount along a common direction specified by $\mathbf{T}$.

When considering multiple translations, we add subscripts to differentiate between them, e.g., $\mathbf{T}_1$ and $\mathbf{T}_2$.

In a pure rotation, a pre-selected point on a rigid body remains fixed relative to the reference configuration. Every pure rotation applied to a rigid body is equivalent to a pure rotation by some angle about a fixed axis through the pre-selected point relative to the reference configuration. We will use the symbol **R** to denote the operation corresponding to such a pure rotation.

$$\xrightarrow{\mathbf{R}} \mathbf{R}(\;)$$

To *apply* or *perform* the operation **R** is to turn the rigid body about the pre-selected point by an amount and about a direction specified by **R**.

When considering multiple rotations, we add subscripts to differentiate between them, e.g., $\mathbf{R}_1$ and $\mathbf{R}_2$.

### 2.1.2 The Identity

The *trivial* pure translation that corresponds to zero net shift is called the *identity translation*. The *trivial* pure rotation that corresponds to a zero net turn is called the *identity rotation*. The final configuration that results after applying the identity rotation is identical to that obtained after applying the identity translation. This observation justifies the use of the symbol **I** to denote both the identity translation and the identity rotation.

$$\xrightarrow{\mathbf{I}} \mathbf{I}(\;) =$$

### 2.1.3 Scaling

If a pure translation **T** corresponds to shifting all the points on a rigid body by a distance $d$ along some fixed direction, let the pure translation

$$\alpha\mathbf{T},\ \alpha \geq 0$$

correspond to shifting all the points on the rigid body by a distance $\alpha d$ along the same fixed direction. If, instead, $\alpha < 0$, then the pure translation $\alpha\mathbf{T}$ corresponds to shifting all the points on the rigid body by a distance $|\alpha|\, d$ in the opposite direction to that of **T**.

If a pure rotation **R** corresponds to a rotation by a given angle $\varphi$ about some fixed direction, let the pure rotation

$$\alpha\mathbf{R},\ \text{for all } \alpha$$

correspond to a rotation by an angle $\alpha\varphi$ about the same fixed direction[1].

[1] Here, a negative angle corresponds to rotating in the opposite direction to a positive angle.

### 2.1.4 Composition

If a pure translation $\mathbf{T}_1$ is applied to a rigid body followed by a pure translation $\mathbf{T}_2$, then the combined operation is represented by the expression

$$\mathbf{T}_2 \circ \mathbf{T}_1.$$

NOTE THE ORDER!

$$\xrightarrow{\mathbf{T}_1} \mathbf{T}_1(\ ) \xrightarrow{\mathbf{T}_2} \mathbf{T}_2 \circ \mathbf{T}_1(\ )$$

The *composition symbol* $\circ$ separates the operations and emphasizes the order in which the operations are applied. If a pure rotation $\mathbf{R}_1$ is applied to a rigid body followed by a pure rotation $\mathbf{R}_2$, then the combined operation is represented by the expression

$$\mathbf{R}_2 \circ \mathbf{R}_1.$$

$$\xrightarrow{\mathbf{R}_1} \mathbf{R}_1(\ ) \xrightarrow{\mathbf{R}_2} \mathbf{R}_2 \circ \mathbf{R}_1(\ )$$

If a pure rotation $\mathbf{R}_1$ is applied to a rigid body followed by a pure translation $\mathbf{T}_1$ that is, in turn, followed by a pure rotation $\mathbf{R}_2$, the combined operation is represented by the expression

$$\mathbf{R}_2 \circ \mathbf{T}_1 \circ \mathbf{R}_1.$$

$$\xrightarrow{\mathbf{R}_1} \mathbf{R}_1(\ ) \xrightarrow{\mathbf{T}_1} \mathbf{T}_1 \circ \mathbf{R}_1(\ ) \xrightarrow{\mathbf{R}_2} \mathbf{R}_2 \circ \mathbf{T}_1 \circ \mathbf{R}_1(\ )$$

It should be clear how to use the composition symbol to represent the combined operation that results from an arbitrary sequence of pure translations and pure rotations.

**Illustration 2.1**

The configuration obtained by first applying a pure translation $\mathbf{T}_1$ followed by a pure translation $\mathbf{T}_2$ is identical to that obtained by first applying $\mathbf{T}_2$ followed by $\mathbf{T}_1$. This observation is represented by the equality

$$\mathbf{T}_1 \circ \mathbf{T}_2 = \mathbf{T}_2 \circ \mathbf{T}_1.$$

$$\begin{array}{ccc} \square & \xrightarrow{\mathbf{T}_1} & \mathbf{T}_1(\square) \\ \downarrow{\scriptstyle \mathbf{T}_2} & & \downarrow{\scriptstyle \mathbf{T}_2} \\ \mathbf{T}_2(\square) & \xrightarrow{\mathbf{T}_1} & \mathbf{T}_2 \circ \mathbf{T}_1(\square) \\ & & = \mathbf{T}_1 \circ \mathbf{T}_2(\square) \end{array}$$

The configuration obtained by first applying a pure translation $\mathbf{T}$ followed by a pure rotation $\mathbf{R}$ is identical to that obtained by first applying $\mathbf{R}$ followed by $\mathbf{T}$. This observation is represented by the equality

$$\mathbf{T} \circ \mathbf{R} = \mathbf{R} \circ \mathbf{T}.$$

$$\begin{array}{ccc} \square & \xrightarrow{\mathbf{T}} & \mathbf{T}(\square) \\ \downarrow{\scriptstyle \mathbf{R}} & & \downarrow{\scriptstyle \mathbf{R}} \\ \mathbf{R}(\square) & \xrightarrow{\mathbf{T}} & \mathbf{R} \circ \mathbf{T}(\square) \\ & & = \mathbf{T} \circ \mathbf{R}(\square) \end{array}$$

The configuration obtained by first applying a pure rotation $\mathbf{R}_1$ followed by a pure rotation $\mathbf{R}_2$ is not generally identical to that obtained by first applying $\mathbf{R}_2$ followed by $\mathbf{R}_1$. This observation is represented by the inequality

$$\mathbf{R}_1 \circ \mathbf{R}_2 \neq \mathbf{R}_2 \circ \mathbf{R}_1.$$

$$\begin{array}{ccccc} \mathbf{R}_2(\square) & \xleftarrow{\mathbf{R}_2} & \square & \xrightarrow{\mathbf{R}_1} & \mathbf{R}_1(\square) \\ & \searrow{\scriptstyle \mathbf{R}_1} & & \swarrow{\scriptstyle \mathbf{R}_2} & \\ & \mathbf{R}_1 \circ \mathbf{R}_2(\square) & \neq & \mathbf{R}_2 \circ \mathbf{R}_1(\square) & \end{array}$$

Now, consider the combined operation that results from applying a sequence of pairs of pure translations and pure rotations:

$$\mathbf{R}_n \circ \mathbf{T}_n \circ \cdots \circ \mathbf{R}_1 \circ \mathbf{T}_1.$$

Using the observations made above, the same operation can be represented by the composition

$$\mathbf{R}_n \circ \mathbf{R}_{n-1} \circ \cdots \circ \mathbf{R}_1 \circ \mathbf{T}_n \circ \mathbf{T}_{n-1} \circ \cdots \circ \mathbf{T}_1,$$

in which the pure translations are applied before the pure rotations. The sequence of translations may be replaced by a single pure translation $\mathbf{T}$, where

$$\mathbf{T} = \mathbf{T}_n \circ \mathbf{T}_{n-1} \circ \cdots \circ \mathbf{T}_1.$$

Similarly, since the pure rotations are assumed to keep the same point fixed, the sequence of rotations may be replaced by a single pure rotation $\mathbf{R}$, where

$$\mathbf{R} = \mathbf{R}_n \circ \mathbf{R}_{n-1} \circ \cdots \circ \mathbf{R}_1.$$

We conclude that

$$\mathbf{R}_n \circ \mathbf{T}_n \circ \cdots \circ \mathbf{R}_1 \circ \mathbf{T}_1 = \mathbf{R} \circ \mathbf{T}.$$

### 2.1.5 Inverses

If a pure translation $\mathbf{T}_1$ is followed by a pure translation $\mathbf{T}_2$ and the combined operation equals the trivial pure translation $\mathbf{I}$, i.e.,

$$\mathbf{T}_2 \circ \mathbf{T}_1 = \mathbf{I},$$

then $\mathbf{T}_2$ is said to be the *inverse* of $\mathbf{T}_1$ and we write

$$\mathbf{T}_2 = (\mathbf{T}_1)^{-1}.$$

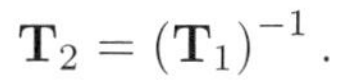

It follows that

$$(\mathbf{T}_1)^{-1} \circ \mathbf{T}_1 = \mathbf{I}.$$

$$\square \xrightarrow{\mathbf{T}_1} \mathbf{T}_1(\square) \xrightarrow{(\mathbf{T}_1)^{-1}} (\mathbf{T}_1)^{-1} \circ \mathbf{T}_1(\square) = \mathbf{I}\,(\square) = \square$$

The inverse of a pure translation $\mathbf{T}$ corresponds to a shift of all points on the rigid body by an equal amount as specified by $\mathbf{T}$ but in the opposite direction, i.e.,

$$(\mathbf{T})^{-1} = (-1)\,\mathbf{T}.$$

If a pure rotation $\mathbf{R}_1$ is followed by a pure rotation $\mathbf{R}_2$ and the combined operation equals the trivial pure rotation $\mathbf{I}$, i.e.,

$$\mathbf{R}_2 \circ \mathbf{R}_1 = \mathbf{I},$$

then $\mathbf{R}_2$ is said to be the inverse of $\mathbf{R}_1$, and we write

$$\mathbf{R}_2 = (\mathbf{R}_1)^{-1}.$$

It follows that

$$(\mathbf{R}_1)^{-1} \circ \mathbf{R}_1 = \mathbf{I}.$$

$$\xrightarrow{\mathbf{R}_1} \mathbf{R}_1(\ ) \xrightarrow{(\mathbf{R}_1)^{-1}} (\mathbf{R}_1)^{-1} \circ \mathbf{R}_1(\ ) = \mathbf{I}\,(\ ) =$$

The inverse of a pure rotation $\mathbf{R}$ corresponds to a rotation of the rigid body about the same axis and by the same amount as specified by $\mathbf{R}$ but in the opposite direction, i.e.,

$$(\mathbf{R})^{-1} = (-1)\,\mathbf{R}.$$

**Illustration 2.2**

Since $\mathbf{R}$ and $\mathbf{R}^{-1}$ correspond to rotations about the same axis, it follows that

$$\mathbf{R} \circ \mathbf{R}^{-1} = \mathbf{R}^{-1} \circ \mathbf{R} = \mathbf{I}.$$

Using the definition of the inverse, we conclude that

$$\left(\mathbf{R}^{-1}\right)^{-1} = \mathbf{R}.$$

### 2.1.6 The Group of Rigid-body Transformations

The collection of all possible pure translations together with the composition operation constitute a *group*.

---

**Definition 2.1** A group is a set $\mathbb{X}$ and a binary operation $\odot$ into $\mathbb{X}$, such that for all $A, B, C \in \mathbb{X}$ :

- *Associativity*: $A \odot (B \odot C) = (A \odot B) \odot C$;
- *Identity*: There exists an element $I \in \mathbb{X}$, such that $I \odot A = A \odot I = A$;
- *Invertibility*: Every element $A$ has an inverse in $\mathbb{X}$ denoted by $A^{-1}$, such that $A^{-1} \odot A = A \odot A^{-1} = I$.

---

Certainly,

$$\mathbf{T}_3 \circ (\mathbf{T}_2 \circ \mathbf{T}_1) = (\mathbf{T}_3 \circ \mathbf{T}_2) \circ \mathbf{T}_1,$$

where $\mathbf{T}_1$, $\mathbf{T}_2$, and $\mathbf{T}_3$ are arbitrary pure translations. Moreover, the identity translation $\mathbf{I}$ satisfies

$$\mathbf{I} \circ \mathbf{T} = \mathbf{T} \circ \mathbf{I} = \mathbf{T},$$

since $\mathbf{I}$ corresponds to the absence of motion. Finally, since the inverse $\mathbf{T}^{-1}$ of a pure translation $\mathbf{T}$ corresponds to a shift of all points on a rigid body by an equal amount as specified by $\mathbf{T}$ but in the opposite direction, we have

$$\mathbf{T}^{-1} \circ \mathbf{T} = \mathbf{T} \circ \mathbf{T}^{-1} = \mathbf{I}.$$

Indeed, the group of pure translations together with the composition operator also satisfy a *commutativity* property

$$\mathbf{T}_1 \circ \mathbf{T}_2 = \mathbf{T}_2 \circ \mathbf{T}_1.$$

This shows that the group of pure translations together with the composition operator is an *Abelian group*.

**Illustration 2.3**

The collection of all pure rotations that keep a pre-selected point on a rigid body fixed relative to the reference configuration together with the composition operator also constitute a group. In fact,

$$\mathbf{R}_1 \circ (\mathbf{R}_2 \circ \mathbf{R}_3) = (\mathbf{R}_1 \circ \mathbf{R}_2) \circ \mathbf{R}_3,$$

$$\mathbf{I} \circ \mathbf{R} = \mathbf{R} \circ \mathbf{I} = \mathbf{R},$$

and

$$\mathbf{R}^{-1} \circ \mathbf{R} = \mathbf{R} \circ \mathbf{R}^{-1} = \mathbf{I}.$$

In contrast to the group of all pure translations with the composition operator, it is **not** generally true that

$$\mathbf{R}_1 \circ \mathbf{R}_2 = \mathbf{R}_2 \circ \mathbf{R}_1,$$

i.e., the group of all pure rotations that keep a pre-selected point on a rigid body fixed relative to the reference configuration together with the composition operator is not Abelian.

(Ex. 2.8 – Ex. 2.11)

## 2.2 Hierarchies

### 2.2.1 Notation

In this text, I consistently denote an observer corresponding to some reference configuration with an upper-case, calligraphic letter, such as

$$\mathcal{A},\ \mathcal{R},\ \text{or}\ \mathcal{W}.$$

The choice of letter is not important, unless you are trying to give the person you are communicating with additional information by a clever choice of letter. If the same letter is to be used to denote separate observers, I use appropriate subscripts to differentiate between the observers, such as $\mathcal{A}_1$, $\mathcal{C}_{\text{laboratory}}$, $\mathcal{W}_{\text{final}}$.

### 2.2.2 Single-body, Single-observer Hierarchy

In spite of the possibility of using a single observer to describe the configuration of a rigid body, it is often convenient to introduce a sequence of intermediate, auxiliary observers. Typically, this reduces the complexity of the pure translations or pure rotations that describe the relative configurations of neighboring observers or that relate the rigid body's configuration to the last auxiliary observer.

We graphically illustrate the collection of observers and the suggested geometric description by using a tree structure, in which the original observer forms the parent node, the auxiliary observers form the internal nodes, the rigid body is found at the leaf node, and each branch represents a combination of a pure translation and a pure rotation.

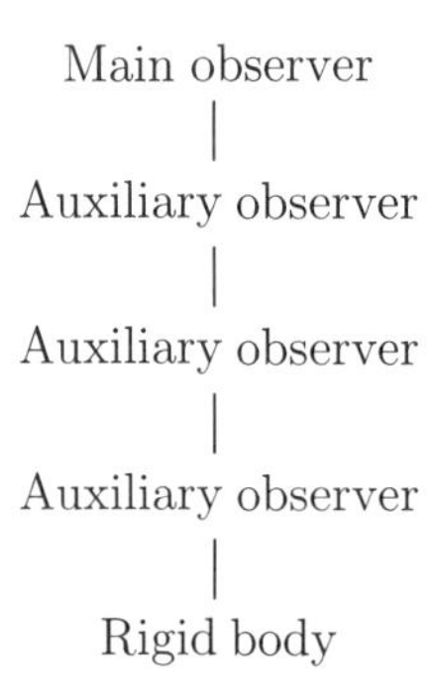


**Illustration 2.4**

The proposed hierarchy is reminiscent of a vertical organizational structure for a company, in which any contact between the employees and the upper echelons of the company takes place through a set of intermediate levels of management. A member at each level of the organization need only be concerned with how to contact his or her superior, ensuring that the executive officer has full knowledge of the status of all employees.

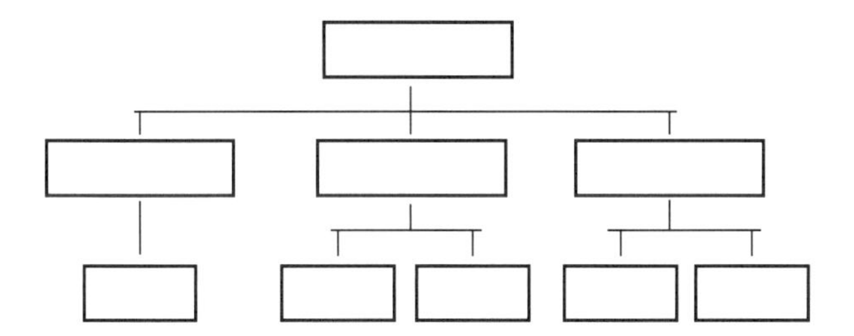

In an actual managerial structure, it is also possible for the upper levels of management to pass information down through the tree to ensure a certain set of actions from the employees. In the case of the tree structure representing the organization of observers, this possibility is excluded, since an observer is limited to observing its environment without controlling it.

### 2.2.3 Single-body, Multiple-observer Hierarchy

> **Auxiliary observers need not be associated with actual physical objects**. Instead, they represent virtual objects, relative to which the configuration of other observers or actual physical objects may be described. Any number of auxiliary observers may be introduced into a description, limited only by their usefulness in simplifying the representation of the observed geometry.

Multiple observers may be introduced for reasons other than a reduction of complexity. Imagine, for example, multiple human observers observing the motion of the same rigid body. Given a choice of point kept fixed by any pure rotation, the rigid body's configuration relative to each observer is uniquely described through a combination of a pure translation and a pure rotation. But that description clearly varies between different observers. The different observers appear to lack a means for communicating their observations to each other, since a statement by one observer that the rigid body is stationary may clash with the observations of other observers.

Let $\mathcal{A}$ and $\mathcal{B}$ be two observers observing the motion of a single block. Let the point kept fixed by pure rotations of the block correspond to the geometric center of the block. Then, the configuration of the block relative to observer $\mathcal{A}$ may be described through a combination of a unique pure translation $\mathbf{T}_\mathcal{A}$ and a unique pure rotation $\mathbf{R}_\mathcal{A}$. Similarly, the configuration of the block relative to observer $\mathcal{B}$ may be described through a combination of a unique pure translation $\mathbf{T}_\mathcal{B}$ and a unique pure rotation $\mathbf{R}_\mathcal{B}$. The following tree structure captures this state of affairs:

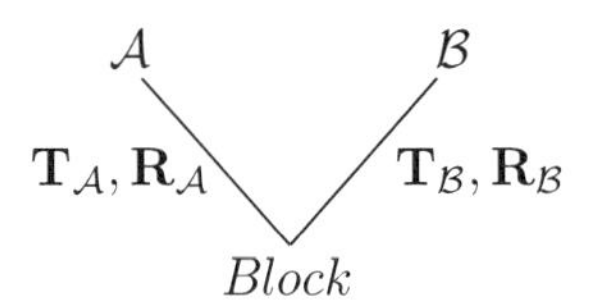


Now, suppose we want to treat observer $\mathcal{B}$ as an auxiliary observer, such that the configuration of the block relative to observer $\mathcal{A}$ is described as a combination of the configuration of the block relative to $\mathcal{B}$ and the configuration of $\mathcal{B}$ relative to $\mathcal{A}$. How could we use the given information to find the unique pure translation $\mathbf{T}_{\mathcal{A}\to\mathcal{B}}$ and unique pure rotation $\mathbf{R}_{\mathcal{A}\to\mathcal{B}}$ relating $\mathcal{B}$ to $\mathcal{A}$?

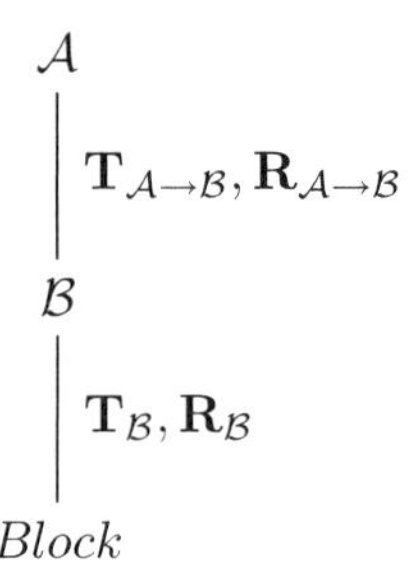


The position and orientation of the block relative to observer $\mathcal{A}$ are uniquely determined by the pure translation $\mathbf{T}_\mathcal{A}$ and the pure rotation $\mathbf{R}_\mathcal{A}$. This implies that a shift given by $\mathbf{T}_\mathcal{A}$, followed by a turn given by $\mathbf{R}_\mathcal{A}$, brings the block from its reference configuration to the actual configuration. The intermediate configuration is characterized by its geometric center coinciding with that of the block in the actual configuration. Similarly, the position and orientation of the block relative to observer $\mathcal{B}$ are uniquely determined by the pure translation $\mathbf{T}_\mathcal{B}$ and the pure rotation $\mathbf{R}_\mathcal{B}$. This implies that a shift given by $\mathbf{T}_\mathcal{B}$, followed by a turn given by $\mathbf{R}_\mathcal{B}$, brings the block from its reference configuration to the actual configuration. Again, the intermediate configuration is characterized by its geometric center coinciding with that of the block in the actual configuration.

$$\mathcal{A} \xrightarrow{\mathbf{T}_\mathcal{A}} \mathbf{T}_\mathcal{A}(\mathcal{A}) \xrightarrow{\mathbf{R}_\mathcal{A}} \mathbf{R}_\mathcal{A} \circ \mathbf{T}_\mathcal{A}(\mathcal{A})$$

$$= \mathbf{R}_\mathcal{B} \circ \mathbf{T}_\mathcal{B}(\mathcal{B}) \xleftarrow{\mathbf{R}_\mathcal{B}} \mathbf{T}_\mathcal{B}(\mathcal{B}) \xleftarrow{\mathbf{T}_\mathcal{B}} \mathcal{B}$$

From this discussion, we conclude that the pure translation $\mathbf{T}_{\mathcal{A}}$ followed by the pure rotation $\mathbf{R}_{\mathcal{A}}$ followed by the pure rotation $(\mathbf{R}_{\mathcal{B}})^{-1}$ followed by the pure translation $(\mathbf{T}_{\mathcal{B}})^{-1}$ bring the block from the reference configuration of $\mathcal{A}$ to the reference configuration of $\mathcal{B}$.

$$\mathcal{A} \xrightarrow{\mathbf{T}_{\mathcal{A}}} \mathbf{T}_{\mathcal{A}}(\mathcal{A}) \xrightarrow{\mathbf{R}_{\mathcal{A}}} \mathbf{R}_{\mathcal{A}} \circ \mathbf{T}_{\mathcal{A}}(\mathcal{A}) \xrightarrow{(\mathbf{R}_{\mathcal{B}})^{-1}}$$
$$\xrightarrow{(\mathbf{R}_{\mathcal{B}})^{-1}} (\mathbf{R}_{\mathcal{B}})^{-1} \circ \mathbf{R}_{\mathcal{A}} \circ \mathbf{T}_{\mathcal{A}}(\mathcal{A}) \xrightarrow{(\mathbf{T}_{\mathcal{B}})^{-1}}$$
$$\xrightarrow{(\mathbf{T}_{\mathcal{B}})^{-1}} (\mathbf{T}_{\mathcal{B}})^{-1} \circ (\mathbf{R}_{\mathcal{B}})^{-1} \circ \mathbf{R}_{\mathcal{A}} \circ \mathbf{T}_{\mathcal{A}}(\mathcal{A}) = \mathcal{B}$$

In other words,

$$\begin{aligned} \mathbf{R}_{\mathcal{A}\to\mathcal{B}} \circ \mathbf{T}_{\mathcal{A}\to\mathcal{B}} &= (\mathbf{T}_{\mathcal{B}})^{-1} \circ (\mathbf{R}_{\mathcal{B}})^{-1} \circ \mathbf{R}_{\mathcal{A}} \circ \mathbf{T}_{\mathcal{A}} \\ &= (\mathbf{R}_{\mathcal{B}})^{-1} \circ \mathbf{R}_{\mathcal{A}} \circ (\mathbf{T}_{\mathcal{B}})^{-1} \circ \mathbf{T}_{\mathcal{A}}, \end{aligned}$$

which implies that

$$\mathbf{T}_{\mathcal{A}\to\mathcal{B}} = (\mathbf{T}_{\mathcal{B}})^{-1} \circ \mathbf{T}_{\mathcal{A}}$$

and

$$\mathbf{R}_{\mathcal{A}\to\mathcal{B}} = (\mathbf{R}_{\mathcal{B}})^{-1} \circ \mathbf{R}_{\mathcal{A}}.$$

The configuration of the block relative to the $\mathcal{A}$ observer is now given by

$$\begin{aligned} \mathbf{R}_{\mathcal{B}} \circ \mathbf{T}_{\mathcal{B}} \circ \mathbf{R}_{\mathcal{A}\to\mathcal{B}} \circ \mathbf{T}_{\mathcal{A}\to\mathcal{B}} &= \mathbf{R}_{\mathcal{B}} \circ \mathbf{T}_{\mathcal{B}} \circ (\mathbf{R}_{\mathcal{B}})^{-1} \circ \mathbf{R}_{\mathcal{A}} \circ (\mathbf{T}_{\mathcal{B}})^{-1} \circ \mathbf{T}_{\mathcal{A}} \\ &= \mathbf{R}_{\mathcal{B}} \circ (\mathbf{R}_{\mathcal{B}})^{-1} \circ \mathbf{R}_{\mathcal{A}} \circ \mathbf{T}_{\mathcal{B}} \circ (\mathbf{T}_{\mathcal{B}})^{-1} \circ \mathbf{T}_{\mathcal{A}} \\ &= \mathbf{R}_{\mathcal{A}} \circ \mathbf{T}_{\mathcal{A}} \end{aligned}$$

as expected.

**Illustration 2.5**

Now, consider three observers $\mathcal{A}$, $\mathcal{B}$, and $\mathcal{C}$ observing the same rigid body as represented by the following tree structure:

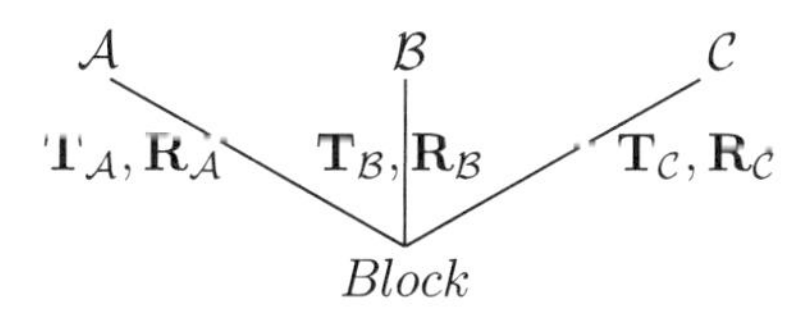


By the same argument as above, we find that

$$\mathbf{T}_{\mathcal{A}\to\mathcal{B}} = (\mathbf{T}_{\mathcal{B}})^{-1} \circ \mathbf{T}_{\mathcal{A}},$$
$$\mathbf{R}_{\mathcal{A}\to\mathcal{B}} = (\mathbf{R}_{\mathcal{B}})^{-1} \circ \mathbf{R}_{\mathcal{A}},$$
$$\mathbf{T}_{\mathcal{B}\to\mathcal{C}} = (\mathbf{T}_{\mathcal{C}})^{-1} \circ \mathbf{T}_{\mathcal{B}},$$

and

$$\mathbf{R}_{\mathcal{B}\to\mathcal{C}} = (\mathbf{R}_{\mathcal{C}})^{-1} \circ \mathbf{R}_{\mathcal{B}}.$$

It follows that the configuration of $\mathcal{C}$ relative to $\mathcal{A}$ is given by

$$\begin{aligned}
\mathbf{R}_{\mathcal{A}\to\mathcal{C}} \circ \mathbf{T}_{\mathcal{A}\to\mathcal{C}} &= \mathbf{R}_{\mathcal{B}\to\mathcal{C}} \circ \mathbf{T}_{\mathcal{B}\to\mathcal{C}} \circ \mathbf{R}_{\mathcal{A}\to\mathcal{B}} \circ \mathbf{T}_{\mathcal{A}\to\mathcal{B}} \\
&= (\mathbf{R}_{\mathcal{C}})^{-1} \circ \mathbf{R}_{\mathcal{B}} \circ (\mathbf{T}_{\mathcal{C}})^{-1} \circ \mathbf{T}_{\mathcal{B}} \\
&\quad \circ (\mathbf{R}_{\mathcal{B}})^{-1} \circ \mathbf{R}_{\mathcal{A}} \circ (\mathbf{T}_{\mathcal{B}})^{-1} \circ \mathbf{T}_{\mathcal{A}} \\
&= (\mathbf{R}_{\mathcal{C}})^{-1} \circ \mathbf{R}_{\mathcal{B}} \circ (\mathbf{R}_{\mathcal{B}})^{-1} \circ \mathbf{R}_{\mathcal{A}} \\
&\quad \circ (\mathbf{T}_{\mathcal{C}})^{-1} \circ \mathbf{T}_{\mathcal{B}} \circ (\mathbf{T}_{\mathcal{B}})^{-1} \circ \mathbf{T}_{\mathcal{A}} \\
&= (\mathbf{R}_{\mathcal{C}})^{-1} \circ \mathbf{R}_{\mathcal{A}} \circ (\mathbf{T}_{\mathcal{C}})^{-1} \circ \mathbf{T}_{\mathcal{A}},
\end{aligned}$$

i.e.,

$$\mathbf{T}_{\mathcal{A}\to\mathcal{C}} = (\mathbf{T}_{\mathcal{C}})^{-1} \circ \mathbf{T}_{\mathcal{A}}$$

and

$$\mathbf{R}_{\mathcal{A}\to\mathcal{C}} = (\mathbf{R}_{\mathcal{C}})^{-1} \circ \mathbf{R}_{\mathcal{A}}$$

as expected.

### 2.2.4 Multiple-body, Multiple-observer Hierarchy

The notion of auxiliary observers is particularly useful when dealing with multiple rigid bodies. Of course, the configuration of the $i$-th rigid body relative to an observer $\mathcal{W}$ could be described by a single pure translation $\mathbf{T}_i$ and a single pure rotation $\mathbf{R}_i$. It may, however, be convenient to introduce multiple auxiliary observers between the main observer $\mathcal{W}$ and each rigid body.

Consider the tree representation below:

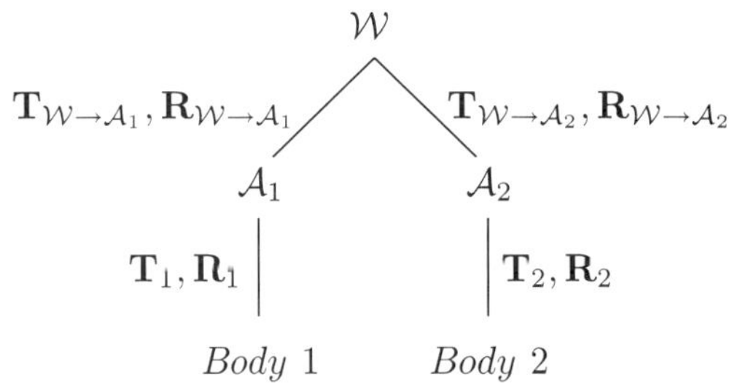

$\mathcal{W}$

$\mathbf{T}_{\mathcal{W}\to\mathcal{A}_1}, \mathbf{R}_{\mathcal{W}\to\mathcal{A}_1}$ $\quad$ $\mathbf{T}_{\mathcal{W}\to\mathcal{A}_2}, \mathbf{R}_{\mathcal{W}\to\mathcal{A}_2}$

$\mathcal{A}_1$ $\quad$ $\mathcal{A}_2$

$\mathbf{T}_1, \mathbf{R}_1$ $\quad$ $\mathbf{T}_2, \mathbf{R}_2$

*Body* 1 $\quad$ *Body* 2

The configuration of each rigid body relative to the main observer $\mathcal{W}$ is decomposed into the configuration of the body relative to an intermediate auxiliary observer and the configuration of the auxiliary observer relative to $\mathcal{W}$. Here,

$$\mathbf{R}_1 \circ \mathbf{T}_1$$

describes the configuration of the first rigid body relative to $\mathcal{A}_1$,

$$\mathbf{R}_2 \circ \mathbf{T}_2$$

describes the configuration of the second rigid body relative to $\mathcal{A}_2$,

$$\mathbf{R}_{\mathcal{W}\to\mathcal{A}_1} \circ \mathbf{T}_{\mathcal{W}\to\mathcal{A}_1}$$

describes the configuration of the $\mathcal{A}_1$ observer relative to $\mathcal{W}$, and

$$\mathbf{R}_{\mathcal{W}\to\mathcal{A}_2} \circ \mathbf{T}_{\mathcal{W}\to\mathcal{A}_2}$$

describes the configuration of the $\mathcal{A}_2$ observer relative to $\mathcal{W}$. It follows that the configuration of the first rigid body relative to $\mathcal{W}$ is given by

$$\mathbf{R}_1 \circ \mathbf{T}_1 \circ \mathbf{R}_{\mathcal{W}\to\mathcal{A}_1} \circ \mathbf{T}_{\mathcal{W}\to\mathcal{A}_1} = \mathbf{R}_1 \circ \mathbf{R}_{\mathcal{W}\to\mathcal{A}_1} \circ \mathbf{T}_1 \circ \mathbf{T}_{\mathcal{W}\to\mathcal{A}_1}$$

and similarly for the second rigid body.

**Illustration 2.6**

We may reorganize the tree structure discussed above to promote the $\mathcal{A}_1$ observer to the main observer according to the tree representation below:

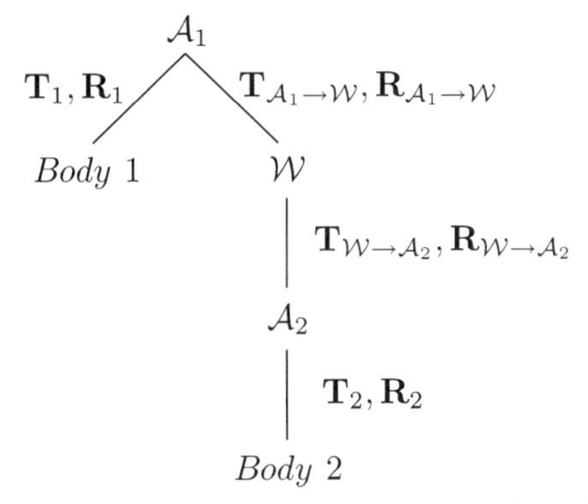


Then, the configuration of the second rigid body relative to $\mathcal{A}_1$ is given by

$$\mathbf{R}_2 \circ \mathbf{T}_2 \circ \mathbf{R}_{\mathcal{W}\to\mathcal{A}_2} \circ \mathbf{T}_{\mathcal{W}\to\mathcal{A}_2} \circ \mathbf{R}_{\mathcal{A}_1\to\mathcal{W}} \circ \mathbf{T}_{\mathcal{A}_1\to\mathcal{W}}$$
$$= \mathbf{R}_2 \circ \mathbf{R}_{\mathcal{W}\to\mathcal{A}_2} \circ \mathbf{R}_{\mathcal{A}_1\to\mathcal{W}} \circ \mathbf{T}_2 \circ \mathbf{T}_{\mathcal{W}\to\mathcal{A}_2} \circ \mathbf{T}_{\mathcal{A}_1\to\mathcal{W}}$$

and so on.

## 2.3 Recommended Methodology

Consider a system with multiple rigid bodies and a multiplicity of possible main observers, relative to which the configuration of each of the rigid bodies can be described. In determining the appropriate course of action to adequately describe the system, two modeling decisions need to be made. First, we must select a single observer to act as the main observer relative to which the configurations of all rigid bodies will ultimately be described. Second, we may introduce any number of auxiliary observers so as to simplify the descriptions of any pure translations and pure rotations corresponding to specific branches in the resulting tree representation. In this section, I shall propose some simple rules of thumb that I recommend you consider and possibly adopt.

### 2.3.1 The Main Observer

The selection of the main observer depends on the purpose of the modeling.

With emphasis on computer animations of multibody systems, the natural main observer corresponds to the internal representation of space within the appropriate computer-graphics application. By relating the configuration of all rigid bodies (and of any auxiliary observers) to this observer, we provide all the information necessary to reproduce the visual appearance of the system within the computer-graphics application. Whichever observer we choose to promote to main observer, the visual representation within the computer-graphics application reflects the positions and orientations of all the rigid bodies relative to this observer.

When considering computer animations, I typically use the letter $\mathcal{W}$ to denote the main observer and refer to it as the *world observer* (hence the choice of "W").

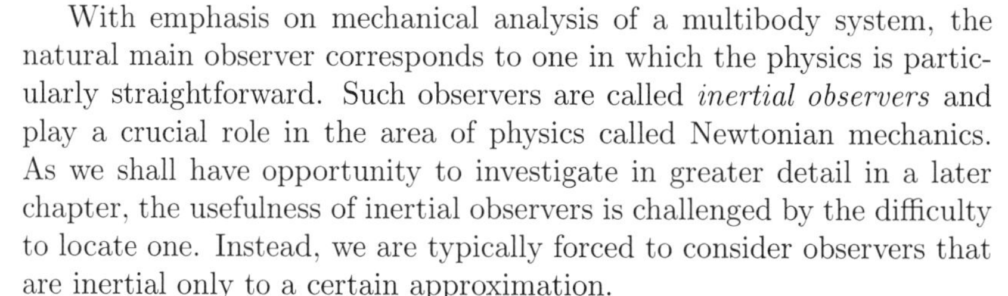

With emphasis on mechanical analysis of a multibody system, the natural main observer corresponds to one in which the physics is particularly straightforward. Such observers are called *inertial observers* and play a crucial role in the area of physics called Newtonian mechanics. As we shall have opportunity to investigate in greater detail in a later chapter, the usefulness of inertial observers is challenged by the difficulty to locate one. Instead, we are typically forced to consider observers that are inertial only to a certain approximation.

When considering mechanical analysis, I typically use the letter $\mathcal{N}$ (as in Newton) to denote the main observer.

### 2.3.2 The Auxiliary Observers

The introduction of auxiliary observers is entirely at the discretion of the modeler.

Whatever the application, the fundamental role of auxiliary observers in the analysis is to reduce the complexity of the mathematics necessary to describe the configuration of any rigid body relative to the main observer.

Although certainly not a necessity, I recommend that the motion of the rigid bodies relative to the main observer be completely described by the motion of the auxiliary observers relative to the main observer. In this fashion, the pure translation and the pure rotation that relate a rigid body to its parent observer are always time-independent. This also implies that at each node in a corresponding tree representation, the branch nodes may be rigid bodies or additional auxiliary observers.

### 2.3.3 Loops

In a tree representation corresponding to a multibody mechanism, auxiliary observers are introduced as intermediate nodes along different branches to reduce the complexity of any given pure translation or pure rotation relating two successive observers along that branch.

A branch segment between two observers $\mathcal{A}$ and $\mathcal{B}$ corresponds to a pair of pure translations $\mathbf{T}_{\mathcal{A}\to\mathcal{B}}$ and $\mathbf{T}_{\mathcal{B}\to\mathcal{A}} = (\mathbf{T}_{\mathcal{A}\to\mathcal{B}})^{-1}$ and a pair of pure rotations $\mathbf{R}_{\mathcal{A}\to\mathcal{B}}$ and $\mathbf{R}_{\mathcal{B}\to\mathcal{A}} = (\mathbf{R}_{\mathcal{A}\to\mathcal{B}})^{-1}$. We say that the two observers $\mathcal{A}$ and $\mathcal{B}$ are *neighbors* and that their configurations are *directly related.*

$$\begin{array}{c} \mathcal{A} \\ \big| \; \mathbf{T}_{\mathcal{A}\to\mathcal{B}}, \mathbf{R}_{\mathcal{A}\to\mathcal{B}} \\ \mathcal{B} \end{array}$$

If the path between two observers $\mathcal{C}$ and $\mathcal{E}$ passes through at least one intermediate node, corresponding to an observer $\mathcal{D}$, then the configurations of the observers $\mathcal{C}$ and $\mathcal{E}$ are *indirectly related.*

$$\begin{array}{c} \mathcal{C} \\ \big| \; \mathbf{T}_{\mathcal{C}\to\mathcal{D}}, \mathbf{R}_{\mathcal{C}\to\mathcal{D}} \\ \mathcal{D} \\ \big| \; \mathbf{T}_{\mathcal{D}\to\mathcal{E}}, \mathbf{R}_{\mathcal{D}\to\mathcal{E}} \\ \mathcal{E} \end{array}$$

As shown above, $\mathbf{T}_{\mathcal{C}\to\mathcal{E}}$ and $\mathbf{R}_{\mathcal{C}\to\mathcal{E}}$ can be computed from the compositions

$$\mathbf{T}_{\mathcal{C}\to\mathcal{E}} = \mathbf{T}_{\mathcal{D}\to\mathcal{E}} \circ \mathbf{T}_{\mathcal{C}\to\mathcal{D}}$$

and

$$\mathbf{R}_{\mathcal{C}\to\mathcal{E}} = \mathbf{R}_{\mathcal{D}\to\mathcal{E}} \circ \mathbf{R}_{\mathcal{C}\to\mathcal{D}}.$$

Similar computations could be employed to relate every observer in the tree structure to every other observer.

It is tempting to suggest the result of these computations by including connections between all nodes in the original tree structure, resulting in a *network* with multiple closed loops. As long as all the pure translations and pure rotations between originally **indirectly related** observers are computed using the above expressions, there is no need for concern.

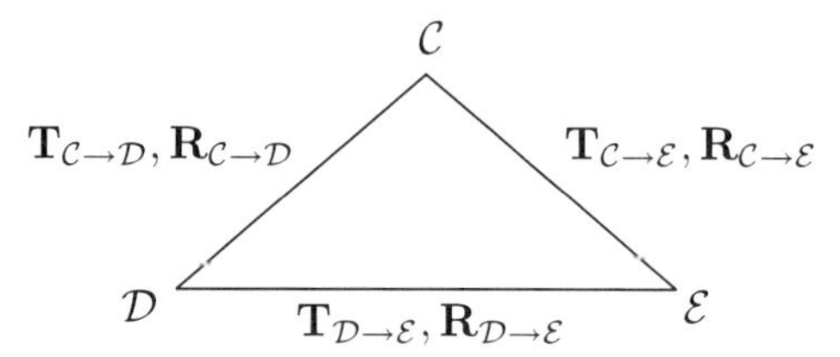


Problems may arise, however, with the network representation if it is employed during the modeling stage. This could occur, for example, as a result of an inconsistent specification of the pure translation and/or pure rotation relating two already indirectly related observers. The possibility of overdetermining the relative configurations of two observers should be excluded in practice. The tree representation, excluding all possible node loops, **cannot** suffer from internal inconsistencies, yet is capable of completely describing all possible configurations of the mechanical system.

(Ex. 2.12 – Ex. 2.13)

## 2.4 Examples

### 2.4.1 A Still Life

Suppose you want to describe the geometry of an assortment of rigid bodies that are stationary relative to you. The tree structure below exemplifies the recommended geometric hierarchy:

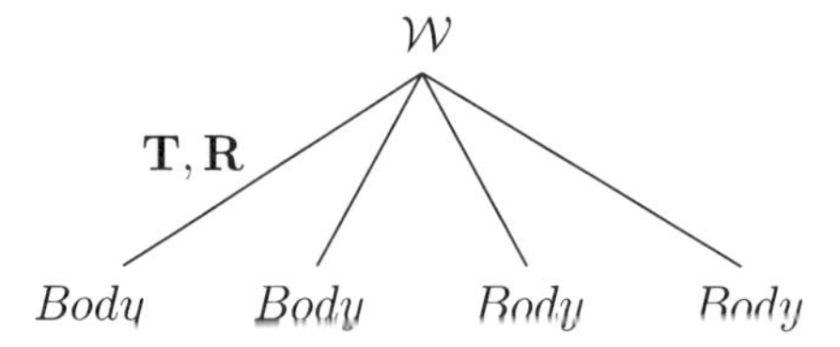


Each rigid body is directly related to the main observer $\mathcal{W}$ through a unique pure translation $\mathbf{T}$ and a pure rotation $\mathbf{R}$. Since these are independent of time, there is no pressing reason to introduce any auxiliary observers. As suggested in the previous chapter, we are certainly free to introduce intermediate observers, with the help of which we can decompose the configuration of any single rigid body relative to $\mathcal{W}$ into more manageable steps.

For example, we may introduce an auxiliary observer $\mathcal{A}$, such that the configuration of the rigid body relative to $\mathcal{A}$ is given by the pure rotation $\mathbf{R}_{\mathcal{A}} = \mathbf{R}$ and the configuration of $\mathcal{A}$ relative to $\mathcal{W}$ is given by the pure translation $\mathbf{T}_{\mathcal{W}\to\mathcal{A}} = \mathbf{T}$. Similarly, we may introduce an auxiliary observer $\mathcal{B}$, such that the configuration of the rigid body relative to $\mathcal{B}$ is given by a pure rotation $\mathbf{R}_{\mathcal{B}}$ about a predetermined axis through the rigid body and the configuration of $\mathcal{B}$ relative to $\mathcal{A}$ is given by a pure rotation $\mathbf{R}_{\mathcal{A}\to\mathcal{B}}$ that aligns this axis through the rigid body with the corresponding axis in the actual configuration of the body.

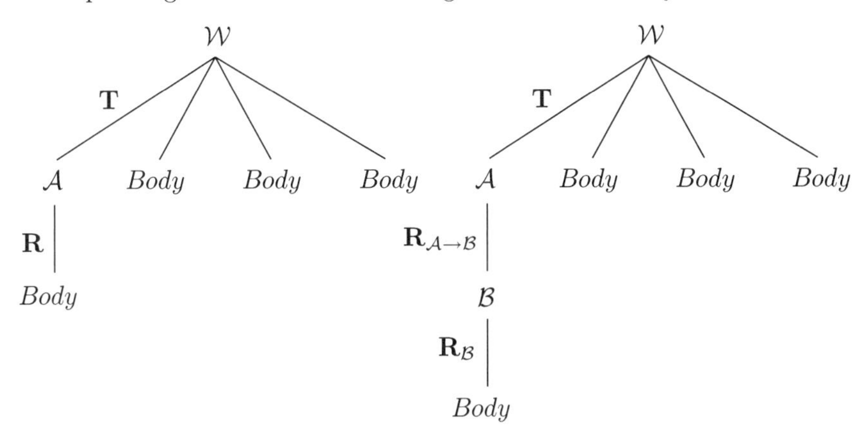


The process of introducing auxiliary observers naturally reaches a conclusion when all pure translations and pure rotations are described in as straightforward a manner as possible. It is never necessary to introduce such observers in the case where all the bodies are stationary relative to the main observer, but it may turn out to be convenient at times. We will have ample opportunity to return to this in greater detail when we develop the quantitative theory of translations and rotations.

## 2.4.2 The Single Moving Rigid Body

We now apply the methodology suggested thus far to the motion of a single rigid body, say a block, relative to a background that is stationary relative to the main observer $\mathcal{W}$. Under the assumption that the block's configuration relative to $\mathcal{W}$ changes with time, the recommended methodology **requires** the introduction of at least one auxiliary observer $\mathcal{A}$, relative to which the block remains stationary. The motion of the block relative to $\mathcal{W}$ is contained within the time-dependent configuration of $\mathcal{A}$ relative to $\mathcal{W}$.

The configuration of $\mathcal{A}$ relative to $\mathcal{W}$ can be uniquely described as a combination of a (possibly time-dependent) pure translation $\mathbf{T}_{\mathcal{W}\to\mathcal{A}}$ and a (possibly time-dependent) pure rotation $\mathbf{R}_{\mathcal{W}\to\mathcal{A}}$. If the orientation of the block relative to $\mathcal{W}$ does not depend on time, then it is appropriate to introduce $\mathcal{A}$ in such a way that $\mathbf{R}_{\mathcal{W}\to\mathcal{A}} = \mathbf{I}$, the identity rotation. Alternatively, if the position of the block relative to $\mathcal{W}$ does not depend on time, then it is appropriate to introduce $\mathcal{A}$ in such a way that $\mathbf{T}_{\mathcal{W}\to\mathcal{A}} = \mathbf{I}$, the identity translation.

**Illustration 2.7**

Suppose you want to describe the *sliding* motion of a puck on the surface of an ice hockey rink.

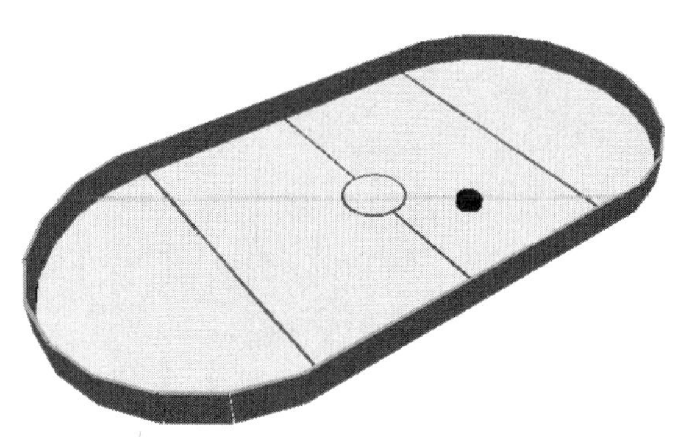

Consider a main observer $\mathcal{W}$, relative to which the ice hockey rink remains stationary. Let its reference configuration coincide with the puck's configuration when it sits at the center of the rink. Introduce an auxiliary observer $\mathcal{A}$ corresponding to a reference configuration that is coincident with the puck at all times. Then, the configuration of $\mathcal{A}$ relative to $\mathcal{W}$ is entirely determined by the pure translation $\mathbf{T}_{\mathcal{W}\to\mathcal{A}}$ corresponding to shifting the block in some direction parallel to the surface of the rink and by some amount that depend on time.

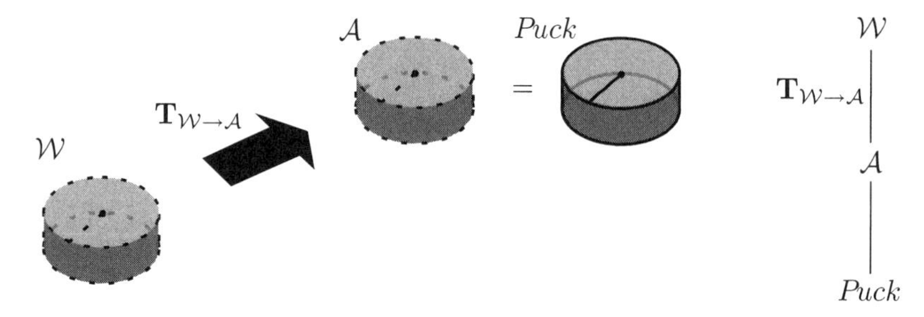


If, in addition to sliding, we also want to consider the *spinning* of the puck about an axis perpendicular to the ice, then the puck's configuration is no longer stationary relative to $\mathcal{A}$. Instead, introduce an auxiliary observer $\mathcal{B}$ corresponding to a reference configuration that is coincident with the puck at all times. The configuration of $\mathcal{B}$ relative to $\mathcal{A}$ is entirely determined by a pure rotation $\mathbf{R}_{\mathcal{A}\to\mathcal{B}}$ corresponding to a rotation about an axis perpendicular to the surface of the rink. As before, the configuration of $\mathcal{A}$ relative to $\mathcal{W}$ is entirely determined by a pure translation $\mathbf{T}_{\mathcal{W}\to\mathcal{A}}$ parallel to the surface of the rink.

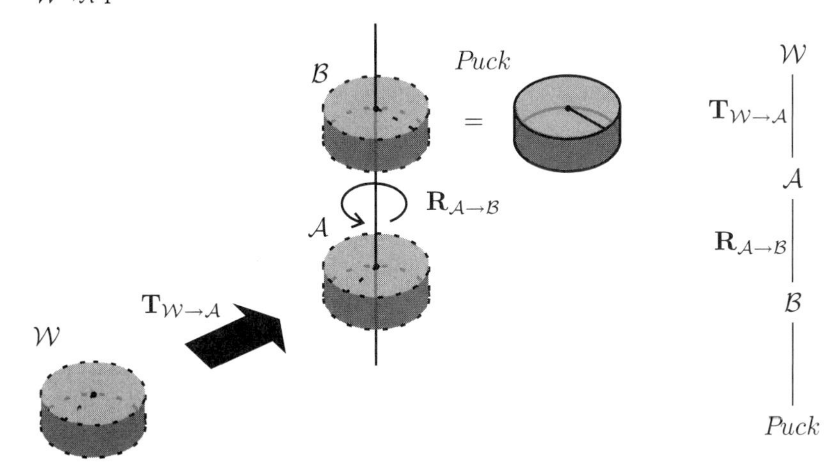


Suppose you want to describe the motion of a small bead sliding on the surface of a sphere.

Consider a main observer $\mathcal{W}$, relative to which the sphere remains stationary. Let its reference configuration be represented by a virtual block whose reference position is at the center of the sphere. Since the bead is so small, we shall disregard changes in its orientation and focus, instead, on describing its position relative to $\mathcal{W}$. Introduce an auxiliary observer $\mathcal{A}$, such that the reference position of its virtual block is at the center of the sphere and such that the actual position of the bead relative to $\mathcal{A}$ is given by a **time-independent** pure translation $\mathbf{T}$ corresponding to a shift by a constant amount (actually, the radius of the sphere) along a fixed direction relative to $\mathcal{A}$.

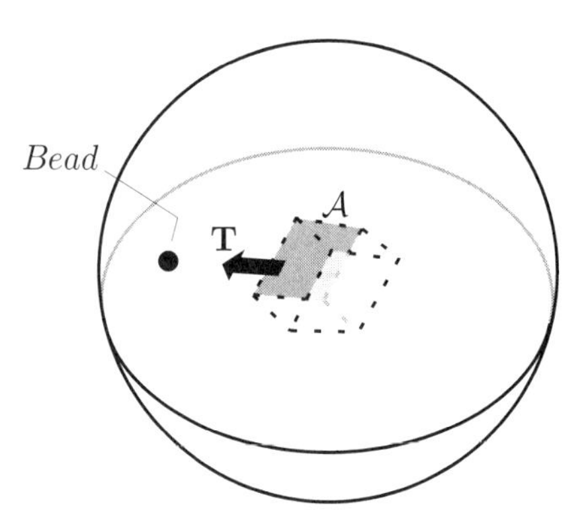


Since the bead can be positioned anywhere on the surface of the sphere, we must be able to accommodate changes in the orientation of $\mathcal{A}$ relative to $\mathcal{W}$ that ensure that the fixed direction corresponding to **T** points from the center of the sphere to the actual location of the bead. Since the reference positions of $\mathcal{W}$ and $\mathcal{A}$ are both at the center of the sphere, it follows that the configuration of $\mathcal{A}$ relative to $\mathcal{W}$ is entirely determined by a **time-dependent** pure rotation $\mathbf{R}_{\mathcal{W}\to\mathcal{A}}$.

Consider two separate axes through the center of the sphere. Then, we may introduce an auxiliary observer $\mathcal{B}$, such that the configuration of $\mathcal{B}$ relative to $\mathcal{W}$ is given by a pure rotation $\mathbf{R}_{\mathcal{W}\to\mathcal{B}}$ about the first axis and the configuration of $\mathcal{A}$ relative to $\mathcal{B}$ is given by a pure rotation $\mathbf{R}_{\mathcal{B}\to\mathcal{A}}$ about the second axis, as suggested in the figure below.

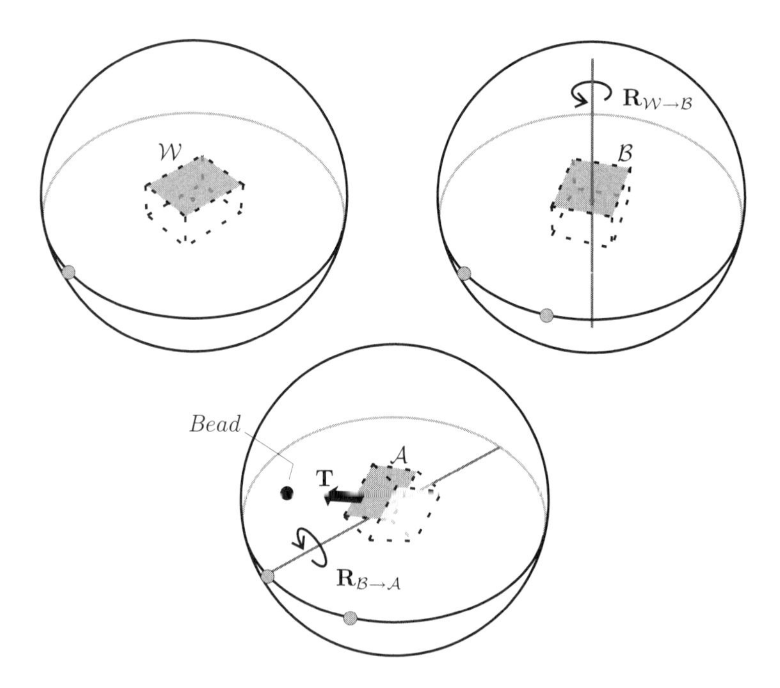

The pure rotation that relates the orientation of $\mathcal{A}$ to that of $\mathcal{W}$ is then given by

$$\mathbf{R}_{\mathcal{W}\to\mathcal{A}} = \mathbf{R}_{\mathcal{B}\to\mathcal{A}} \circ \mathbf{R}_{\mathcal{W}\to\mathcal{B}}.$$

Thus, the configuration of the bead is given by the composition

$$\mathbf{T} \circ \mathbf{R}_{\mathcal{W}\to\mathcal{A}} = \mathbf{T} \circ \mathbf{R}_{\mathcal{B}\to\mathcal{A}} \circ \mathbf{R}_{\mathcal{W}\to\mathcal{B}},$$

where any changes with time are contained within the pure rotations $\mathbf{R}_{\mathcal{B}\to\mathcal{A}}$ and $\mathbf{R}_{\mathcal{W}\to\mathcal{B}}$.

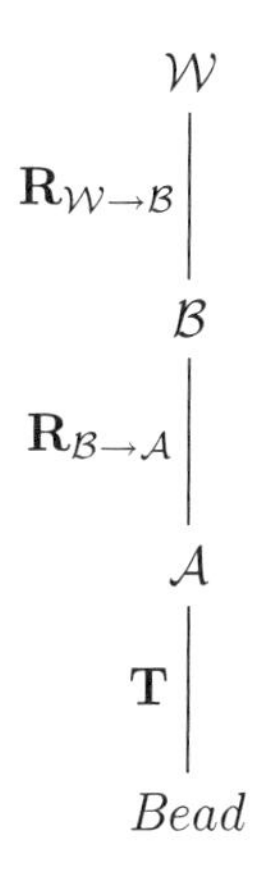


### 2.4.3 Mechanical Joints

**Illustration 2.8**

Suppose you want to describe the motion of two rigid rods that are joined at a hinge joint, allowing each rod to rotate relative to the other rod about a direction fixed relative to the two rods.

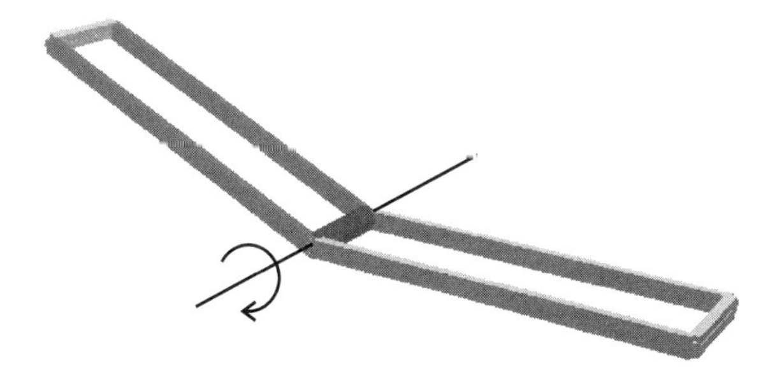

Let $\mathcal{W}$ denote the main observer, relative to which the motion of the pair of rods will be described. Since both rods are free to move relative to $\mathcal{W}$, introduce two auxiliary observers $\mathcal{A}_1$ and $\mathcal{A}_2$, such that the configurations of rods 1 and 2 are stationary relative to $\mathcal{A}_1$ and $\mathcal{A}_2$, respectively. A first attempt to describe this geometry is contained within the tree structure below, wherein each of the auxiliary observers is directly related to the main observer through a unique pure translation $\mathbf{T}_{\mathcal{W}\to\mathcal{A}_i}$ and a pure rotation $\mathbf{R}_{\mathcal{W}\to\mathcal{A}_i}$.

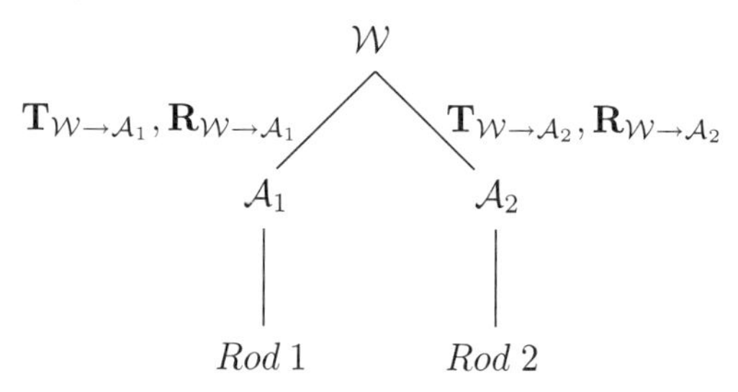


In contrast, you may reorganize the tree structure to suggest the presence of the hinge joint and the severe restriction on the relative configuration of the two rods as shown below.

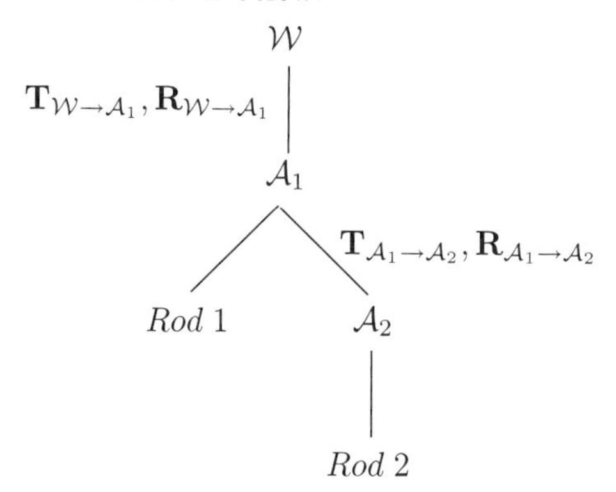


Here, the configuration of the second auxiliary observer $\mathcal{A}_2$ relative to the first auxiliary observer $\mathcal{A}_1$ is given by the pure translation

$$\mathbf{T}_{\mathcal{A}_1\to\mathcal{A}_2} = \mathbf{T}_{\mathcal{W}\to\mathcal{A}_2} \circ \left(\mathbf{T}_{\mathcal{W}\to\mathcal{A}_1}\right)^{-1}$$

and the pure rotation

$$\mathbf{R}_{\mathcal{A}_1\to\mathcal{A}_2} = \mathbf{R}_{\mathcal{W}\to\mathcal{A}_2} \circ \left(\mathbf{R}_{\mathcal{W}\to\mathcal{A}_1}\right)^{-1}.$$

We note that in the present arrangement, the auxiliary observer $\mathcal{A}_1$ fills two purposes. On the one hand, it acts to contain all the time-dependent changes in the configuration of the first rod relative to $\mathcal{W}$. On the other hand, it acts to decompose the configuration of $\mathcal{A}_2$ relative to $\mathcal{W}$. In its latter role, it is particularly useful if the pure translation $\mathbf{T}_{\mathcal{A}_1\to\mathcal{A}_2}$ and the pure rotation $\mathbf{R}_{\mathcal{A}_1\to\mathcal{A}_2}$ are simple to describe.

Consider choosing $\mathcal{A}_1$ and $\mathcal{A}_2$, such that the corresponding reference positions coincide with the location of the hinge joint.

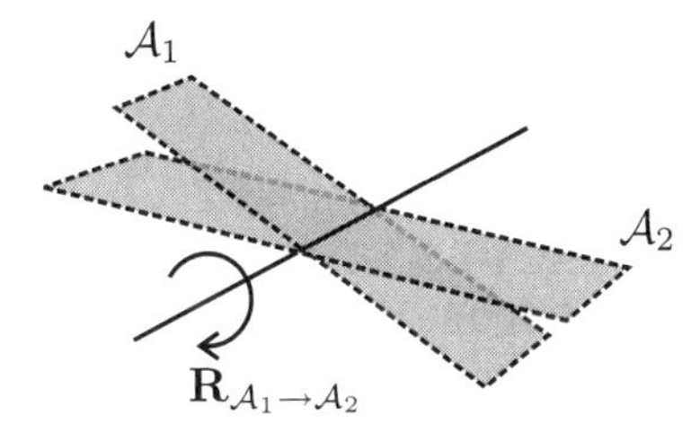


This choice respects the requirement that the rods be stationary relative to the corresponding auxiliary observer. We may now choose the reference orientations of $\mathcal{A}_1$ and $\mathcal{A}_2$, such that $\mathbf{T}_{\mathcal{A}_1 \to \mathcal{A}_2} = \mathbf{I}$ and $\mathbf{R}_{\mathcal{A}_1 \to \mathcal{A}_2}$ corresponds to a rotation about the hinge axis, a dramatic reduction in the complexity of the geometry description.

If the configuration of a rigid body relative to another rigid body is easier to describe than the configuration of the rigid body relative to some observer $\mathcal{O}$, consider a hierarchy that reflects this observation. For example, if $\mathcal{A}_1$ and $\mathcal{A}_2$ are the auxiliary observers relative to which the two rigid bodies are stationary, then introduce $\mathcal{A}_1$ (or $\mathcal{A}_2$) as the parent observer of $\mathcal{A}_2$ (or $\mathcal{A}_1$) rather than directly relating both of these to $\mathcal{O}$ as shown below.

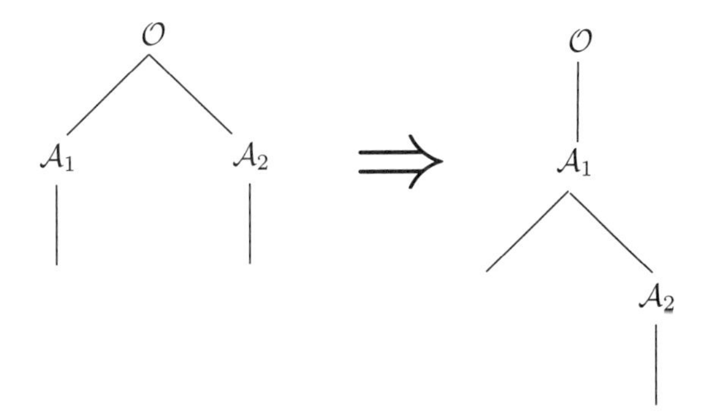


Suppose, for example, that you want to describe the motion of the trolley that rolls along the jib of a tower crane and from which the lifting cable and hook are suspended, and the motion of the jib and counterweight assembly that may rotate relative to the tower.

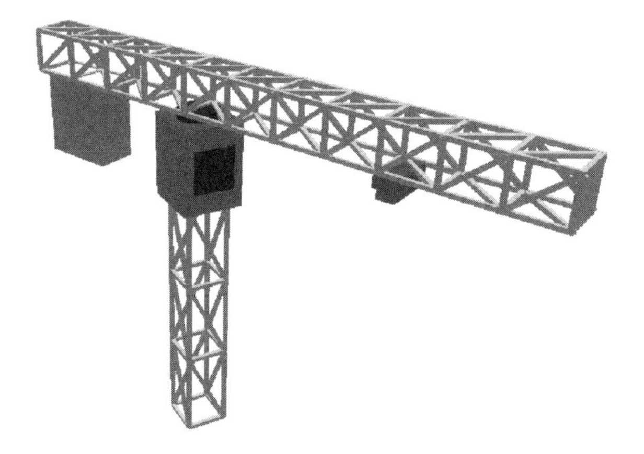

Let $\mathcal{W}$ denote the main observer, relative to which the tower remains stationary, such that the reference position of $\mathcal{W}$ coincides with the joint between the main jib and the tower. Introduce two auxiliary observers $\mathcal{T}$ and $\mathcal{J}$, relative to which the trolley and the main jib remain stationary, such that the reference position of $\mathcal{J}$ coincides with that of $\mathcal{W}$. The tree structure below reflects the hierarchy inherent in the mechanical construction.

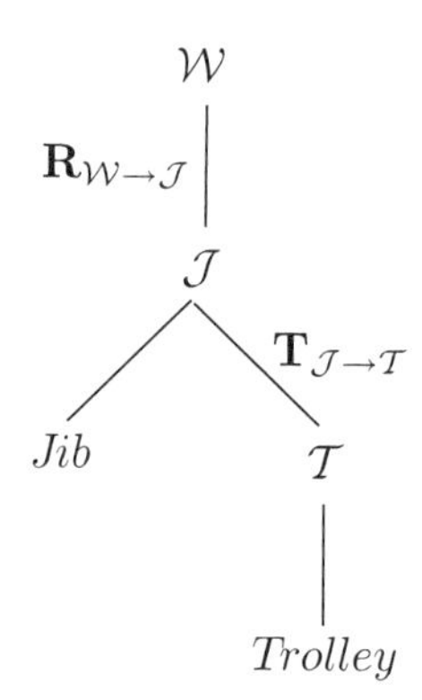


We may now choose the reference orientations of $\mathcal{W}$, $\mathcal{J}$, and $\mathcal{T}$, such that the configuration of $\mathcal{J}$ relative to $\mathcal{W}$ is determined by a pure rotation $\mathbf{R}_{\mathcal{W}\to\mathcal{J}}$ about an axis parallel to the tower, while the configuration of $\mathcal{T}$ relative to $\mathcal{J}$ is determined by a pure translation $\mathbf{T}_{\mathcal{J}\to\mathcal{T}}$ along the direction of the main jib.

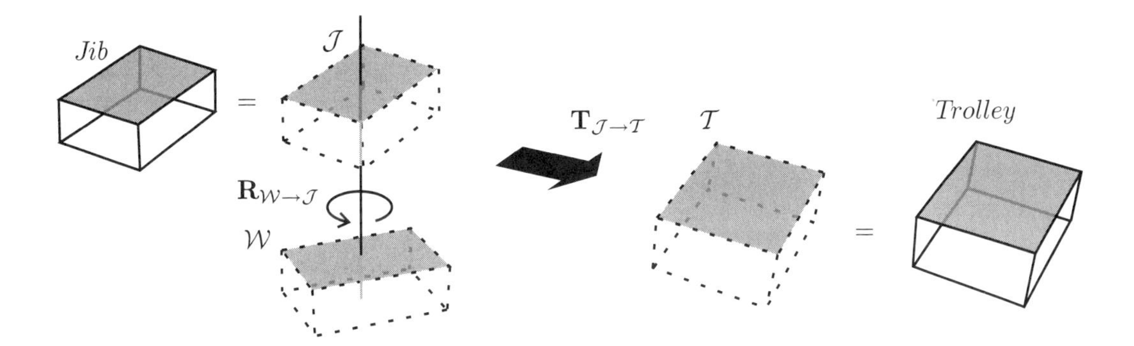


## 2.4.4 A Bicycle

Suppose you want to describe the motion of a simple bicycle consisting of a rear wheel, a front wheel, a steering shaft, and a frame.

In particular, the rear wheel may rotate only about a fixed direction relative to the frame; the front wheel may rotate only about a fixed direction relative to the steering shaft; and the steering shaft may rotate only about a fixed direction relative to the frame.

Let $\mathcal{W}$ denote the main observer, relative to which the motion of the bicycle will be described. Introduce four auxiliary observers $\mathcal{A}_{\text{rear wheel}}$, $\mathcal{A}_{\text{front wheel}}$, $\mathcal{A}_{\text{steering}}$, and $\mathcal{A}_{\text{frame}}$, relative to which the rear wheel, front wheel, steering shaft, and frame, respectively, are stationary. In particular, let the reference positions of $\mathcal{A}_{\text{rear wheel}}$ and $\mathcal{A}_{\text{frame}}$ coincide with the center of the rear wheel and let the reference positions of $\mathcal{A}_{\text{front wheel}}$ and $\mathcal{A}_{\text{steering}}$ coincide with the center of the front wheel.

With a suitable choice of reference orientations, the configuration of $\mathcal{A}_{\text{rear wheel}}$ relative to $\mathcal{A}_{\text{frame}}$ is described in terms of a pure rotation $\mathbf{R}_{\mathcal{A}_{\text{frame}}\to\mathcal{A}_{\text{rear wheel}}}$ about an axis perpendicular to the rear wheel. Similarly, the configuration of $\mathcal{A}_{\text{front wheel}}$ relative to $\mathcal{A}_{\text{steering}}$ is described in terms of a pure rotation $\mathbf{R}_{\mathcal{A}_{\text{steering}}\to\mathcal{A}_{\text{front wheel}}}$ about an axis perpendicular to the front wheel. Finally, the configuration of $\mathcal{A}_{\text{frame}}$ relative to

$\mathcal{A}_{\text{steering}}$ is described in terms of a pure translation $\mathbf{T}_{\mathcal{A}_{\text{steering}} \to \mathcal{A}_{\text{frame}}}$ along the axis through the centers of the rear and front wheels, followed by a rotation $\mathbf{R}_{\mathcal{A}_{\text{steering}} \to \mathcal{A}_{\text{frame}}}$ about an axis parallel to the axis of rotation of the shaft relative to the frame.

Any of the four observer hierarchies below reflects the inherent mechanical design.

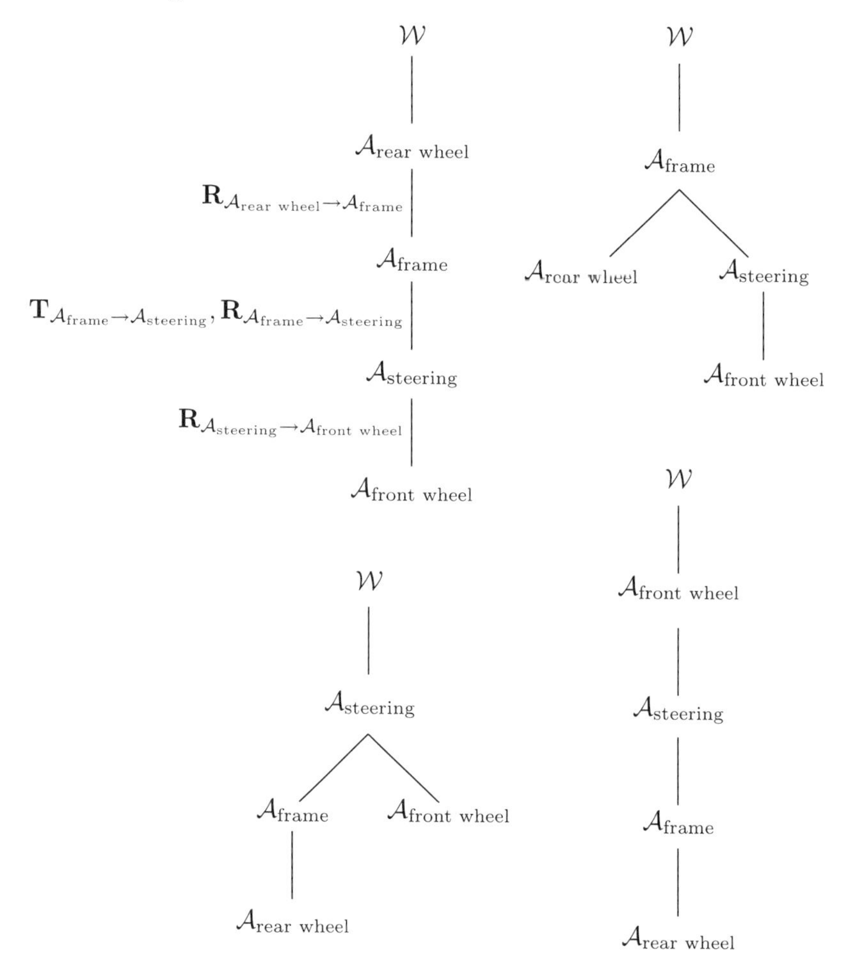


In each case, the configuration of the uppermost auxiliary observer relative to the main observer $\mathcal{W}$ is given by a unique pure translation and pure rotation. Each of the four hierarchies appears superior from a modeling perspective to the "flat" hierarchy, in which all the parts of the bicycle are directly related to the main observer.

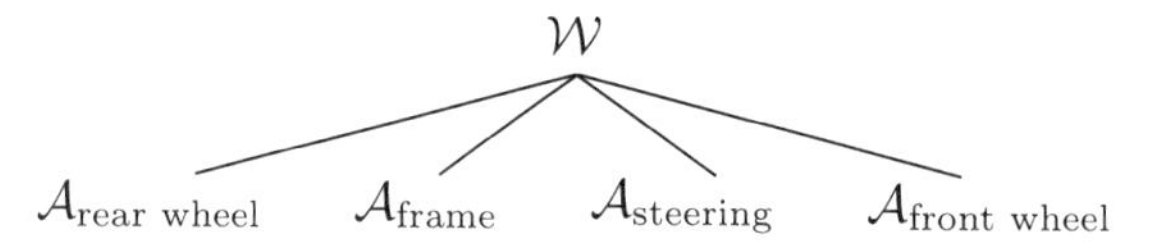

### 2.4.5 A Desk Lamp

Suppose you want to describe the motion of a desk lamp consisting of a base, an upper beam, a middle beam, a lower beam, a bracket, and a lamp shade.

In particular, the upper, middle, and lower beams may rotate only about a fixed direction relative to the base; the bracket may rotate only about a fixed direction relative to the upper and lower beams; and the lamp shade may rotate only about a fixed direction relative to the bracket.

In contrast to the previous examples, however, these rotations are not all independent. Instead, the hinge joints connecting the bracket to the upper and lower beams constrain the rotations of the lower and upper beams relative to the base and of the bracket relative to the beams, in such a way as to sustain the connection for all time. Similarly, the spur gears attached to the horizontal bars supporting the upper and middle beams constrain changes in the rotations of the upper and middle beams relative to the base, in such a way as to respect the impenetrability of the gear teeth as shown in the figure on the next page.

Let $\mathcal{W}$ denote the main observer, relative to which the motion of the lamp will be described. Introduce six auxiliary observers $\mathcal{A}_{\text{base}}$, $\mathcal{A}_{\text{upper beam}}$, $\mathcal{A}_{\text{middle beam}}$, $\mathcal{A}_{\text{lower beam}}$, $\mathcal{A}_{\text{bracket}}$, and $\mathcal{A}_{\text{shade}}$, relative to which the base, upper beam, middle beam, lower beam, bracket, and lamp shade, respectively, are stationary. In particular, let the reference position of $\mathcal{A}_{\text{base}}$ coincide with the point on the top of the base centered between the vertical posts; let the reference positions of $\mathcal{A}_{\text{upper beam}}$,

$\mathcal{A}_{\text{middle beam}}$, and $\mathcal{A}_{\text{lower beam}}$ coincide with the centers of the horizontal bars supporting the corresponding beam; let the reference position of $\mathcal{A}_{\text{bracket}}$ coincide with the far end point of the lower beam, and let the reference position of $\mathcal{A}_{\text{shade}}$ coincide with the hinge joint about which the lamp shade rotates.

With a suitable choice of reference orientations of the auxiliary observers, the configurations of $\mathcal{A}_{\text{upper beam}}$, $\mathcal{A}_{\text{middle beam}}$, and $\mathcal{A}_{\text{lower beam}}$ relative to $\mathcal{A}_{\text{base}}$ are described in terms of time-independent pure translations $\mathbf{T}_{\mathcal{A}_{\text{base}}\to\mathcal{A}_{\text{upper beam}}}$, $\mathbf{T}_{\mathcal{A}_{\text{base}}\to\mathcal{A}_{\text{middle beam}}}$, $\mathbf{T}_{\mathcal{A}_{\text{base}}\to\mathcal{A}_{\text{lower beam}}}$ along directions parallel to the vertical posts, and time-dependent pure rotations $\mathbf{R}_{\mathcal{A}_{\text{base}}\to\mathcal{A}_{\text{upper beam}}}$, $\mathbf{R}_{\mathcal{A}_{\text{base}}\to\mathcal{A}_{\text{middle beam}}}$, $\mathbf{R}_{\mathcal{A}_{\text{base}}\to\mathcal{A}_{\text{lower beam}}}$ about axes parallel to the horizontal bars. Similarly, the configuration of $\mathcal{A}_{\text{bracket}}$ relative to $\mathcal{A}_{\text{lower beam}}$ is described in terms of a time-independent pure translation $\mathbf{T}_{\mathcal{A}_{\text{lower beam}}\to\mathcal{A}_{\text{bracket}}}$ along a direction parallel to the lower beam and a time-dependent pure rotation $\mathbf{R}_{\mathcal{A}_{\text{lower beam}}\to\mathcal{A}_{\text{bracket}}}$ about an axis parallel to the horizontal bars. Finally, the configuration of $\mathcal{A}_{\text{shade}}$ relative to $\mathcal{A}_{\text{bracket}}$ is described in terms of a time-independent pure translation $\mathbf{T}_{\mathcal{A}_{\text{bracket}}\to\mathcal{A}_{\text{shade}}}$ and a time-dependent pure rotation $\mathbf{R}_{\mathcal{A}_{\text{bracket}}\to\mathcal{A}_{\text{shade}}}$ about an axis parallel to the horizontal bars.

The observer hierarchy below reflects the inherent mechanical design.

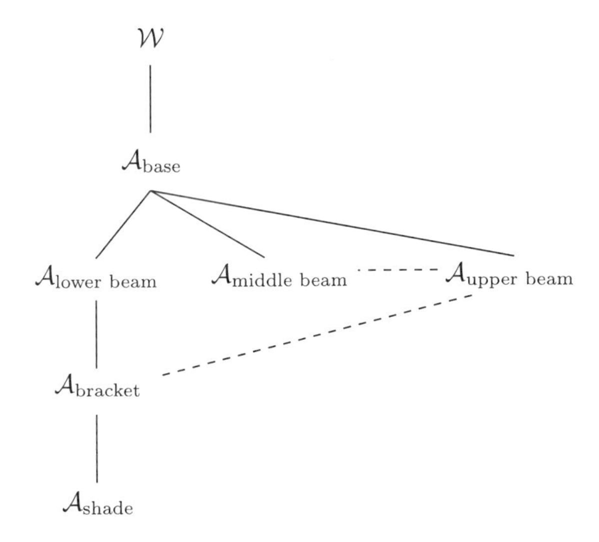


As noted above, the pure rotations contained within this hierarchy are not all independent. Instead, the closed loops formed by the base, upper and lower beams, and the bracket, on the one hand, and by the base and the upper and middle beams (through the spur gears), on the other hand, constrain the relative orientations and changes in these relative orientations, respectively, between different observers. Although it might appear natural to introduce a direct connection between the upper beam and the bracket and/or the upper and middle beams in the tree hierarchy above, the accepted methodology explicitly prohibits the creation of such loops in the observer hierarchy. Instead, we may choose to suggest the existence of constraints on the relative configurations by the use of dashed lines in the tree hierarchy as was done in the figure above.

## 2.5 MAMBO

(Ex. 2.14 – Ex. 2.15)

The computer-graphics application MAMBO interfaces with graphical subroutines within the computer operating system to represent an arbitrarily complex array of rigid bodies in a three-dimensional environment. Detail about the geometric description of the multibody mechanism is provided to the application through a MAMBO *geometry description* (a MAMBO .geo file). The MAMBO online reference manual contains a complete description of the .geo-file grammar.

**Illustration 2.9**

The simplest, grammatically correct MAMBO geometry description is given by the single statement:

```
MODULE World {
}
```

corresponding to the tree structure

*World*

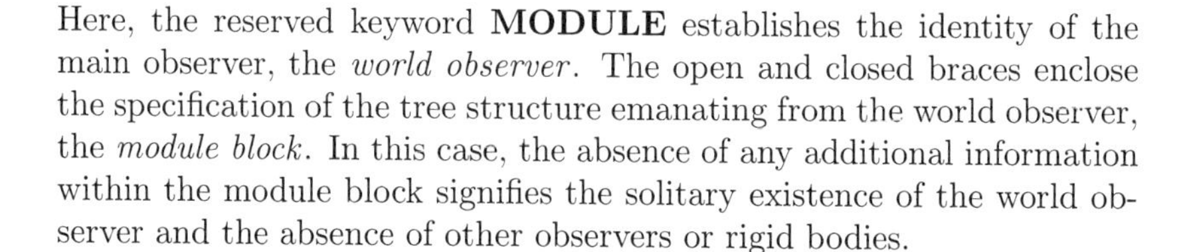

Here, the reserved keyword **MODULE** establishes the identity of the main observer, the *world observer*. The open and closed braces enclose the specification of the tree structure emanating from the world observer, the *module block*. In this case, the absence of any additional information within the module block signifies the solitary existence of the world observer and the absence of other observers or rigid bodies.

The reserved keyword **BODY** represents auxiliary observers. Any number of auxiliary observers may be contained within a module or body block, e.g.,

```
MODULE World {
   BODY Vehicle {
      BODY Rightfrontwheel {
      }
      BODY Leftfrontwheel {
      }
      BODY Rightrearwheel {
      }
      BODY Leftrearwheel {
      }
   }
   BODY Driver {
   }
}
```

In this example, the configuration of the driver observer is directly related to the world observer. The configurations of each of the four wheel observers, however, are only indirectly related to the world observer via the vehicle observer.

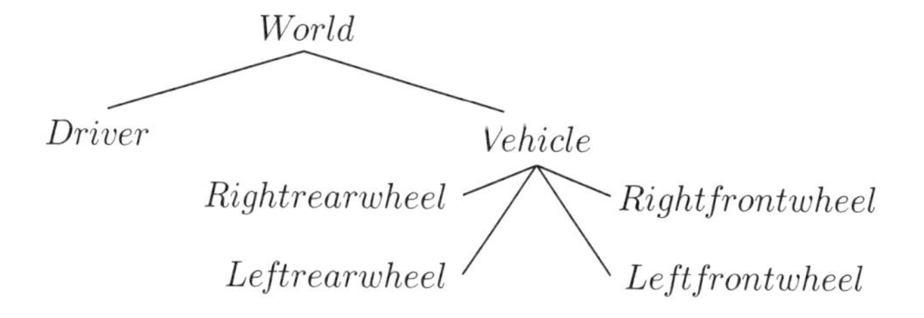

**Illustration 2.10**

Rigid bodies may be introduced at any level in the tree structure, i.e., within any observer block. MAMBO is shipped with an object library containing geometric primitives such as spheres, cylinders, and rectangular blocks. Each such primitive is represented within the MAMBO .geo file as an object block, as shown in the following geometry description:

```
MODULE World {
   BODY Vehicle {
      BLOCK {
      }
      BODY Rightfrontwheel {
         CYLINDER {
         }
      }
      BODY Leftfrontwheel {
         CYLINDER {
         }
      }
      BODY Rightrearwheel {
         CYLINDER {
         }
      }
      BODY Leftrearwheel {
         CYLINDER {
         }
      }
   }
   BODY Driver {
      SPHERE {
      }
   }
}
```

Loading[2] this geometry description into MAMBO will display the block, cylinders, and sphere, but will hardly represent the multibody mechanism you had in mind. (Try it!) A more appealing representation requires a specification of the pure translations and pure rotations that relate the configurations of rigid bodies to their parent observers and so on.

[2]You must first load a MAMBO motion description (a MAMBO .dyn file). An empty text file with the .dyn extension will do at this point (see the MAMBO reference manual).

The notion of observers and the corresponding tree representations is represented within the MAMBO computer-algebra toolbox through the definition of a global variable `GlobalObserverDeclarations`. Changes to this variable initiated by the user are made possible through the procedures `DeclareObservers` and `DefineNeighbors`.

The declaration of an observer using the `DeclareObservers` procedure appends the global variable `GlobalObserverDeclarations` to include the name associated with the observer.

**Illustration 2.11**

In the following extract from a MAMBO toolbox session, the content of the global variable `GlobalObserverDeclarations` is displayed immediately after invoking the `Restart` procedure and following the declaration of the three observers $\mathcal{A}$, $\mathcal{B}$, and $\mathcal{C}$.

```
> Restart():
> print(GlobalObserverDeclarations);
```

table([
])

```
> DeclareObservers(A,B,C):
> print(GlobalObserverDeclarations);
```

table([
$B = \{\}$
$C = \{\}$
$A = \{\}$
])

Here, prior to the declaration of any observers, the global variable is an empty table (see Appendix A for more detail on MAPLE data structures). After the observers have been declared, `GlobalObserverDeclarations` contains three entries with labels given by the names of the observers and associated empty sets. The empty sets signify the independence of the three observers with no information about the relative configuration of the different observers.

That the configuration of an observer is modeled as **directly related** to the configuration of a different observer is established by invoking the `DefineNeighbors` procedure.

**Illustration 2.12**

In the continuation below of the previous MAMBO toolbox session, `DefineNeighbors` establishes the $\mathcal{B}$ observer as a neighbor of the $\mathcal{A}$ and $\mathcal{C}$ observers. The resulting modification to `GlobalObserverDeclarations` is also shown.

```
> DefineNeighbors([A,B],[B,C]):
> print(GlobalObserverDeclarations);
```

$$\begin{array}{l}\text{table}([\\ \quad B = \{A, C\}\\ \quad C = \{B\}\\ \quad A = \{B\}\\ \quad ])\end{array}$$

Although no new observer labels have been added to the global variable `GlobalObserverDeclarations`, the associated, previously empty sets now contain the names of all observers that neighbor the observer specified by the label.

The MAMBO toolbox utility `GeometryOutput` can be used to generate tree representations of the information stored in `GlobalObserverDeclarations` in a format suitable for export into a MAMBO geometry description (i.e., a MAMBO .geo file). Given the name of an already declared observer to serve as the main observer, `GeometryOutput` generates a tree structure incorporating all observers in `GlobalObserverDeclarations` whose configuration can be directly or indirectly related to that of the main observer.

**Illustration 2.13**

We continue with the same MAMBO toolbox session as in the previous illustration. Here, we illustrate the output that results from different calls to the `GeometryOutput` utility.

```
> GeometryOutput(main=A);

MODULE A {
   BODY B {
      BODY C {
      }
   }
}
```

```
> GeometryOutput(main=B);

MODULE B {
   BODY A {
   }
   BODY C {
   }
}
```

```
> GeometryOutput(main=C);

MODULE C {
   BODY B {
      BODY A {
      }
   }
}
```

With the command

```
> GeometryOutput(main=A,filename="hierarchy.geo");
```

the output is directed to the file by the name `hierarchy.geo` within the current working directory[3]. More detail on the optional arguments to `GeometryOutput` may be found in Appendix B.

[3] The current working directory may be accessed and set using the `currentdir` command in Maple.

## 2.6 Exercises

**Exercise 2.1** As shown in the text, the successive composition of a sequence of pure translations is equivalent to a single pure translation. Similarly, sequences of pure translations that differ only in the order in which individual pure translations are applied are equivalent. What other mathematical objects and associated operations are you familiar with that exhibit the same set of properties?

**Exercise 2.2** As shown in the text, the successive composition of a sequence of pure rotations that keep the same point on the block fixed is equivalent to a single pure rotation. In contrast with pure translations, however, sequences of pure rotations that differ only in the order in which individual pure rotations are applied are generally not equivalent. What other mathematical objects and associated operations are you familiar with that exhibit the same set of properties?

**Exercise 2.3** Consider two pure rotations $\mathbf{R}_1$ and $\mathbf{R}_2$. Under what conditions on $\mathbf{R}_1$ and $\mathbf{R}_2$ does the equality

$$\mathbf{R}_1 \circ \mathbf{R}_2 = \mathbf{R}_2 \circ \mathbf{R}_1$$

hold?

**Exercise 2.4** Show that

$$\left(\mathbf{T}^{-1}\right)^{-1} = \mathbf{T}.$$

[Hint: $\mathbf{T} \circ \mathbf{T}^{-1} = \mathbf{T}^{-1} \circ \mathbf{T} = \mathbf{I}$.]

**Exercise 2.5** Reduce the following sequences of pure translations and pure rotations to a single combination of a pure translation and a pure rotation:

a) $(\mathbf{T}_1)^{-1} \circ \mathbf{R}_1 \circ \mathbf{T}_1$
b) $\mathbf{R}_2 \circ \mathbf{T}_1 \circ \mathbf{T}_1 \circ (\mathbf{R}_2)^{-1}$
c) $\mathbf{T}_1 \circ \mathbf{R}_1 \circ \mathbf{R}_1 \circ \mathbf{T}_1$
d) $\mathbf{R}_1 \circ \mathbf{T}_1 \circ (-1)\,\mathbf{R}_1$
e) $\mathbf{T}_2 \circ (\mathbf{T}_1)^{-1} \circ \mathbf{R}_2 \circ 2\mathbf{R}_2 \circ (\mathbf{T}_2)^{-1}$
f) $(-1)\,\mathbf{T}_1 \circ 2\mathbf{I} \circ 3\mathbf{R}_1 \circ (\mathbf{R}_1)^{-1}$

**Exercise 2.6** Show that the collection of all combinations of pure translations and pure rotations, together with the composition operator, constitutes a group.

**Exercise 2.7** Show that the group of all combinations of pure translations and pure rotations, together with the composition operator, is not Abelian.

**Exercise 2.8** Let $\mathcal{A}$ and $\mathcal{B}$ be two observers. Draw the tree structures that correspond to treating $\mathcal{A}$ or $\mathcal{B}$, respectively, as the main observer. Denote each branch with the pure translation and pure rotation that relate the observers.

**Exercise 2.9** Let $\mathcal{A}$, $\mathcal{B}$, and $\mathcal{C}$ be three observers. Draw the possible tree structures that correspond to treating $\mathcal{A}$, $\mathcal{B}$, or $\mathcal{C}$, respectively, as the main observer. Denote each branch with the pure translation and pure rotation that relate the observers.

**Exercise 2.10** Consider the tree structure below. Draw the equivalent tree structure corresponding to letting the $\mathcal{A}_2$ observer be the main observer and find the pure translation and the pure rotation that describe the configuration of the first rigid body relative to $\mathcal{A}_2$.

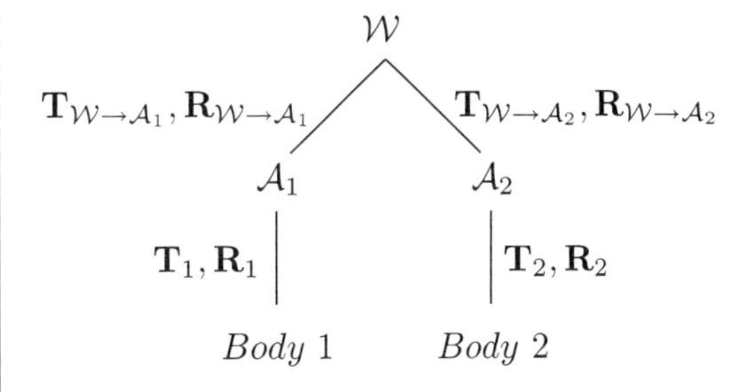


**Exercise 2.11** Consider the tree structure in the previous exercise. Draw the equivalent tree structure corresponding to eliminating the $\mathcal{A}_2$ observer and relating the configuration of the second rigid body directly to $\mathcal{W}$.

**Exercise 2.12** For each of the mechanisms below, introduce a main observer and auxiliary observers, and draw the corresponding tree structures including symbols for the pure translation and pure rotation that correspond to each branch.

a) A trombone
b) A unicycle
c) A pair of plyers
d) A backhoe
e) A wooden labyrinth game
f) A padlock
g) An electric fan
h) A Ferris wheel

**Exercise 2.13** For each of the mechanisms below, introduce a main observer and auxiliary observers, and draw the corresponding tree structures including symbols for the pure translation and pure rotation that correspond to each branch. Identify constraints on the relative configurations of observers in the tree and highlight these in your diagram.

a) A marionette
b) The action in a piano
c) A gymnast on the Roman rings
d) A reel lawn mower
e) A car wheel assembly
f) A cuckoo clock

**Exercise 2.14** Use the MAMBO toolbox to generate the MAMBO geometry description corresponding to the observer hierachy below.

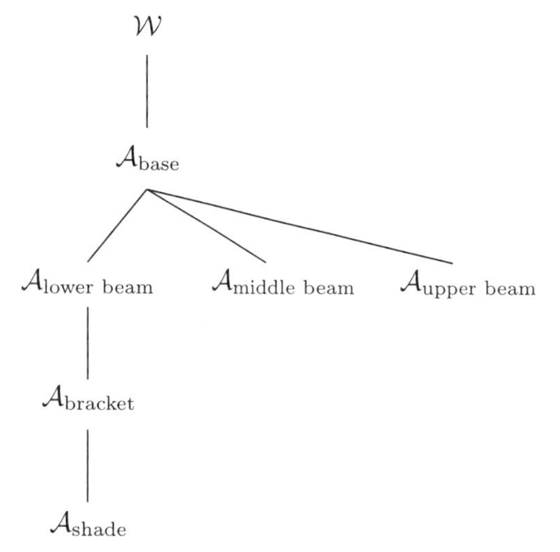


**Solution.**

```
> Restart():
> DeclareObservers(W,Base,
> LowerBeam,Bracket,Shade,
> MiddleBeam,UpperBeam):
> DefineNeighbors([W,Base],
> [Base,LowerBeam],
> [LowerBeam,Bracket],
> [Bracket,Shade],
> [Base,MiddleBeam],
> [Base,UpperBeam]):
> GeometryOutput(main=W);
```

**Exercise 2.15** Use the MAMBO toolbox to generate the MAMBO geometry descriptions corresponding to each of the observer hierarchies below.

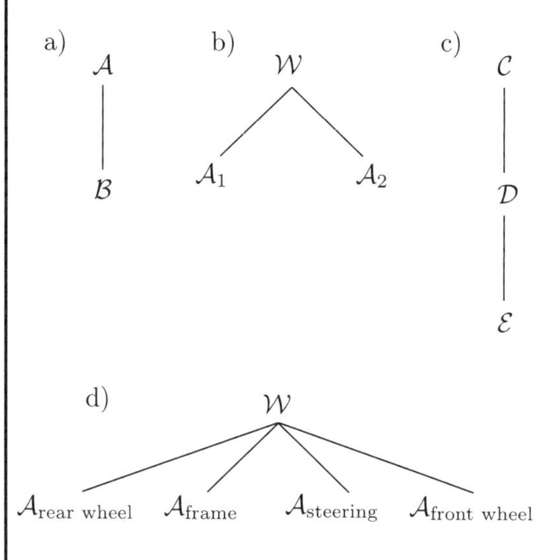


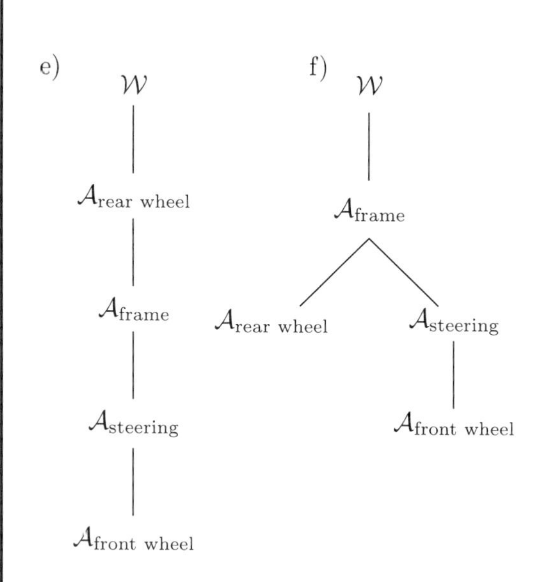

## Summary of notation

The symbols $\mathbf{T}$ and $\mathbf{R}$ were used in this chapter to denote arbitrary pure translations and pure rotations. The same notation, but with subscripts (e.g., $\mathbf{T}_1$ or $\mathbf{R}_2$) was used to distinguish between multiple pure translations or pure rotations. For example, $\mathbf{T}_{\mathcal{A}}$ denoted a pure translation relating the configuration of a rigid body relative to the observer $\mathcal{A}$. Similarly, $\mathbf{T}_{\mathcal{A}\to\mathcal{B}}$ denoted the pure translation that relates the configuration of the observer $\mathcal{B}$ relative to $\mathcal{A}$.

The symbol $\mathbf{I}$ was used in this chapter to denote the identity translation as well as the identity rotation, since these both correspond to absence of motion.

The symbol $\alpha$ (*alpha*) was used in this chapter to denote a scalar multiple of a pure translation or pure rotation, as in $\alpha\mathbf{T}$ or $\alpha\mathbf{R}$.

The symbol $\circ$ was used in this chapter to denote a composition of pure translations and pure rotations, as in $\mathbf{T} \circ \mathbf{R}$.

The superscript $^{-1}$ was used in this chapter to denote an inverse of a pure translation or a pure rotation, as in $\mathbf{T}^{-1}$.

Upper-case, calligraphic letters, such as $\mathcal{A}$, $\mathcal{B}$, and $\mathcal{W}$ were used in this chapter to denote observers.

## Summary of terminology

An observer *hierarchy* is a tree structure with a main observer as the parent node, auxiliary observers as internal nodes, and rigid bodies as leaf nodes. (Page 50)

With emphasis on computer animations of multibody systems, the main observer is called the *world observer*. (Page 56)

With emphasis on mechanical analysis of multibody systems, the main observer is an *inertial observer*. (Page 56)

Each *branch* in an observer hierarchy corresponds to a pair of unique translations $\mathbf{T}$ and $\mathbf{T}^{-1}$ and a pair of unique rotations $\mathbf{R}$ and $\mathbf{R}^{-1}$. (Page 57)

Observers at nodes that are connected by a path that passes through no other nodes are said to be *directly related*. (Page 57)

Observers at nodes that are connected by a path that passes through at least one other node are said to be *indirectly related*. (Page 57)

In MAMBO, the reserved keyword **MODULE** represents the main observer. (Page 72)

(Page 72) In MAMBO, the reserved keyword **BODY** represents an auxiliary observer.

(Page 72) In MAMBO, the *module block* contains information about all auxiliary observers and rigid bodies.

(Page 72) In MAMBO, a *body block* contains information about all descendant auxiliary observers and rigid bodies.

(Page 73) In MAMBO, the reserved keywords **BLOCK**, **SPHERE**, and **CYLINDER** represent rigid bodies with the shape of a rectangular block, a sphere, and a cylinder, respectively.

(Page 74) In the MAMBO toolbox, the global variable `GlobalObserverDeclarations` contains the names of all declared observers and information about their neighbors.

(Page 74) In the MAMBO toolbox, the procedure `DeclareObservers` appends `GlobalObserverDeclarations` with any number of observer names.

(Page 75) In the MAMBO toolbox, the procedure `DefineNeighbors` appends `GlobalObserverDeclarations` with information about directly related observers.

(Page 75) In the MAMBO toolbox, the procedure `GeometryOutput` generates a MAMBO geometry description with the main observer corresponding to some declared observer.

# Chapter 3

# Translations

*wherein the reader learns of:*

- *The association between points, separations, position vectors, and translations;*
- *The mathematics of vectors and translations;*
- *The use of computer-algebra software to expedite computations.*

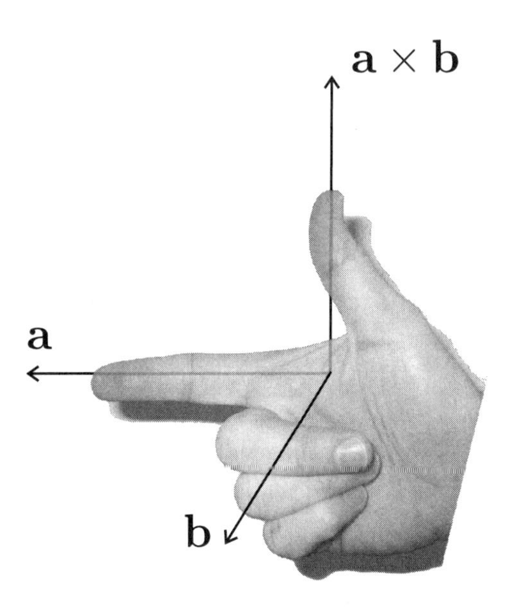

## Practicum

The previous chapters were intended to develop your *intuition* for three-dimensional geometries and the structure of multibody mechanisms. The present chapter is intended to provide you with a *language* to communicate a three-dimensional geometry and the structure of a multibody mechanism. In order to master this language, you must learn to recognize and correctly interpret its individual words and sentences. You must learn to shun grammatical aberrations. You must become fluent.

Follow along with the text in this chapter with a pen and a piece of paper. Write, draw, and speak the objects and the operations discussed here. Use the numerical examples in the exercise section to solidify your transition from the concrete to the abstract. Become proficient with the relevant MAMBO toolbox procedures. You will be amply rewarded for your efforts.

## 3.1 Points

### 3.1.1 Notation

To denote points in space, I consistently use upper-case, italicized letters, e.g.,

$$A,\ R,\ X,$$

and so on. The point in space that coincides with the tip of your nose may be denoted by $N$. The point in space that coincides with the center of the Earth may be denoted by $E$. The choice of letter is not important, unless you are trying to give the person you are communicating with additional information by a clever choice of letter. For example, if a point is to be used to represent the reference position of a specific observer, you may prefer to denote it by the same letter that was used to denote the observer. To distinguish between multiple points that use the same letter, I include appropriate subscripts, e.g., $A_1$, $R_{\text{reference}}$, and $C_{\text{world}}$.

To graphically represent a point in space, this text consistently uses a tiny circular dot. For later reference, it is a good idea to place the corresponding letter adjacent to the dot.

$R$

$A$

$B_{\text{ball}}$

### 3.1.2 Common Misconceptions

> Points are **not** numerical constructs. They represent geometrical features of space, but are **not directly** associated with numbers.
>
> A point is a point is a point!

It **does not** make sense to refer to a point as a combination of three numbers. The assertion that the point $A$ is given by the triplet $(1, -0.5, 0)$ is nonsensical without additional information. The notion of a point's coordinates presupposes the existence of a coordinate system. Equivalently, the position of a point may be reduced to a triplet of numbers only with respect to a specific observer.

For example, denote by $B$ some arbitrary point on the virtual block corresponding to the observer $\mathcal{B}$ and consider a coordinate system with origin at $B$ and axes parallel to the edges of the virtual block. The position of a point $A$ relative to $\mathcal{B}$ is uniquely specified by its coordinates with respect to this coordinate system. Clearly, these coordinates depend on the choice of observer.

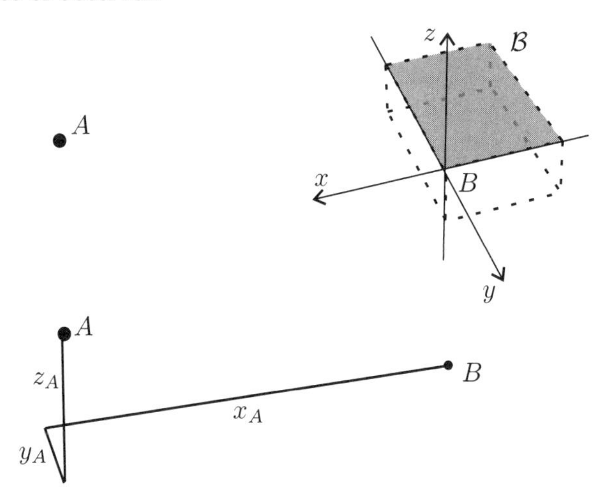


My personal preference is to consistently refer to a point with the corresponding letter and avoid all mention of its coordinates relative to some observer. The notion of coordinate systems and coordinates of a point, however, is quite common and standard and should be understood. I will describe this in detail in Chapter 4.

(Ex. 3.1 – Ex. 3.5)

## 3.2 Separations

### 3.2.1 Notation

The *separation* from point $A$ to point $B$ is the straight-line segment from $A$ to $B$ and is denoted by $\overrightarrow{AB}$. Note that the arrow signifies the direction of the separation from $A$ to $B$ and that the letters corresponding to the points are both represented in the notation. Other examples of separations are

$$\overrightarrow{RS},\ \overrightarrow{BF},\ \text{and}\ \overrightarrow{AG}.$$

In all cases, the notation for a separation involves the two points (using upper-case, italicized letters) connected by the separation and a superscripted arrow indicating the direction of the separation.

The separation $\overrightarrow{AB}$ uniquely describes the location of the point $B$ relative to the point $A$. The separation documents the shortest path to traverse from point $A$ to point $B$.

To graphically represent a separation in space, this text consistently uses an actual arrow from one point to another. For later reference, it is a good idea to place the corresponding combination of letters and superscripted arrow adjacent to the actual arrow.

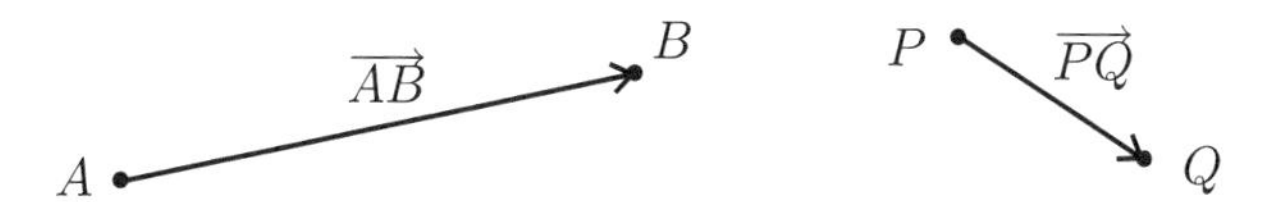

There is no unique measure of the *length* of a separation. Equivalently, there is no unique measure of the *distance* between two points. Instead, all statements regarding measures of length and distance are relative to some accepted standard, say a platinum bar in a sealed vault in Paris[1].

### 3.2.2 Common Misconceptions

> Separations are **not** numerical constructs. They represent geometrical features of space, but are not directly associated with numbers.
>
> A separation is a separation is a separation!

It **does not** make sense to refer to a separation as a combination of three or more numbers. The separation is uniquely determined by the two points it connects. Since we are unable to ascribe unique numbers to points, the same follows for separations. Given an observer, we could certainly describe a separation by the coordinates of each of the points relative to the coordinate system with origin at some stationary point relative to the observer and axes parallel to the edges of the virtual block corresponding to the observer. As before, these numbers would depend on the choice of observer.

My personal preference is to consistently refer to a separation by the corresponding combination of letters and superscripted arrow and avoid all mention of the coordinates of its constituent points relative to some observer.

### 3.2.3 Algebra of Separations

The separation $\overrightarrow{AB}$ documents the shortest path to traverse from point $A$ to point $B$.

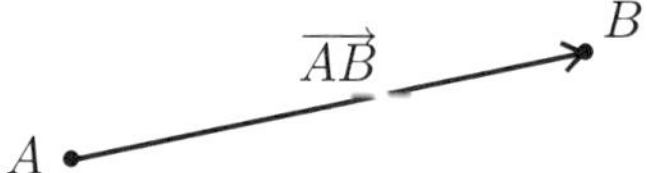


Motion along the path given by $\overrightarrow{AB}$ is denoted by the symbol

$$\overset{\overrightarrow{AB}}{\curvearrowright}.$$

[1]The SI system unit of length, the *meter*, was originally introduced as a ten-millionth of the distance between the north pole and the equator along the Paris meridian and represented by the distance between notches on a platinum-iridium bar. The current definition is in terms of the distance traveled by light in vacuum in 1/299 792 458 seconds.

It makes sense to write

$$A \overset{\overrightarrow{AB}}{\curvearrowright} B,$$

suggesting that motion from $A$ along $\overrightarrow{AB}$ brings one to $B$.

**Illustration 3.1**

That the shortest path from $B$ to $C$ is given by the separation $\overrightarrow{BC}$ implies that

$$B \overset{\overrightarrow{BC}}{\curvearrowright} C.$$

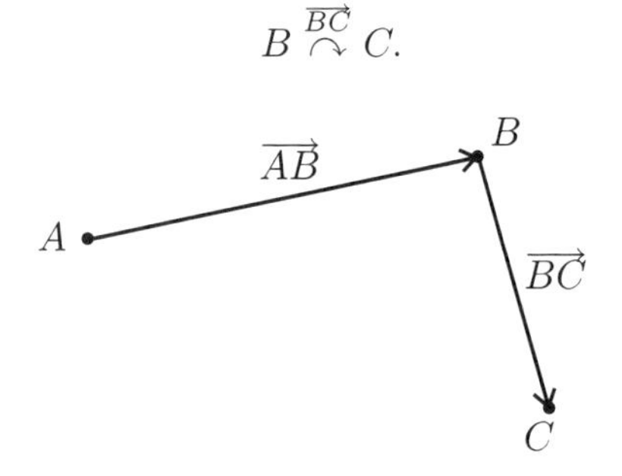


Since from before

$$A \overset{\overrightarrow{AB}}{\curvearrowright} B,$$

we conclude that

$$A \overset{\overrightarrow{AB}}{\curvearrowright} \overset{\overrightarrow{BC}}{\curvearrowright} C,$$

i.e., that motion from $A$ along $\overrightarrow{AB}$ and subsequently along $\overrightarrow{BC}$ brings one to $C$.

But since

$$A \overset{\overrightarrow{AC}}{\curvearrowright} C,$$

the motion described by the composition $\overset{\overrightarrow{AB}}{\curvearrowright} \overset{\overrightarrow{BC}}{\curvearrowright}$ produces the same outcome as that described by $\overset{\overrightarrow{AC}}{\curvearrowright}$. While the latter motion is along the shortest path between $A$ and $C$, the former is, in general, not. Although the end results are the same for the two motions, we refrain from suggesting that they are equal, in light of this marked difference in actual path.

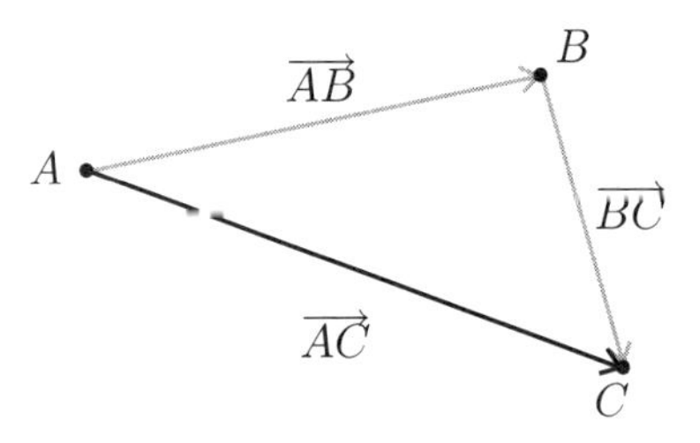

Any number of expressions of the form $\overset{\overrightarrow{AB}}{\curvearrowright}$ may be combined, provided that the endpoint of one such motion coincides with the starting point of the next motion. Thus, the expression

$$\overset{\overrightarrow{AD}}{\curvearrowright}\ \overset{\overrightarrow{DO}}{\curvearrowright}\ \overset{\overrightarrow{OX}}{\curvearrowright}$$

makes sense and results in the same final displacement as the motion

$$\overset{\overrightarrow{AX}}{\curvearrowright}\,.$$

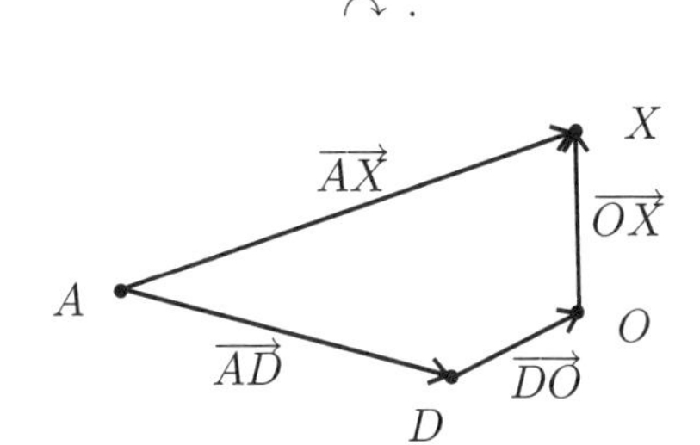


On the other hand, the expression

$$\overset{\overrightarrow{AD}}{\curvearrowright}\ \overset{\overrightarrow{AV}}{\curvearrowright}\ \overset{\overrightarrow{DV}}{\curvearrowright}$$

does **not** make sense.

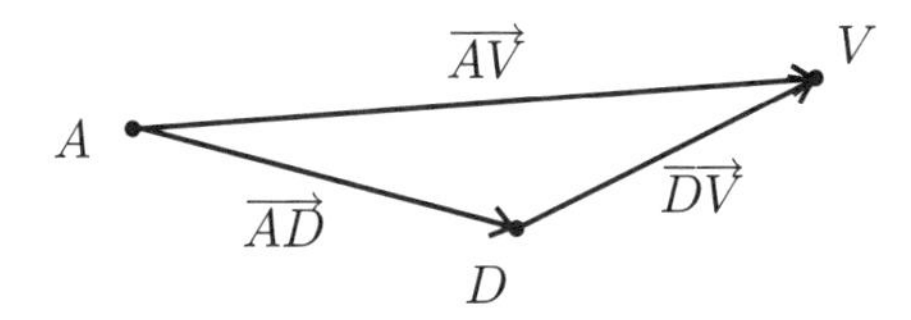


### 3.2.4 Affine Space

We may parametrize the points on the separation $\overrightarrow{PQ}$ from $P$ to $Q$ by the real numbers in the interval $[0, 1]$, such that 0 corresponds to $P$, 1 corresponds to $Q$, and $0 \leq \gamma \leq 1$ corresponds to the point on $\overrightarrow{PQ}$ whose distance from $P$ is a fraction $\gamma$ of the distance between $P$ and $Q$.

It is convenient to think of the notation $\overrightarrow{PQ}$ as a function of the real numbers in $[0, 1]$, such that $\overrightarrow{PQ}\,(0) = P$ and $\overrightarrow{PQ}\,(1) = Q$. It follows that $\overrightarrow{PQ}\left(\frac{1}{2}\right)$ is the *midpoint* on the separation, halfway between $P$ and $Q$. From the definition of a separation, it follows that

$$\overrightarrow{PQ}\,(\gamma) = \overrightarrow{QP}\,(1 - \gamma)$$

for any $\gamma \in [0, 1]$.

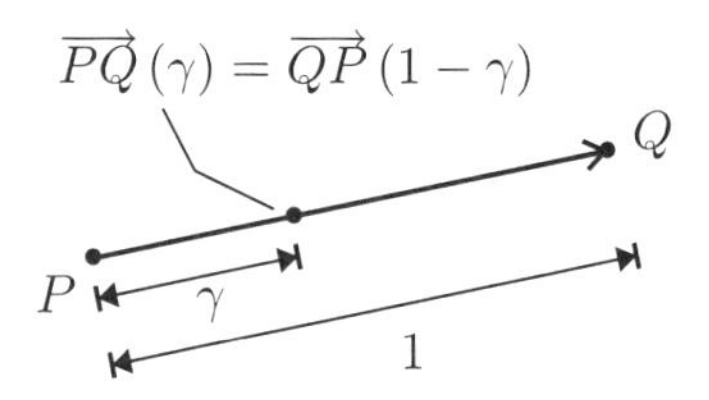


**Illustration 3.2**

Given any four points $P$, $Q$, $R$, and $S$, the condition that

$$\overrightarrow{PS}\left(\frac{1}{2}\right) = \overrightarrow{QR}\left(\frac{1}{2}\right) = M$$

implies that the separations $\overrightarrow{PQ}$ and $\overrightarrow{RS}$ have equal length, are parallel, and have the same heading.

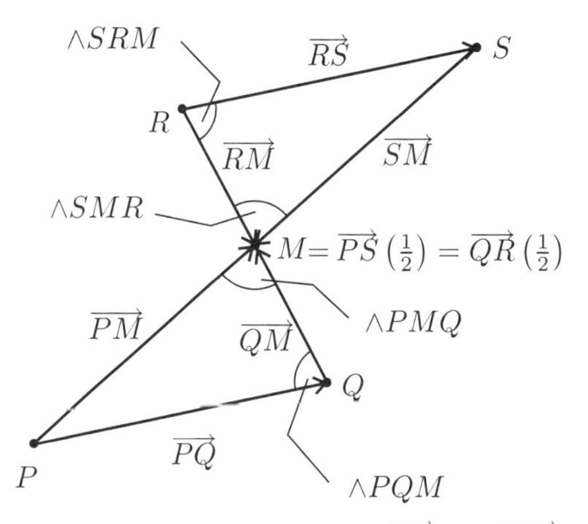


Indeed, the lengths of the separations $\overrightarrow{SM}$ and $\overrightarrow{PM}$ are equal. Similarly, the lengths of the separations $\overrightarrow{QM}$ and $\overrightarrow{RM}$ are equal. Finally, the angle $\wedge PMQ$ equals the angle $\wedge SMR$. Since the triangles $PQM$ and $SRM$ have two sides and one angle in common, they are congruent. It follows that the lengths of the separations $\overrightarrow{PQ}$ and $\overrightarrow{RS}$ are equal and that the angles $\wedge PQM$ and $\wedge SRM$ are equal. From Euclidean geometry, the latter observation implies that $\overrightarrow{PQ}$ and $\overrightarrow{RS}$ are parallel.

The converse to the statement in the illustration is also true: if two separations $\overrightarrow{PQ}$ and $\overrightarrow{RS}$ have equal length, are parallel, and have the same heading, then

$$\overrightarrow{PS}\left(\frac{1}{2}\right) = \overrightarrow{QR}\left(\frac{1}{2}\right).$$

If, in addition, $\overrightarrow{RS}$ and $\overrightarrow{TU}$ have equal length, are parallel, and have the same heading, then

$$\overrightarrow{ST}\left(\frac{1}{2}\right) = \overrightarrow{RU}\left(\frac{1}{2}\right).$$

But this implies that $\overrightarrow{PQ}$ and $\overrightarrow{TU}$ have equal length, are parallel, and have the same heading, i.e.,

$$\overrightarrow{PU}\left(\frac{1}{2}\right) = \overrightarrow{QT}\left(\frac{1}{2}\right).$$

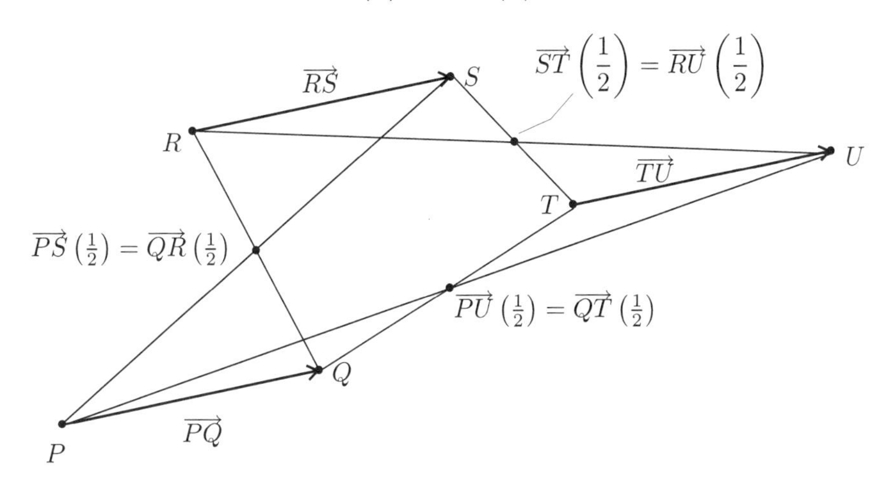


These properties show that the set of points together with the function $\rightarrow$ is an *affine space*.

**Definition 3.1** A set of elements $\mathbb{A}$ together with a ternary function[2]

$$\rightarrow : \mathbb{A} \times \mathbb{A} \times [0,1] \rightarrow \mathbb{A}$$

is called an affine space if for any elements $P, Q \in \mathbb{A}$ :

- *Uniqueness:* $\overrightarrow{PQ}(0) = P$ and $\overrightarrow{PQ}(1) = Q$;
- *Symmetry:* $\overrightarrow{QP}\left(\frac{1}{2}\right) = \overrightarrow{PQ}\left(\frac{1}{2}\right)$;
- *Transitivity:* $\overrightarrow{PS}\left(\frac{1}{2}\right) = \overrightarrow{QR}\left(\frac{1}{2}\right)$ and $\overrightarrow{ST}\left(\frac{1}{2}\right) = \overrightarrow{RU}\left(\frac{1}{2}\right)$ implies that $\overrightarrow{PU}\left(\frac{1}{2}\right) = \overrightarrow{QT}\left(\frac{1}{2}\right)$.

[2]To truly qualify as an affine space, the function $\rightarrow$ must satisfy a number of additional conditions. The conditions listed here guarantee that an equivalence relation can be introduced on $\mathbb{A}$. The additional conditions guarantee that the equivalence relation supports the construction of a *vector space* (see Section 3.3.2 for further discussion).

As $\gamma \in [0, 1]$ varies from 0 to 1, the function $\overrightarrow{PQ}(\gamma)$ maps out a path from $P$ to $Q$. While this path is a straight-line segment in the flat space that we appear to experience, this need not generally be the case.

Consider, as an example, a sphere of radius 1 centered at the origin of a Cartesian coordinate system and restrict attention to the set of points on the sphere for which their $z$-coordinate is greater than zero. There is a unique straight line through the origin intersecting each point in this set. We can identify each such point with the unique point of intersection of the corresponding straight line with the $z = 1$ plane.

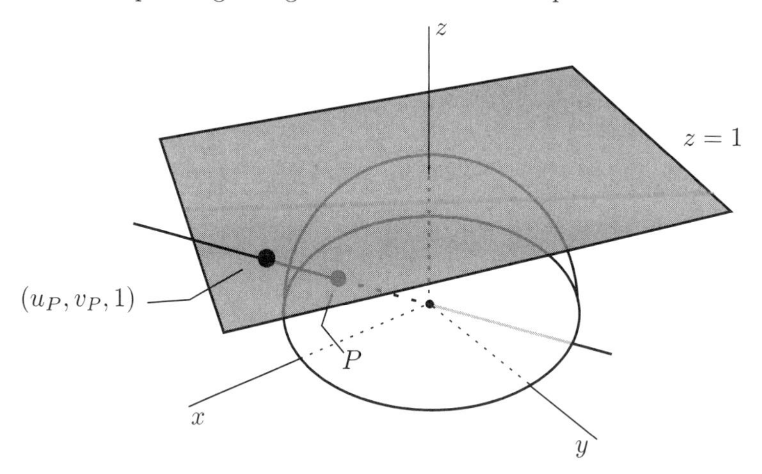


To each point $P$ on the hemisphere, there thus corresponds a unique triplet of coordinates $(u_P, v_P, 1)$, and vice versa. We represent this observation by the statement

$$P \leftrightarrow (u_P, v_P, 1).$$

Let $P$ and $Q$ be two points on the hemisphere, such that

$$P \leftrightarrow (u_P, v_P, 1) \text{ and } Q \leftrightarrow (u_Q, v_Q, 1).$$

Then, define the function $\overrightarrow{\ }$, such that

$$\overrightarrow{PQ}(\gamma) \leftrightarrow (u_P + \gamma(u_Q - u_P), v_P + \gamma(v_Q - v_P), 1).$$

Clearly,

$$\overrightarrow{PQ}(0) = P,$$

since

$$\overrightarrow{PQ}(0) \leftrightarrow (u_P + 0 * (u_Q - u_P), v_P + 0 * (v_Q - v_P), 1) = (u_P, v_P, 1).$$

Moreover,

$$\overrightarrow{PQ}(1) \leftrightarrow (u_P + 1 * (u_Q - u_P), v_P + 1 * (v_Q - v_P), 1) = (u_Q, v_Q, 1),$$

i.e.,

$$\overrightarrow{PQ}(1) = Q.$$

The path through the points $P$ and $Q$ described by the function $\overrightarrow{PQ}(\gamma)$ is an arc segment of the great circle[3] through $P$ and $Q$.

As in the case of separations in flat space, $\overrightarrow{PQ}$ describes the shortest path from $P$ to $Q$. It is no longer the case, however, that $\gamma$ denotes the ratio between the distance from $P$ to $\overrightarrow{PQ}(\gamma)$ and the distance[4] from $P$ to $Q$. For example, if $P$ lies at the top of the hemisphere and $Q$ near the $z = 0$ plane, then

$$\overrightarrow{PQ}\left(\frac{1}{2}\right) \leftrightarrow \left(\frac{u_Q}{2}, \frac{v_Q}{2}, 1\right),$$

which can be made arbitrarily close to the $z = 0$ plane by picking $u_Q$ and/or $v_Q$ arbitrarily large.

**Illustration 3.3**

The function $\rightarrow$ on the hemisphere satisfies the symmetry property, since

$$\begin{aligned}(u_P + \gamma (u_Q - u_P), v_P + \gamma (v_Q - v_P), 1) \\ = (u_Q + (1 - \gamma)(u_P - u_Q), v_Q + (1 - \gamma)(v_P - v_Q), 1).\end{aligned}$$

Furthermore,

$$\begin{aligned}\left(u_P + \frac{1}{2}(u_S - u_P), v_P + \frac{1}{2}(v_S - v_P), 1\right) \\ = \left(u_Q + \frac{1}{2}(u_R - u_Q), v_Q + \frac{1}{2}(v_R - v_Q), 1\right),\end{aligned}$$

$$\begin{aligned}\left(u_S + \frac{1}{2}(u_T - u_S), v_S + \frac{1}{2}(v_T - v_S), 1\right) \\ = \left(u_R + \frac{1}{2}(u_U - u_R), v_R + \frac{1}{2}(v_U - v_R), 1\right)\end{aligned}$$

imply that

$$\begin{aligned}u_P &= -u_S + u_Q + u_R, \\ v_P &= -v_S + v_Q + v_R, \\ u_U &= u_S + u_T - u_R,\end{aligned}$$

[3]The corresponding path on the $z = 1$ plane is a straight-line segment from $(u_P, v_P, 1)$ to $(u_Q, v_Q, 1)$.

[4]Unless we define the distance between two points on the hemisphere to equal the distance between the corresponding points on the $z = 1$ plane.

2

and

$$v_U = v_S + v_T - v_R.$$

Substitution shows that

$$\left(u_P + \frac{1}{2}(u_U - u_P), v_P + \frac{1}{2}(v_U - v_P), 1\right) = \left(u_Q + \frac{1}{2}(u_T - u_Q), v_Q + \frac{1}{2}(v_T - v_Q), 1\right),$$

i.e., that $\rightarrow$ satisfies the transitivity property.

(Ex. 3.6 – Ex. 3.27)

## 3.3 Vectors

### 3.3.1 Equivalent Separations

**Illustration 3.4**

In a pure translation, all points on a rigid body are shifted by an equal amount along a common direction. Let $A$ and $B$ be any two points on the rigid body and denote by $A_r$ and $B_r$ the corresponding locations in space that coincide with $A$ and $B$, respectively, when the rigid body is in the reference configuration. Similarly, let $A_f$ and $B_f$ denote the points in space that coincide with $A$ and $B$, respectively, when the rigid body is in the final configuration.

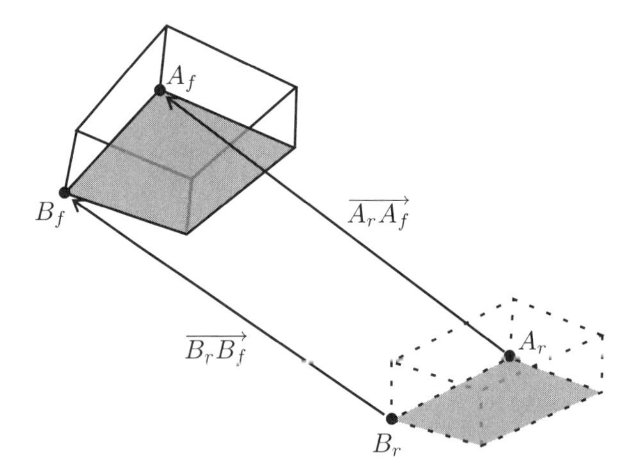

By the definition of a pure translation, the separations $\overrightarrow{A_r A_f}$ and $\overrightarrow{B_r B_f}$:

- Have equal length;
- Are parallel;
- Have the same heading.

In fact, all separations from points in the reference configuration to the corresponding points in the final configuration satisfy these three conditions. An equivalent observation is to suggest that all motions of the form

$$\overset{\overrightarrow{A_r A_f}}{\curvearrowright}$$

are identical in distance and in direction, and differ only in the choice of points that are involved.

When two separations $\overrightarrow{PQ}$ and $\overrightarrow{RS}$ have equal length, are parallel, and have the same heading, they are said to be *equivalent* and we write

$$\overrightarrow{PQ} \sim \overrightarrow{RS}.$$

Clearly, $\overrightarrow{PQ} \nsim \overrightarrow{QP}$, since the separations $\overrightarrow{PQ}$ and $\overrightarrow{QP}$ differ in heading.

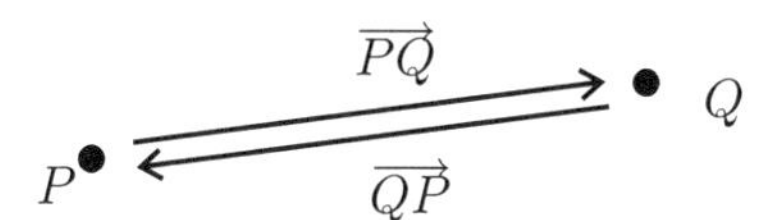


Given a separation $\overrightarrow{RS}$, there exists a separation $\overrightarrow{PQ}$ for every point $P$ in space, such that $\overrightarrow{PQ} \sim \overrightarrow{RS}$. We can use this observation to make sense of statements like

$$P \overset{\overrightarrow{RS}}{\curvearrowright} Q,$$

which are *a priori* nonsensical, since the separation $\overrightarrow{RS}$ connects $R$ and $S$, **not** $P$ and $Q$. If, instead, we choose to interpret the expression $\overset{\overrightarrow{RS}}{\curvearrowright}$ to suggest a motion along any separation that is equivalent to $\overrightarrow{RS}$, then

$$P \overset{\overrightarrow{RS}}{\curvearrowright} Q$$

is true if and only if $\overrightarrow{PQ} \sim \overrightarrow{RS}$.

To emphasize this interpretation, consider the notation

$$\overset{\left[\overrightarrow{RS}\right]}{\curvearrowright},$$

where $\left[\overrightarrow{RS}\right]$ represents the infinite collection of separations that are equivalent to $\overrightarrow{RS}$. In particular, the notation $\overrightarrow{PQ} \in \left[\overrightarrow{RS}\right]$ means that the separation $\overrightarrow{PQ}$ is in that collection, i.e., that $\overrightarrow{PQ} \sim \overrightarrow{RS}$. The statement

$$P \overset{\left[\overrightarrow{RS}\right]}{\curvearrowright} Q$$

is true if and only if $\overrightarrow{PQ} \in \left[\overrightarrow{RS}\right]$.

### 3.3.2 Equivalence Classes

The result from Exercise 3.6 shows that the relation between two equivalent separations is an example of an *equivalence relation*:

**Definition 3.2** An equivalence relation $\sim$ on a set $\mathbb{F}$ is a property of pairs of elements for which the following holds true:

- *Reflexivity:* $x \sim x$;
- *Symmetry:* $x \sim y \Rightarrow y \sim x$;
- *Transitivity:* $x \sim y$ and $y \sim z \Rightarrow x \sim z$.

The subset of elements equivalent to $x$ is commonly denoted by $[x]$ and is called an *equivalence class*. The equivalence relation generates a set of equivalence classes on $\mathbb{F}$. This set is denoted by $\mathbb{F}/\sim$ and is called the *quotient set*.

The equivalence class in flat space of all separations equivalent to the separation $\overrightarrow{PQ}$ is the collection $\left[\overrightarrow{PQ}\right]$. Each such equivalence class in flat space corresponds to a unique pure translation, and vice versa. It follows that the quotient set corresponding to the equivalence relation $\sim$ on the set of separations in flat space is the set of pure translations.

Two separations $\overrightarrow{PQ}$ and $\overrightarrow{RS}$ in flat space are equivalent if and only if they have the same length, are parallel, and have the same heading. But in the previous section, this was shown to be true if and only if[5]

$$\overrightarrow{PS}\left(\frac{1}{2}\right) = \overrightarrow{QR}\left(\frac{1}{2}\right).$$

In a general affine space, we can use this as the definition of equivalence. The implications in a general affine space may no longer be interpretable in terms of the separations having equal length, being parallel, and having the same heading as was the case in flat space.

### Illustration 3.5

We must show that the definition of equivalence in a general affine space satisfies the conditions for an equivalence relation. From the conditions on the $\overrightarrow{\ }$ function in an affine space, we recall that

$$\overrightarrow{PQ}\left(\frac{1}{2}\right) = \overrightarrow{QP}\left(\frac{1}{2}\right).$$

But this implies that the separation $\overrightarrow{PQ}$ is equivalent to itself, confirming that the property of equivalence satisfies the reflexivity condition. Moreover, if

$$\overrightarrow{PS}\left(\frac{1}{2}\right) = \overrightarrow{QR}\left(\frac{1}{2}\right),$$

then it follows that

$$\overrightarrow{SP}\left(\frac{1}{2}\right) = \overrightarrow{RQ}\left(\frac{1}{2}\right).$$

But this implies that, if the separation $\overrightarrow{PQ}$ is equivalent to the separation $\overrightarrow{RS}$, then the separation $\overrightarrow{RS}$ is equivalent to the separation $\overrightarrow{PQ}$ as required by the symmetry condition.

Finally, transitivity follows if

$$\overrightarrow{PS}\left(\frac{1}{2}\right) = \overrightarrow{QR}\left(\frac{1}{2}\right) \text{ and } \overrightarrow{RU}\left(\frac{1}{2}\right) = \overrightarrow{ST}\left(\frac{1}{2}\right)$$

imply that

$$\overrightarrow{PU}\left(\frac{1}{2}\right) = \overrightarrow{QT}\left(\frac{1}{2}\right).$$

[5]See the previous section for a definition of this notation.

But this is the transitivity condition on the function $\rightarrow$ in a general affine space. We conclude that the definition of equivalence introduced here truly satisfies the conditions for an equivalence relation.

### 3.3.3 Position Vectors

A collection of separations $\left[\overrightarrow{PQ}\right]$ that are equivalent to the separation $\overrightarrow{PQ}$ is called a *position vector from $P$ to $Q$*, and is generally denoted by $\mathbf{r}^{PQ}$ or $\bar{r}^{PQ}$.

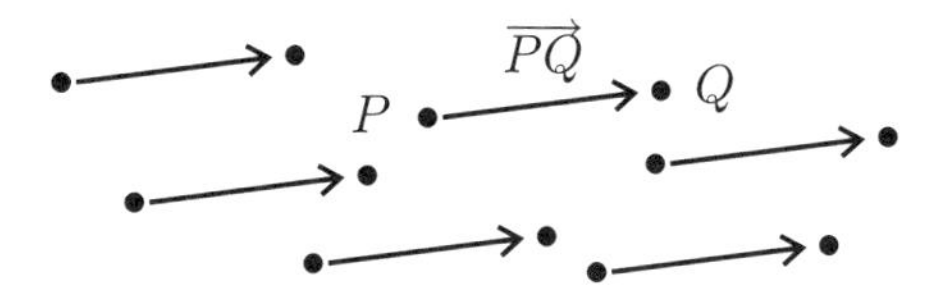


In this text, I consistently use the first version of the notation, whereas I recommend that you use the second form when writing on paper or blackboard. The notation for a position vector is a lower-case "r" (bold-faced or with a bar above as described) followed by a superscript involving two points. Examples of position vectors are

$$\mathbf{r}^{BD} \text{ or } \bar{r}^{BD}, \mathbf{r}^{QJ} \text{ or } \bar{r}^{QJ}, \text{ and } \mathbf{r}^{P_1P_2} \text{ or } \bar{r}^{P_1P_2}.$$

The position vector $\mathbf{r}^{BE}$ **does not** equal the separation $\overrightarrow{BE}$, but contains it. $\overrightarrow{BE}$ is said to be a *representation* of the position vector $\mathbf{r}^{BE}$ or to *represent* the position vector $\mathbf{r}^{BE}$.

To graphically depict a position vector, this text consistently uses an arbitrary separation that represents the position vector. For later reference, it is a good idea to place the corresponding symbol adjacent to the separation.

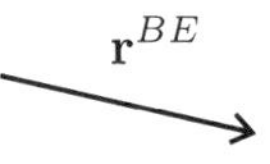


**Illustration 3.6**

A separation is uniquely associated with two points in space. Thus, $\overrightarrow{AG}$ is entirely determined by the location of the points $A$ and $G$, respectively. A position vector, however, can be defined without referring to any points. In fact, a position vector is entirely determined by the *length*, *direction*, and *heading* of any of its representations. Thus, unless we are particularly interested in using a position vector to describe the collection of separations equivalent to the separation between two **specific** points, we can omit the superscript in the notation for the position vector. When this is the case, we just refer to the position vector as a *vector*.

Reserving the letter "r" for position vectors, vectors in general are denoted by a lower-case letter (bold-faced or with a bar above it as described previously), e.g.,

$$\mathbf{v} \text{ or } \overline{v},\ \mathbf{w} \text{ or } \overline{w}, \text{ and } \mathbf{x} \text{ or } \overline{x}.$$

Since all the representations of a vector have the same *length*, *direction*, and *heading*, it is customary to ascribe these characteristics to the vector. For example, the length of a vector is the length of any of its representations. The length of the vector $\mathbf{v}$ is denoted by $\|\mathbf{v}\|$.

> Every pure translation corresponds to a unique vector.
> Every vector corresponds to a unique translation.

In particular, the vector $\mathbf{v}$ corresponds to the pure translation $\mathbf{T}$ that shifts all the points on a rigid body by an amount and in a direction given by the length, direction, and heading of $\mathbf{v}$. Similarly, given a pure translation $\mathbf{T}$, we can construct the corresponding vector by collecting all the separations $\overrightarrow{A_r A_f}$ between points in the reference and final configurations.

## 3.3.4 Common Misconceptions

A vector **does not** have a location in space. In fact, a vector does not even exist in space as a single object. A vector is a collection of infinitely many equivalent separations, each of which does have a location in space. It **does not** make sense to suggest that a vector can be moved freely in space, since it cannot have a location in space in the first place. It is the selection of different separations to **represent** the vector that suggests the idea of moving the vector. A vector **is not** an arrow between two points.

A vector **is not** a column matrix of numbers. Certainly, the terminology "vector" is common in linear-algebra texts to refer to such matrices. Although a possible source of confusion, the dual usage of this terminology is justified by the similarity between the properties of the vectors introduced here and those of column matrices.

> The vectors introduced here represent geometrical features of space, and are **not directly** associated with numbers.
>
> A vector is a vector is a vector!

### 3.3.5 Algebra of Vectors

**The identity translation**

**Illustration 3.7**

Every pure translation corresponds to a unique vector. As a special case, the identity translation **I** corresponds to the absence of motion. We denote the corresponding vector by $\mathbf{0}$ and refer to it as the *zero vector*.

The zero vector $\mathbf{0}$ contains all separations equivalent to the separation $\overrightarrow{PP}$. In fact, $\overrightarrow{AB} \in \mathbf{0}$ if and only if the points $A$ and $B$ coincide. Every separation representing the zero vector has zero length. In other words, $\|\mathbf{0}\| = 0$.

We graphically represent the zero vector by a point and the symbol $\mathbf{0}$ adjacent to the point.

$$\bullet^{\mathbf{0}}$$

The statement

$$P \overset{\mathbf{0}}{\curvearrowright} P$$

is true for all $P$.

**Scaling of translations**

**Illustration 3.8**

Let $\overrightarrow{PQ}$ and $\overrightarrow{RS}$ be two separations that have the same direction and heading, but such that the length of $\overrightarrow{RS}$ is twice that of $\overrightarrow{PQ}$. It follows that $\overrightarrow{PQ} \nsim \overrightarrow{RS}$.

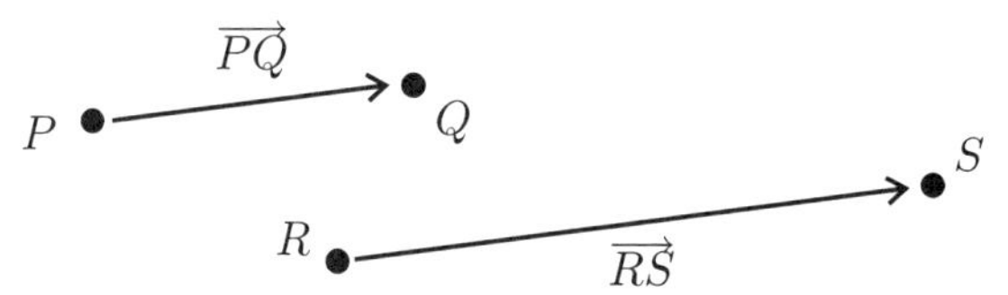


Since the two separations connect different points, there is no straightforward way to relate these separations. This is not true, however, for the corresponding vectors. Indeed, the vector $\left[\overrightarrow{RS}\right]$ consists of all separations that have the same direction and heading as the separations in the vector $\left[\overrightarrow{PQ}\right]$ but twice the length. It makes sense to express this observation by the statement

$$2\left[\overrightarrow{PQ}\right] = \left[\overrightarrow{RS}\right].$$

Define the *multiplication of the vector* $\mathbf{v}$ *with a scalar* (i.e., real number) $\alpha \geq 0$ as the vector $\alpha\mathbf{v}$, such that any representation of $\alpha\mathbf{v}$ has the same heading and direction as any representation of $\mathbf{v}$, but is longer by a factor of $\alpha$ (or shorter if $\alpha < 1$). For a negative scalar $\alpha$, any representation of the vector $\alpha\mathbf{v}$ has the same direction and length as any representation of $|\alpha|\,\mathbf{v}$, but opposite heading.

**Illustration 3.9**

In Chapter 2, we considered scaling a pure translation $\mathbf{T}$ by a real number $\alpha$ to yield the pure translation $\alpha\mathbf{T}$. In fact, if $\alpha \geq 0$, then all separations $\overrightarrow{A_r A_f}$ between points in the reference and final configurations under the pure translation $\alpha\mathbf{T}$ are a factor of $\alpha$ longer than the corresponding separations for the pure translation $\mathbf{T}$. Similarly, if $\alpha < 0$, the corresponding separations are a factor of $|\alpha|$ longer but have opposite heading.

It follows that if $\mathbf{v}$ is the vector corresponding to a pure translation $\mathbf{T}$, then $\alpha\mathbf{v}$ is the vector corresponding to the pure translation $\alpha\mathbf{T}$, and vice versa.

There is no difference between expressions like $(5+1)\,\mathbf{v}$ and $\mathbf{v}\,(5+1)$. Both of these represent the vector $6\mathbf{v}$.

### Compositions of translations

If the pure translation $\mathbf{T}_1$ corresponds to the position vector $\mathbf{r}^{AB}$ and the translation $\mathbf{T}_2$ corresponds to the position vector $\mathbf{r}^{BC}$, then what is the position vector corresponding to the combined translation

$$\mathbf{T}_2 \circ \mathbf{T}_1?$$

The motion

$$\overset{\mathbf{r}^{AB}}{\curvearrowright}\ \overset{\mathbf{r}^{BC}}{\curvearrowright}$$

yields the same end result as the motion

$$\overset{\mathbf{r}^{AC}}{\curvearrowright},$$

where the separations $\overrightarrow{AB}$, $\overrightarrow{BC}$, and $\overrightarrow{AC}$ form the three sides of a triangle.

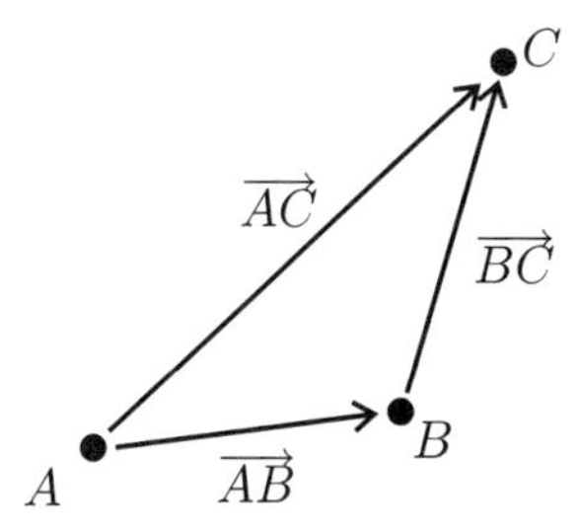

The translation $\mathbf{T}_2 \circ \mathbf{T}_1$ corresponds to the position vector $\mathbf{r}^{AC}$ and we write

$$\mathbf{r}^{AB} + \mathbf{r}^{BC} = \mathbf{r}^{AC}$$

and call $\mathbf{r}^{AC}$ the *sum* of the two position vectors $\mathbf{r}^{AB}$ and $\mathbf{r}^{BC}$.

**Illustration 3.10**

We can illustrate the concept of adding vectors by drawing triangles whose sides are given by representations of the vectors involved. For example, the figure below shows that the sum

$$\left[\overrightarrow{QR}\right] + \left[\overrightarrow{PS}\right]$$

equals the position vector

$$\left[\overrightarrow{QT}\right].$$

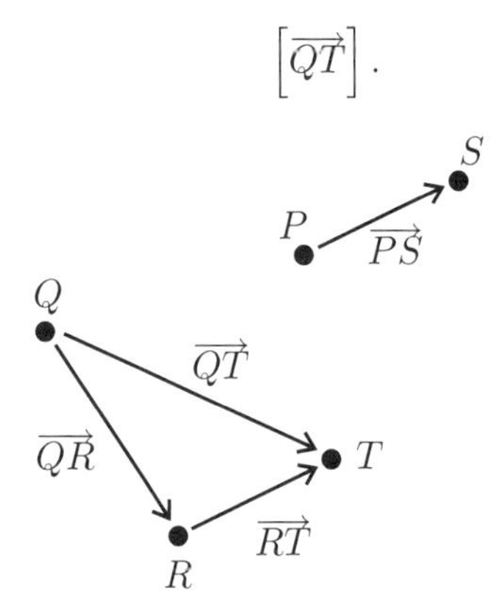


Similarly, the following diagram shows that

$$\left[\overrightarrow{PS}\right] + \left[\overrightarrow{QR}\right] = \left[\overrightarrow{PT'}\right].$$

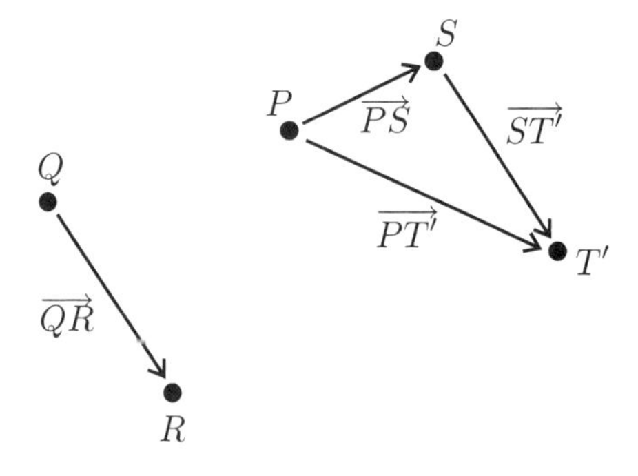


But,

$$\overrightarrow{QT} \sim \overrightarrow{PT'},$$

i.e.,

$$\left[\overrightarrow{QR}\right] + \left[\overrightarrow{PS}\right] = \left[\overrightarrow{PS}\right] + \left[\overrightarrow{QR}\right].$$

You are used to seeing the + symbol used in contexts where the order of summation is immaterial to the result of the operation. The illustration shows that the order in which vectors are added is immaterial to the result of the summation. This is a restatement of an observation made in the first two chapters about the order of successive pure translations.

### Inverses of translations

If the vector $\mathbf{v}$ corresponds to the translation $\mathbf{T}$, then the vector $(-1)\,\mathbf{v}$ corresponds to the translation $\mathbf{T}^{-1}$, since $\mathbf{T}^{-1}$ corresponds to a shift of all points by the same amount as described by $\mathbf{T}$ but in the opposite direction.

Since

$$\mathbf{T}^{-1} \circ \mathbf{T} = \mathbf{I}\,,$$

we conclude that

$$\mathbf{v} + (-1)\,\mathbf{v} = \mathbf{0}\,.$$

It makes sense to define subtraction of vectors by the formula

$$\mathbf{v} - \mathbf{w} = \mathbf{v} + (-1)\,\mathbf{w}.$$

Moreover, it is standard notation to write

$$-\mathbf{v}$$

instead of $(-1)\,\mathbf{v}$.

### Vector products

Vectors (and the corresponding pure translations) may be compared using their lengths and heading. Given two non-zero vectors $\mathbf{a}$ and $\mathbf{b}$, the angle $\theta\,(\mathbf{a}, \mathbf{b})$ between the vectors is a measure of the difference in heading. The extreme cases

$$\theta\,(\mathbf{a}, \mathbf{b}) = 0°,\ 90°,\ \text{and}\ 180°$$

are particularly useful in applications. Here, $\theta\,(\mathbf{a}, \mathbf{b}) = 0°$ implies that the vectors $\mathbf{a}$ and $\mathbf{b}$ are parallel and have the same heading. They must therefore differ only in length, say

$$\mathbf{b} = \alpha\mathbf{a}$$

for some real number $\alpha > 0$.

If, instead, $\theta(\mathbf{a}, \mathbf{b}) = 180^\circ$, the two vectors $\mathbf{a}$ and $\mathbf{b}$ are parallel but have the opposite heading. In analogy with the $\theta(\mathbf{a}, \mathbf{b}) = 0^\circ$ case, it follows that

$$\mathbf{b} = \alpha \mathbf{a}$$

for some real number $\alpha < 0$.

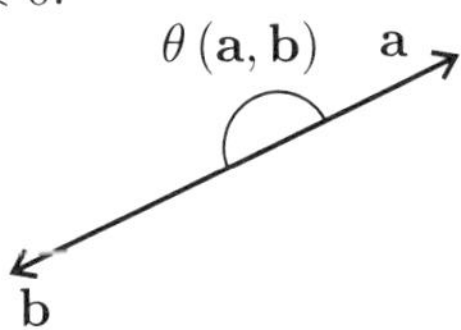


Finally, if $\theta(\mathbf{a}, \mathbf{b}) = 90^\circ$, the two vectors $\mathbf{a}$ and $\mathbf{b}$ are perpendicular.

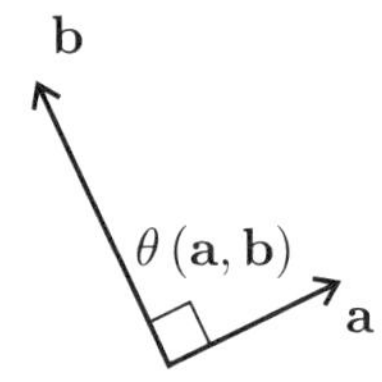


From the properties of the trigonometric functions sine and cosine, it follows that

$$\sin \theta(\mathbf{a}, \mathbf{b}) = 0$$

corresponds to the case when the vectors are parallel, while

$$\cos \theta(\mathbf{a}, \mathbf{b}) = 0$$

corresponds to the case when the vectors are perpendicular. To detect whether two given vectors are parallel or perpendicular, it would be convenient to be able to easily compute the quantities $\sin \theta(\mathbf{a}, \mathbf{b})$ and $\cos \theta(\mathbf{a}, \mathbf{b})$. This is made possible through the definition of two vector products.

**Illustration 3.11**

Consider the triangle with sides corresponding to separations representing the vectors $\mathbf{a}$, $\mathbf{b}$, and $\mathbf{a} + \mathbf{b}$. From the figure on the following page, we see that

$$\begin{aligned} \|\mathbf{a} + \mathbf{b}\| &= \|\mathbf{a}\| \cos \phi_1 + \|\mathbf{b}\| \cos \phi_2, \\ \|\mathbf{a} + \mathbf{b}\| \cos \phi_1 &= \|\mathbf{a}\| + \|\mathbf{b}\| \cos \theta(\mathbf{a}, \mathbf{b}), \end{aligned}$$

and

$$\|\mathbf{a}+\mathbf{b}\| \cos \phi_2 = \|\mathbf{b}\| + \|\mathbf{a}\| \cos \theta (\mathbf{a}, \mathbf{b}) .$$

Multiply the second equation by $\|\mathbf{a}\|$ and the third equation by $\|\mathbf{b}\|$ and use the first equation to obtain

$$\|\mathbf{a}\|^2 + \|\mathbf{b}\|^2 + 2 \|\mathbf{a}\| \|\mathbf{b}\| \cos \theta (\mathbf{a}, \mathbf{b}) = \|\mathbf{a}+\mathbf{b}\|^2 .$$

This statement is known as the *cosine theorem.*

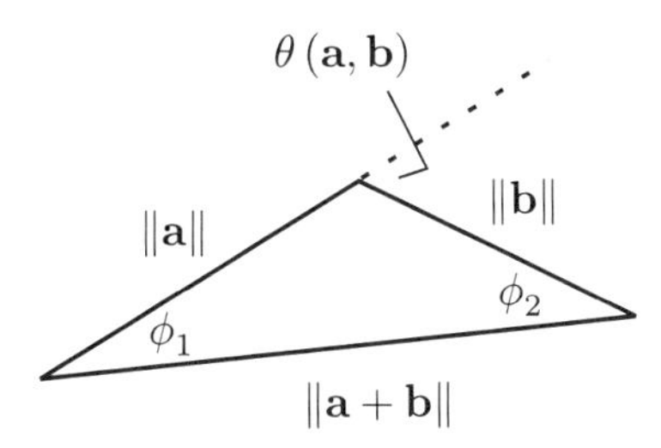


The innocuous quantity

$$\|\mathbf{a}\| \|\mathbf{b}\| \cos \theta (\mathbf{a}, \mathbf{b}) \stackrel{def}{=} \mathbf{a} \bullet \mathbf{b}$$

that appears in the illustration is called the *dot product* of the two vectors $\mathbf{a}$ and $\mathbf{b}$.

The dot product finds widespread use in the remainder of this text and should be well understood.

Since $\|\mathbf{0}\| = 0$, it follows that

$$\mathbf{0} \bullet \mathbf{a} = \mathbf{a} \bullet \mathbf{0} = 0$$

for any vector $\mathbf{a}$. In fact, since $\|\mathbf{v}\| > 0$ for $\mathbf{v} \neq \mathbf{0}$,

$$\mathbf{a} \bullet \mathbf{b} = 0$$

if and only if $\mathbf{a} = \mathbf{0}$, $\mathbf{b} = \mathbf{0}$, or $\theta (\mathbf{a}, \mathbf{b}) = 90°$.

To compute the dot product as defined here requires knowledge of the lengths of the vectors and the angle between them. As we shall see in the next section, it is possible to compute the dot product without direct knowledge of this angle. As such, the dot product is a tool for detecting whether two vectors are perpendicular!

Since

$$\mathbf{a} \bullet \mathbf{b} = \|\mathbf{a}\| \|\mathbf{b}\| \cos \theta (\mathbf{a}, \mathbf{b}) = \|\mathbf{a}\| \left[\|\mathbf{b}\| \cos \theta (\mathbf{a}, \mathbf{b})\right] ,$$

we see that the dot product between $\mathbf{a}$ and $\mathbf{b}$ amounts to multiplying the length of $\mathbf{a}$ with the length of the projection of $\mathbf{b}$ onto $\mathbf{a}$, $\|\mathbf{b}\| \cos \theta (\mathbf{a}, \mathbf{b})$.

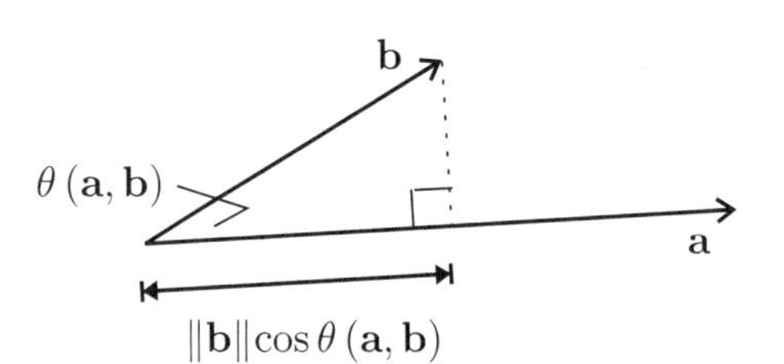


**Illustration 3.12**

Let **a**, **b**, and **c** be three arbitrary vectors as in the figure below. If you consider **a** and **c** to be parallel to some plane, the vector **b** is **not** necessarily parallel to this plane. The image is one in three dimensions.

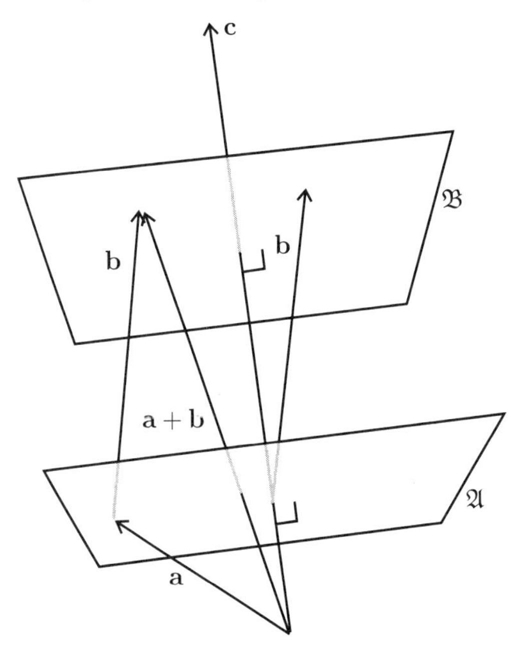


Now, imagine two planes $\mathfrak{A}$ and $\mathfrak{B}$, perpendicular to the separation representing **c** and intersecting the end points of the separations representing **a** and **b**, respectively. It follows that

$$\|\mathbf{a}\| \cos \theta (\mathbf{a}, \mathbf{c})$$

is the distance from the starting point of the separation representing **a** to $\mathfrak{A}$. Similarly,

$$\|\mathbf{b}\| \cos \theta (\mathbf{b}, \mathbf{c})$$

is the distance between $\mathfrak{A}$ and $\mathfrak{B}$. Their sum equals the distance from the starting point of the separation representing **a** to the plane $\mathfrak{B}$. From the figure, it is now clear that

$$\|\mathbf{a}+\mathbf{b}\| \cos\theta\,(\mathbf{a}+\mathbf{b},\mathbf{c}) = \|\mathbf{a}\| \cos\theta\,(\mathbf{a},\mathbf{c}) + \|\mathbf{b}\| \cos\theta\,(\mathbf{b},\mathbf{c})\,.$$

Multiplication with $\|\mathbf{c}\|$ gives

$$\|\mathbf{a}+\mathbf{b}\|\,\|\mathbf{c}\| \cos\theta\,(\mathbf{a}+\mathbf{b},\mathbf{c}) = \|\mathbf{a}\|\,\|\mathbf{c}\| \cos\theta\,(\mathbf{a},\mathbf{c}) + \|\mathbf{b}\|\,\|\mathbf{c}\| \cos\theta\,(\mathbf{b},\mathbf{c})\,.$$

The result of the illustration implies that

$$\mathbf{c} \bullet (\mathbf{a}+\mathbf{b}) = \mathbf{c} \bullet \mathbf{a} + \mathbf{c} \bullet \mathbf{b}.$$

Combined with the observation that (show this!)

$$\alpha\,(\mathbf{a} \bullet \mathbf{b}) = (\alpha \mathbf{a}) \bullet \mathbf{b},$$

it follows that the dot product is *linear*. It is this property that enables its computation without resorting to the definition above.

In analogy with the definition of the dot product, consider the scalar quantity

$$\|\mathbf{a}\|\,\|\mathbf{b}\| \sin\theta\,(\mathbf{a},\mathbf{b})\,.$$

As with the dot product, this quantity is zero if and only if $\mathbf{a}=\mathbf{0}$, $\mathbf{b}=\mathbf{0}$, or $\theta\,(\mathbf{a},\mathbf{b}) = 0°$ or $180°$. Since the angle $\theta\,(\mathbf{a},\mathbf{b})$ lies between $0°$ and $180°$, this quantity is $\geq 0$. As we shall see in the next section, it is possible to compute this quantity without direct knowledge of the angle $\theta\,(\mathbf{a},\mathbf{b})$. As such, it is a tool for detecting whether two vectors are parallel!

In fact, the formula presented in the next section does not merely compute the product above. Instead, it generates a vector, called the *cross product* of the two vectors. In particular, the cross product $\mathbf{a}\times\mathbf{b}$ between the vectors $\mathbf{a}$ and $\mathbf{b}$ is defined as the vector whose length equals

$$\|\mathbf{a}\|\,\|\mathbf{b}\| \sin\theta\,(\mathbf{a},\mathbf{b})$$

and whose direction is perpendicular to both $\mathbf{a}$ and $\mathbf{b}$, pointing in the direction of the right-hand thumb when the fingers curl from $\mathbf{a}$ to $\mathbf{b}$.

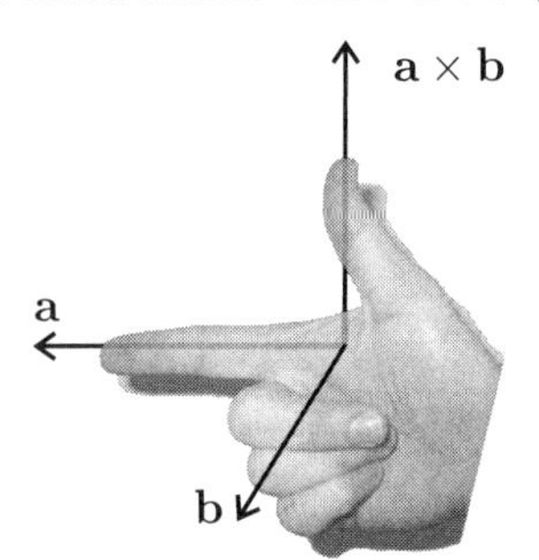


The cross product finds widespread use in the remainder of this text and should be well understood.

**Illustration 3.13**

Consider the parallelogram whose sides correspond to representations of the two vectors $\mathbf{a}$ and $\mathbf{b}$. Then, if we use the side corresponding to $\mathbf{a}$ as base, we find the height to be

$$\|\mathbf{b}\| \sin \theta (\mathbf{a}, \mathbf{b})$$

and the area of the parallelogram is

$$\|\mathbf{a}\| \, \|\mathbf{b}\| \sin \theta (\mathbf{a}, \mathbf{b}) .$$

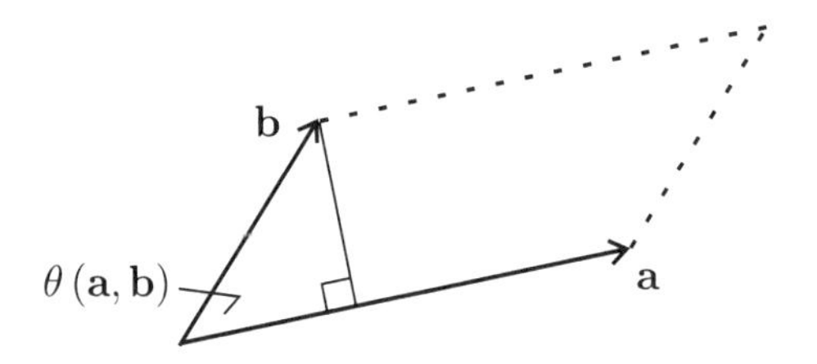


### 3.3.6 Vector Space

The definitions of multiplication of a vector with a scalar and the addition of two vectors makes the set of all (position) vectors a *vector space.*

**Definition 3.3** A vector space $\mathbb{V}$ is a set of elements with well-defined addition and scaling operations, such that given two elements $\mathbf{v}_1, \mathbf{v}_2 \in \mathbb{V}$

$$\mathbf{v}_1 + \mathbf{v}_2 \in \mathbb{V}, \text{ and } k\mathbf{v}_1 \in \mathbb{V} \text{ for any } k \in \mathbb{R}.$$

Moreover, the operations satisfy the following conditions:

- *Commutativity:* $\mathbf{v}_1 + \mathbf{v}_2 = \mathbf{v}_2 + \mathbf{v}_1, \forall \mathbf{v}_1, \mathbf{v}_2 \in \mathbb{V}$;
- *Associativity:* $\mathbf{v}_1 + (\mathbf{v}_2 + \mathbf{v}_3) = (\mathbf{v}_1 + \mathbf{v}_2) + \mathbf{v}_3$, and $k_1 (k_2 \mathbf{v}_1) = (k_1 k_2) \mathbf{v}_1, \forall \mathbf{v}_1, \mathbf{v}_2, \mathbf{v}_3 \in \mathbb{V}$, and $\forall k_1, k_2 \in \mathbb{R}$;
- *Distributivity:* $(k_1 + k_2) \mathbf{v}_1 = k_1 \mathbf{v}_1 + k_2 \mathbf{v}_1$, and $k_1 (\mathbf{v}_1 + \mathbf{v}_2) = k_1 \mathbf{v}_1 + k_1 \mathbf{v}_2, \forall \mathbf{v}_1, \mathbf{v}_2 \in \mathbb{V}$, and $\forall k_1, k_2 \in \mathbb{R}$;
- There exists a *zero element* $\mathbf{0}$, such that $\mathbf{v} + \mathbf{0} = \mathbf{v}$ and $\mathbf{v} + (-1)\mathbf{v} = \mathbf{0}$;
- *Multiplication by* 1 leaves a vector unchanged: $1\mathbf{v} = \mathbf{v}$.

2

**Illustration 3.14**

The elements of a vector space together with the addition operation constitute an Abelian group. Here, the identity element on $\mathbb{V}$ is the zero element $\mathbf{0}$, since

$$\mathbf{v} + \mathbf{0} = \mathbf{0} + \mathbf{v} = \mathbf{v}.$$

Moreover, the inverse of an element $\mathbf{v}$ is the element $-\mathbf{v}$, since

$$\mathbf{v} + (-\mathbf{v}) = (-\mathbf{v}) + \mathbf{v} = \mathbf{0}.$$

Associativity and commutativity follow from the same properties on the vector space.

It is possible to generate a vector space from a general affine space. In a general affine space, two separations $\overrightarrow{PQ}$ and $\overrightarrow{RS}$ are said to be equivalent, i.e., $\overrightarrow{PQ} \sim \overrightarrow{RS}$, if

$$\overrightarrow{PS}\left(\frac{1}{2}\right) = \overrightarrow{QR}\left(\frac{1}{2}\right).$$

The relation $\sim$ is an equivalence relation and we may, consequently, consider the corresponding quotient set on the space of separations. Multiplication of an equivalence class with a scalar and addition of equivalence classes may now be introduced, provided that the $\rightarrow$ function satisfies a number of additional properties that respect the properties of a vector space. To identify these additional conditions on $\rightarrow$ is a nice exercise for the particularly inquisitive.

The length of a vector $\|\mathbf{v}\|$ was defined as the length of any separation representing the vector. It can be shown that the property of length satisfies the properties of a *norm* on a vector space.

**Definition 3.4** A norm $\|\cdot\|$ is a real-valued function on a vector space $\mathbb{V}$, such that for all $\mathbf{v}_1, \mathbf{v}_2 \in \mathbb{V}$, and all $\alpha \in \mathbb{R}$:

- *Positive definiteness:* $\|\mathbf{v}_1\| > 0$ unless $\mathbf{v}_1 = \mathbf{0}$, for which $\|\mathbf{0}\| = 0$;
- *Homogeneity:* $\|\alpha \mathbf{v}_1\| = |\alpha| \cdot \|\mathbf{v}_1\|$;
- *Triangle inequality:* $\|\mathbf{v}_1 + \mathbf{v}_2\| \leq \|\mathbf{v}_1\| + \|\mathbf{v}_2\|$.

2

Certainly $\|\mathbf{0}\| = 0$, since every separation $\overrightarrow{AA}$ representing the zero vector has zero length. Indeed, if $\|\mathbf{v}\| = 0$, then all separations representing $\mathbf{v}$ must have zero length. But this implies that $\overrightarrow{AA} \in \mathbf{v}$ for some point $A$, i.e., $\mathbf{v} = \mathbf{0}$. For all other vectors, the length must be a positive quantity, confirming that the length of a position vector satisfies the positive definiteness property.

The homogeneity property is an immediate consequence of the definition of the multiplication of a position vector by a scalar. Finally, the triangle inequality states that the length of one side in a triangle is always less than or equal to the sum of the lengths of the other sides.

The dot product introduced on the collection of all position vectors is an example of an *inner product*.

---

**Definition 3.5** An inner product $\bullet$ is an operation on pairs of elements of a vector space $\mathbb{V}$ into the reals, such that for all $\mathbf{v}_1, \mathbf{v}_2, \mathbf{v}_3 \in \mathbb{V}$ and all $\alpha, \beta \in \mathbb{R}$:

- *Positive definiteness:* $\mathbf{v}_1 \bullet \mathbf{v}_1 > 0$ unless $\mathbf{v}_1 = \mathbf{0}$, for which $\mathbf{0} \bullet \mathbf{0} = 0$;
- *Symmetry:* $\mathbf{v}_1 \bullet \mathbf{v}_2 = \mathbf{v}_2 \bullet \mathbf{v}_1$;
- *Linearity:* $(\alpha \mathbf{v}_1 + \beta \mathbf{v}_2) \bullet \mathbf{v}_3 = \alpha (\mathbf{v}_1 \bullet \mathbf{v}_3) + \beta (\mathbf{v}_2 \bullet \mathbf{v}_3)$.

---

Given an inner product, we may define a function $f(\mathbf{v})$ on the vector space by

$$f(\mathbf{v}) = \sqrt{\mathbf{v} \bullet \mathbf{v}}.$$

It follows that

$$f(\mathbf{v}) > 0$$

unless $\mathbf{v} = \mathbf{0}$ and

$$f(\mathbf{0}) = \sqrt{\mathbf{0} \bullet \mathbf{0}} = \sqrt{0} = 0,$$

i.e., that $f$ is a positive definite function.

Moreover,

$$f(\alpha \mathbf{v}) = \sqrt{(\alpha \mathbf{v}) \bullet (\alpha \mathbf{v})} = \sqrt{\alpha^2 \mathbf{v} \bullet \mathbf{v}} = |\alpha| \cdot f(\mathbf{v}),$$

showing that $f$ is a homogeneous function.

**Illustration 3.15**

The result in Exercise 3.26 shows that for an arbitrary inner product,

$$|\mathbf{v} \bullet \mathbf{w}| \leq \sqrt{\mathbf{v} \bullet \mathbf{v}} \sqrt{\mathbf{w} \bullet \mathbf{w}}.$$

It follows that

$$\begin{aligned} f^2(\mathbf{v}+\mathbf{w}) &= (\mathbf{v}+\mathbf{w}) \bullet (\mathbf{v}+\mathbf{w}) \\ &= \mathbf{v} \bullet \mathbf{v} + 2\mathbf{v} \bullet \mathbf{w} + \mathbf{w} \bullet \mathbf{w} \\ &= f^2(\mathbf{v}) + f^2(\mathbf{w}) + 2\mathbf{v} \bullet \mathbf{w} \\ &\leq f^2(\mathbf{v}) + f^2(\mathbf{w}) + 2|\mathbf{v} \bullet \mathbf{w}| \\ &\leq f^2(\mathbf{v}) + f^2(\mathbf{w}) + 2f(\mathbf{v}) \cdot f(\mathbf{w}) \\ &= [f(\mathbf{v}) + f(\mathbf{w})]^2 . \end{aligned}$$

Since the quantities being squared on both sides of the equality are $\geq 0$, we can take the square root to obtain

$$f(\mathbf{v}+\mathbf{w}) \leq f(\mathbf{v}) + f(\mathbf{w}) .$$

The function $f$ is said to be *subadditive.*

The properties of the function $f$ show that it qualifies as a norm on the vector space. The subadditivity of $f$ is just a restatement of the triangle inequality. We have found that every inner product automatically generates a norm.

## 3.4 Bases

(Ex. 3.28 – Ex. 3.72)

**Illustration 3.16**

Let $\mathbf{a}_1 \neq \mathbf{0}$ be a vector and $P$ some point in space. Then, if

$$P \overset{v_1\mathbf{a}_1}{\curvearrowright} Q$$

for some scalar $v_1$, the point $Q$ lies on a straight line through $P$ that is parallel to $\mathbf{a}_1$. In fact, every point on this straight line corresponds to some value for the coefficient $v_1$. We say that the vector $\mathbf{a}_1$ *spans the straight line.*

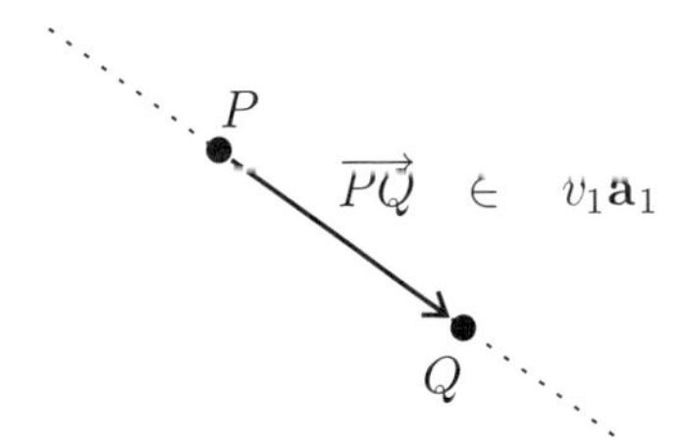


Let $\mathbf{a}_2$ be a second vector that is not parallel to the line spanned by $\mathbf{a}_1$, i.e., such that

$$\mathbf{a}_2 \neq \beta\mathbf{a}_1$$

for all values of $\beta$. Then,

$$P \xrightarrow{v_1\mathbf{a}_1+v_2\mathbf{a}_2} Q$$

for some scalars $v_1$ and $v_2$ implies that $Q$ lies in a plane through $P$ that is parallel to $\mathbf{a}_1$ and $\mathbf{a}_2$. In fact, every point in this plane corresponds to some value for the coefficients $v_1$ and $v_2$. We say that the vectors $\mathbf{a}_1$ and $\mathbf{a}_2$ *span the plane.*

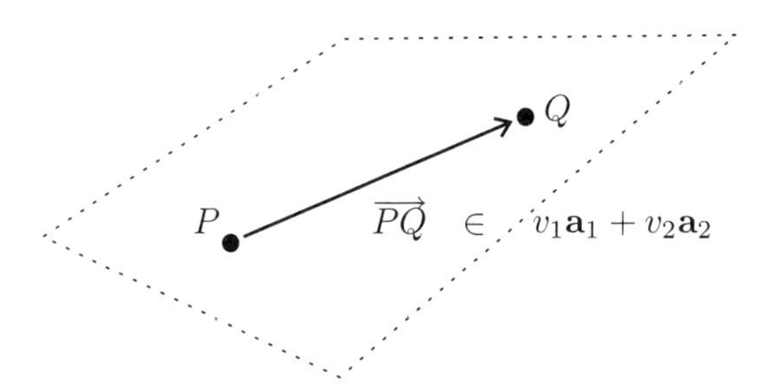


Finally, let $\mathbf{a}_3$ be a third vector that is not parallel to the plane spanned by $\mathbf{a}_1$ and $\mathbf{a}_2$, i.e., such that

$$\mathbf{a}_3 \neq \beta_1\mathbf{a}_1 + \beta_2\mathbf{a}_2$$

for all values of $\beta_1$ and $\beta_2$. Then, for every point in space, it is possible to find some scalars $v_1$, $v_2$, and $v_3$, such that

$$P \xrightarrow{v_1\mathbf{a}_1+v_2\mathbf{a}_2+v_3\mathbf{a}_3} Q.$$

We say that the vectors $\mathbf{a}_1$, $\mathbf{a}_2$, and $\mathbf{a}_3$ *span all of space.*

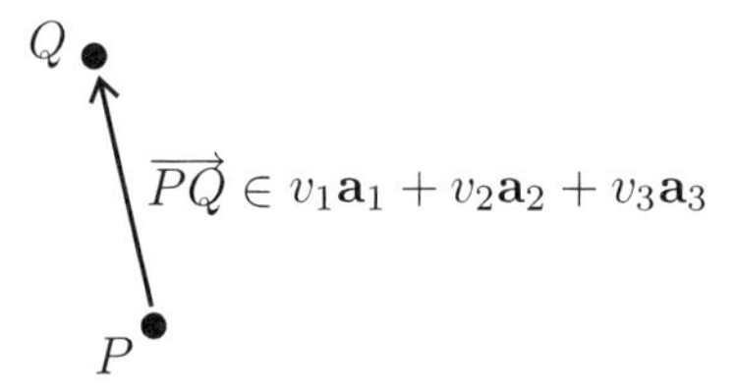

A set of three vectors $\{\mathbf{a}_1, \mathbf{a}_2, \mathbf{a}_3\}$ that span all of space is called a *basis of space* and the vectors are known as *basis vectors.*

Let $\{\mathbf{a}_1, \mathbf{a}_2, \mathbf{a}_3\}$ be a basis of space. For every pair of points $P$ and $Q$, it is possible to find scalars $v_1$, $v_2$, and $v_3$, such that

$$P \xrightarrow{v_1\mathbf{a}_1+v_2\mathbf{a}_2+v_3\mathbf{a}_3} Q.$$

Equivalently, for every pair of points $P$ and $Q$, it is possible to find scalars $v_1$, $v_2$, and $v_3$, such that

$$\left[\overrightarrow{PQ}\right] = v_1\mathbf{a}_1 + v_2\mathbf{a}_2 + v_3\mathbf{a}_3.$$

In particular, for every vector $\mathbf{v}$, it is possible to find scalars $v_1$, $v_2$, and $v_3$, such that

$$\mathbf{v} = v_1\mathbf{a}_1 + v_2\mathbf{a}_2 + v_3\mathbf{a}_3.$$

The result of Exercise 3.29 shows that the coefficients $v_1$, $v_2$, and $v_3$ are unique. These coefficients are known as the *coordinates of the vector relative to the basis* $\{\mathbf{a}_1, \mathbf{a}_2, \mathbf{a}_3\}$. We say that the vector *is expressed relative to the basis* $\{\mathbf{a}_1, \mathbf{a}_2, \mathbf{a}_3\}$.

### 3.4.1 Orthonormal Bases

Let $\{\mathbf{a}_1, \mathbf{a}_2, \mathbf{a}_3\}$ be a basis of space. For every vector $\mathbf{v}$, there exists a unique set of coordinates $v_1$, $v_2$, and $v_3$ of the vector relative to the basis, such that

$$\mathbf{v} = v_1\mathbf{a}_1 + v_2\mathbf{a}_2 + v_3\mathbf{a}_3.$$

Let $\mathbf{v}$ and $\mathbf{w}$ be two arbitrary vectors with coordinates $v_1$, $v_2$, and $v_3$ and $w_1$, $w_2$, and $w_3$, respectively, relative to the basis $\{\mathbf{a}_1, \mathbf{a}_2, \mathbf{a}_3\}$. Now, consider the dot product

$$\begin{aligned}
\mathbf{v} \bullet \mathbf{w} &= (v_1\mathbf{a}_1 + v_2\mathbf{a}_2 + v_3\mathbf{a}_3) \bullet (w_1\mathbf{a}_1 + w_2\mathbf{a}_2 + w_3\mathbf{a}_3) \\
&= v_1 w_1 (\mathbf{a}_1 \bullet \mathbf{a}_1) + v_1 w_2 (\mathbf{a}_1 \bullet \mathbf{a}_2) + v_1 w_3 (\mathbf{a}_1 \bullet \mathbf{a}_3) \\
&\quad + v_2 w_1 (\mathbf{a}_2 \bullet \mathbf{a}_1) + v_2 w_2 (\mathbf{a}_2 \bullet \mathbf{a}_2) + v_2 w_3 (\mathbf{a}_2 \bullet \mathbf{a}_3) \\
&\quad + v_3 w_1 (\mathbf{a}_3 \bullet \mathbf{a}_1) + v_3 w_2 (\mathbf{a}_3 \bullet \mathbf{a}_2) + v_3 w_3 (\mathbf{a}_3 \bullet \mathbf{a}_3),
\end{aligned}$$

where the second equality follows from the linearity of the dot product. Rewrite this sum as a matrix product:

$$\mathbf{v} \bullet \mathbf{w} = \begin{pmatrix} v_1 & v_2 & v_3 \end{pmatrix} \begin{pmatrix} \mathbf{a}_1 \bullet \mathbf{a}_1 & \mathbf{a}_1 \bullet \mathbf{a}_2 & \mathbf{a}_1 \bullet \mathbf{a}_3 \\ \mathbf{a}_2 \bullet \mathbf{a}_1 & \mathbf{a}_2 \bullet \mathbf{a}_2 & \mathbf{a}_2 \bullet \mathbf{a}_3 \\ \mathbf{a}_3 \bullet \mathbf{a}_1 & \mathbf{a}_3 \bullet \mathbf{a}_2 & \mathbf{a}_3 \bullet \mathbf{a}_3 \end{pmatrix} \begin{pmatrix} w_1 \\ w_2 \\ w_3 \end{pmatrix}.$$

It follows that the value of the dot product $\mathbf{v} \bullet \mathbf{w}$ is determined once the matrix

$$\begin{pmatrix} \mathbf{a}_1 \bullet \mathbf{a}_1 & \mathbf{a}_1 \bullet \mathbf{a}_2 & \mathbf{a}_1 \bullet \mathbf{a}_3 \\ \mathbf{a}_2 \bullet \mathbf{a}_1 & \mathbf{a}_2 \bullet \mathbf{a}_2 & \mathbf{a}_2 \bullet \mathbf{a}_3 \\ \mathbf{a}_3 \bullet \mathbf{a}_1 & \mathbf{a}_3 \bullet \mathbf{a}_2 & \mathbf{a}_3 \bullet \mathbf{a}_3 \end{pmatrix}$$

is known.

From Exercise 3.16, we recall that

$$\mathbf{a}_1 \bullet \mathbf{a}_2 = \mathbf{a}_2 \bullet \mathbf{a}_1,$$

$$\mathbf{a}_1 \bullet \mathbf{a}_3 = \mathbf{a}_3 \bullet \mathbf{a}_1,$$

and

$$\mathbf{a}_2 \bullet \mathbf{a}_3 = \mathbf{a}_3 \bullet \mathbf{a}_2.$$

It follows that the matrix

$$\begin{pmatrix} \mathbf{a}_1 \bullet \mathbf{a}_1 & \mathbf{a}_1 \bullet \mathbf{a}_2 & \mathbf{a}_1 \bullet \mathbf{a}_3 \\ \mathbf{a}_2 \bullet \mathbf{a}_1 & \mathbf{a}_2 \bullet \mathbf{a}_2 & \mathbf{a}_2 \bullet \mathbf{a}_3 \\ \mathbf{a}_3 \bullet \mathbf{a}_1 & \mathbf{a}_3 \bullet \mathbf{a}_2 & \mathbf{a}_3 \bullet \mathbf{a}_3 \end{pmatrix}$$

is *symmetric*.

**Illustration 3.17**

The result of Exercise 3.36 shows that for every basis $\{\mathbf{a}_1, \mathbf{a}_2, \mathbf{a}_3\}$ there exist independent angles $\theta_1$, $\theta_2$, and $\theta_3$, such that the matrix

$$\begin{pmatrix} \mathbf{a}_1 \bullet \mathbf{a}_1 & \mathbf{a}_1 \bullet \mathbf{a}_2 & \mathbf{a}_1 \bullet \mathbf{a}_3 \\ \mathbf{a}_2 \bullet \mathbf{a}_1 & \mathbf{a}_2 \bullet \mathbf{a}_2 & \mathbf{a}_2 \bullet \mathbf{a}_3 \\ \mathbf{a}_3 \bullet \mathbf{a}_1 & \mathbf{a}_3 \bullet \mathbf{a}_2 & \mathbf{a}_3 \bullet \mathbf{a}_3 \end{pmatrix}$$

takes the form

$$\begin{pmatrix} \|\mathbf{a}_1\|^2 & \|\mathbf{a}_1\| \|\mathbf{a}_2\| c_1 & \|\mathbf{a}_1\| \|\mathbf{a}_3\| c_2 \\ \|\mathbf{a}_1\| \|\mathbf{a}_2\| c_1 & \|\mathbf{a}_2\|^2 & \|\mathbf{a}_2\| \|\mathbf{a}_3\| (s_1 s_2 c_3 + c_1 c_2) \\ \|\mathbf{a}_1\| \|\mathbf{a}_3\| c_2 & \|\mathbf{a}_2\| \|\mathbf{a}_3\| (s_1 s_2 c_3 + c_1 c_2) & \|\mathbf{a}_3\|^2 \end{pmatrix},$$

where

$$c_i = \cos\theta_i, s_i = \sin\theta_i$$

and

$$\sin\theta_1, \sin\theta_2, \sin\theta_3 \neq 0.$$

It follows that

$$\cos\theta(\mathbf{a}_1, \mathbf{a}_2) = \frac{\mathbf{a}_1 \bullet \mathbf{a}_2}{\|\mathbf{a}_1\| \|\mathbf{a}_2\|} = \cos\theta_1,$$
$$\cos\theta(\mathbf{a}_1, \mathbf{a}_3) = \frac{\mathbf{a}_1 \bullet \mathbf{a}_3}{\|\mathbf{a}_1\| \|\mathbf{a}_3\|} = \cos\theta_2,$$

and

$$\cos\theta(\mathbf{a}_2, \mathbf{a}_3) = \frac{\mathbf{a}_2 \bullet \mathbf{a}_3}{\|\mathbf{a}_2\| \|\mathbf{a}_3\|} = \sin\theta_1 \sin\theta_2 \cos\theta_3 + \cos\theta_1 \cos\theta_2.$$

Moreover,

$$\begin{aligned} \mathbf{v} \bullet \mathbf{v} = {} & (\|\mathbf{a}_1\| v_1 + \|\mathbf{a}_2\| v_2 \cos\theta_1 + \|\mathbf{a}_3\| v_3 \cos\theta_2)^2 \\ & + (\|\mathbf{a}_2\| v_2 \sin\theta_1 + \|\mathbf{a}_3\| v_3 \sin\theta_2 \cos\theta_3)^2 \\ & + (\|\mathbf{a}_3\| v_3 \sin\theta_2 \sin\theta_3)^2, \end{aligned}$$

which is positive for all $\mathbf{v} \neq \mathbf{0}$.

A basis of space is said to be *orthonormal* if

$$\begin{pmatrix} \mathbf{a}_1 \bullet \mathbf{a}_1 & \mathbf{a}_1 \bullet \mathbf{a}_2 & \mathbf{a}_1 \bullet \mathbf{a}_3 \\ \mathbf{a}_2 \bullet \mathbf{a}_1 & \mathbf{a}_2 \bullet \mathbf{a}_2 & \mathbf{a}_2 \bullet \mathbf{a}_3 \\ \mathbf{a}_3 \bullet \mathbf{a}_1 & \mathbf{a}_3 \bullet \mathbf{a}_2 & \mathbf{a}_3 \bullet \mathbf{a}_3 \end{pmatrix}$$

equals the identity matrix. It follows that

$$\begin{aligned} \mathbf{v} \bullet \mathbf{w} &= \begin{pmatrix} v_1 & v_2 & v_3 \end{pmatrix} \begin{pmatrix} \mathbf{a}_1 \bullet \mathbf{a}_1 & \mathbf{a}_1 \bullet \mathbf{a}_2 & \mathbf{a}_1 \bullet \mathbf{a}_3 \\ \mathbf{a}_2 \bullet \mathbf{a}_1 & \mathbf{a}_2 \bullet \mathbf{a}_2 & \mathbf{a}_2 \bullet \mathbf{a}_3 \\ \mathbf{a}_3 \bullet \mathbf{a}_1 & \mathbf{a}_3 \bullet \mathbf{a}_2 & \mathbf{a}_3 \bullet \mathbf{a}_3 \end{pmatrix} \begin{pmatrix} w_1 \\ w_2 \\ w_3 \end{pmatrix} \\ &= \begin{pmatrix} v_1 & v_2 & v_3 \end{pmatrix} \begin{pmatrix} 1 & 0 & 0 \\ 0 & 1 & 0 \\ 0 & 0 & 1 \end{pmatrix} \begin{pmatrix} w_1 \\ w_2 \\ w_3 \end{pmatrix} \\ &= \begin{pmatrix} v_1 & v_2 & v_3 \end{pmatrix} \begin{pmatrix} w_1 \\ w_2 \\ w_3 \end{pmatrix} \\ &= v_1 w_1 + v_2 w_2 + v_3 w_3, \end{aligned}$$

where the $v$'s and $w$'s are the coordinates of $\mathbf{v}$ and $\mathbf{w}$ relative to an orthonormal basis $\{\mathbf{a}_1, \mathbf{a}_2, \mathbf{a}_3\}$.

**Illustration 3.18**

Let $\{\mathbf{a}_1, \mathbf{a}_2, \mathbf{a}_3\}$ be an orthonormal basis of space. Then

$$\begin{pmatrix} \mathbf{a}_1 \bullet \mathbf{a}_1 & \mathbf{a}_1 \bullet \mathbf{a}_2 & \mathbf{a}_1 \bullet \mathbf{a}_3 \\ \mathbf{a}_2 \bullet \mathbf{a}_1 & \mathbf{a}_2 \bullet \mathbf{a}_2 & \mathbf{a}_2 \bullet \mathbf{a}_3 \\ \mathbf{a}_3 \bullet \mathbf{a}_1 & \mathbf{a}_3 \bullet \mathbf{a}_2 & \mathbf{a}_3 \bullet \mathbf{a}_3 \end{pmatrix} = \begin{pmatrix} 1 & 0 & 0 \\ 0 & 1 & 0 \\ 0 & 0 & 1 \end{pmatrix}$$

shows that

$$\|\mathbf{a}_i\| = \sqrt{\mathbf{a}_i \bullet \mathbf{a}_i} = 1, \; i = 1, 2, 3,$$

i.e., the basis vectors have unit length. Moreover,

$$\mathbf{a}_1 \bullet \mathbf{a}_2 = \mathbf{a}_1 \bullet \mathbf{a}_3 = \mathbf{a}_2 \bullet \mathbf{a}_3 = 0,$$

i.e., the basis vectors are mutually perpendicular.

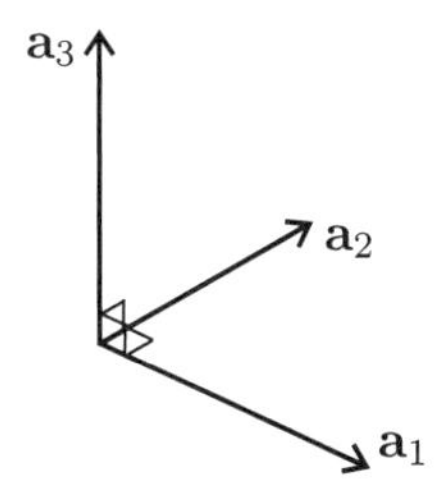


The length of a vector $\mathbf{v}$ is given by

$$\|\mathbf{v}\| = \sqrt{\mathbf{v} \bullet \mathbf{v}} = \sqrt{v_1^2 + v_2^2 + v_3^2},$$

where the $v_i$'s are the coordinates of $\mathbf{v}$ relative to the orthonormal basis. This statement is equivalent to the Pythagorean theorem in three dimensions.

The formula

$$\mathbf{v} \bullet \mathbf{w} = v_1 w_1 + v_2 w_2 + v_3 w_3$$

offers a straightforward method for computing the dot product between arbitrary vectors without appealing to the original geometric definition of $\mathbf{v} \bullet \mathbf{w}$ in terms of the lengths of $\mathbf{v}$ and $\mathbf{w}$ and the intermediate angle $\theta(\mathbf{v}, \mathbf{w})$. Instead, we find that

$$\begin{aligned} \cos\theta(\mathbf{v}, \mathbf{w}) &= \frac{\mathbf{v} \bullet \mathbf{w}}{\|\mathbf{v}\| \, \|\mathbf{w}\|} \\ &= \frac{v_1 w_1 + v_2 w_2 + v_3 w_3}{\sqrt{v_1^2 + v_2^2 + v_3^2}\sqrt{w_1^2 + w_2^2 + w_3^2}}. \end{aligned}$$

In particular, we conclude that two non-zero vectors $\mathbf{v}$ and $\mathbf{w}$ are perpendicular if and only if

$$v_1 w_1 + v_2 w_2 + v_3 w_3 = 0,$$

where the $v_i$'s and $w_i$'s are the coordinates of $\mathbf{v}$ and $\mathbf{w}$ relative to some orthonormal basis.

The result of Exercise 3.47 shows that if $\{\mathbf{a}_1, \mathbf{a}_2, \mathbf{a}_3\}$ is an orthonormal basis and

$$\mathbf{v} = v_1 \mathbf{a}_1 + v_2 \mathbf{a}_2 + v_3 \mathbf{a}_3,$$

then the dot product $\mathbf{a}_i \bullet \mathbf{v}$ equals the $i$-th coordinate of the vector $\mathbf{v}$ relative to the basis $\{\mathbf{a}_1, \mathbf{a}_2, \mathbf{a}_3\}$, i.e.

$$\mathbf{a}_i \bullet \mathbf{v} = v_i.$$

### 3.4.2 Notation

From this point on, all coordinate descriptions will be stated relative to orthonormal bases. We will find it algebraically convenient to organize the basis vectors of a basis $\{\mathbf{a}_1, \mathbf{a}_2, \mathbf{a}_3\}$ into a row matrix denoted by a lower-case, unsubscripted letter:

$$a \stackrel{\text{def}}{=} \begin{pmatrix} \mathbf{a}_1 & \mathbf{a}_2 & \mathbf{a}_3 \end{pmatrix}.$$

When referring to this matrix, we simply speak of the orthonormal basis $a$. To distinguish between different orthonormal bases that use the same letter, I include appropriate superscripts within parentheses to the right of the symbol of the basis, e.g., $a^{(1)}$, $b^{(\text{block})}$, and so on. The same superscripts are then added to the basis vectors, e.g.,

$$b^{(r)} = \begin{pmatrix} \mathbf{b}_1^{(r)} & \mathbf{b}_2^{(r)} & \mathbf{b}_3^{(r)} \end{pmatrix}.$$

With a slight stretch of normal matrix multiplication, any vector $\mathbf{v}$ can be expanded as

$$\begin{aligned} \mathbf{v} &= \mathbf{a}_1 (\mathbf{a}_1 \bullet \mathbf{v}) + \mathbf{a}_2 (\mathbf{a}_2 \bullet \mathbf{v}) + \mathbf{a}_3 (\mathbf{a}_3 \bullet \mathbf{v}) \\ &= \begin{pmatrix} \mathbf{a}_1 & \mathbf{a}_2 & \mathbf{a}_3 \end{pmatrix} \begin{pmatrix} \mathbf{a}_1 \bullet \mathbf{v} \\ \mathbf{a}_2 \bullet \mathbf{v} \\ \mathbf{a}_3 \bullet \mathbf{v} \end{pmatrix} = a\, {}^a v, \end{aligned}$$

where the column matrix

$${}^a v \stackrel{def}{=} \begin{pmatrix} \mathbf{a}_1 \bullet \mathbf{v} \\ \mathbf{a}_2 \bullet \mathbf{v} \\ \mathbf{a}_3 \bullet \mathbf{v} \end{pmatrix} \stackrel{def}{=} \begin{pmatrix} {}^a v_1 \\ {}^a v_2 \\ {}^a v_3 \end{pmatrix}$$

contains the coordinates of the vector $\mathbf{v}$ relative to the orthonormal basis $a$. The matrices $a$ and ${}^a v$ are multiplied with each other following the standard rules of matrix multiplication, in spite of the non-standard nature of their components.

${}^a v$ is called the *matrix representation of the vector* $\mathbf{v}$ *relative to the orthonormal basis* $a$. The letter (lower-case, italicized) used in the notation ${}^a v$ for the matrix representation agrees with the letter used to denote the corresponding vector. The left superscript, in turn, specifies the orthonormal basis, relative to which the vector is expressed. Other examples are

$${}^b w \text{ and } {}^e (\mathbf{u}_{\text{base}}),$$

where the latter expression refers to the matrix representation of the vector $\mathbf{u}_{\text{base}}$ relative to the orthonormal basis $e$.

A further generalization of notation allows one to consider dot products of vectors with matrices of vectors, such as

$$a^T \bullet \mathbf{v} = \begin{pmatrix} \mathbf{a}_1 \\ \mathbf{a}_2 \\ \mathbf{a}_3 \end{pmatrix} \bullet \mathbf{v} \stackrel{def}{=} \begin{pmatrix} \mathbf{a}_1 \bullet \mathbf{v} \\ \mathbf{a}_2 \bullet \mathbf{v} \\ \mathbf{a}_3 \bullet \mathbf{v} \end{pmatrix} = {}^a v;$$

dot products of matrices of vectors, such as

$$\begin{aligned} a^T \bullet b &= \begin{pmatrix} \mathbf{a}_1 \\ \mathbf{a}_2 \\ \mathbf{a}_3 \end{pmatrix} \bullet \begin{pmatrix} \mathbf{b}_1 & \mathbf{b}_2 & \mathbf{b}_3 \end{pmatrix} \\ &\stackrel{def}{=} \begin{pmatrix} \mathbf{a}_1 \bullet \mathbf{b}_1 & \mathbf{a}_1 \bullet \mathbf{b}_2 & \mathbf{a}_1 \bullet \mathbf{b}_3 \\ \mathbf{a}_2 \bullet \mathbf{b}_1 & \mathbf{a}_2 \bullet \mathbf{b}_2 & \mathbf{a}_2 \bullet \mathbf{b}_3 \\ \mathbf{a}_3 \bullet \mathbf{b}_1 & \mathbf{a}_3 \bullet \mathbf{b}_2 & \mathbf{a}_3 \bullet \mathbf{b}_3 \end{pmatrix}; \end{aligned}$$

and similarly for cross products[6]. In the formulae above, the $^T$ superscript denotes the matrix transpose.

### 3.4.3 Common Misconceptions

It is important to recognize that $a$ is a $1 \times 3$ matrix of **vectors** and $^a v$ is a $3 \times 1$ matrix of **numbers**. $^a v$ is **not** a vector. $^a v$ is **not** the vector $\mathbf{v}$ expressed in the $a$ basis. $^a v$ is simply a matrix. It **cannot** be used alone to represent the vector $\mathbf{v}$. Any description of $\mathbf{v}$ using $^a v$ **must** include mention of the corresponding orthonormal basis $a$.

The expression

$$\mathbf{v} = a \,{}^a v$$

is true, as is shown by matrix multiplication. The expression

$$\mathbf{v} = {}^a v$$

is false, however, since the left-hand side is a vector (i.e., a collection of infinitely many equivalent separations) and the right-hand side is a matrix of numbers.

There is nothing wrong with an expression like

$$b \,{}^a v.$$

This is **not** the vector $\mathbf{v}$, since the expression mixes the matrix representation of $\mathbf{v}$ relative to $a$ with the row matrix $b$. The expression does evaluate to a vector, however, whose matrix representation relative to the orthonormal basis $b$ is given by $^a v$.

### 3.4.4 Handedness

Let $\{\mathbf{a}_1, \mathbf{a}_2, \mathbf{a}_3\}$ be an orthonormal basis. Then, the basis vectors $\mathbf{a}_1$, $\mathbf{a}_2$, and $\mathbf{a}_3$ are mutually perpendicular and of unit length. Since

$$\|-\mathbf{a}_i\| = \|\mathbf{a}_i\| = 1,\ i = 1, 2, 3,$$

it follows that $\{\pm\mathbf{a}_1, \pm\mathbf{a}_2, \pm\mathbf{a}_3\}$ is an orthonormal basis for any combination of plus and minus signs. There are eight such combinations, each corresponding to a different orthonormal basis.

[6] Note that care must be taken to account for the antisymmetry of the cross product when applying the transpose operator to expressions involving cross products, e.g., $\left(a^T \times a\right)^T = -a^T \times \left(a^T\right)^T = -a^T \times a$.

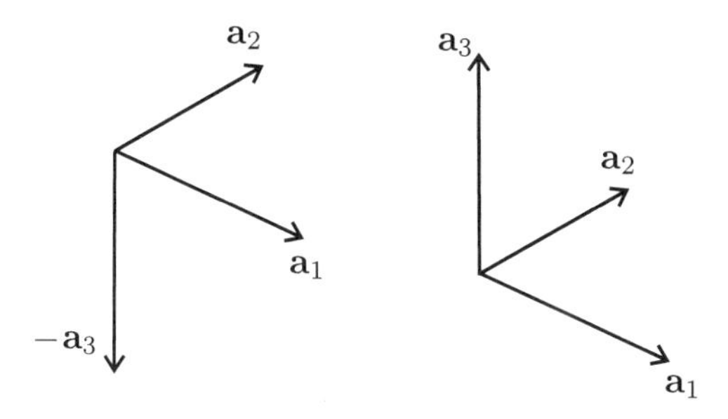


**Illustration 3.19**

Consider changing the signs on two of the basis vectors and compare the orthonormal bases $\{\mathbf{a}_1, \mathbf{a}_2, \mathbf{a}_3\}$ and $\{-\mathbf{a}_1, -\mathbf{a}_2, \mathbf{a}_3\}$. Imagine the basis vectors $\mathbf{a}_1$, $\mathbf{a}_2$, and $\mathbf{a}_3$ fixed to a rigid body. Then, a rotation of the rigid body half a turn about an axis parallel to $\mathbf{a}_3$ will make these vectors coincide with the basis vectors $-\mathbf{a}_1$, $-\mathbf{a}_2$, and $\mathbf{a}_3$. Here, the cross products between the first and second basis vectors, respectively, in each of the bases yield identical results:

$$\mathbf{a}_1 \times \mathbf{a}_2 = (-\mathbf{a}_1) \times (-\mathbf{a}_2).$$

In contrast, there is no rotation of the rigid body that will make the basis vectors $\mathbf{a}_1$, $\mathbf{a}_2$, and $\mathbf{a}_3$ coincide with the basis vectors $-\mathbf{a}_1$, $\mathbf{a}_2$, and $\mathbf{a}_3$. Here, the cross products between the first and second basis vectors, respectively, in each of the bases yield different results:

$$\mathbf{a}_1 \times \mathbf{a}_2 = -(-\mathbf{a}_1) \times \mathbf{a}_2 \neq (-\mathbf{a}_1) \times \mathbf{a}_2.$$

An orthonormal basis is said to be *right-handed* if

$$\mathbf{a}_1 \times \mathbf{a}_2 = \mathbf{a}_3.$$

By the definition of the cross product, this implies that $\mathbf{a}_3$ points in the direction of the right-hand thumb when the fingers curl from $\mathbf{a}_1$ to $\mathbf{a}_2$. It is straightforward to see that for a right-handed, orthonormal basis

$$\mathbf{a}_2 \times \mathbf{a}_3 = \mathbf{a}_1$$

and

$$\mathbf{a}_3 \times \mathbf{a}_1 = \mathbf{a}_2,$$

each of which can be taken to define a right-handed, orthonormal basis.

If, instead,

$$\mathbf{a}_1 \times \mathbf{a}_2 = -\mathbf{a}_3,$$

then the orthonormal basis is *left-handed*.

**Illustration 3.20**

Let the orthonormal basis $\{\mathbf{a}_1, \mathbf{a}_2, \mathbf{a}_3\}$ be right-handed. Then, since

$$(-\mathbf{a}_1) \times (-\mathbf{a}_2) = \mathbf{a}_1 \times \mathbf{a}_2 = \mathbf{a}_3,$$

the orthonormal basis $\{-\mathbf{a}_1, -\mathbf{a}_2, \mathbf{a}_3\}$ is also right-handed.

On the other hand, the orthonormal basis $\{-\mathbf{a}_1, \mathbf{a}_2, \mathbf{a}_3\}$ is left-handed, since

$$(-\mathbf{a}_1) \times \mathbf{a}_2 = -(\mathbf{a}_1 \times \mathbf{a}_2) = -\mathbf{a}_3.$$

The result from Exercise 3.58 shows that the cross product between two arbitrary vectors $\mathbf{v}$ and $\mathbf{w}$ may be computed from the expression

$$\begin{aligned} \mathbf{v} \times \mathbf{w} &= \mathbf{a}_1 \begin{vmatrix} v_2 & v_3 \\ w_2 & w_3 \end{vmatrix} - \mathbf{a}_2 \begin{vmatrix} v_1 & v_3 \\ w_1 & w_3 \end{vmatrix} + \mathbf{a}_3 \begin{vmatrix} v_1 & v_2 \\ w_1 & w_2 \end{vmatrix} \\ &= \begin{vmatrix} \mathbf{a}_1 & \mathbf{a}_2 & \mathbf{a}_3 \\ v_1 & v_2 & v_2 \\ w_1 & w_2 & w_3 \end{vmatrix}, \end{aligned}$$

where the $v_i$'s and $w_i$'s are the coordinates of the two vectors relative to the right-handed, orthonormal basis $\{\mathbf{a}_1, \mathbf{a}_2, \mathbf{a}_3\}$. The last determinant is computed using the standard rules from matrix algebra, in spite of the mixed nature of the entries of the matrix.

### 3.4.5 Generating New Bases

Let $a$ be a right-handed orthonormal basis and let $\mathbf{v} \neq \mathbf{0}$ be some vector that is not parallel to $\mathbf{a}_1$. Since

$$\left\| \frac{\mathbf{v}}{\|\mathbf{v}\|} \right\| = \frac{1}{\|\mathbf{v}\|} \|\mathbf{v}\| = 1,$$

it follows that

$$\mathbf{b}_1 = \frac{\mathbf{v}}{\|\mathbf{v}\|}$$

is a vector of unit length that is parallel to $\mathbf{v}$.

Since the cross product between two non-zero, non-parallel vectors results in a vector perpendicular to the two vectors, it follows that

$$\mathbf{b}_2 = \frac{\mathbf{b}_1 \times \mathbf{a}_1}{\|\mathbf{b}_1 \times \mathbf{a}_1\|}$$

is a vector of unit length that is perpendicular to $\mathbf{b}_1$. Finally, the vector

$$\mathbf{b}_3 = \mathbf{b}_1 \times \mathbf{b}_2$$

is perpendicular to $\mathbf{b}_1$ and $\mathbf{b}_2$ and has length

$$\|\mathbf{b}_3\| = \|\mathbf{b}_1\| \, \|\mathbf{b}_2\| \sin\theta\left(\mathbf{b}_1, \mathbf{b}_2\right) = 1,$$

since $\theta\left(\mathbf{b}_1, \mathbf{b}_2\right) = 90°$. We conclude that the set $\{\mathbf{b}_1, \mathbf{b}_2, \mathbf{b}_3\}$ forms a right-handed, orthonormal basis.

**Illustration 3.21**

Let $a$ be a right-handed, orthonormal basis and consider the vector

$$\mathbf{v} = a \begin{pmatrix} {}^a v_1 \\ {}^a v_2 \\ {}^a v_3 \end{pmatrix},$$

such that $\mathbf{v}$ is not parallel to $\mathbf{a}_1$. Then,

$$\mathbf{b}_1 = \frac{\mathbf{v}}{\|\mathbf{v}\|} = a \begin{pmatrix} \frac{{}^a v_1}{\sqrt{{}^a v_1^2 + {}^a v_2^2 + {}^a v_3^2}} \\ \frac{{}^a v_2}{\sqrt{{}^a v_1^2 + {}^a v_2^2 + {}^a v_3^2}} \\ \frac{{}^a v_3}{\sqrt{{}^a v_1^2 + {}^a v_2^2 + {}^a v_3^2}} \end{pmatrix},$$

$$\mathbf{b}_2 = \frac{\mathbf{b}_1 \times \mathbf{a}_1}{\|\mathbf{b}_1 \times \mathbf{a}_1\|} = a \begin{pmatrix} 0 \\ \frac{{}^a v_3}{\sqrt{{}^a v_2^2 + {}^a v_3^2}} \\ -\frac{{}^a v_2}{\sqrt{{}^a v_2^2 + {}^a v_3^2}} \end{pmatrix},$$

and, finally,

$$\mathbf{b}_3 = \mathbf{b}_1 \times \mathbf{b}_2 = a \begin{pmatrix} -\frac{\sqrt{{}^a v_2^2 + {}^a v_3^2}}{\sqrt{{}^a v_1^2 + {}^a v_2^2 + {}^a v_3^2}} \\ \frac{{}^a v_1 {}^a v_2}{\sqrt{{}^a v_1^2 + {}^a v_2^2 + {}^a v_3^2}\sqrt{{}^a v_2^2 + {}^a v_3^2}} \\ \frac{{}^a v_1 {}^a v_3}{\sqrt{{}^a v_1^2 + {}^a v_2^2 + {}^a v_3^2}\sqrt{{}^a v_2^2 + {}^a v_3^2}} \end{pmatrix}$$

The cross product $\mathbf{b}_1 \times \mathbf{a}_1$ used to generate $\mathbf{b}_2$ was arbitrary. Any vector (except one parallel to $\mathbf{b}_1$) could take the place of $\mathbf{a}_1$. Each such choice would lead to a different right-handed, orthonormal basis.

2

### 3.4.6 Linear Combinations and Bases

In a general vector space, the associative property of addition of vectors allows you to consider adding several vectors without worrying about the order in which the operation is performed. Given a collection of vectors $\mathbf{v}$, $\mathbf{w}$, and $\mathbf{x}$, we can multiply them by the real numbers 2, $-3.4$, and 1, respectively, and add up the resulting vectors to obtain

$$2\mathbf{v} - 3.4\mathbf{w} + \mathbf{x}.$$

This is an example of a *linear combination* of the vectors $\mathbf{v}$, $\mathbf{w}$, and $\mathbf{x}$.

---

**Definition 3.6** Let $\{\mathbf{v}_1, \mathbf{v}_2, \ldots, \mathbf{v}_n\}$ be a set of arbitrary vectors and let $\{\alpha_1, \alpha_2, \ldots, \alpha_n\}$ be a set of arbitrary real numbers. Then, the vector sum

$$\alpha_1\mathbf{v}_1 + \alpha_2\mathbf{v}_2 + \ldots + \alpha_n\mathbf{v}_n$$

is called a *linear combination* of the vectors $\{\mathbf{v}_1, \mathbf{v}_2, \ldots, \mathbf{v}_n\}$. The set of all possible linear combinations of these vectors is denoted by

$$\text{span}\,\{\mathbf{v}_1, \mathbf{v}_2, \ldots, \mathbf{v}_n\}$$

and is said to be *spanned* by the $\mathbf{v}_i$'s.

---

**Illustration 3.22**

Let $\mathbf{v}_1$ and $\mathbf{v}_2$ be two vectors and assume that

$$2\mathbf{v}_1 - \mathbf{v}_2 = \mathbf{v}_1 + 3\mathbf{v}_2.$$

Then, it follows that

$$\mathbf{v}_1 - 4\mathbf{v}_2 = \mathbf{0}.$$

More generally, if two linear combinations of the vectors $\mathbf{v}_1$ and $\mathbf{v}_2$ give the same result

$$\alpha_1\mathbf{v}_1 + \alpha_2\mathbf{v}_2 = \beta_1\mathbf{v}_1 + \beta_2\mathbf{v}_2,$$

then the linear combination

$$(\beta_1 - \alpha_1)\,\mathbf{v}_1 + (\beta_2 - \alpha_2)\,\mathbf{v}_2 = \mathbf{0}$$

even though at least one of the coefficients $\beta_1 - \alpha_1$ and $\beta_2 - \alpha_2$ is different from zero.

If it is not possible to find coefficients $\gamma_1$ and $\gamma_2$, such that the linear combination

$$\gamma_1 \mathbf{v}_1 + \gamma_2 \mathbf{v}_2 = \mathbf{0}$$

other than $\gamma_1 = \gamma_2 = 0$, then every linear combination of the two vectors must give a different result.

If the only linear combination of a set of vectors $\{\mathbf{v}_1, \mathbf{v}_2, \ldots, \mathbf{v}_n\}$

$$\alpha_1 \mathbf{v}_1 + \alpha_2 \mathbf{v}_2 + \cdots + \alpha_n \mathbf{v}_n$$

that results in the zero vector $\mathbf{0}$ is the one with $\alpha_1 = \alpha_2 = \cdots = \alpha_n = 0$, then the vectors are said to be *linearly independent*.

If every vector can be written as a linear combination of a set of linearly independent vectors $\{\mathbf{v}_1, \mathbf{v}_2, \ldots, \mathbf{v}_n\}$, then the vectors are said to form a *basis* of the vector space. It follows that $\text{span}\{\mathbf{v}_1, \mathbf{v}_2, \ldots, \mathbf{v}_n\}$ is the entire space. In analogy to the result of Exercise 3.29, the coefficients in such a linear combination are unique. These coefficients are called the coordinates of the vector relative to the basis. The number of vectors in a basis of a vector space is the *dimension* of the vector space.

Let $\{\mathbf{v}_1, \mathbf{v}_2, \ldots, \mathbf{v}_n\}$ be a basis of an $n$-dimensional vector space. By linearity, it follows that the inner product between two arbitrary vectors $\mathbf{a}$ and $\mathbf{b}$ becomes

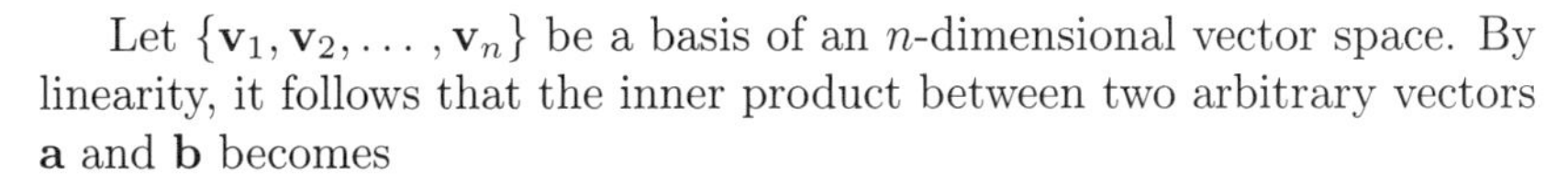

$$\mathbf{a} \bullet \mathbf{b} = \left( \sum_i a_i \mathbf{v}_i \right) \bullet \left( \sum_j b_j \mathbf{v}_j \right) = \sum_i \sum_j a_i b_j \left( \mathbf{v}_i \bullet \mathbf{v}_j \right),$$

where the $a_i$'s and $b_j$'s are the coordinates of $\mathbf{a}$ and $\mathbf{b}$ relative to the basis $\{\mathbf{v}_1, \ldots, \mathbf{v}_n\}$. Clearly, the value of the inner product is determined once the $n^2$ products $\mathbf{v}_i \bullet \mathbf{v}_j$ have been chosen for all $i, j \in \{1, 2, \ldots, n\}$. An *orthonormal basis* is obtained when

$$\mathbf{v}_i \bullet \mathbf{v}_j = \delta_{ij} = \begin{cases} 1 & \text{when } i = j \\ 0 & \text{when } i \neq j \end{cases},$$

where $\delta_{ij}$ is the so-called *Kronecker delta*. With respect to an orthonormal basis,

$$\mathbf{a} \bullet \mathbf{b} = \sum_i \sum_j a_i b_j \delta_{ij} = \sum_i a_i b_i.$$

## 3.5 The MAMBO Toolbox

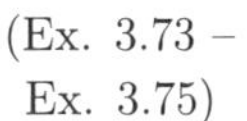

(Ex. 3.73 – Ex. 3.75)

The MAMBO toolbox contains a complete set of procedures and functions to define and operate on vectors. Vectors in the MAMBO toolbox are specified by the names of right-handed, orthonormal bases and the corresponding matrix representations.

**Illustration 3.23**

In the following MAMBO toolbox session, the right-handed, orthonormal basis $a$ is used to define the vector

$$\mathbf{v} = a \begin{pmatrix} 1 \\ 2 \\ 0 \end{pmatrix}.$$

```
> Restart():
> DeclareTriads(a):
> v:=MakeTranslations(a,1,2,0);
```

$$\begin{aligned} v := \text{table}([ \\ & 1 = \text{table}([ \\ & \text{“Triad”} = a \\ & \text{“Coordinates”} = [1, 2, 0] \\ & ]) \\ & \text{“Size”} = 1 \\ & \text{“Type”} = \text{“Vector”} \\ & ]) \end{aligned}$$

The `MakeTranslations` procedure will only return a valid MAMBO vector if the first argument corresponds to the name of a right-handed, orthonormal basis that has been previously declared with the `DeclareTriads` procedure. Invoking the `DeclareTriads` procedure appends the global variable `GlobalTriadDeclarations`[7].

A MAMBO vector is a MAPLE table with at least three entries, namely a `Type`, a `Size`, and a nested table containing the name of a right-handed, orthonormal basis and the corresponding matrix representation.

We recall the use of the `eval` command to access and display the content of a MAPLE table as illustrated by the following statements:

```
> v;
```

$$v$$

```
> eval(v);
```

[7]More detail on the `GlobalTriadDeclarations` variable in Chapter 5.

table([
1 = table([
"Coordinates" = [1, 2, 0]
"Triad" = $a$
])
"Type" = "Vector"
"Size" = 1
])

---

The simplest vector is the zero vector **0**, whose matrix representation relative to any arbitrary basis $b$ is

$$^{b}(\mathbf{0}) = \begin{pmatrix} 0 \\ 0 \\ 0 \end{pmatrix}.$$

In the MAMBO toolbox, we can generate the zero vector using the `MakeTranslations` procedure or more directly using the `NullVector` procedure:

```
> MakeTranslations(a,0,0,0);
```

table([
"Size" = 0
"Type" = "Vector"
])

```
> NullVector();
```

table([
"Size" = 0
"Type" = "Vector"
])

Since the zero vector has the same matrix representation relative to all bases, there is no need to include the name of a right-handed, orthonormal basis and the corresponding matrix representation within the table structure. This is also reflected in the `Size` entry.

The MAMBO toolbox contains functions to compute:

- The multiplication of a vector with a scalar: `&**`;
- The sum of two vectors: `&++`;
- The difference between two vectors: `&--`;
- The dot product between two vectors: `&oo`;
- The cross product between two vectors: `&xx`;
- The length of a vector: `VectorLength`.

The `&**` function is demonstrated in the following MAMBO toolbox statement:

```
>  2 &** v;
```

table([
1 = table([
"Coordinates" = [2, 4, 0]
"Triad" = $a$
])
"Type" = "Vector"
"Size" = 1
])

To multiply a MAMBO vector with a negative scalar, it is important to place the scalar within parentheses:

```
>  (-1) &** v;
```

table([
1 = table([
"Coordinates" = [−1, −2, 0]
"Triad" = $a$
])
"Type" = "Vector"
"Size" = 1
])

Parentheses should also be used if the scalar is expressed as a product of several scalars, as in

```
>  (1/VectorLength(v)) &** v;
```

table([
"Type" = "Vector"
1 = table([
"Coordinates" = $\left[\frac{1}{5}\sqrt{5},\ \frac{2}{5}\sqrt{5},\ 0\right]$
"Triad" = $a$
])
"Size" = 1
])

where the vector **v** is multiplied by the inverse of its length (obtained with the `VectorLength` function) to yield a vector parallel to **v** but of unit length.

We may add two vectors with the `&++` function. If the two vectors are expressed relative to the same right-handed, orthonormal basis, the `&++` function returns a MAMBO vector with `Size` equal to 1 and matrix representation given by the sum of the matrix representations of the two vectors.

```
>  w:=MakeTranslations(a,-1,q,0):
>  w &++ v;
```

table([
1 = table([
"Triad" = $a$
"Coordinates" = $[0,\ q+2,\ 0]$
])
"Size" = 1
"Type" = "Vector"
])

If the two vectors are expressed relative to different right-handed, orthonormal bases, the `&++` function returns a MAMBO vector with `Size` equal to 2 and containing two nested tables, each of which contains the name of the basis, relative to which the corresponding vector is expressed and the associated matrix representation.

```
> DeclareTriads(b):
> u:=MakeTranslations(b,r,r-s,s):
> u &++ v;
```

table([
1 = table([
"Triad" = $b$
"Coordinates" = $[r, r - s, s]$
])
2 = table([
"Triad" = $a$
"Coordinates" = $[1, 2, 0]$
])
"Size" = 2
"Type" = "Vector"
])

**Illustration 3.24**

A MAMBO vector with multiple nested tables can be generated by the `MakeTranslations` procedure.

```
> MakeTranslations([a,1,2,0],[b,r,r-s,s]);
```

table([
1 = table([
"Coordinates" = $[1, 2, 0]$
"Triad" = $a$
])
2 = table([
"Coordinates" = $[r, r - s, s]$
"Triad" = $b$
])
"Type" = "Vector"
"Size" = 2
])

Note the use of the square brackets to separate the distinct components.

The `&--` function is defined using the `&**` and `&++` functions as suggested in the definition of vector subtraction. Some additional examples of the use of these functions are given in the MAMBO toolbox statements below.

```
> (v &++ u) &++ v;
```

table([
1 = table([
"Coordinates" = [2, 4, 0]
"Triad" = $a$
])
2 = table([
"Coordinates" = [$r$, $r - s$, $s$]
"Triad" = $b$
])
"Type" = "Vector"
"Size" = 2
])

```
> v &++ u &++ v;
```

table([
1 = table([
"Coordinates" = [2, 4, 0]
"Triad" = $a$
])
2 = table([
"Coordinates" = [$r$, $r - s$, $s$]
"Triad" = $b$
])
"Type" = "Vector"
"Size" = 2
])

```
> v &++ NullVector();
```

$$\begin{array}{l} \text{table}([ \\ \quad 1 = \text{table}([ \\ \quad \text{“Coordinates”} = [1,\ 2,\ 0] \\ \quad \text{“Triad”} = a \\ \quad ]) \\ \quad \text{“Type”} = \text{“Vector”} \\ \quad \text{“Size”} = 1 \\ \quad ]) \end{array}$$

```
> v &-- v;
```

$$\begin{array}{l} \text{table}([ \\ \quad \text{“Type”} = \text{“Vector”} \\ \quad \text{“Size”} = 0 \\ \quad ]) \end{array}$$

```
> v &++ u &-- v;
```

$$\begin{array}{l} \text{table}([ \\ \quad 1 = \text{table}([ \\ \quad \text{“Coordinates”} = [r,\ r-s,\ s] \\ \quad \text{“Triad”} = b \\ \quad ]) \\ \quad \text{“Type”} = \text{“Vector”} \\ \quad \text{“Size”} = 1 \\ \quad ]) \end{array}$$

Note how the `Size` of the MAMBO vector adapts to accommodate changes in the number of nested tables.

The `&oo` and `&xx` functions implement the formulae derived in the previous section for the dot product and cross product as confirmed by the results of the following MAMBO toolbox session:

```
> Restart():
> DeclareTriads(a):
> v:=MakeTranslations(a,v1,v2,v3):
> w:=MakeTranslations(a,w1,w2,w3):

> v &oo w;
```

$$v1\, w1 + v2\, w2 + v3\, w3$$

```
> v &xx w;
```

table([
"Type" = "Vector"
1 = table([
"Coordinates" = $[-w2\, v3 + w3\, v2,\ -w3\, v1 + v3\, w1,\ w2\, v1 - v2\, w1]$
"Triad" = $a$
])
"Size" = 1
])

Here, $a$ is declared as a right-handed, orthonormal basis prior to the definition of the vectors $\mathbf{v}$ and $\mathbf{w}$.

**Illustration 3.25**

Let $a$ be a right-handed, orthonormal basis. Then, the following sequence of MAMBO toolbox statements generates an alternative right-handed, orthonormal basis $b$, such that $\mathbf{b}_1$ is parallel to the vector

$$a \begin{pmatrix} 1 \\ 0 \\ -2 \end{pmatrix}.$$

```
> Restart():
> DeclareTriads(a):
> a1:=MakeTranslations(a,1):
> v:=MakeTranslations(a,1,0,-2):
> b1:=(1/VectorLength(v)) &** v;
```

$b1$ := table([
1 = table([
"Triad" = $a$
"Coordinates" = $\left[\frac{1}{5}\sqrt{5},\ 0,\ -\frac{2}{5}\sqrt{5}\right]$
])
"Size" = 1
"Type" = "Vector"
])

```
> b2:=(1/VectorLength(b1 &xx a1)) &** (b1 &xx a1);
```

$$b2 := \text{table}([$$
$$1 = \text{table}([$$
$$\text{"Triad"} = a$$
$$\text{"Coordinates"} = [0, -1, 0]$$
$$])$$
$$\text{"Size"} = 1$$
$$\text{"Type"} = \text{"Vector"}$$
$$])$$

```
> b3:=b1 &xx b2;
```

$$b3 := \text{table}([$$
$$1 = \text{table}([$$
$$\text{"Triad"} = a$$
$$\text{"Coordinates"} = \left[-\frac{2}{5}\sqrt{5},\ 0,\ -\frac{1}{5}\sqrt{5}\right]$$
$$])$$
$$\text{"Size"} = 1$$
$$\text{"Type"} = \text{"Vector"}$$
$$])$$

Here, the shorthand form of the `MakeTranslations` procedure is used to generate the basis vector $\mathbf{a}_1$. That the basis $b$ is orthonormal follows from

```
> matrix(3,3,(i,j)->cat(b,i) &oo cat(b,j));
```

$$\begin{bmatrix} 1 & 0 & 0 \\ 0 & 1 & 0 \\ 0 & 0 & 1 \end{bmatrix}$$

corresponding to the matrix

$$\begin{pmatrix} \mathbf{b}_1 \bullet \mathbf{b}_1 & \mathbf{b}_1 \bullet \mathbf{b}_2 & \mathbf{b}_1 \bullet \mathbf{b}_3 \\ \mathbf{b}_2 \bullet \mathbf{b}_1 & \mathbf{b}_2 \bullet \mathbf{b}_2 & \mathbf{b}_2 \bullet \mathbf{b}_3 \\ \mathbf{b}_3 \bullet \mathbf{b}_1 & \mathbf{b}_3 \bullet \mathbf{b}_2 & \mathbf{b}_3 \bullet \mathbf{b}_3 \end{pmatrix}.$$

Finally, that the basis is right-handed is confirmed by

```
> b1 &oo (b2 &xx b3);
```

$$1$$

## 3.6 Exercises

**Exercise 3.1** Show that the motion

$$\underset{\curvearrowright}{\overrightarrow{AB}}\ \underset{\curvearrowright}{\overrightarrow{BA}}$$

produces the same outcome as the motion $\underset{\curvearrowright}{\overrightarrow{AA}}$, but whereas the latter involves no motion whatsoever, the former may involve arbitrarily large displacements.

**Exercise 3.2** When is

$$\underset{\curvearrowright}{\overrightarrow{AB}}\ \underset{\curvearrowright}{\overrightarrow{BC}} = \underset{\curvearrowright}{\overrightarrow{AC}}?$$

**Exercise 3.3** For each of the sequences of motions below, i) determine whether the expression makes sense and ii) find the separation between the starting and ending points when it exists.

a) $\underset{\curvearrowright}{\overrightarrow{P_1P_2}}\ \underset{\curvearrowright}{\overrightarrow{P_3P_4}}$ b) $\underset{\curvearrowright}{\overrightarrow{P_1P_2}}\ \underset{\curvearrowright}{\overrightarrow{P_2P_3}}\ \underset{\curvearrowright}{\overrightarrow{P_3P_1}}$

c) $\underset{\curvearrowright}{\overrightarrow{P_2P_1}}\ \underset{\curvearrowright}{\overrightarrow{P_2P_3}}\ \underset{\curvearrowright}{\overrightarrow{P_3P_4}}$ d) $\underset{\curvearrowright}{\overrightarrow{P_2P_1}}\ \underset{\curvearrowright}{\overrightarrow{P_1P_3}}\ \underset{\curvearrowright}{\overrightarrow{P_3P_1}}$

**Exercise 3.4** Consider the affine space of points on the upper hemisphere of a sphere of unit radius introduced following Definition 3.1. Each point on the hemisphere corresponds to a unique straight line through the center of the sphere. Equivalently, every straight line through the center of the sphere that is not perpendicular to the straight line through the poles of the sphere corresponds to a unique point on the upper hemisphere.

Consider all the straight lines that intersect at some point in space. Denote one of these lines by $L$. Now, eliminate from this collection all the straight lines that are perpendicular to $L$. Use the above observation to construct an affine space of the remaining set of straight lines.

**Exercise 3.5** Consider the collection of all straight lines from the previous exercise that intersect at some point in space and are not perpendicular to the line labeled by $L$. Each such straight line corresponds to a unique plane perpendicular to the straight line. Use this observation to construct an affine space of the corresponding set of planes.

**Exercise 3.6** Show that

1. $\overrightarrow{PQ} \sim \overrightarrow{PQ}$;

2. $\overrightarrow{PQ} \sim \overrightarrow{RS}$ implies that $\overrightarrow{RS} \sim \overrightarrow{PQ}$;

3. $\overrightarrow{PQ} \sim \overrightarrow{RS}$ and $\overrightarrow{RS} \sim \overrightarrow{TU}$ imply that $\overrightarrow{PQ} \sim \overrightarrow{TU}$.

**Exercise 3.7** Consider the points on the surface of a sphere. Define the relation $\sim$ so that two points $P$ and $Q$ on the sphere satisfy $P \sim Q$ if and only if they lie on the same line through the center of the sphere. Show that $\sim$ is an equivalence relation. Characterize the corresponding equivalence classes and the quotient set.

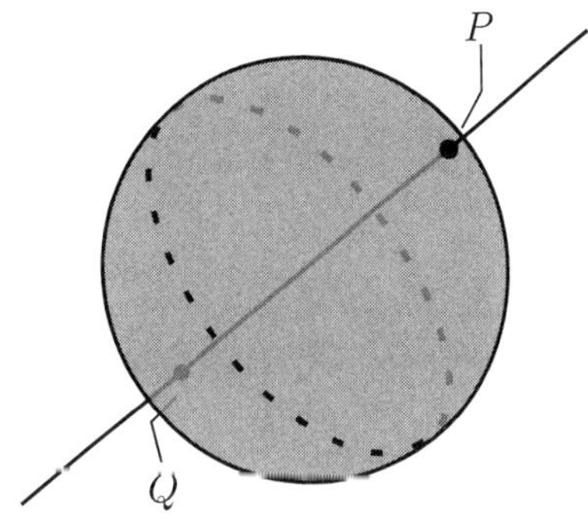


**Solution.** Let $P$, $Q$, and $R$ be points on the sphere's surface. Clearly, $P \sim P$, i.e., $\sim$ is reflexive. Moreover, $P \sim Q$ implies that $Q \sim P$, i.e., $\sim$ is symmetric. Finally, if $P \sim Q$ and $Q \sim R$, then $P \sim R$, i.e., $\sim$ is transitive. Thus, $\sim$ is an equivalence relation.

The equivalence classes corresponding to $\sim$ consist of diametrically opposite points on the sphere's surface.

The quotient set is conveniently represented by the set of straight lines through the center of the sphere.

**Exercise 3.8** Consider the affine space of points on the upper hemisphere as introduced following Definition 3.1. Consider the equivalence relation $\sim$, such that

$$\overrightarrow{PQ} \sim \overrightarrow{RS}$$
if and only if
$$\overrightarrow{PS}\left(\frac{1}{2}\right) = \overrightarrow{QR}\left(\frac{1}{2}\right).$$

Characterize the resulting equivalence classes. Can you define addition and scalar multiplication on the quotient set to make it a vector space?

**Exercise 3.9** Show that $\mathbf{v}+\mathbf{0} = \mathbf{0}+\mathbf{v} = \mathbf{v}$ for all position vectors $\mathbf{v}$.

**Solution.** That $\mathbf{v}+\mathbf{0} = \mathbf{0}+\mathbf{v}$ follows from the result in Illustration 3.10. Let the separations $\overrightarrow{PQ}$ and $\overrightarrow{QQ}$ represent the vectors $\mathbf{v}$ and $\mathbf{0}$, respectively. Then,

$$\mathbf{v}+\mathbf{0} = \left[\overrightarrow{PQ}\right] + \left[\overrightarrow{QQ}\right] = \left[\overrightarrow{PQ}\right] = \mathbf{v}.$$

**Exercise 3.10** Show that $\mathbf{v}-\mathbf{v} = \mathbf{0}$ for all position vectors $\mathbf{v}$.

**Solution.** Let the separation $\overrightarrow{PQ}$ represent the vector $\mathbf{v}$. Then,

$$\begin{aligned}\mathbf{v}-\mathbf{v} &= \mathbf{v}+(-1)\,\mathbf{v} \\ &= \left[\overrightarrow{PQ}\right] + \left[\overrightarrow{QP}\right] = \left[\overrightarrow{PP}\right] = \mathbf{0}.\end{aligned}$$

**Exercise 3.11** Show that $(\mathbf{u}+\mathbf{v})+\mathbf{w} = \mathbf{u}+(\mathbf{v}+\mathbf{w})$ for all position vectors $\mathbf{u}$, $\mathbf{v}$, and $\mathbf{w}$.

**Exercise 3.12** Show that $(\alpha_1+\alpha_2)\,\mathbf{v} = \alpha_1\mathbf{v}+\alpha_2\mathbf{v}$ for all scalars $\alpha_1$ and $\alpha_2$ and all position vectors $\mathbf{v}$.

**Exercise 3.13** Show that $a\,(\mathbf{v}+\mathbf{w}) = \alpha\mathbf{v}+\alpha\mathbf{w}$ for all position vectors $\mathbf{v}$ and $\mathbf{w}$ and any scalar $\alpha$.

**Exercise 3.14** Show that the cosine theorem implies that

$$\|\mathbf{v}+\mathbf{w}\| \leq \|\mathbf{v}\| + \|\mathbf{w}\|$$

for all position vectors $\mathbf{v}$ and $\mathbf{w}$.

**Solution.** The cosine theorem states that

$$\begin{aligned}\|\mathbf{v}+\mathbf{w}\|^2 &= \|\mathbf{v}\|^2 + \|\mathbf{w}\|^2 \\ &\quad +2\,\|\mathbf{v}\|\,\|\mathbf{w}\|\cos\theta\,(\mathbf{v},\mathbf{w}) \\ &\leq \|\mathbf{v}\|^2 + \|\mathbf{w}\|^2 + 2\,\|\mathbf{v}\|\,\|\mathbf{w}\| \\ &= (\|\mathbf{v}\| + \|\mathbf{w}\|)^2,\end{aligned}$$

since $\cos\theta\,(\mathbf{v},\mathbf{w}) \leq 1$. Since the quantities being squared on both sides of the inequality are positive, it follows that

$$\|\mathbf{v}+\mathbf{w}\| \leq \|\mathbf{v}\| + \|\mathbf{w}\|.$$

**Exercise 3.15** Show that $-\|\mathbf{v}-\mathbf{w}\| \leq \|\mathbf{v}\| - \|\mathbf{w}\| \leq \|\mathbf{v}-\mathbf{w}\|$ for all position vectors $\mathbf{v}$ and $\mathbf{w}$.

**Solution.** From the previous exercise, we have

$$\begin{aligned}\|\mathbf{v}\| &= \|\mathbf{v}-\mathbf{w}+\mathbf{w}\| \leq \|\mathbf{v}-\mathbf{w}\| + \|\mathbf{w}\| \\ &\Longrightarrow \|\mathbf{v}\| - \|\mathbf{w}\| \leq \|\mathbf{v}-\mathbf{w}\|.\end{aligned}$$

Similarly,

$$\begin{aligned}\|\mathbf{w}\| &= \|\mathbf{w}-\mathbf{v}+\mathbf{v}\| \leq \|\mathbf{w}-\mathbf{v}\| + \|\mathbf{v}\| \\ &\Rightarrow \|\mathbf{v}\| - \|\mathbf{w}\| \geq -\|\mathbf{w}-\mathbf{v}\|.\end{aligned}$$

But,

$$\begin{aligned}\|\mathbf{w}-\mathbf{v}\| &= \|(-1)\,(\mathbf{v}-\mathbf{w})\| \\ &= |(-1)|\,\|\mathbf{v}-\mathbf{w}\| = \|\mathbf{v}-\mathbf{w}\|,\end{aligned}$$

i.e.,

$$-\|\mathbf{v}-\mathbf{w}\| \leq \|\mathbf{v}\| - \|\mathbf{w}\| \leq \|\mathbf{v}-\mathbf{w}\|.$$

**Exercise 3.16** Show that $\mathbf{v} \bullet \mathbf{w} = \mathbf{w} \bullet \mathbf{v}$ for all position vectors $\mathbf{v}$ and $\mathbf{w}$.

**Exercise 3.17** Show that $\|\mathbf{v}\| = \sqrt{\mathbf{v} \bullet \mathbf{v}}$ for all position vectors $\mathbf{v}$.

**Exercise 3.18** Show that $\mathbf{v} \bullet \mathbf{v} > 0$ unless $\mathbf{v} = \mathbf{0}$, for which $\mathbf{0} \bullet \mathbf{0} = 0$.

**Exercise 3.19** Use the dot product to show that

$$\|\mathbf{v}+\mathbf{w}\|^2 + \|\mathbf{v}-\mathbf{w}\|^2 = 2\|\mathbf{v}\|^2 + 2\|\mathbf{w}\|^2.$$

Interpret the result geometrically.

**Solution.** Using the dot product, we find $\|\mathbf{v}+\mathbf{w}\|^2 + \|\mathbf{v}-\mathbf{w}\|^2$

$$\begin{aligned}
&= (\mathbf{v}+\mathbf{w}) \bullet (\mathbf{v}+\mathbf{w}) + (\mathbf{v}-\mathbf{w}) \bullet (\mathbf{v}-\mathbf{w}) \\
&= \mathbf{v} \bullet \mathbf{v} + \mathbf{v} \bullet \mathbf{w} + \mathbf{w} \bullet \mathbf{v} + \mathbf{w} \bullet \mathbf{w} \\
&\quad + \mathbf{v} \bullet \mathbf{v} - \mathbf{v} \bullet \mathbf{w} - \mathbf{w} \bullet \mathbf{v} + \mathbf{w} \bullet \mathbf{w} \\
&= 2\mathbf{v} \bullet \mathbf{v} + 2\mathbf{w} \bullet \mathbf{w} \\
&= 2\|\mathbf{v}\|^2 + 2\|\mathbf{w}\|^2,
\end{aligned}$$

where the second equality follows from linearity.

Now, consider the parallelogram spanned by the two vectors $\mathbf{v}$ and $\mathbf{w}$.

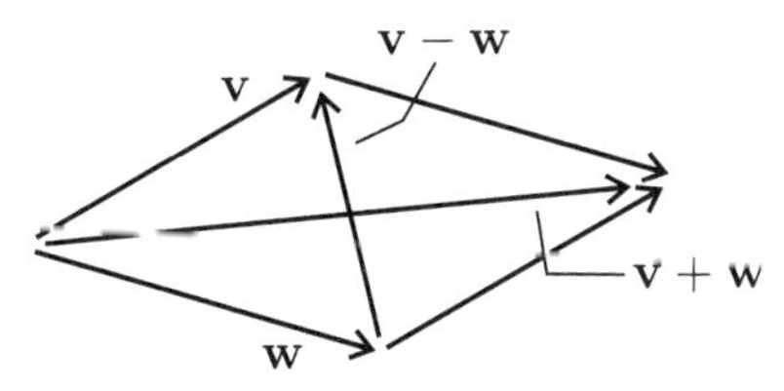


The vectors $\mathbf{v}+\mathbf{w}$ and $\mathbf{v}-\mathbf{w}$ then correspond to the diagonals in the parallelogram. The statement above implies that the sum of the squares of the lengths of the diagonals in a parallelogram equals twice the sum of the squares of the lengths of the sides of the parallelogram.

**Exercise 3.20** Show that $\mathbf{w} \times \mathbf{v} = -\mathbf{v} \times \mathbf{w}$.

[Hint: $\theta(\mathbf{v},\mathbf{w}) = \theta(\mathbf{w},\mathbf{v})$.]

**Exercise 3.21** Show that $\mathbf{v} \times \mathbf{v} = \mathbf{0}$.

**Exercise 3.22** Recall that $\|\mathbf{v} \times \mathbf{w}\|$ equals the **area** of the parallelogram spanned by the vectors $\mathbf{v}$ and $\mathbf{w}$. What quantity does $|\mathbf{u} \bullet (\mathbf{v} \times \mathbf{w})|$ equal for some arbitrary vector $\mathbf{u}$?

[Hint: Consider the volume of the prism whose edges are separations representing the three vectors $\mathbf{u}$, $\mathbf{v}$, and $\mathbf{w}$.]

**Exercise 3.23** Show that $\mathbf{v} \bullet (\mathbf{v} \times \mathbf{w}) = 0$.

**Solution.** The vector $\mathbf{v} \times \mathbf{w}$ is perpendicular to both $\mathbf{v}$ and $\mathbf{w}$.

Thus, the angle between the vectors $\mathbf{v}$ and $\mathbf{v} \times \mathbf{w}$ is $90°$ and $\cos 90° = 0$, which proves the claim.

**Exercise 3.24** Show that

$$\mathbf{v} = (\mathbf{v} \bullet \mathbf{n})\,\mathbf{n} + \mathbf{n} \times (\mathbf{v} \times \mathbf{n})$$

for any vector $\mathbf{v}$ and any vector $\mathbf{n}$ of unit length.

**Solution.** Consider the right triangle, for which $\mathbf{v}$ represents the hypotenuse and the adjacent side lies along the vector $\mathbf{n}$.

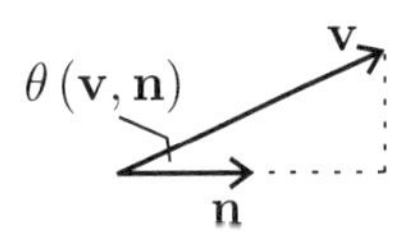


Then the length of the adjacent side is

$$\begin{aligned}
&\|\mathbf{v}\| \cos\theta(\mathbf{v},\mathbf{n}) \\
&\qquad = \|\mathbf{v}\|\,\|\mathbf{n}\| \cos\theta(\mathbf{v},\mathbf{n}) = \mathbf{v} \bullet \mathbf{n},
\end{aligned}$$

since $\|\mathbf{n}\| = 1$. Indeed, the vector corresponding to the adjacent side is given by this length multiplied by the unit vector $\mathbf{n}$.

Furthermore, the length of the opposite side equals

$$\begin{aligned}\|\mathbf{v}\| \sin\theta(\mathbf{v},\mathbf{n}) &= \|\mathbf{v}\| \|\mathbf{n}\| \|\mathbf{n}\| \sin\theta(\mathbf{v},\mathbf{n}) \\ &= \|\mathbf{v}\times\mathbf{n}\| \|\mathbf{n}\| \sin\frac{\pi}{2} \\ &= \|\mathbf{n}\times(\mathbf{v}\times\mathbf{n})\|,\end{aligned}$$

since $\mathbf{n}$ is perpendicular to $\mathbf{v}\times\mathbf{n}$. Also, the vector $\mathbf{n}\times(\mathbf{v}\times\mathbf{n})$ is parallel to the opposite side of the triangle pointing from the right-angled corner toward the hypotenuse. The statement of the problem now follows by considering the vector sum of the two vectors corresponding to the adjacent and opposite sides of the triangle.

**Exercise 3.25** Find the general solution to the equation $\mathbf{u}\bullet(\mathbf{v}\times\mathbf{w}) = 0$, where $\mathbf{v}$ and $\mathbf{w}$ are two given, linearly independent vectors.

**Solution.** The vector $\mathbf{v}\times\mathbf{w}$ is perpendicular to both $\mathbf{v}$ and $\mathbf{w}$. In other words, $\mathbf{v}\times\mathbf{w}$ is normal to any plane spanned by $\mathbf{v}$ and $\mathbf{w}$.

The equation

$$\mathbf{u}\bullet(\mathbf{v}\times\mathbf{w}) = 0$$

implies that $\mathbf{u}$ is perpendicular to $\mathbf{v}\times\mathbf{w}$, i.e., that $\mathbf{u}$ is parallel to any plane spanned by $\mathbf{v}$ and $\mathbf{w}$. It follows that $\mathbf{u}\in\operatorname{span}\{\mathbf{v},\mathbf{w}\}$, or

$$\mathbf{u} = \alpha\mathbf{v} + \beta\mathbf{w}$$

for some pair of scalars $\alpha$ and $\beta$.

**Exercise 3.26** Let $\mathbf{v}$ and $\mathbf{w}$ be two arbitrary vectors in a vector space with an inner product $\bullet$. Prove the Cauchy-Schwarz inequality:

$$|\mathbf{v}\bullet\mathbf{w}| \leq \sqrt{\mathbf{v}\bullet\mathbf{v}}\sqrt{\mathbf{w}\bullet\mathbf{w}}.$$

**Solution.** Suppose that

$$\mathbf{w}\bullet\mathbf{w} = 0.$$

It follows that

$$\mathbf{v}\bullet\mathbf{w} = 0$$

and thus

$$0 = |\mathbf{v}\bullet\mathbf{w}| \leq \sqrt{\mathbf{v}\bullet\mathbf{v}}\sqrt{\mathbf{w}\bullet\mathbf{w}} = 0$$

confirming the validity of the inequality for this special case.

Suppose, instead, that

$$\mathbf{w}\bullet\mathbf{w} \neq 0.$$

Positive definiteness implies that

$$\begin{aligned}0 &\leq (\mathbf{v}-\alpha\mathbf{w})\bullet(\mathbf{v}-\alpha\mathbf{w}) \\ &= \mathbf{v}\bullet\mathbf{v} - 2\alpha\mathbf{v}\bullet\mathbf{w} + \alpha^2\mathbf{w}\bullet\mathbf{w}\end{aligned}$$

for all $\alpha$. If, in particular

$$\alpha = \frac{\mathbf{v}\bullet\mathbf{w}}{\mathbf{w}\bullet\mathbf{w}},$$

it follows that

$$\begin{aligned}0 &\leq \mathbf{v}\bullet\mathbf{v} - 2\alpha\mathbf{v}\bullet\mathbf{w} + \alpha^2\mathbf{w}\bullet\mathbf{w} \\ &= \mathbf{v}\bullet\mathbf{v} - 2\frac{\mathbf{v}\bullet\mathbf{w}}{\mathbf{w}\bullet\mathbf{w}}\mathbf{v}\bullet\mathbf{w} \\ &\quad + \left(\frac{\mathbf{v}\bullet\mathbf{w}}{\mathbf{w}\bullet\mathbf{w}}\right)^2\mathbf{w}\bullet\mathbf{w} \\ &= \mathbf{v}\bullet\mathbf{v} - \frac{(\mathbf{v}\bullet\mathbf{w})^2}{\mathbf{w}\bullet\mathbf{w}},\end{aligned}$$

i.e.,

$$(\mathbf{v}\bullet\mathbf{w})^2 \leq (\mathbf{v}\bullet\mathbf{v})(\mathbf{w}\bullet\mathbf{w})$$

and the claim follows by taking square roots on both sides.

**Exercise 3.27** Show that the Cauchy-Schwarz inequality is true by construction in the case of the dot product on the vector space of translations.

**Exercise 3.28** Let $\{\mathbf{a}_1,\mathbf{a}_2,\mathbf{a}_3\}$ be a basis of space. Show that

$$v_1\mathbf{a}_1 + v_2\mathbf{a}_2 + v_3\mathbf{a}_3 = \mathbf{0}$$

is true if and only if

$$v_1 = v_2 = v_3 = 0.$$

**Solution.** Since

$$0\mathbf{a} = \mathbf{0} \text{ and } \mathbf{0} + \mathbf{0} = \mathbf{0}$$

for any vector $\mathbf{v}$, the assertion that

$$v_1 = v_2 = v_3 = 0 \Rightarrow v_1\mathbf{a}_1 + v_2\mathbf{a}_2 + v_3\mathbf{a}_3 = \mathbf{0}$$

is trivial.

Assume, instead, that at least one of the scalars $v_1$, $v_2$, or $v_3$ is non-zero, e.g., $v_3$. Then,

$$v_1\mathbf{a}_1 + v_2\mathbf{a}_2 + v_3\mathbf{a}_3 = \mathbf{0}$$

only if

$$\mathbf{a}_3 = -\frac{1}{v_3}\left(v_1\mathbf{a}_1 + v_2\mathbf{a}_2\right).$$

But this implies that $\mathbf{a}_3$ is parallel to the plane spanned by $\mathbf{a}_1$ and $\mathbf{a}_2$, in contradiction with the assumption that the vectors form a basis.

If $v_3 = 0$, but $v_2 \neq 0$, then

$$v_1\mathbf{a}_1 + v_2\mathbf{a}_2 + v_3\mathbf{a}_3 = \mathbf{0}$$

only if

$$\mathbf{a}_2 = -\frac{v_1}{v_2}\mathbf{a}_1.$$

But this implies that $\mathbf{a}_2$ is parallel to the line spanned by $\mathbf{a}_1$, in contradiction with the assumption that the vectors form a basis.

Finally, if $v_3 = v_2 = 0$, but $v_1 \neq 0$, then

$$v_1\mathbf{a}_1 + v_2\mathbf{a}_2 + v_3\mathbf{a}_3 = \mathbf{0}$$

only if

$$\mathbf{a}_1 = \mathbf{0}.$$

Again, this contradicts the assumptions that the vectors form a basis and the claim follows.

**Exercise 3.29** Let $\{\mathbf{a}_1, \mathbf{a}_2, \mathbf{a}_3\}$ be a basis of space. Show that

$$v_1\mathbf{a}_1 + v_2\mathbf{a}_2 + v_3\mathbf{a}_3 = w_1\mathbf{a}_1 + w_2\mathbf{a}_2 + w_3\mathbf{a}_3$$

is possibly only if

$$v_1 = w_1,\ v_2 = w_2, \text{ and } v_3 = w_3.$$

**Exercise 3.30** Let $\{\mathbf{a}_1, \mathbf{a}_2, \mathbf{a}_3\}$ be a basis of space. Show that an alternative basis is given by the vectors

$$\begin{aligned} \mathbf{b}_1 &= 2\mathbf{a}_1 - \mathbf{a}_3 \\ \mathbf{b}_2 &= -\mathbf{a}_2 + 3\mathbf{a}_3 \\ \mathbf{b}_3 &= \mathbf{a}_1 + \mathbf{a}_3 \end{aligned}$$

**Solution.** The vectors $\mathbf{b}_1$, $\mathbf{b}_2$, and $\mathbf{b}_3$ constitute a basis if

$$\mathbf{b}_2 - \beta\mathbf{b}_1 \neq \mathbf{0}$$

for all values of $\beta$ and

$$\mathbf{b}_3 - \beta_1\mathbf{b}_1 - \beta_2\mathbf{b}_2 \neq \mathbf{0}$$

for all values of $\beta_1$ and $\beta_2$.

Here,

$$\mathbf{b}_2 - \beta\mathbf{b}_1 = -2\beta\mathbf{a}_1 - \mathbf{a}_2 + (3 + \beta)\,\mathbf{a}_3.$$

But, by a previous exercise, this cannot equal the zero vector for any choice of $\beta$. Similarly,

$$\begin{aligned} \mathbf{b}_3 - \beta_1\mathbf{b}_1 - \beta_2\mathbf{b}_2 &= (1 - 2\beta_1)\,\mathbf{a}_1 + \beta_2\mathbf{a}_2 \\ &\quad + (1 + \beta_1 - 3\beta_2)\,\mathbf{a}_3. \end{aligned}$$

But, by a previous exercise, this equals the zero vector only if

$$1 - 2\beta_1 = \beta_2 = 1 + \beta_1 - 3\beta_2 = 0,$$

which is not possible for any choice of $\beta_1$ and $\beta_2$. We conclude that $\{\mathbf{b}_1, \mathbf{b}_2, \mathbf{b}_3\}$ is a basis of space.

**Exercise 3.31** Let $\{\mathbf{a}_1, \mathbf{a}_2, \mathbf{a}_3\}$ be a basis of space. For each of the sets of vectors below, determine whether they constitute an alternative basis of space.

a) $\mathbf{b}_1 = 2\mathbf{a}_2 + \mathbf{a}_3$, $\mathbf{b}_2 = -2\mathbf{a}_2 + \mathbf{a}_3$, $\mathbf{b}_3 = 3\mathbf{a}_3$

b) $\mathbf{b}_1 = 3\mathbf{a}_1$, $\mathbf{b}_2 = \mathbf{a}_1$, $\mathbf{b}_3 = \mathbf{a}_1 + \mathbf{a}_2$

c) $\mathbf{b}_1 = -\mathbf{a}_2 - \mathbf{a}_3$, $\mathbf{b}_2 = \mathbf{a}_2 - \mathbf{a}_3$, $\mathbf{b}_3 = \mathbf{a}_3$

d) $\mathbf{b}_1 = \mathbf{a}_2 + 2\mathbf{a}_3$, $\mathbf{b}_2 = \mathbf{a}_1$, $\mathbf{b}_3 = \mathbf{a}_3$

e) $\mathbf{b}_1 = \mathbf{a}_1 + \mathbf{a}_2 + \mathbf{a}_3$, $\mathbf{b}_2 = \mathbf{a}_1 - \mathbf{a}_2 + \mathbf{a}_3$, $\mathbf{b}_3 = 2\mathbf{a}_1 - 2\mathbf{a}_3$

f) $\mathbf{b}_1 = \mathbf{a}_1 + \mathbf{a}_3$, $\mathbf{b}_2 = 2\mathbf{a}_2 + \mathbf{a}_3$, $\mathbf{b}_3 = -2\mathbf{a}_1 - \mathbf{a}_2 + 2\mathbf{a}_3$

**Exercise 3.32** Let ${}^a v_1$, ${}^a v_2$, and ${}^a v_3$ be the coordinates of a vector $\mathbf{v}$ relative to the basis $\{\mathbf{a}_1, \mathbf{a}_2, \mathbf{a}_3\}$. Find the coordinates ${}^b v_1$, ${}^b v_2$, and ${}^b v_3$ of the vector $\mathbf{v}$ relative to the basis $\{\mathbf{b}_1, \mathbf{b}_2, \mathbf{b}_3\}$ where

$$\begin{aligned} \mathbf{b}_1 &= 2\mathbf{a}_1 - \mathbf{a}_3, \\ \mathbf{b}_2 &= -\mathbf{a}_2 + 3\mathbf{a}_3, \\ \mathbf{b}_3 &= \mathbf{a}_1 + \mathbf{a}_3. \end{aligned}$$

**Solution.** Since ${}^a v_1, {}^a v_2$, and ${}^a v_3$ are the coordinates of $\mathbf{v}$ relative to the basis $\{\mathbf{a}_1, \mathbf{a}_2, \mathbf{a}_3\}$, we can write

$$\mathbf{v} = {}^a v_1 \mathbf{a}_1 + {}^a v_2 \mathbf{a}_2 + {}^a v_3 \mathbf{a}_3.$$

But, from the definition of the vectors $\mathbf{b}_1$, $\mathbf{b}_2$, and $\mathbf{b}_3$ we find

$$\begin{aligned} \mathbf{a}_1 &= \frac{1}{3}\mathbf{b}_1 + \frac{1}{3}\mathbf{b}_3, \\ \mathbf{a}_2 &= -\mathbf{b}_1 - \mathbf{b}_2 + 2\mathbf{b}_3, \\ \mathbf{a}_3 &= -\frac{1}{3}\mathbf{b}_1 + \frac{2}{3}\mathbf{b}_3. \end{aligned}$$

Substitution into the expression for $\mathbf{v}$ then yields

$$\begin{aligned} \mathbf{v} &= {}^a v_1 \left(\frac{1}{3}\mathbf{b}_1 + \frac{1}{3}\mathbf{b}_3\right) \\ &\quad + {}^a v_2 \left(-\mathbf{b}_1 - \mathbf{b}_2 + 2\mathbf{b}_3\right) \\ &\quad + {}^a v_3 \left(-\frac{1}{3}\mathbf{b}_1 + \frac{2}{3}\mathbf{b}_3\right) \\ &= \left(\frac{{}^a v_1}{3} - {}^a v_2 - \frac{{}^a v_3}{3}\right)\mathbf{b}_1 - {}^a v_2 \mathbf{b}_2 \\ &\quad + \left(\frac{{}^a v_1}{3} + 2\,{}^a v_2 + \frac{2\,{}^a v_3}{3}\right)\mathbf{b}_3. \end{aligned}$$

The coordinates ${}^b v_1$, ${}^b v_2$, and ${}^b v_3$ of $\mathbf{v}$ relative to the basis $\{\mathbf{b}_1, \mathbf{b}_2, \mathbf{b}_3\}$ are therefore

$$\begin{aligned} {}^b v_1 &= \frac{{}^a v_1}{3} - {}^a v_2 - \frac{{}^a v_3}{3}, \\ {}^b v_2 &= -\,{}^a v_2, \\ {}^b v_3 &= \frac{{}^a v_1}{3} + 2\,{}^a v_2 + \frac{2\,{}^a v_3}{3}. \end{aligned}$$

**Exercise 3.33** Let ${}^a v_1$, ${}^a v_2$, and ${}^a v_3$ be the coordinates of a vector $\mathbf{v}$ relative to the basis $\{\mathbf{a}_1, \mathbf{a}_2, \mathbf{a}_3\}$. Find the coordinates ${}^b v_1$, ${}^b v_2$, and ${}^b v_3$ of the vector $\mathbf{v}$ relative to each of the bases below.

a) $\mathbf{b}_1 = \mathbf{a}_1 + \mathbf{a}_3$, $\mathbf{b}_2 = 2\mathbf{a}_2 + \mathbf{a}_3$, $\mathbf{b}_3 = -2\mathbf{a}_1 - \mathbf{a}_2$

b) $\mathbf{b}_1 = -\mathbf{a}_1$, $\mathbf{b}_2 = 2\mathbf{a}_2 + \mathbf{a}_3$, $\mathbf{b}_3 = -2\mathbf{a}_1 + 2\mathbf{a}_3$

c) $\mathbf{b}_1 = \mathbf{a}_1 + \mathbf{a}_3$, $\mathbf{b}_2 = 2\mathbf{a}_2 - \mathbf{a}_3$, $\mathbf{b}_3 = -2\mathbf{a}_1 - \mathbf{a}_3$

d) $\mathbf{b}_1 = \mathbf{a}_1 + 2\mathbf{a}_3$, $\mathbf{b}_2 = 2\mathbf{a}_2 + \mathbf{a}_3$, $\mathbf{b}_3 = -2\mathbf{a}_1 + 2\mathbf{a}_3$

e) $\mathbf{b}_1 = \mathbf{a}_3$, $\mathbf{b}_2 = \mathbf{a}_2$, $\mathbf{b}_3 = -2\mathbf{a}_1$

f) $\mathbf{b}_1 = 2\mathbf{a}_1 - \mathbf{a}_3$, $\mathbf{b}_2 = 2\mathbf{a}_2 + \mathbf{a}_3$, $\mathbf{b}_3 = -2\mathbf{a}_1 - \mathbf{a}_2 + 2\mathbf{a}_3$

**Exercise 3.34** Let $\{\mathbf{a}_1, \mathbf{a}_2, \mathbf{a}_3\}$ be a basis of space. Find the coordinates relative to

$\{\mathbf{a}_1, \mathbf{a}_2, \mathbf{a}_3\}$ of the vector $5\mathbf{v}$, where

a) $\mathbf{v} = \frac{1}{3}\mathbf{a}_1 + \frac{1}{3}\mathbf{a}_2 - \mathbf{a}_3$ b) $\mathbf{v} = -3\mathbf{a}_3$
c) $\mathbf{v} = 2\mathbf{a}_1 + \mathbf{a}_2$ d) $\mathbf{v} = -\mathbf{a}_1 + 2\mathbf{a}_3$
e) $\mathbf{v} = \mathbf{a}_1 + \frac{1}{5}\mathbf{a}_2 - \mathbf{a}_3$ f) $\mathbf{v} = \mathbf{a}_1 - \mathbf{a}_3$

**Exercise 3.35** Let $\{\mathbf{a}_1, \mathbf{a}_2, \mathbf{a}_3\}$ be a basis of space. Find the coordinates relative to $\{\mathbf{a}_1, \mathbf{a}_2, \mathbf{a}_3\}$ of the vector $\mathbf{v} + \mathbf{w}$, where

a) $\begin{array}{l}\mathbf{v} = \mathbf{a}_1 - \mathbf{a}_3 \\ \mathbf{w} = \mathbf{a}_1 + \mathbf{a}_3\end{array}$ b) $\begin{array}{l}\mathbf{v} = 2\mathbf{a}_1 + \mathbf{a}_2 \\ \mathbf{w} = -\mathbf{a}_1 + \mathbf{a}_3\end{array}$

c) $\begin{array}{l}\mathbf{v} = -3\mathbf{a}_3 \\ \mathbf{w} = -2\mathbf{a}_1 + \frac{1}{3}\mathbf{a}_3\end{array}$ d) $\begin{array}{l}\mathbf{v} = -\mathbf{a}_1 + \frac{2}{3}\mathbf{a}_3 \\ \mathbf{w} = -2\mathbf{a}_1 + \frac{2}{3}\mathbf{a}_2\end{array}$

e) $\begin{array}{l}\mathbf{v} = \frac{1}{3}\mathbf{a}_1 + \frac{1}{3}\mathbf{a}_2 - \mathbf{a}_3 \\ \mathbf{w} = -2\mathbf{a}_1 + \frac{1}{3}\mathbf{a}_3\end{array}$

f) $\begin{array}{l}\mathbf{v} = \frac{4}{3}\mathbf{a}_1 + \frac{1}{3}\mathbf{a}_2 - \frac{1}{3}\mathbf{a}_3 \\ \mathbf{w} = -\frac{2}{3}\mathbf{a}_1 - \frac{4}{3}\mathbf{a}_3\end{array}$

**Exercise 3.36** Let $\{\mathbf{a}_1, \mathbf{a}_2, \mathbf{a}_3\}$ be a basis of space. Show that there exist independent angles $\theta_1$, $\theta_2$, and $\theta_3$, such that

$$\mathbf{a}_1 \bullet \mathbf{a}_2 = \|\mathbf{a}_1\| \, \|\mathbf{a}_2\| \cos\theta_1,$$
$$\mathbf{a}_1 \bullet \mathbf{a}_3 = \|\mathbf{a}_1\| \, \|\mathbf{a}_3\| \cos\theta_2,$$

and

$$\mathbf{a}_2 \bullet \mathbf{a}_3 = \|\mathbf{a}_2\| \, \|\mathbf{a}_3\| \left( \begin{array}{c} \sin\theta_1 \sin\theta_2 \cos\theta_3 \\ + \cos\theta_1 \cos\theta_2 \end{array} \right),$$

where

$$\sin\theta_1, \sin\theta_2, \sin\theta_3 \neq 0.$$

[Hint: Let $\theta_3$ be the angle between the plane spanned by the vectors $\mathbf{a}_1$ and $\mathbf{a}_2$ and the plane spanned by the vectors $\mathbf{a}_1$ and $\mathbf{a}_3$.]

**Exercise 3.37** Let $\{\mathbf{a}_1, \mathbf{a}_2, \mathbf{a}_3\}$ be a basis of space. Use the result of the previous exercise to show that

$$\left[ (\mathbf{a}_1 \bullet \mathbf{a}_1)(\mathbf{a}_2 \bullet \mathbf{a}_2) - (\mathbf{a}_1 \bullet \mathbf{a}_2)^2 \right] \cdot \left[ (\mathbf{a}_1 \bullet \mathbf{a}_1)(\mathbf{a}_3 \bullet \mathbf{a}_3) - (\mathbf{a}_1 \bullet \mathbf{a}_3)^2 \right] - \left[ (\mathbf{a}_1 \bullet \mathbf{a}_1)(\mathbf{a}_2 \bullet \mathbf{a}_3) - (\mathbf{a}_1 \bullet \mathbf{a}_2)(\mathbf{a}_1 \bullet \mathbf{a}_3) \right]^2 > 0.$$

**Exercise 3.38** Let $\{\mathbf{a}_1, \mathbf{a}_2, \mathbf{a}_3\}$ be a basis of space, such that

$$\begin{pmatrix} \mathbf{a}_1 \bullet \mathbf{a}_1 & \mathbf{a}_1 \bullet \mathbf{a}_2 & \mathbf{a}_1 \bullet \mathbf{a}_3 \\ \mathbf{a}_2 \bullet \mathbf{a}_1 & \mathbf{a}_2 \bullet \mathbf{a}_2 & \mathbf{a}_2 \bullet \mathbf{a}_3 \\ \mathbf{a}_3 \bullet \mathbf{a}_1 & \mathbf{a}_3 \bullet \mathbf{a}_2 & \mathbf{a}_3 \bullet \mathbf{a}_3 \end{pmatrix} = \begin{pmatrix} 1 & 0 & \frac{1}{4} \\ 0 & \frac{1}{2} & \frac{1}{3} \\ \frac{1}{4} & \frac{1}{3} & 1 \end{pmatrix}.$$

Evaluate the dot product between the two vectors

$$\mathbf{v} = 3\mathbf{a}_1 - \mathbf{a}_3 \text{ and } \mathbf{w} = -\mathbf{a}_2 + 2\mathbf{a}_3.$$

**Solution.** From the linearity of the dot product, we have

$$\begin{aligned} \mathbf{v} \bullet \mathbf{w} &= (3\mathbf{a}_1 - \mathbf{a}_3) \bullet (-\mathbf{a}_2 + 2\mathbf{a}_3) \\ &= -3\mathbf{a}_1 \bullet \mathbf{a}_2 + 6\mathbf{a}_1 \bullet \mathbf{a}_3 \\ &\quad + \mathbf{a}_3 \bullet \mathbf{a}_2 - 2\mathbf{a}_3 \bullet \mathbf{a}_3 \\ &= -3 * 0 + 6 * \frac{1}{4} + \frac{1}{3} - 2 * 1 \\ &= -\frac{1}{6}. \end{aligned}$$

**Exercise 3.39** Let $\{\mathbf{a}_1, \mathbf{a}_2, \mathbf{a}_3\}$ be a basis of space. Evaluate the dot product between the vectors

$$\mathbf{v} = 3\mathbf{a}_1 - \mathbf{a}_3 \text{ and } \mathbf{w} = -\mathbf{a}_2 + 2\mathbf{a}_3,$$

when

$$\begin{pmatrix} \mathbf{a}_1 \bullet \mathbf{a}_1 & \mathbf{a}_1 \bullet \mathbf{a}_2 & \mathbf{a}_1 \bullet \mathbf{a}_3 \\ \mathbf{a}_2 \bullet \mathbf{a}_1 & \mathbf{a}_2 \bullet \mathbf{a}_2 & \mathbf{a}_2 \bullet \mathbf{a}_3 \\ \mathbf{a}_3 \bullet \mathbf{a}_1 & \mathbf{a}_3 \bullet \mathbf{a}_2 & \mathbf{a}_3 \bullet \mathbf{a}_3 \end{pmatrix} =$$

$$\text{a)} \begin{pmatrix} 1 & 0 & \frac{1}{4} \\ 0 & 2 & \frac{\sqrt{3}}{2} \\ \frac{1}{4} & \frac{\sqrt{3}}{2} & 1 \end{pmatrix}$$

$$\text{b)} \begin{pmatrix} 5 & \frac{1}{\sqrt{2}} & \frac{\sqrt{5}}{\sqrt{2}} \\ \frac{1}{\sqrt{2}} & \frac{1}{5} & \frac{1}{2\sqrt{5}} \\ \frac{\sqrt{5}}{\sqrt{2}} & \frac{1}{2\sqrt{5}} & 1 \end{pmatrix}$$

$$\text{c)} \begin{pmatrix} \frac{1}{2} & 0 & 0 \\ 0 & \frac{1}{2} & \frac{1}{2\sqrt{2}} \\ 0 & \frac{1}{2\sqrt{2}} & \frac{1}{2} \end{pmatrix}$$

$$\text{d)} \begin{pmatrix} 2 & \frac{1}{\sqrt{2}} & 0 \\ \frac{1}{\sqrt{2}} & 1 & 0 \\ 0 & 0 & \frac{1}{2} \end{pmatrix}$$

$$\text{e)} \begin{pmatrix} 1 & -\sqrt{2} & \frac{1}{\sqrt{2}} \\ -\sqrt{2} & 4 & -1 \\ \frac{1}{\sqrt{2}} & -1 & 1 \end{pmatrix}$$

$$\text{f)} \begin{pmatrix} 1 & -\frac{1}{2} & -\frac{1}{\sqrt{2}} \\ -\frac{1}{2} & \frac{1}{2} & \frac{1}{2\sqrt{2}} \\ -\frac{1}{\sqrt{2}} & \frac{1}{2\sqrt{2}} & 1 \end{pmatrix}$$

**Exercise 3.40** Let $\{\mathbf{a}_1, \mathbf{a}_2, \mathbf{a}_3\}$ be a basis of space. Show that the columns of the matrix

$$\begin{pmatrix} \mathbf{a}_1 \bullet \mathbf{a}_1 & \mathbf{a}_1 \bullet \mathbf{a}_2 & \mathbf{a}_1 \bullet \mathbf{a}_3 \\ \mathbf{a}_2 \bullet \mathbf{a}_1 & \mathbf{a}_2 \bullet \mathbf{a}_2 & \mathbf{a}_2 \bullet \mathbf{a}_3 \\ \mathbf{a}_3 \bullet \mathbf{a}_1 & \mathbf{a}_3 \bullet \mathbf{a}_2 & \mathbf{a}_3 \bullet \mathbf{a}_3 \end{pmatrix}$$

are linearly independent.

[Hint: Use the result from Exercise 3.36 and show that the determinant of the matrix must be non-zero.]

**Exercise 3.41** Let $\{\mathbf{a}_1, \mathbf{a}_2, \mathbf{a}_3\}$ be a basis of space, such that

$$\begin{pmatrix} \mathbf{a}_1 \bullet \mathbf{a}_1 & \mathbf{a}_1 \bullet \mathbf{a}_2 & \mathbf{a}_1 \bullet \mathbf{a}_3 \\ \mathbf{a}_2 \bullet \mathbf{a}_1 & \mathbf{a}_2 \bullet \mathbf{a}_2 & \mathbf{a}_2 \bullet \mathbf{a}_3 \\ \mathbf{a}_3 \bullet \mathbf{a}_1 & \mathbf{a}_3 \bullet \mathbf{a}_2 & \mathbf{a}_3 \bullet \mathbf{a}_3 \end{pmatrix} = \begin{pmatrix} 1 & 0 & \frac{1}{4} \\ 0 & \frac{1}{2} & \frac{1}{3} \\ \frac{1}{4} & \frac{1}{3} & 1 \end{pmatrix}.$$

Show that this guarantees that $\mathbf{v} \bullet \mathbf{v} > 0$ for any non-zero vector $\mathbf{v}$.

**Solution.** Let $v_1$, $v_2$, and $v_3$ be the coordinates of the vector $\mathbf{v}$ relative to the basis $\{\mathbf{a}_1, \mathbf{a}_2, \mathbf{a}_3\}$. Since $\mathbf{v} \neq \mathbf{0}$, the result of a previous exercise shows that not all the coordinates can equal zero.

From

$$\mathbf{v} = v_1 \mathbf{a}_1 + v_2 \mathbf{a}_2 + v_3 \mathbf{a}_3$$

it follows that

$$\begin{aligned} \mathbf{v} \bullet \mathbf{v} &= v_1^2 \mathbf{a}_1 \bullet \mathbf{a}_1 + v_2^2 \mathbf{a}_2 \bullet \mathbf{a}_2 + v_3^2 \mathbf{a}_3 \bullet \mathbf{a}_3 \\ &\quad +2v_1 v_2 \mathbf{a}_1 \bullet \mathbf{a}_2 + 2v_1 v_3 \mathbf{a}_1 \bullet \mathbf{a}_3 \\ &\quad +2v_2 v_3 \mathbf{a}_2 \bullet \mathbf{a}_3 \\ &= v_1^2 + \frac{1}{2}v_2^2 + v_3^2 + \frac{1}{2}v_1 v_3 + \frac{2}{3}v_2 v_3 \\ &= \left(v_1 + \frac{1}{4}v_3\right)^2 + \frac{1}{2}\left(v_2 + \frac{2}{3}v_3\right)^2 \\ &\quad + \frac{103}{144}v_3^2, \end{aligned}$$

where the last equality follows from completing the squares. Since not all the coordinates can equal zero, the last expression must be greater than zero.

**Exercise 3.42** Let $\{\mathbf{a}_1, \mathbf{a}_2, \mathbf{a}_3\}$ be a basis of space. Each of the matrices below is a possible candidate for

$$\begin{pmatrix} \mathbf{a}_1 \bullet \mathbf{a}_1 & \mathbf{a}_1 \bullet \mathbf{a}_2 & \mathbf{a}_1 \bullet \mathbf{a}_3 \\ \mathbf{a}_2 \bullet \mathbf{a}_1 & \mathbf{a}_2 \bullet \mathbf{a}_2 & \mathbf{a}_2 \bullet \mathbf{a}_3 \\ \mathbf{a}_3 \bullet \mathbf{a}_1 & \mathbf{a}_3 \bullet \mathbf{a}_2 & \mathbf{a}_3 \bullet \mathbf{a}_3 \end{pmatrix}.$$

In each case, show that this guarantees that $\mathbf{v} \bullet \mathbf{v} > 0$ for any arbitrary vector $\mathbf{v}$.

$$\text{a)} \begin{pmatrix} 1 & 0 & \frac{1}{4} \\ 0 & 2 & \frac{\sqrt{3}}{2} \\ \frac{1}{4} & \frac{\sqrt{3}}{2} & 1 \end{pmatrix}$$

$$\text{b)} \begin{pmatrix} 5 & \frac{1}{\sqrt{2}} & \frac{\sqrt{5}}{\sqrt{2}} \\ \frac{1}{\sqrt{2}} & \frac{1}{5} & \frac{1}{2\sqrt{5}} \\ \frac{\sqrt{5}}{\sqrt{2}} & \frac{1}{2\sqrt{5}} & 1 \end{pmatrix}$$

c) $\begin{pmatrix} \frac{1}{2} & 0 & 0 \\ 0 & \frac{1}{2} & \frac{1}{2\sqrt{2}} \\ 0 & \frac{1}{2\sqrt{2}} & \frac{1}{2} \end{pmatrix}$

d) $\begin{pmatrix} 2 & \frac{1}{\sqrt{2}} & 0 \\ \frac{1}{\sqrt{2}} & 1 & 0 \\ 0 & 0 & \frac{1}{2} \end{pmatrix}$

e) $\begin{pmatrix} 1 & -\sqrt{2} & \frac{1}{\sqrt{2}} \\ -\sqrt{2} & 4 & -1 \\ \frac{1}{\sqrt{2}} & -1 & 1 \end{pmatrix}$

f) $\begin{pmatrix} 1 & -\frac{1}{2} & -\frac{1}{\sqrt{2}} \\ -\frac{1}{2} & \frac{1}{2} & \frac{1}{2\sqrt{2}} \\ -\frac{1}{\sqrt{2}} & \frac{1}{2\sqrt{2}} & 1 \end{pmatrix}$

**Exercise 3.43** Let $\{\mathbf{a}_1, \mathbf{a}_2, \mathbf{a}_3\}$ be an orthonormal basis of space. For each pair of vectors $\mathbf{v}$ and $\mathbf{w}$ below, evaluate the dot product $\mathbf{v} \bullet \mathbf{w}$.

a) $\mathbf{v} = \mathbf{a}_1 - \mathbf{a}_3$, $\mathbf{w} = \mathbf{a}_1 + \mathbf{a}_3$

b) $\mathbf{v} = -3\mathbf{a}_3$, $\mathbf{w} = -2\mathbf{a}_1 + \frac{1}{3}\mathbf{a}_3$

c) $\mathbf{v} = -\mathbf{a}_1 + \frac{2}{3}\mathbf{a}_3$, $\mathbf{w} = -2\mathbf{a}_1 + \frac{2}{3}\mathbf{a}_2$

d) $\mathbf{v} = 2\mathbf{a}_1 + \mathbf{a}_2$, $\mathbf{w} = -\mathbf{a}_1 + \mathbf{a}_3$

e) $\mathbf{v} = \frac{1}{3}\mathbf{a}_1 + \frac{1}{3}\mathbf{a}_2 - \mathbf{a}_3$, $\mathbf{w} = -2\mathbf{a}_1 + \frac{1}{3}\mathbf{a}_3$

f) $\mathbf{v} = \frac{4}{3}\mathbf{a}_1 + \frac{1}{3}\mathbf{a}_2 - \frac{1}{3}\mathbf{a}_3$, $\mathbf{w} = -\frac{2}{3}\mathbf{a}_1 - \frac{4}{3}\mathbf{a}_3$

**Exercise 3.44** Let $\{\mathbf{a}_1, \mathbf{a}_2, \mathbf{a}_3\}$ be an orthonormal basis of space. Find the angle $\theta(\mathbf{v}, \mathbf{w})$ between the vectors $\mathbf{v} = \mathbf{a}_1 - \mathbf{a}_3$ and $\mathbf{w} = 2\mathbf{a}_1 - \mathbf{a}_2 + 3\mathbf{a}_3$.

**Solution.** From the definition of the dot product, it follows that

$$\cos\theta(\mathbf{v}, \mathbf{w}) = \frac{\mathbf{v} \bullet \mathbf{w}}{\|\mathbf{v}\| \|\mathbf{w}\|}$$

$$= \frac{1 * 2 + 0 * (-1) + (-1) * 3}{\sqrt{1^2 + 0^2 + (-1)^2}\sqrt{2^2 + (-1)^2 + 3^2}}$$

$$= -\frac{1}{\sqrt{28}}$$

and thus

$$\theta(\mathbf{v}, \mathbf{w}) = \arccos\left(-\frac{1}{\sqrt{28}}\right) \approx 100.9°.$$

**Exercise 3.45** Let $\{\mathbf{a}_1, \mathbf{a}_2, \mathbf{a}_3\}$ be an orthonormal basis of space. For each pair of vectors $\mathbf{v}$ and $\mathbf{w}$ below, find the angle $\theta(\mathbf{v}, \mathbf{w})$.

a) $\mathbf{v} = \mathbf{a}_1 - \mathbf{a}_3$, $\mathbf{w} = \mathbf{a}_1 + \mathbf{a}_3$

b) $\mathbf{v} = -3\mathbf{a}_3$, $\mathbf{w} = -2\mathbf{a}_1 + \frac{1}{3}\mathbf{a}_3$

c) $\mathbf{v} = -\mathbf{a}_1 + \frac{2}{3}\mathbf{a}_3$, $\mathbf{w} = -2\mathbf{a}_1 + \frac{2}{3}\mathbf{a}_2$

d) $\mathbf{v} = 2\mathbf{a}_1 + \mathbf{a}_2$, $\mathbf{w} = -\mathbf{a}_1 + \mathbf{a}_3$

e) $\mathbf{v} = \frac{1}{3}\mathbf{a}_1 + \frac{1}{3}\mathbf{a}_2 - \mathbf{a}_3$, $\mathbf{w} = -2\mathbf{a}_1 + \frac{1}{3}\mathbf{a}_3$

f) $\mathbf{v} = \frac{4}{3}\mathbf{a}_1 + \frac{1}{3}\mathbf{a}_2 - \frac{1}{3}\mathbf{a}_3$, $\mathbf{w} = -\frac{2}{3}\mathbf{a}_1 - \frac{4}{3}\mathbf{a}_3$

**Exercise 3.46** Let $\{\mathbf{a}_1, \mathbf{a}_2, \mathbf{a}_3\}$ be an orthonormal basis of space. For each of the vectors $\mathbf{v}$ and $\mathbf{w}$ below, determine whether they are perpendicular.

a) $\mathbf{v} = \mathbf{a}_1 - \mathbf{a}_3$, $\mathbf{w} = \mathbf{a}_1 + \mathbf{a}_3$

b) $\mathbf{v} = -3\mathbf{a}_3$, $\mathbf{w} = -2\mathbf{a}_1 + \frac{1}{3}\mathbf{a}_3$

c) $\mathbf{v} = -\mathbf{a}_1 + \frac{2}{3}\mathbf{a}_3$, $\mathbf{w} = -2\mathbf{a}_1 + \frac{2}{3}\mathbf{a}_2$

d) $\mathbf{v} = 2\mathbf{a}_1 + \mathbf{a}_2$, $\mathbf{w} = -\mathbf{a}_1 + \mathbf{a}_3$

e) $\mathbf{v} = \frac{1}{3}\mathbf{a}_1 + \frac{1}{3}\mathbf{a}_2 - \mathbf{a}_3$, $\mathbf{w} = -2\mathbf{a}_1 + \frac{1}{3}\mathbf{a}_3$

f) $\mathbf{v} = \frac{4}{3}\mathbf{a}_1 + \frac{1}{3}\mathbf{a}_2 - \frac{1}{3}\mathbf{a}_3$, $\mathbf{w} = -\frac{2}{3}\mathbf{a}_1 - \frac{4}{3}\mathbf{a}_3$

**Exercise 3.47** Let $\{\mathbf{a}_1, \mathbf{a}_2, \mathbf{a}_3\}$ be an orthonormal basis of space. Show that

$$\mathbf{a}_i \bullet \mathbf{v} = v_i, \; i = 1, 2, 3,$$

and thus that

$$\mathbf{v} = \mathbf{a}_1 (\mathbf{a}_1 \bullet \mathbf{v}) + \mathbf{a}_2 (\mathbf{a}_2 \bullet \mathbf{v}) + \mathbf{a}_3 (\mathbf{a}_3 \bullet \mathbf{v}),$$

where $v_i$ is the $i$-th coordinate of $\mathbf{v}$ relative to the given basis.

**Solution.** By definition,

$$\mathbf{v} = v_1\mathbf{a}_1 + v_2\mathbf{a}_2 + v_3\mathbf{a}_3,$$

where $v_1$, $v_2$, and $v_3$ are the coordinates of the vector $\mathbf{v}$ relative to the basis $\{\mathbf{a}_1, \mathbf{a}_2, \mathbf{a}_3\}$. Then,

$$\begin{aligned} \mathbf{a}_1 \bullet \mathbf{v} &= \mathbf{a}_1 \bullet (v_1\mathbf{a}_1 + v_2\mathbf{a}_2 + v_3\mathbf{a}_3) \\ &= v_1(\mathbf{a}_1 \bullet \mathbf{a}_1) + v_2(\mathbf{a}_1 \bullet \mathbf{a}_2) \\ &\quad + v_3(\mathbf{a}_1 \bullet \mathbf{a}_3) \\ &= v_1, \end{aligned}$$

$$\begin{aligned} \mathbf{a}_2 \bullet \mathbf{v} &= \mathbf{a}_2 \bullet (v_1\mathbf{a}_1 + v_2\mathbf{a}_2 + v_3\mathbf{a}_3) \\ &= v_1(\mathbf{a}_2 \bullet \mathbf{a}_1) + v_2(\mathbf{a}_2 \bullet \mathbf{a}_2) \\ &\quad + v_3(\mathbf{a}_2 \bullet \mathbf{a}_3) \\ &= v_2, \end{aligned}$$

and

$$\begin{aligned} \mathbf{a}_3 \bullet \mathbf{v} &= \mathbf{v}_3 \bullet (v_1\mathbf{a}_1 + v_2\mathbf{a}_2 + v_3\mathbf{a}_3) \\ &= v_1(\mathbf{a}_3 \bullet \mathbf{a}_1) + v_2(\mathbf{a}_3 \bullet \mathbf{a}_2) \\ &\quad + v_3(\mathbf{a}_3 \bullet \mathbf{a}_3) \\ &= v_3 \end{aligned}$$

as claimed.

**Exercise 3.48** Show that

$$\mathbf{v} = a\ {}^a v = \left({}^a v\right)^T a^T,$$

where $^T$ denotes the matrix transpose.

**Exercise 3.49** Let $a$ be an orthonormal basis. Compute $a^T \bullet a$.

**Solution.** Since $a$ is orthonormal, we find

$$\begin{aligned} a^T \bullet a &= \begin{pmatrix} \mathbf{a}_1 \\ \mathbf{a}_2 \\ \mathbf{a}_3 \end{pmatrix} \bullet \begin{pmatrix} \mathbf{a}_1 & \mathbf{a}_2 & \mathbf{a}_3 \end{pmatrix} \\ &= \begin{pmatrix} \mathbf{a}_1 \bullet \mathbf{a}_1 & \mathbf{a}_1 \bullet \mathbf{a}_2 & \mathbf{a}_1 \bullet \mathbf{a}_3 \\ \mathbf{a}_2 \bullet \mathbf{a}_1 & \mathbf{a}_2 \bullet \mathbf{a}_2 & \mathbf{a}_2 \bullet \mathbf{a}_3 \\ \mathbf{a}_3 \bullet \mathbf{a}_1 & \mathbf{a}_3 \bullet \mathbf{a}_2 & \mathbf{a}_3 \bullet \mathbf{a}_3 \end{pmatrix} \\ &= \begin{pmatrix} 1 & 0 & 0 \\ 0 & 1 & 0 \\ 0 & 0 & 1 \end{pmatrix}. \end{aligned}$$

**Exercise 3.50** Let $a$ be an orthonormal basis. Compute $a \bullet a^T$.

**Exercise 3.51** Let $a$ be an orthonormal basis. Find an expression for the dot product between two vectors $\mathbf{v}$ and $\mathbf{w}$ in terms of their matrix representations, ${}^a v$ and ${}^a w$.

**Solution.** Using the result of the previous exercises, we find

$$\begin{aligned} \mathbf{v} \bullet \mathbf{w} &= \left({}^a v^T a^T\right) \bullet \left(a\ {}^a w\right) \\ &= {}^a v^T \left(a^T \bullet a\right)\ {}^a w \\ &= {}^a v^T \begin{pmatrix} 1 & 0 & 0 \\ 0 & 1 & 0 \\ 0 & 0 & 1 \end{pmatrix} {}^a w \\ &= {}^a v^T\ {}^a w, \end{aligned}$$

where the second equality follows from the linearity of the dot product. Thus, the dot product between two vectors reduces to a matrix multiplication of the matrix representations of the vectors (with a transpose suitably inserted).

**Exercise 3.52** Let $\{\mathbf{a}_1, \mathbf{a}_2, \mathbf{a}_3\}$ be an orthonormal basis. Show that four of the eight orthonormal bases $\{\pm\mathbf{a}_1, \pm\mathbf{a}_2, \pm\mathbf{a}_3\}$ are right-handed and four are left-handed.

**Exercise 3.53** Let $a$ be a right-handed, orthonormal basis. Show that

$$a^T \times a = \begin{pmatrix} \mathbf{0} & \mathbf{a}_3 & -\mathbf{a}_2 \\ -\mathbf{a}_3 & \mathbf{0} & \mathbf{a}_1 \\ \mathbf{a}_2 & -\mathbf{a}_1 & \mathbf{0} \end{pmatrix}.$$

**Exercise 3.54** Let $a$ be a right-handed, orthonormal basis. Show that $a \bullet \left(a^T \times a\right) = \left(\begin{array}{ccc} 0 & 0 & 0 \end{array}\right)$.

**Solution.** We have

$$a \bullet \left(a^T \times a\right) = a \bullet \left(\begin{array}{ccc} \mathbf{0} & \mathbf{a}_3 & -\mathbf{a}_2 \\ -\mathbf{a}_3 & \mathbf{0} & \mathbf{a}_1 \\ \mathbf{a}_2 & -\mathbf{a}_1 & \mathbf{0} \end{array}\right)$$

$$\begin{aligned} &= \left(\begin{array}{c} \mathbf{a}_1 \bullet \mathbf{0} - \mathbf{a}_2 \bullet \mathbf{a}_3 + \mathbf{a}_3 \bullet \mathbf{a}_2 \\ \mathbf{a}_1 \bullet \mathbf{a}_3 + \mathbf{a}_2 \bullet \mathbf{0} - \mathbf{a}_3 \bullet \mathbf{a}_1 \\ -\mathbf{a}_1 \bullet \mathbf{a}_2 + \mathbf{a}_2 \bullet \mathbf{a}_1 + \mathbf{a}_3 \bullet \mathbf{0} \end{array}\right)^T \\ &= \left(\begin{array}{ccc} 0 & 0 & 0 \end{array}\right), \end{aligned}$$

thus proving the claim.

**Exercise 3.55** Let $a$ be a right-handed, orthonormal basis. Compute the product $-\frac{1}{2} a \times \left(a^T \times a\right)$.

**Exercise 3.56** Let $a$ be a left-handed orthonormal basis. Compute the product $-\frac{1}{2} a \times \left(a^T \times a\right)$.

**Exercise 3.57** Show that a triad $a$ is right-handed if and only if

$$\mathbf{a}_i \bullet \left(\mathbf{a}_j \times \mathbf{a}_k\right) = 1,$$

where $i, j, k$ is any subsequence of three consecutive numbers from the sequence $1, 2, 3, 1, 2$;

$$\mathbf{a}_i \bullet \left(\mathbf{a}_j \times \mathbf{a}_k\right) = -1,$$

where $i, j, k$ is any subsequence of three consecutive numbers from the sequence $3, 2, 1, 3, 2$; and

$$\mathbf{a}_i \bullet \left(\mathbf{a}_j \times \mathbf{a}_k\right) = 0$$

for all other choices of $i$, $j$, and $k$.

**Exercise 3.58** Let $a$ be a right-handed, orthonormal basis. Show that the cross product between two arbitrary vectors $\mathbf{v}$ and $\mathbf{w}$ can be computed from the formula

$$\begin{aligned} \mathbf{v} \times \mathbf{w} &= \mathbf{a}_1 \left|\begin{array}{cc} v_2 & v_3 \\ w_2 & w_3 \end{array}\right| - \mathbf{a}_2 \left|\begin{array}{cc} v_1 & v_3 \\ w_1 & w_3 \end{array}\right| \\ &\quad + \mathbf{a}_3 \left|\begin{array}{cc} v_1 & v_2 \\ w_1 & w_2 \end{array}\right| \\ &\overset{def}{=} \left|\begin{array}{ccc} \mathbf{a}_1 & \mathbf{a}_2 & \mathbf{a}_3 \\ v_1 & v_2 & v_3 \\ w_1 & w_2 & w_3 \end{array}\right|, \end{aligned}$$

where the $v_i$'s and $w_i$'s are the coordinates of the vectors $\mathbf{v}$ and $\mathbf{w}$, respectively, relative to the basis $a$.

[Hint: Let $r$ denote the cross product of the vectors $v$ and $w$, i.e.,

$$\mathbf{r} = \mathbf{v} \times \mathbf{w}.$$

Use the fact that $\mathbf{r}$ is perpendicular to both $\mathbf{v}$ and $\mathbf{w}$ to conclude that

$$\begin{aligned} r_1 v_1 + r_2 v_2 + r_3 v_3 &= 0, \\ r_1 w_1 + r_2 w_2 + r_3 w_3 &= 0. \end{aligned}$$

Use the fact that

$$\begin{aligned} \|\mathbf{r}\|^2 &= \|\mathbf{v}\|^2 \|\mathbf{w}\|^2 \sin^2 \theta(\mathbf{v}, \mathbf{w}) \\ &= \|\mathbf{v}\|^2 \|\mathbf{w}\|^2 \left(1 - \cos^2 \theta(\mathbf{v}, \mathbf{w})\right) \\ &= \|\mathbf{v}\|^2 \|\mathbf{w}\|^2 - (\mathbf{v} \bullet \mathbf{w})^2 \end{aligned}$$

to show that

$$\begin{aligned} r_1^2 + r_2^2 + r_3^2 = &\left(v_1^2 + v_2^2 + v_3^2\right) \\ &\cdot \left(w_1^2 + w_2^2 + w_3^2\right) \\ &- \left(v_1 w_1 + v_2 w_2 + v_3 w_3\right)^2. \end{aligned}$$

Now solve the first two equations for $r_1$ and $r_2$ in terms of $r_3$ and substitute the result into the last equation. Solve the resulting equation for $r_3$ in terms of the $v_i$'s and $w_i$'s and substitute the result into the expressions for $r_1$ and $r_2$. You should get two possible solutions. Select the one that agrees with the assumption that $a$ is right-handed.]

**Exercise 3.59** Let $a$ be a right-handed, orthonormal basis. Show that

$$\mathbf{v} \times \mathbf{w} = {}_a\begin{pmatrix} 0 & -{}^a v_3 & {}^a v_2 \\ {}^a v_3 & 0 & -{}^a v_1 \\ -{}^a v_2 & {}^a v_1 & 0 \end{pmatrix} {}^a w.$$

**Exercise 3.60** Let $a$ be a right-handed, orthonormal basis. For each pair of vectors $\mathbf{v}$ and $\mathbf{w}$ below, compute the cross product $\mathbf{v} \times \mathbf{w}$.

a) $\mathbf{v} = \mathbf{a}_1 - \mathbf{a}_3$, $\mathbf{w} = \mathbf{a}_1 + \mathbf{a}_3$

b) $\mathbf{v} = -3\mathbf{a}_3$, $\mathbf{w} = -2\mathbf{a}_1 + \frac{1}{3}\mathbf{a}_3$

c) $\mathbf{v} = -\mathbf{a}_1 + \frac{2}{3}\mathbf{a}_3$, $\mathbf{w} = -2\mathbf{a}_1 + \frac{2}{3}\mathbf{a}_2$

d) $\mathbf{v} = 2\mathbf{a}_1 + \mathbf{a}_2$, $\mathbf{w} = -\mathbf{a}_1 + \mathbf{a}_3$

e) $\mathbf{v} = \frac{1}{3}\mathbf{a}_1 + \frac{1}{3}\mathbf{a}_2 - \mathbf{a}_3$, $\mathbf{w} = -2\mathbf{a}_1 + \frac{1}{3}\mathbf{a}_3$

f) $\mathbf{v} = \frac{4}{3}\mathbf{a}_1 + \frac{1}{3}\mathbf{a}_2 - \frac{1}{3}\mathbf{a}_3$, $\mathbf{w} = -\frac{2}{3}\mathbf{a}_1 - \frac{4}{3}\mathbf{a}_3$

**Exercise 3.61** Consider applying a pure rotation to a block in its reference configuration corresponding to a half turn about an edge through a given corner on the block, followed by a pure rotation corresponding to a quarter turn about a different edge through the same corner as shown in the figure below.

Show that the final configuration is related to the reference configuration by a single pure rotation about an axis through the corner making an angle of $\theta_1 = 45°$ with the first edge and $\theta_2 = 90°$ with the second edge.

**Solution.** Denote the corner kept fixed by the pure rotations $A$ and introduce a right-handed, orthonormal basis $a$, such that the first edge is parallel to $\mathbf{a}_3$ and the second edge is parallel to $\mathbf{a}_1$ as shown in the figure. Let $B$ and $C$ correspond to two other points in the block, such that

$$\mathbf{r}^{AB_{\text{reference}}} = \mathbf{a}_3 \text{ and } \mathbf{r}^{AC_{\text{reference}}} = -\mathbf{a}_1,$$

where the $_{\text{reference}}$ subscript refers to points in the reference configuration.

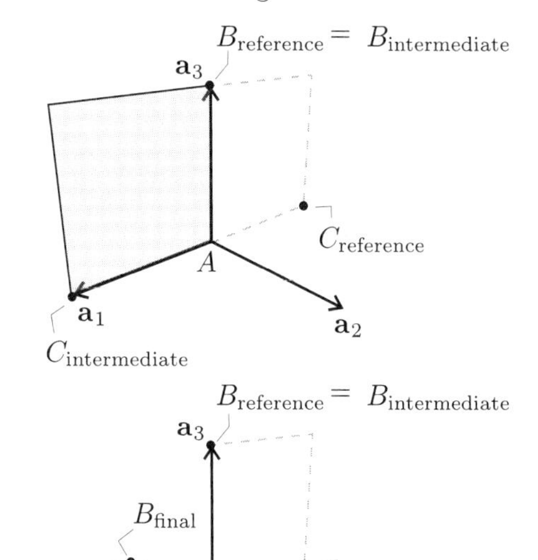




From the figure it follows that

$$\mathbf{r}^{AB_{\text{intermediate}}} = \mathbf{a}_3 \text{ and } \mathbf{r}^{AC_{\text{intermediate}}} = \mathbf{a}_1,$$

where the $_{\text{intermediate}}$ subscript refers to points in the intermediate configuration. Finally,

$$\mathbf{r}^{AB_{\text{final}}} = -\mathbf{a}_2 \text{ and } \mathbf{r}^{AC_{\text{final}}} = \mathbf{a}_1,$$

where the $_{\text{final}}$ subscript refers to points in the final configuration. Since the point $A$ is kept fixed by the pure rotations, it follows that the final configuration is related to the reference configuration by a single pure rotation keeping $A$ fixed. From Exercise 1.8, we recall that every pure rotation is equivalent to a rotation about a unique axis through the point kept fixed. It follows that the vectors $\mathbf{r}^{B_{\text{reference}}B_{\text{final}}}$ and $\mathbf{r}^{C_{\text{reference}}C_{\text{final}}}$ must be perpendicular to the axis of rotation, i.e., that the axis of rotation is parallel to the vector

$$\mathbf{r}^{B_{\text{reference}}B_{\text{final}}} \times \mathbf{r}^{C_{\text{reference}}C_{\text{final}}},$$

provided that this vector is non-zero. Here,

$$\begin{aligned}
\mathbf{r}^{B_{\text{reference}}B_{\text{final}}} &\times \mathbf{r}^{C_{\text{reference}}C_{\text{final}}} \\
&= \left(\mathbf{r}^{AB_{\text{final}}} - \mathbf{r}^{AB_{\text{reference}}}\right) \\
&\quad \times \left(\mathbf{r}^{AC_{\text{final}}} - \mathbf{r}^{AC_{\text{reference}}}\right) \\
&= (-\mathbf{a}_2 - \mathbf{a}_3) \times 2\mathbf{a}_1 = 2\mathbf{a}_3 - 2\mathbf{a}_2,
\end{aligned}$$

i.e., the axis of rotation is parallel to the vector $2\mathbf{a}_3 - 2\mathbf{a}_2$. Indeed,

$$\cos\theta_1 = \frac{(2\mathbf{a}_3 - 2\mathbf{a}_2) \bullet \mathbf{a}_1}{\|2\mathbf{a}_3 - 2\mathbf{a}_2\| \; \|\mathbf{a}_1\|} = 0$$

and

$$\cos\theta_1 = \frac{(2\mathbf{a}_3 - 2\mathbf{a}_2) \bullet \mathbf{a}_3}{\|2\mathbf{a}_3 - 2\mathbf{a}_2\| \; \|\mathbf{a}_3\|} = \frac{1}{\sqrt{2}},$$

from which the claim follows.

**Exercise 3.62** Consider applying a pure rotation to a block in its reference configuration corresponding to a half turn about an edge through a given corner on a block followed by a pure rotation by an angle $\theta$ about a different edge through the same corner. The final configuration is related to the reference configuration by a single pure rotation about an axis through the corner making an angle $\phi$ with the first edge and perpendicular to the second edge. Show that

$$\phi = \frac{|\theta|}{2}.$$

**Exercise 3.63** Consider applying a pure rotation to a block in its reference configuration corresponding to a half turn about some axis through a given corner on a block followed by a pure rotation corresponding to a quarter turn about a different axis through the same corner making an angle $\theta$ with the first axis. The final configuration is related to the reference configuration by a single pure rotation about an axis through the corner making an angle $\phi_1$ with the first axis and $\phi_2$ with the second axis. Show that

$$\cos\phi_1 = \frac{1}{\sqrt{1+\sin^2\theta}}$$

and

$$\cos\phi_2 = \frac{\cos\theta}{\sqrt{1+\sin^2\theta}}.$$

[Hint: Let the first axis be parallel to the basis vector $\mathbf{a}_3$ of a right-handed, orthonormal basis and let the second axis be parallel to the vector $\sin\theta\mathbf{a}_1 + \cos\theta\mathbf{a}_3$.]

**Exercise 3.64** Let $a$ be a right-handed, orthonormal basis. For each of the vectors $\mathbf{v}$ below, find an alternative right-handed, orthonormal basis $b$, such that $\mathbf{b}_1$ is parallel to $\mathbf{v}$.

a) $\mathbf{v} = \mathbf{a}_1 - \mathbf{a}_3$
b) $\mathbf{v} = -3\mathbf{a}_3$
c) $\mathbf{v} = -\mathbf{a}_1 + \frac{2}{3}\mathbf{a}_2 - \mathbf{a}_3$
d) $\mathbf{v} = 2\mathbf{a}_1 + \mathbf{a}_2$
e) $\mathbf{v} = \frac{1}{3}\mathbf{a}_1 + \frac{1}{3}\mathbf{a}_2 - \mathbf{a}_3$
f) $\mathbf{v} = \frac{4}{3}\mathbf{a}_1 + \frac{1}{3}\mathbf{a}_2 - \frac{1}{3}\mathbf{a}_3$

**Exercise 3.65** Let $a$ be a right-handed, orthonormal basis. For each of the vectors $\mathbf{v}$ below, find an alternative right-handed, orthonormal basis $b$, such that $\mathbf{b}_2$ is parallel to $\mathbf{v}$.

a) $\mathbf{v} = \mathbf{a}_1 - \mathbf{a}_3$
b) $\mathbf{v} = -3\mathbf{a}_3$
c) $\mathbf{v} = -\mathbf{a}_1 + \frac{2}{3}\mathbf{a}_2 - \mathbf{a}_3$
d) $\mathbf{v} = 2\mathbf{a}_1 + \mathbf{a}_2$
e) $\mathbf{v} = \frac{1}{3}\mathbf{a}_1 + \frac{1}{3}\mathbf{a}_2 - \mathbf{a}_3$
f) $\mathbf{v} = \frac{4}{3}\mathbf{a}_1 + \frac{1}{3}\mathbf{a}_2 - \frac{1}{3}\mathbf{a}_3$

**Exercise 3.66** Let $a$ be a right-handed, orthonormal basis. For each of the vectors $\mathbf{v}$ below, find an alternative right-handed, orthonormal basis $b$, such that $\mathbf{b}_3$ is parallel to $\mathbf{v}$.

a) $\mathbf{v} = \mathbf{a}_1 - \mathbf{a}_3$
b) $\mathbf{v} = -3\mathbf{a}_3$
c) $\mathbf{v} = -\mathbf{a}_1 + \frac{2}{3}\mathbf{a}_2 - \mathbf{a}_3$
d) $\mathbf{v} = 2\mathbf{a}_1 + \mathbf{a}_2$
e) $\mathbf{v} = \frac{1}{3}\mathbf{a}_1 + \frac{1}{3}\mathbf{a}_2 - \mathbf{a}_3$
f) $\mathbf{v} = \frac{4}{3}\mathbf{a}_1 + \frac{1}{3}\mathbf{a}_2 - \frac{1}{3}\mathbf{a}_3$

**Exercise 3.67** Let $\{\mathbf{v}_1, \mathbf{v}_2, \dots, \mathbf{v}_n\}$ be a set of arbitrary vectors in a vector space $\mathbb{V}$. Show that $\text{span}\{\mathbf{v}_1, \mathbf{v}_2, \dots, \mathbf{v}_n\}$ is a subset of $\mathbb{V}$ and that $\text{span}\{\mathbf{v}_1, \mathbf{v}_2, \dots, \mathbf{v}_n\}$ is a vector space in its own right.

**Solution.** Any linear combination in $\text{span}\{\mathbf{v}_1, \mathbf{v}_2, \dots, \mathbf{v}_n\}$ may be expressed as

$$(((\alpha_1\mathbf{v}_1 + \alpha_2\mathbf{v}_2) + \alpha_3\mathbf{v}_3) + \cdots)\cdots + \alpha_n\mathbf{v}_n.$$

Each sum of two vectors yields a vector in $\mathbb{V}$ and therefore the linear combination must be a vector in $\mathbb{V}$, confirming that $\text{span}\{\mathbf{v}_1, \mathbf{v}_2, \dots, \mathbf{v}_n\}$ is a subset of $\mathbb{V}$.

The set $\text{span}\{\mathbf{v}_1, \mathbf{v}_2, \dots, \mathbf{v}_n\}$ contains the zero vector $\mathbf{0}$, since

$$0\mathbf{v}_1 + 0\mathbf{v}_2 + \cdots + 0\mathbf{v}_n = \mathbf{0}.$$

Moreover, if $\mathbf{v} \in \text{span}\{\mathbf{v}_1, \mathbf{v}_2, \dots, \mathbf{v}_n\}$, then there exists real numbers $\alpha_1, \dots, \alpha_n$, such that

$$\mathbf{v} = \alpha_1\mathbf{v}_1 + \alpha_2\mathbf{v}_2 + \cdots + \alpha_n\mathbf{v}_n.$$

But this implies that

$$\begin{aligned} -\mathbf{v} &= -(\alpha_1\mathbf{v}_1 + \alpha_2\mathbf{v}_2 + \cdots + \alpha_n\mathbf{v}_n) \\ &= -\alpha_1\mathbf{v}_1 - \alpha_2\mathbf{v}_2 - \cdots - \alpha_n\mathbf{v}_n, \end{aligned}$$

i.e., $-\mathbf{v} \in \text{span}\{\mathbf{v}_1, \mathbf{v}_2, \dots, \mathbf{v}_n\}$.

It is straightforward to show that if $\mathbf{v}$ and $\mathbf{w}$ are two vectors in $\text{span}\{\mathbf{v}_1, \mathbf{v}_2, \dots, \mathbf{v}_n\}$, then

$$\mathbf{v} + \mathbf{w} \in \text{span}\{\mathbf{v}_1, \mathbf{v}_2, \dots, \mathbf{v}_n\}$$

and

$$\alpha\mathbf{v} \in \text{span}\{\mathbf{v}_1, \mathbf{v}_2, \dots, \mathbf{v}_n\}$$

for any real number $\alpha$.

**Exercise 3.68** Show that the vectors

$$\mathbf{v} = 2\mathbf{v}_1 - 3\mathbf{v}_2, \ \mathbf{w} = -\mathbf{v}_1 + 1.5\mathbf{v}_2$$

are linearly dependent.

**Solution.** The vectors $\mathbf{v}$ and $\mathbf{w}$ are linearly dependent if there exists a pair of scalars $\beta_1$ and $\beta_2$ (not both zero), such that

$$\beta_1\mathbf{v} + \beta_2\mathbf{w} = \mathbf{0}.$$

In this case, a solution to this equation is given by $\beta_1 = 1$ and $\beta_2 = 2$, since

$$1\mathbf{v} + 2\mathbf{w} = (2\mathbf{v}_1 - 3\mathbf{v}_2) + (-2\mathbf{v}_1 + 3\mathbf{v}_2) = \mathbf{0}.$$

**Exercise 3.69** Show that the set of vectors $\{\mathbf{0}, \mathbf{v}_1, \dots, \mathbf{v}_n\}$ is linearly dependent for any vectors $\mathbf{v}_1$ through $\mathbf{v}_n$.

**Exercise 3.70** Show that if a vector space has a basis with $n$ vectors, then any set of $n+1$ vectors is linearly dependent.

**Exercise 3.71** Show that if a vector space has a basis with $n$ vectors, then no set of $n-1$ vectors will span the whole vector space.

**Exercise 3.72** Show that if a vector space has a basis with $n$ vectors, then every basis of the vector space has $n$ basis vectors.

**Exercise 3.73** Let $a$ be a right-handed, orthonormal basis. Consider the vector

$$\mathbf{v} = a\begin{pmatrix} 2 \\ 1 \\ 4 \end{pmatrix}$$

and a vector $\mathbf{n}$ of unit length parallel to the vector $\mathbf{w} = \mathbf{a}_1 + \mathbf{a}_2 - 2\mathbf{a}_3$. Let $\mathbf{v} = \mathbf{v}_{\|} + \mathbf{v}_{\perp}$ be a decomposition of the vector into a component parallel to $\mathbf{n}$ and one perpendicular to $\mathbf{n}$. Use the MAMBO toolbox to find these components.

**Solution.** From a previous problem, we have

$$\mathbf{v} = (\mathbf{v} \bullet \mathbf{n})\,\mathbf{n} + \mathbf{n} \times (\mathbf{v} \times \mathbf{n}),$$

where $\mathbf{v}$ and $\mathbf{n}$ are some arbitrary vectors, such that $\|\mathbf{n}\| = 1$. Using the following MAMBO toolbox statements:

```
> Restart():
> a:='a':
> DeclareTriads(a):
> v:=MakeTranslations(a,2,1,4):
> w:=MakeTranslations(a,1,1,-2):
> n:=(1/VectorLength(w)) &** w:
> vpar:=(v &oo n) &** n;
> vperp:=n &xx (v &xx n);
```

we find

$$\mathbf{v}_{\|} = a \begin{pmatrix} -\frac{5}{6} \\ -\frac{5}{6} \\ \frac{5}{3} \end{pmatrix} \text{ and } \mathbf{v}_{\perp} = a \begin{pmatrix} \frac{17}{6} \\ \frac{11}{6} \\ \frac{7}{3} \end{pmatrix}.$$

Finally, we confirm the truth of the formula:

```
> v &-- vpar &-- vperp;
```

table([
"Type" = "Vector"
"Size" = 0
])

**Exercise 3.74** Use the MAMBO toolbox to show that

$$\mathbf{a} \times (\mathbf{b} \times \mathbf{c}) = \mathbf{b}\,(\mathbf{a} \bullet \mathbf{c}) - \mathbf{c}\,(\mathbf{a} \bullet \mathbf{b}).$$

**Solution.** The following MAMBO toolbox statements confirm the claim by performing explicit coordinate calculations:

```
> Restart():
> n:='n':
> DeclareTriads(n):
> a:=MakeTranslations(n,a1,a2,a3):
> b:=MakeTranslations(n,b1,b2,b3):
> c:=MakeTranslations(n,c1,c2,c3):
> a &xx (b &xx c)
> &-- ((a &oo c) &** b)
> &++ ((a &oo b) &** c);
```

table([
"Type" = "Vector"
"Size" = 0
])

Note the placement of parentheses to ensure that the multiplication with a scalar is computed prior to any addition or subtraction of vectors.

**Exercise 3.75** Repeat Exercises 3.43, 3.45, 3.46, 3.60, 3.64, 3.65, and 3.66 using the MAMBO toolbox.

### Summary of notation

Upper-case, italicized letters, such as $A$, $R$, and $X$, were used in this chapter to denote arbitrary points in space. The same notation, but with subscripts, e.g., $A_1$ or $R_{\text{reference}}$, was used to distinguish between multiple points.

Pairs of upper-case, italicized letters with a superscripted arrow, such as $\overrightarrow{AB}$ and $\overrightarrow{PQ}$, were used in this chapter to denote arbitrary separations between points.

The symbol $\sim$ was used in this chapter to express the equivalence between two separations.

Bracketed separations, such as $\left[\overrightarrow{AB}\right]$ and $\left[\overrightarrow{PQ}\right]$, were used in this chapter to denote collections of equivalent separations.

Lower-case, bold-faced r's with superscripted pairs of upper-case, italicized letters, such as $\mathbf{r}^{AB}$ and $\mathbf{r}^{PQ}$, were used in this chapter to denote the position vectors corresponding to the collections of equivalent separations $\left[\overrightarrow{AB}\right]$ and $\left[\overrightarrow{PQ}\right]$. The equivalent notation for use on a blackboard or paper was $\bar{r}^{AB}$ and $\bar{r}^{PQ}$. The superscripted, upper-case letters were omitted when referring to a general vector, such as $\mathbf{v}$ or $\mathbf{w}$.

Curved arrows with superscripted separations or position vectors, such as $\overset{\overrightarrow{AB}}{\curvearrowright}$ and $\overset{\mathbf{r}^{AB}}{\curvearrowright}$, were used in this chapter to denote a motion along a separation or in a direction and by a distance corresponding to the direction and length of a position vector.

The symbol $\|\cdot\|$ was used in this chapter to denote the length of a vector.

The symbol $\mathbf{0}$ was used in this chapter to denote the zero vector.

The symbol $+$ was used in this chapter to denote vector addition.

The symbol $-$ was used in this chapter to denote vector subtraction and the unary multiplication of a vector with the number $-1$.

The symbol $\theta(\mathbf{a}, \mathbf{b})$ was used in this chapter to denote the angle between the vectors $\mathbf{a}$ and $\mathbf{b}$.

The symbol $\bullet$ was used in this chapter to denote the vector dot product.

The symbol $\times$ was used in this chapter to denote the vector cross product.

Lower-case, unsubscripted letters, such as $a$ and $b$, were used in this chapter to denote $1 \times 3$ matrices with entries given by the basis vectors of an orthonormal basis.

Lower-case letters with a left superscript, such as ${}^a v$ and ${}^b w$, were used in this chapter to denote $3 \times 1$ matrices with entries given by the coordinates of a vector relative to an orthonormal basis.

The symbol $\delta_{ij}$ was used in this chapter to denote the Kronecker delta, such that $\delta_{ij}$ equals 1 if $i = j$ and 0 otherwise.

## Summary of terminology

The *separation* from point $A$ to point $B$ is the straight-line segment from $A$ to $B$. (Page 84)

Two separations are said to be *equivalent* if they have equal length, are parallel, and have the same heading. (Page 93)

The *position vector* $\mathbf{r}^{AB}$ is the collection of all separations that are equivalent to the separation $\overrightarrow{AB}$. (Page 96)

The separation $\overrightarrow{AB}$ is a *representation* of the position vector $\mathbf{r}^{AB}$. (Page 96)

The *length*, *direction*, and *heading* of a vector equals the length, direction, and heading of any one of its separations. (Page 97)

Every *pure translation* corresponds to a unique vector. Every vector corresponds to a unique pure translation. (Page 97)

The *zero vector* $\mathbf{0}$ corresponds to the identity translation. (Page 98)

The *multiplication* of a vector $\mathbf{v}$ with a scalar $\alpha$ is a vector $\alpha\mathbf{v}$ with length equal to $|\alpha|$ times the length of $\mathbf{v}$ and that is parallel ($\alpha > 0$) or antiparallel ($\alpha < 0$) to $\mathbf{v}$. (Page 99)

The *sum* of two vectors corresponds to the composition of the corresponding translations. (Page 100)

The *dot product* of two vectors is a **real number** equal to the product of the lengths of the two vectors and **cosine** of the angle between the vectors. (Page 103)

The *cross product* of two vectors is a **vector** with length equal to the product of the lengths of the two vectors and **sine** of the angle between the vectors and direction given by the **right-hand rule.** (Page 105)

A vector is said to *span a line* if the separation between any two points on the line represents some scalar multiple of the vector. (Page 109)

A pair of non-parallel vectors is said to *span a plane* if the separation between any two points on the plane represents some linear combination of the vectors. (Page 110)

(Page 110) Three vectors that are not parallel to the same plane are said to be a *basis of space.*

(Page 110) The vectors in a basis of space are called *basis vectors.*

(Page 111) To *express* a vector relative to a basis is to write it as a sum of multiples of the basis vectors.

(Page 111) The coefficients in front of the basis vectors in an expression of a vector relative to a basis are called the *coordinates* of the vector relative to the basis.

(Page 113) A basis is *orthonormal* if the basis vectors are of unit length and mutually perpendicular.

(Page 115) The $3 \times 1$ matrix with entries equal to the coordinates of a vector relative to an orthonormal basis is called the *matrix representation* of the vector relative to the basis.

(Page 117) An orthonormal basis is said to be *right-handed* if the cross product of the first two vectors equals the last vector and *left-handed* otherwise.

(Page 122) In the MAMBO toolbox, the global variable `GlobalTriadDeclarations` contains the names of all declared right-handed, orthonormal bases.

(Page 122) In the MAMBO toolbox, the procedure `DeclareTriads` appends `GlobalTriadDeclarations` with any number of basis labels.

(Page 122) In the MAMBO toolbox, the procedure `MakeTranslations` is used to define an arbitrary vector.

(Page 123) In the MAMBO toolbox, the procedure `NullVector` is used to define the zero vector.

(Page 124) In the MAMBO toolbox, the procedure `&**` is used to compute a multiplication of a vector with a scalar.

(Page 125) In the MAMBO toolbox, the procedure `VectorLength` is used to compute the length of a vector.

(Page 125) In the MAMBO toolbox, the procedure `&++` is used to compute the sum of two vectors.

(Page 127) In the MAMBO toolbox, the procedure `&--` is used to compute the difference between two vectors.

(Page 128) In the MAMBO toolbox, the procedure `&oo` is used to compute the vector dot product.

(Page 129) In the MAMBO toolbox, the procedure `&xx` is used to compute the vector cross product.

# Chapter 4

# Positions

*wherein the reader learns of:*

- *Using position vectors to describe the position of a rigid body or observer relative to another observer;*
- *Using configuration coordinates to describe time-dependent positions;*
- *Configuration constraints and their implications on the allowable configurations of a mechanism;*
- *Animation of a multibody mechanism.*

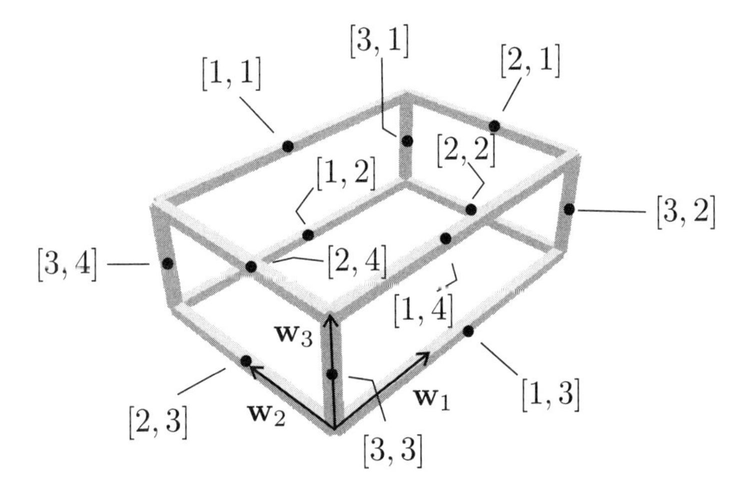

Practicum

As you complete this chapter, you will be able to generate a complicated multibody mechanism within the MAMBO application and to view the effects of changes in geometric parameters or configuration coordinates. Take advantage of the opportunities offered through MAMBO to visualize the significance of constraints and singularities.

Try examining the number of degrees of freedom of everyday mechanisms in your surroundings. Attempt to introduce configuration coordinates and formulate constraints that correspond to limitations on the allowable configurations of your mechanisms.

# 4.1 Review

## 4.1.1 Reference Points

Recall the following observations from Chapter 1:

- The configuration of a rigid body relative to a reference configuration is uniquely described through a combination of a pure translation and a pure rotation, given the selection of a specific point on the body that is kept fixed by the pure rotation;
- The pure translation is given by a shift of all points on the body from the reference configuration to an intermediate configuration, in such a way that the selected point coincides with the corresponding point in the final configuration;
- The magnitude of the translation is the distance between the corresponding points in the reference and final configurations, respectively. The direction of the translation is given by the straight line through the two points.

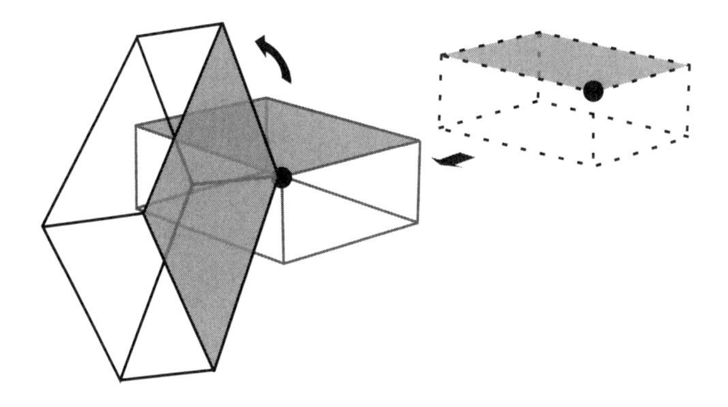

**Illustration 4.1**
The relative configuration of two observers $\mathcal{A}$ and $\mathcal{B}$ may be represented by the configuration of the virtual block corresponding to $\mathcal{A}$ relative to the reference configuration corresponding to $\mathcal{B}$.

The position and orientation of the virtual block relative to the reference configuration can be uniquely described through a combination of a pure translation and a pure rotation given the selection of a specific point on the virtual block that is kept fixed by the pure rotation.

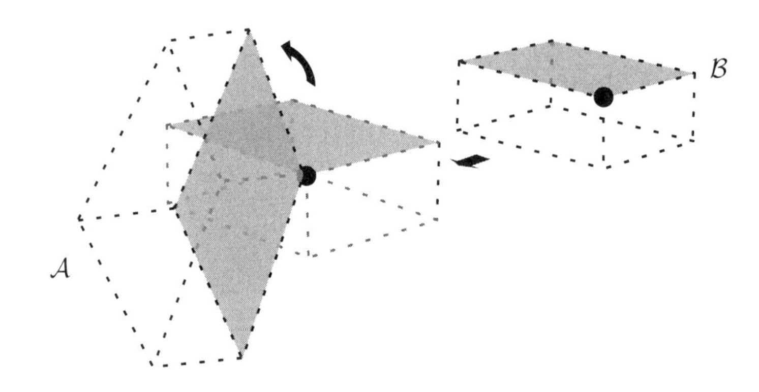


The pure translation is given by a shift of all points on the virtual block from the reference configuration of $\mathcal{B}$ to an intermediate configuration, in such a way that the selected point coincides with the corresponding point in the reference configuration of $\mathcal{A}$.

The magnitude of the translation is the distance between the corresponding points in the two reference configurations. The direction of the translation is given by the straight line through the two points.

The point about which the pure rotation takes place is called the *reference point* of the rigid body or of the observer. The reference point is a point fixed on the rigid body or fixed relative to the reference configuration of the observer.

> There is no preferred choice of point to qualify as the reference point of a given rigid body or observer.

When describing the configuration of rigid bodies, it is common to choose points that correspond to some geometrical feature. For example, a natural choice of reference point of a sphere is at the sphere's center. In the case of a rectangular block, we may select the geometric center or any of the eight corners. In the absence of geometrical features, such as corners or symmetries, to base the selection of reference point on, any point will do.

If two observers $\mathcal{A}$ and $\mathcal{B}$ share the same reference point, then the configuration of $\mathcal{B}$ relative to $\mathcal{A}$ is described through a pure rotation $\mathbf{R}_{\mathcal{A}\to\mathcal{B}}$ but no translation. In other words,

$$\mathbf{T}_{\mathcal{A}\to\mathcal{B}} = \mathbf{I}.$$

Conversely, if

$$\mathbf{T}_{\mathcal{A}\to\mathcal{B}} = \mathbf{I},$$

then the reference points of the two observers $\mathcal{A}$ and $\mathcal{B}$ coincide.

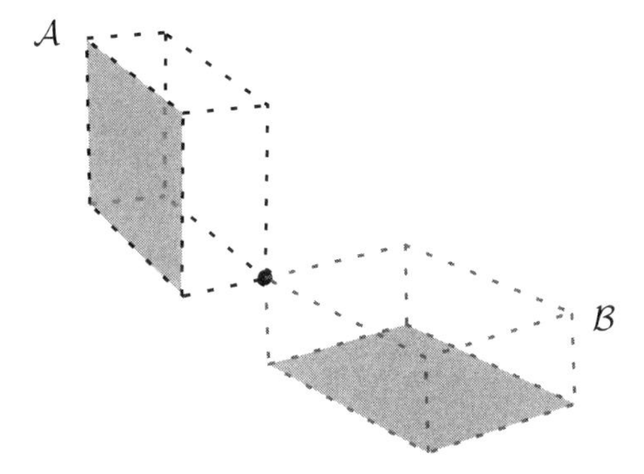


Similarly, if the reference point of a rigid body, say a sphere, coincides with the reference point of an observer $\mathcal{A}$, then the rigid body's configuration relative to $\mathcal{A}$ is described through a pure rotation $\mathbf{R}_{\mathcal{A}}$, but no translation. In other words,

$$\mathbf{T}_{\mathcal{A}} = \mathbf{I}.$$

**Illustration 4.2**

Suppose the configuration of an observer $\mathcal{B}$ relative to an observer $\mathcal{A}$ is given by a non-trivial pure translation, but no rotation, i.e.,

$$\mathbf{T}_{\mathcal{A}\to\mathcal{B}} \neq \mathbf{I}, \mathbf{R}_{\mathcal{A}\to\mathcal{B}} = \mathbf{I}.$$

Then, the reference points of $\mathcal{A}$ and $\mathcal{B}$ do not coincide. The pure translation $\mathbf{T}_{\mathcal{A}\to\mathcal{B}}$ contains the information necessary to shift the position of all points of the reference configuration of $\mathcal{A}$ so that they coincide with the corresponding points of the reference configuration of $\mathcal{B}$. The translation $\mathbf{T}_{\mathcal{A}\to\mathcal{B}}$ is uniquely determined by the location of the reference points of $\mathcal{A}$ and $\mathcal{B}$.

### 4.1.2 Translations

If the reference points of two observers $\mathcal{A}$ and $\mathcal{B}$ are denoted by $A$ and $B$, respectively, then the pure translation $\mathbf{T}_{\mathcal{A}\to\mathcal{B}}$ is uniquely determined by the separation $\overrightarrow{AB}$. In fact, the translation corresponds to a shift of all points by a distance given by the length of $\overrightarrow{AB}$ and in a direction parallel to and with the same heading as $\overrightarrow{AB}$. The separation $\overrightarrow{P_rP_f}$ from the initial location $P_r$ of some arbitrary point to its final location $P_f$ after the application of the pure translation:

- Has the same length as $\overrightarrow{AB}$;

- Is parallel to $\overrightarrow{AB}$;
- Has the same heading as $\overrightarrow{AB}$.

It follows that $\overrightarrow{P_rP_f}$ is equivalent to $\overrightarrow{AB}$, i.e., $\overrightarrow{P_rP_f} \sim \overrightarrow{AB}$. Clearly, the pure translation $\mathbf{T}_{\mathcal{A}\to\mathcal{B}}$ generates infinitely many separations $\overrightarrow{P_rP_f}$, each of which is equivalent to $\overrightarrow{AB}$. In the previous chapter, we concluded that the pure translation corresponds to the collection of all separations equivalent to $\overrightarrow{AB}$, i.e., the position vector $\mathbf{r}^{AB} = \left[\overrightarrow{AB}\right]$.

In the previous chapter, we developed algebraic operations on position vectors that corresponded to the operations on pure translations introduced in Chapter 2. For example, the correspondences

$$\mathbf{v}_1 \leftrightarrow \mathbf{T}_1, \mathbf{v}_2 \leftrightarrow \mathbf{T}_2$$

imply that

$$\alpha\mathbf{v}_1 + \beta\mathbf{v}_2 \leftrightarrow \beta\mathbf{T}_2 \circ \alpha\mathbf{T}_1,$$

where $\alpha$ and $\beta$ are any real numbers. We also introduced two vector products, namely the dot product $\bullet$ and the cross product $\times$, with which we can detect when two pure translations are perpendicular or parallel, respectively.

The vector formalism reduces to straightforward matrix algebra when all vectors are expressed relative to right-handed, orthonormal bases. If $a = \begin{pmatrix} \mathbf{a}_1 & \mathbf{a}_2 & \mathbf{a}_3 \end{pmatrix}$ is a right-handed, orthonormal basis, and if ${}^a v$ and ${}^a w$ are the matrix representations of two vectors $\mathbf{v}$ and $\mathbf{w}$ relative to $a$, then

$$\begin{aligned} \alpha\mathbf{v} &= \alpha\left(a\,{}^a v\right) \\ &= a\left(\alpha\,{}^a v\right), \end{aligned}$$

$$\begin{aligned} \mathbf{v} \pm \mathbf{w} &= a\,{}^a v \pm a\,{}^a w \\ &= a\left({}^a v \pm {}^a w\right), \end{aligned}$$

$$\begin{aligned} \mathbf{v} \bullet \mathbf{w} &= \left({}^a v\right)^T {}^a w \\ &= \left({}^a w\right)^T {}^a v, \end{aligned}$$

and

$$\mathbf{v} \times \mathbf{w} = \begin{vmatrix} \mathbf{a}_1 & \mathbf{a}_2 & \mathbf{a}_3 \\ {}^a v_1 & {}^a v_2 & {}^a v_3 \\ {}^a w_1 & {}^a w_2 & {}^a w_3 \end{vmatrix}.$$

## 4.2 Examples

(Ex. 4.1 – Ex. 4.14)

In all the examples below, all vectors will be expressed relative to a common, right-handed, orthonormal basis

$$w = \begin{pmatrix} \mathbf{w}_1 & \mathbf{w}_2 & \mathbf{w}_3 \end{pmatrix}.$$

In Chapter 6, we will allow for multiple right-handed, orthonormal bases, but will have to forego that pleasure until we have developed the mathematics needed to convert between matrix representations relative to different bases.

### 4.2.1 A Still Life

Suppose you want to describe the geometry of a wireframe representation of a rectangular block, as depicted below.

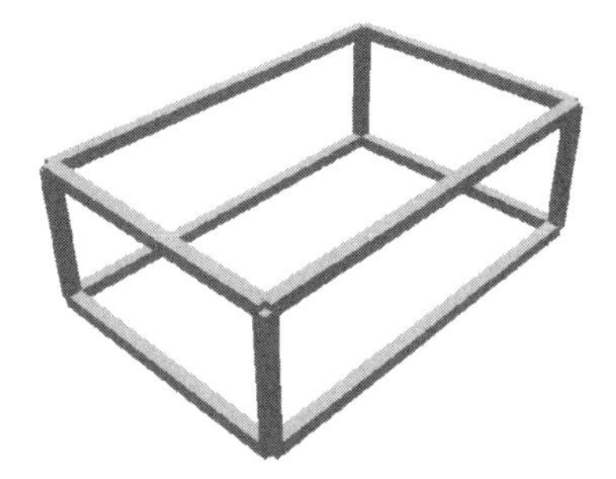

The wireframe structure can be decomposed into 12 rigid edges, four of which are parallel to the $\mathbf{w}_1$ basis vector, four of which are parallel to the $\mathbf{w}_2$ basis vector, and four of which are parallel to the $\mathbf{w}_3$ basis vector. The edges will be labeled by pairs of integers $[i, j]$, corresponding to the $j$-th edge parallel to the $i$-th basis vector of $w$ as indicated in the figure.

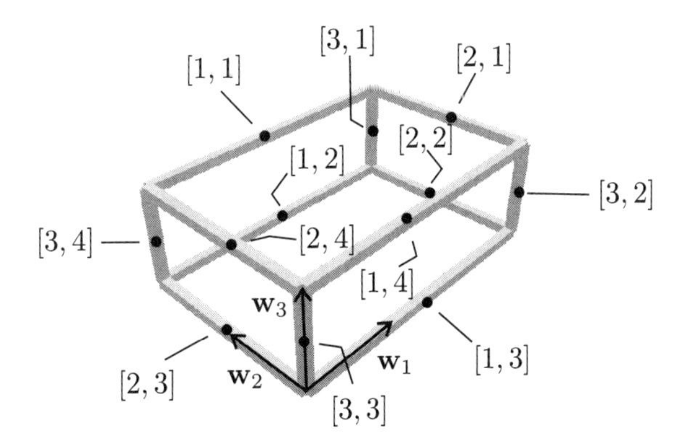

Introduce a main observer $\mathcal{W}$ with reference point $W$ at the center of the block and let the reference point $E_{[i,j]}$ of the $[i,j]$-th edge correspond to the geometric center of the edge.

The configuration of the $[i,j]$-th edge relative to $\mathcal{W}$ is then given by a pure translation $\mathbf{T}_{[i,j]}$ corresponding to the position vector

$$\mathbf{r}^{WE_{[i,j]}}.$$

Specifically, we find

$$\mathbf{r}^{WE_{[1,1]}} = w\begin{pmatrix} 0 \\ l_2/2 \\ l_3/2 \end{pmatrix}, \quad \mathbf{r}^{WE_{[1,2]}} = w\begin{pmatrix} 0 \\ l_2/2 \\ -l_3/2 \end{pmatrix},$$

$$\mathbf{r}^{WE_{[1,3]}} = w\begin{pmatrix} 0 \\ -l_2/2 \\ -l_3/2 \end{pmatrix}, \quad \mathbf{r}^{WE_{[1,4]}} = w\begin{pmatrix} 0 \\ -l_2/2 \\ l_3/2 \end{pmatrix},$$

$$\mathbf{r}^{WE_{[2,1]}} = w\begin{pmatrix} l_1/2 \\ 0 \\ l_3/2 \end{pmatrix}, \quad \mathbf{r}^{WE_{[2,2]}} = w\begin{pmatrix} l_1/2 \\ 0 \\ -l_3/2 \end{pmatrix},$$

$$\mathbf{r}^{WE_{[2,3]}} = w\begin{pmatrix} -l_1/2 \\ 0 \\ -l_3/2 \end{pmatrix}, \quad \mathbf{r}^{WE_{[2,4]}} = w\begin{pmatrix} -l_1/2 \\ 0 \\ l_3/2 \end{pmatrix},$$

$$\mathbf{r}^{WE_{[3,1]}} = w\begin{pmatrix} l_1/2 \\ l_2/2 \\ 0 \end{pmatrix}, \quad \mathbf{r}^{WE_{[3,2]}} = w\begin{pmatrix} l_1/2 \\ -l_2/2 \\ 0 \end{pmatrix},$$

$$\mathbf{r}^{WE_{[3,3]}} = w\begin{pmatrix} -l_1/2 \\ -l_2/2 \\ 0 \end{pmatrix}, \quad \text{and } \mathbf{r}^{WE_{[3,4]}} = w\begin{pmatrix} -l_1/2 \\ l_2/2 \\ 0 \end{pmatrix}.$$

**Illustration 4.3**

Suppose you want to describe the geometry of the arrangement of spheres shown below.

The spheres in the bottom layer are resting on a plane parallel to the $\mathbf{w}_1$ and $\mathbf{w}_2$ basis vectors and $\mathbf{w}_3$ points away from this plane toward the upper sphere.

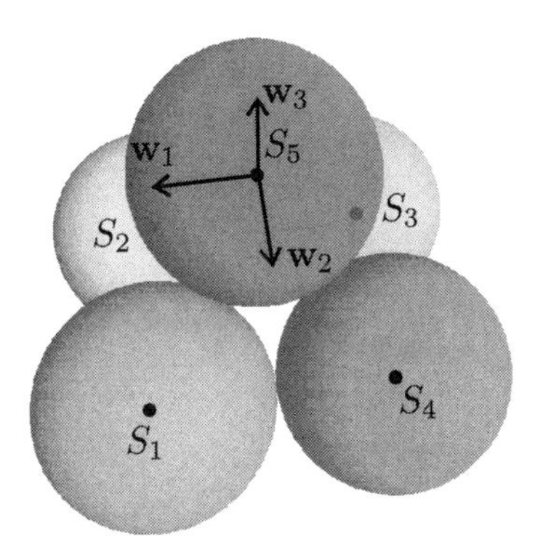


Introduce a main observer $\mathcal{W}$ with reference point $W$ at the center of the upper sphere and let the reference point $S_i$ of the $i$-th sphere correspond to its geometric center. Then, the configuration of the $i$-th sphere relative to $\mathcal{W}$ is given by a pure translation $\mathbf{T}_i$ corresponding to the position vector

$$\mathbf{r}^{WS_i}.$$

Specifically,

$$\mathbf{r}^{WS_1} = w\begin{pmatrix} R \\ R \\ -h \end{pmatrix}, \mathbf{r}^{WS_2} = w\begin{pmatrix} R \\ -R \\ -h \end{pmatrix},$$

$$\mathbf{r}^{WS_3} = w\begin{pmatrix} -R \\ -R \\ -h \end{pmatrix}, \mathbf{r}^{WS_4} = w\begin{pmatrix} -R \\ R \\ -h \end{pmatrix}, \text{ and } \mathbf{r}^{WS_5} = \mathbf{0},$$

where $R$ is the radius of the spheres and $h$ is the height of the center of the upper sphere above the centers of the spheres in the bottom layer.

The height $h$ can be related to the radius $R$ by requiring that

$$\left\|\mathbf{r}^{WS_1}\right\| = \left\|\mathbf{r}^{WS_2}\right\| = \left\|\mathbf{r}^{WS_3}\right\| = \left\|\mathbf{r}^{WS_4}\right\| = 2R.$$

This is equivalent to the equation

$$\sqrt{2R^2 + h^2} = 2R,$$

which implies that

$$h = \sqrt{2}R.$$

In the case of a time-independent configuration of a rigid body, there is no immediate need to introduce auxiliary observers, although it may at times be convenient.

### 4.2.2 The Single Moving Rigid Body

When time-dependent changes take place in the configuration of a rigid body relative to the main observer, the recommended methodology requires the introduction of at least one auxiliary observer between the rigid body and the main observer. Specifically, the auxiliary observer is introduced in such a way that the rigid body remains stationary relative to the auxiliary observer, while the motion of the auxiliary observer relative to the main observer contains the entire time-dependence.

Suppose, for example, that you want to describe the geometry of a single, freely moving rigid body. Introduce a main observer $\mathcal{W}$ with reference point $W$ somewhere in space. Introduce an auxiliary observer $\mathcal{A}$, relative to which the rigid body is stationary and with reference point $A$ coinciding with some arbitrary point on the rigid body.

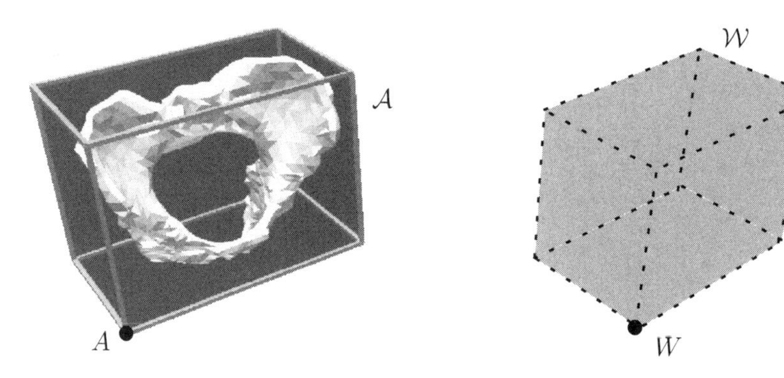


Assume for simplicity that the orientation of the rigid body relative to $\mathcal{W}$ is described by the identity rotation. Then the configuration of the observer $\mathcal{A}$ relative to $\mathcal{W}$ is given by the pure translation $\mathbf{T}_{\mathcal{W}\to\mathcal{A}}$ corresponding to the position vector

$$\mathbf{r}^{WA}.$$

Since the rigid body's position is unrestricted, we can write

$$\mathbf{r}^{WA} = w \begin{pmatrix} q_1 \\ q_2 \\ q_3 \end{pmatrix},$$

where $q_1$, $q_2$, and $q_3$ are time-dependent quantities that uniquely specify the matrix representation of the position vector relative to the $w$ basis. These quantities are called *configuration coordinates*, since they provide information about the configuration of the rigid body relative to the main observer as a function of time.

The configuration coordinates $q_1$, $q_2$, and $q_3$ are the coordinates of the vector $\mathbf{r}^{WA}$ relative to the $w$ basis. In fact, given a coordinate system with origin at $W$ and axes parallel to the basis vectors $\mathbf{w}_1$, $\mathbf{w}_2$, and $\mathbf{w}_3$, the configuration coordinates are the *Cartesian coordinates* of the point $A$ with respect to this coordinate system.

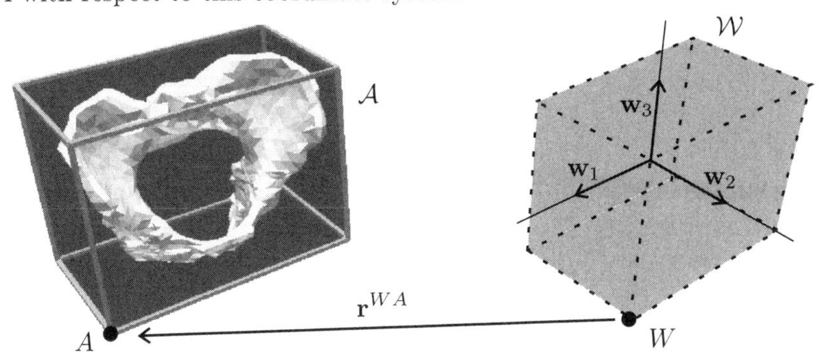


We may express this observation by the notation

$$^{\mathcal{W}}A = \begin{pmatrix} q_1 \\ q_2 \\ q_3 \end{pmatrix},$$

where the left-hand side denotes the *coordinate representation of the point $A$ relative to the observer $\mathcal{W}$*.

**Illustration 4.4**

Consider the quantities $\tilde{q}_1$, $\tilde{q}_2$, and $\tilde{q}_3$ introduced in the figure below.

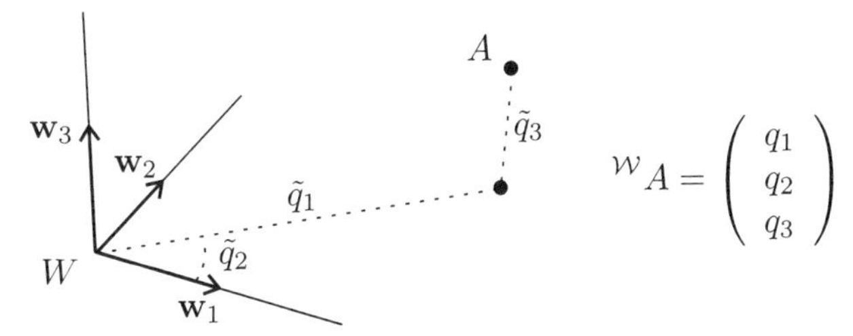


$$^{\mathcal{W}}A = \begin{pmatrix} q_1 \\ q_2 \\ q_3 \end{pmatrix}$$

It follows that

$$q_1 = \tilde{q}_1 \cos \tilde{q}_2,$$
$$q_2 = \tilde{q}_1 \sin \tilde{q}_2,$$

and

$$q_3 = \tilde{q}_3.$$

Consequently, the position vector $\mathbf{r}^{WA}$ can be written as

$$\mathbf{r}^{WA} = w \begin{pmatrix} \tilde{q}_1 \cos \tilde{q}_2 \\ \tilde{q}_1 \sin \tilde{q}_2 \\ \tilde{q}_3 \end{pmatrix}.$$

Similarly,

$$^{\mathcal{W}}A = \begin{pmatrix} \tilde{q}_1 \cos \tilde{q}_2 \\ \tilde{q}_1 \sin \tilde{q}_2 \\ \tilde{q}_3 \end{pmatrix}.$$

The new configuration coordinates $\tilde{q}_1$, $\tilde{q}_2$, and $\tilde{q}_3$ that have replaced $q_1$, $q_2$, and $q_3$ are known as the *polar coordinates* of the point $A$ with respect to the coordinate system with origin at $W$ and axes parallel to the basis vectors $\mathbf{w}_1$, $\mathbf{w}_2$, and $\mathbf{w}_3$.

The example in the illustration represented the configuration of the rigid body relative to $\mathcal{W}$ using an alternative set of three configuration coordinates. But what is the significance of the number **three**? Is it possible to use **more than three** configuration coordinates to describe the configuration of the rigid body relative to $\mathcal{W}$? Is it possible to use **fewer than three**?

Consider the vector

$$\mathbf{r}^{WA} = w \begin{pmatrix} \tilde{q}_1 \cos \tilde{q}_2 \\ \tilde{q}_1 \sin \tilde{q}_3 \\ \tilde{q}_4 \end{pmatrix}.$$

Here, four configuration coordinates $\tilde{q}_1$, $\tilde{q}_2$, $\tilde{q}_3$, and $\tilde{q}_4$ are used to represent the vector relative to the $w$ basis. How are these related to the Cartesian coordinates introduced above? By identifying the coordinates of the vector $\mathbf{r}^{WA}$, we find

$$\tilde{q}_1 \cos \tilde{q}_2 = q_1,$$
$$\tilde{q}_1 \sin \tilde{q}_3 = q_2,$$

and

$$\tilde{q}_4 = q_3,$$

which implies that

$$\tilde{q}_1 = s,$$
$$\tilde{q}_2 = \arccos \frac{q_1}{s},$$
$$\tilde{q}_3 = \arcsin \frac{q_2}{s},$$

and

$$\tilde{q}_4 = q_3,$$

where $s$ is an arbitrary number, such that $s \geq |q_1|, |q_2|$. It follows that every choice of values for the coordinates $q_1$, $q_2$, and $q_3$ corresponds to infinitely many choices of values for the coordinates $\tilde{q}_1$, $\tilde{q}_2$, $\tilde{q}_3$, and $\tilde{q}_4$. We say that the configuration coordinates $\tilde{q}_1$, $\tilde{q}_2$, $\tilde{q}_3$, and $\tilde{q}_4$ are a *redundant set.*

**Illustration 4.5**

If we were to assert that

$$\mathbf{r}^{WA} = w \begin{pmatrix} 0 \\ \tilde{q}_1 + \tilde{q}_2 \\ \tilde{q}_3 - \tilde{q}_4 \end{pmatrix},$$

we would effectively be constraining the rigid body's position, such that the reference point $A$ would lie in the plane spanned by $\mathbf{w}_2$ and $\mathbf{w}_3$ through $W$. If our aim was to describe an entirely free rigid body, this formulation would be erroneous. The variables $\tilde{q}_1$, $\tilde{q}_2$, $\tilde{q}_3$, and $\tilde{q}_4$ are an insufficient set of configuration coordinates.

Now, consider using only two variables $\tilde{q}_1$ and $\tilde{q}_2$ to describe the position vector $\mathbf{r}^{WA}$, such that

$$\mathbf{r}^{WA} = w \begin{pmatrix} a_1\tilde{q}_1 + a_2\tilde{q}_2 \\ a_3\tilde{q}_1 + a_4\tilde{q}_2 \\ a_5\tilde{q}_1 + a_6\tilde{q}_2 \end{pmatrix},$$

for some constants $a_1$, $a_2$, $a_3$, $a_4$, $a_5$, and $a_6$. This implies that

$$\begin{aligned} q_1 &= a_1\tilde{q}_1 + a_2\tilde{q}_2, \\ q_2 &= a_3\tilde{q}_1 + a_4\tilde{q}_2, \\ q_3 &= a_5\tilde{q}_1 + a_6\tilde{q}_2, \end{aligned}$$

where $q_1$, $q_2$, and $q_3$ are the Cartesian coordinates from above. This is a system of three equations in two unknowns ($\tilde{q}_1$ and $\tilde{q}_2$) and can only be solved if the third equation is linearly dependent on the first two. Since this is not generally the case, the two variables $\tilde{q}_1$ and $\tilde{q}_2$ **cannot** be used to describe a general configuration of the rigid body relative to $\mathcal{W}$.

The smallest number of configuration coordinates required to completely describe all possible positions of the rigid body relative to $\mathcal{W}$ is three! We say that the rigid body, **in the absence of rotation**, has three *geometric degrees of freedom.*

### 4.2.3 Constraints

When the position of a rigid body can be described by fewer than three configuration coordinates, the rigid body's configuration is said to be *constrained*. In the absence of rotation, a constrained rigid body has fewer than three geometric degrees of freedom.

1

Suppose you want to describe the motion of a puck sliding on an ice hockey rink.

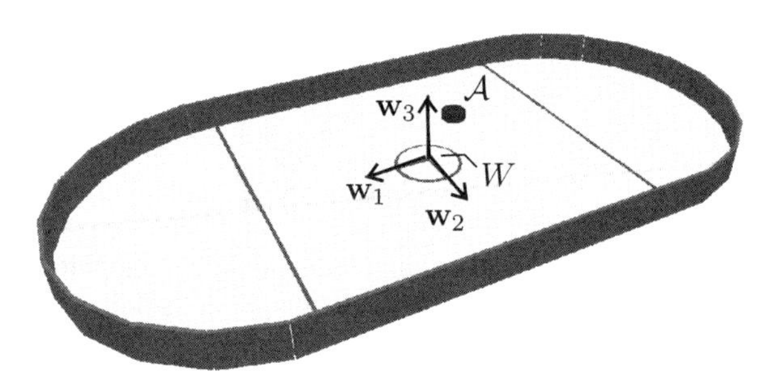


Here, the plane of the ice is parallel to the vectors $\mathbf{w}_1$ and $\mathbf{w}_2$. The vector $\mathbf{w}_3$ points away from the ice in the direction of the center of the puck. Following the corresponding example in Chapter 2, we introduce a main observer $\mathcal{W}$, relative to which the ice hockey rink is stationary and with reference point $W$ at the center of the rink. Since the puck's position relative to $\mathcal{W}$ will change with time, we introduce an auxiliary observer $\mathcal{A}$, relative to which the puck is stationary and with reference point $A$ at the center of the puck.

The configuration of the puck relative to the observer $\mathcal{A}$ is described by the identity translation and the identity rotation. The configuration of $\mathcal{A}$ relative to $\mathcal{W}$, on the other hand, is described by a pure translation $\mathbf{T}_{\mathcal{W}\to\mathcal{A}}$ corresponding to the position vector

$$\mathbf{r}^{WA}.$$

Since the puck is restricted to positions on the ice, we conclude that

$$\mathbf{r}^{WA} = w \begin{pmatrix} q_1 \\ q_2 \\ \frac{h}{2} \end{pmatrix} \text{ or } {}^{\mathcal{W}}A = \begin{pmatrix} q_1 \\ q_2 \\ \frac{h}{2} \end{pmatrix},$$

where $h$ is the height of the puck. It follows that the puck has only two geometric degrees of freedom. The restriction on the puck's position is a constraint on the puck's configuration.

In the absence of the constraint on the puck's configuration, the pure translation $\mathbf{T}_{\mathcal{W}\to\mathcal{A}}$ could be described by the position vector

$$\mathbf{r}^{WA} = w \begin{pmatrix} \tilde{q}_1 \\ \tilde{q}_2 \\ \tilde{q}_3 \end{pmatrix},$$

where $\tilde{q}_1$, $\tilde{q}_2$, and $\tilde{q}_3$ are the Cartesian coordinates introduced above. A quick comparison between the two expressions for $\mathbf{r}^{WA}$ shows that the constrained mechanism is obtained from the unconstrained mechanism by requiring that

$$\tilde{q}_3 = \frac{h}{2}.$$

This equation in the configuration coordinate is called a *configuration constraint*.

**Illustration 4.6**

Suppose you want to describe the motion of a bead sliding on the surface of a sphere.

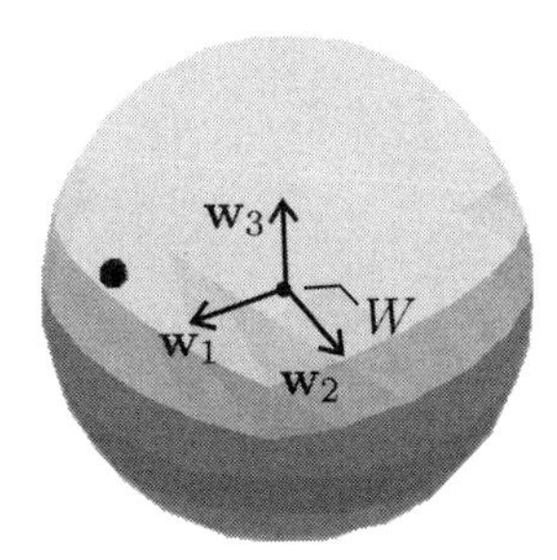


Introduce a main observer $\mathcal{W}$ with reference point $W$ at the center of the sphere. Since the bead's position relative to $\mathcal{W}$ will change with time, we introduce an auxiliary observer $\mathcal{A}$, relative to which the bead is stationary and with reference point $A$ at the center of the bead. (The observer $\mathcal{A}$ is different from the observer $\mathcal{A}$ introduced in Chapter 2.)

The configuration of the observer $\mathcal{A}$ relative to $\mathcal{W}$ is then given by a pure translation $\mathbf{T}_{\mathcal{W}\to\mathcal{A}}$ corresponding to the position vector

$$\mathbf{r}^{WA}.$$

Using *spherical coordinates*, the position vector may be written as

$$\mathbf{r}^{WA} = w \begin{pmatrix} R\sin q_1 \cos q_2 \\ R\sin q_1 \sin q_2 \\ R\cos q_1 \end{pmatrix} \text{ or } {}^{\mathcal{W}}A = \begin{pmatrix} R\sin q_1 \cos q_2 \\ R\sin q_1 \sin q_2 \\ R\cos q_1 \end{pmatrix},$$

where $R$ is the radius of the sphere. It follows that the bead has only two geometric degrees of freedom. The restriction on the bead's position is a constraint on the bead's configuration.

Alternatively, we may use Cartesian coordinates to describe the position vector $\mathbf{r}^{WA}$:

$$\mathbf{r}^{WA} = w \begin{pmatrix} \tilde{q}_1 \\ \tilde{q}_2 \\ \tilde{q}_3 \end{pmatrix}.$$

The constraint on the bead's configuration, however, implies that the length of this vector must equal the radius of the sphere, i.e.,

$$\left\|\mathbf{r}^{WA}\right\| = \sqrt{\tilde{q}_1^2 + \tilde{q}_2^2 + \tilde{q}_3^2} = R.$$

The only allowable values for the configuration coordinates $\tilde{q}_1$, $\tilde{q}_2$, and $\tilde{q}_3$ are those that satisfy this configuration constraint.

### 4.2.4 The Implicit Function Theorem

The configuration constraint on the configuration coordinates $\tilde{q}_1$, $\tilde{q}_2$, and $\tilde{q}_3$ in the previous illustration

$$\sqrt{\tilde{q}_1^2 + \tilde{q}_2^2 + \tilde{q}_3^2} = R$$

can be solved for $\tilde{q}_3$ in terms of $\tilde{q}_1$ and $\tilde{q}_2$:

$$\tilde{q}_3 = \pm\sqrt{R^2 - \tilde{q}_1^2 - \tilde{q}_2^2},$$

where the plus or minus sign reflects whether the bead is on the upper or lower hemisphere, respectively.

The constraint equation is satisfied for $\tilde{q}_1 = \tilde{q}_2 = 0$ and $\tilde{q}_3 = R$. Now, consider arbitrary values of $\tilde{q}_1$ and $\tilde{q}_2$ near 0. Then, there exists a unique value for $\tilde{q}_3$ near $R$ that satisfies the constraint equation. We may comfortably argue that $\tilde{q}_3$ near $R$ is a **function**[1] of $\tilde{q}_1$ and $\tilde{q}_2$ near 0.

The constraint equation is also satisfied for $\tilde{q}_1 = R$ and $\tilde{q}_2 = \tilde{q}_3 = 0$. Now, consider arbitrary values of $\tilde{q}_1$ near $R$ and $\tilde{q}_2$ near 0. In contrast to the previous case, it is not always possible to find a value for $\tilde{q}_3$ near 0 that satisfies the constraint equation. Clearly, if

$$\tilde{q}_1^2 + \tilde{q}_2^2 > R,$$

then the quantity under the radical is negative and no real solution of the constraint equation exists for $\tilde{q}_3$. If, instead,

$$\tilde{q}_1^2 + \tilde{q}_2^2 = R,$$

$\tilde{q}_3 = 0$ is a unique solution to the constraint equation. Finally, if

$$\tilde{q}_1^2 + \tilde{q}_2^2 < R,$$

there are two possible values for $\tilde{q}_3$ near 0 that satisfy the constraint equation.

[1]Recall that a function must be single-valued. The $\pm$ in front of the square root is of concern in this respect. That the value of $\tilde{q}_3$ near $R$ is unique guarantees that the function is single-valued.

While it is theoretically possible to eliminate $\tilde{q}_3$ from the expression for the position vector $\mathbf{r}^{WA}$ when $\tilde{q}_1, \tilde{q}_2 \approx 0$ and $\tilde{q}_3 \approx R$, this is not possible for $\tilde{q}_1 \approx R$ and $\tilde{q}_2, \tilde{q}_3 \approx 0$. This observation is consistent with the predictions of the *implicit function theorem.*

**Theorem:** Let $f$ be a function of $n$ real configuration coordinates and let

$$q_1 = q_{1,0}, q_2 = q_{2,0}, \dots, \text{ and } q_n = q_{n,0}$$

be a solution to the constraint equation

$$f(q_1, q_2, \dots, q_n) = 0.$$

Consider $q_1 \approx q_{1,0}, \dots, q_{i-1} \approx q_{i-1,0}, q_{i+1} \approx q_{i+1,0}, \dots, q_n \approx q_{n,0}$. Then, if

$$\frac{\partial f}{\partial q_i}(q_{1,0}, q_{2,0}, \dots, q_{n,0}) \neq 0,$$

there exists a unique value for $q_i \approx q_{i,0}$ that satisfies the constraint equation.

If

$$\frac{\partial f}{\partial q_i}(q_{1,0}, q_{2,0}, \dots, q_{n,0}) = 0,$$

the choice of values

$$q_1 = q_{1,0}, q_2 = q_{2,0}, \dots, \text{ and } q_n = q_{n,0}$$

is said to be *singular relative to* $q_i$. Otherwise the choice of values is said to be *regular relative to* $q_i$.

**Illustration 4.7**
In the case of the bead,

$$f(\tilde{q}_1, \tilde{q}_2, \tilde{q}_3) = \sqrt{\tilde{q}_1^2 + \tilde{q}_2^2 + \tilde{q}_3^2} - R.$$

It follows that

$$\frac{\partial f}{\partial \tilde{q}_3} = \frac{\tilde{q}_3}{\sqrt{\tilde{q}_1^2 + \tilde{q}_2^2 + \tilde{q}_3^2}}$$

and thus

$$\frac{\partial f}{\partial \tilde{q}_3}(0, 0, R) = 1,$$

2

whereas

$$\frac{\partial f}{\partial \tilde{q}_3}(R,0,0) = 0.$$

In the former case, the implicit function theorem guarantees that $\tilde{q}_3$ can be eliminated in terms of $\tilde{q}_1$ and $\tilde{q}_2$, whereas no such guarantee is offered in the second case. This agrees with the observations made above.

Even though it is not possible to express $\tilde{q}_3$ near 0 as a function of $\tilde{q}_1$ near $R$ and $\tilde{q}_2$ near 0, the implicit function theorem shows that it is possible to express $\tilde{q}_1$ near $R$ as a function of $\tilde{q}_2$ and $\tilde{q}_3$ near 0. In fact,

$$\frac{\partial f}{\partial \tilde{q}_1}(R,0,0) = 1 \neq 0.$$

Thus, while $\tilde{q}_1 = R$, $\tilde{q}_2 = \tilde{q}_3 = 0$ is singular relative to $\tilde{q}_3$, it is regular relative to $\tilde{q}_1$. The result of Exercise 4.10 shows that all choices of values for the configuration coordinates $\tilde{q}_1$, $\tilde{q}_2$, and $\tilde{q}_3$ are regular relative to at least one of the configuration coordinates.

The terminology and methodology introduced here carries over to the case of multiple constraints.

**Theorem:** Let $f_1, f_2, \ldots, f_m$ be $m$ functions of $n \geq m$ real configuration coordinates and let

$$q_1 = q_{1,0}, q_2 = q_{2,0}, \ldots, \text{ and } q_n = q_{n,0}$$

be a solution to the constraint equations

$$\begin{aligned} f_1(q_1, q_2, \ldots, q_n) &= 0, \\ f_2(q_1, q_2, \ldots, q_n) &= 0, \\ &\vdots \\ f_m(q_1, q_2, \ldots, q_n) &= 0. \end{aligned}$$

Consider $q_{m+1} \approx q_{m+1,0}, \ldots, q_n \approx q_{n,0}$. Then, if

$$\begin{vmatrix} \frac{\partial f_1}{\partial q_1} & \cdots & \frac{\partial f_1}{\partial q_m} \\ \vdots & \ddots & \vdots \\ \frac{\partial f_m}{\partial q_1} & \cdots & \frac{\partial f_m}{\partial q_m} \end{vmatrix} (q_{1,0}, q_{2,0}, \ldots, q_{n,0}) \neq 0,$$

there exists a unique choice of values for $q_1 \approx q_{1,0}, \ldots, q_m \approx q_{m,0}$ that satisfies the constraint equations.

**Illustration 4.8**

Suppose you want to describe the motion of a bead that is restricted to move on the intersection between the surfaces of two spheres, such that the separation $\overrightarrow{S_1 S_2}$ from the center of the first sphere to the center of the second sphere is contained in the position vector

$$\mathbf{r}^{S_1 S_2} = w \begin{pmatrix} \frac{3R}{2} \\ 0 \\ 0 \end{pmatrix},$$

where $R$ is the spheres' radius.

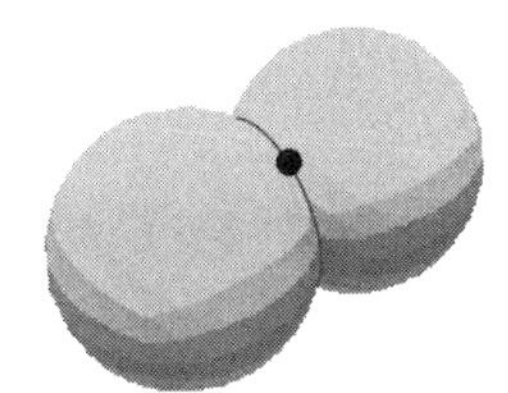

Introduce a main observer $\mathcal{W}$ with reference point $W$ at the center of the first sphere, i.e., such that

$$\mathbf{r}^{WS_1} = \mathbf{0} \text{ and } \mathbf{r}^{WS_2} = w \begin{pmatrix} \frac{3R}{2} \\ 0 \\ 0 \end{pmatrix}.$$

Let $\mathcal{A}$ be an auxiliary observer, relative to which the bead is stationary, and with reference point $A$ at the center of the bead. In the absence of the constraint on the bead's motion, the configuration of $\mathcal{A}$ relative to $\mathcal{W}$ is described by the pure translation $\mathbf{T}_{\mathcal{W}\to\mathcal{A}}$ corresponding to the position vector

$$\mathbf{r}^{WA} = w \begin{pmatrix} q_1 \\ q_2 \\ q_3 \end{pmatrix}.$$

The requirement that the bead lies on the intersection between the spherical surfaces implies that

$$\left\|\mathbf{r}^{S_1 A}\right\| = \left\|\mathbf{r}^{S_2 A}\right\| = R.$$

But, since

$$\mathbf{r}^{S_1 A} = \mathbf{r}^{WA} \text{ and } \mathbf{r}^{S_2 A} = \mathbf{r}^{S_2 W} + \mathbf{r}^{WA} = -\mathbf{r}^{WS_2} + \mathbf{r}^{WA},$$

we find

$$f_1(q_1, q_2, q_3) = \sqrt{q_1^2 + q_2^2 + q_3^2} - R = 0$$

2

and

$$f_2(q_1, q_2, q_3) = \sqrt{q_1^2 - 3Rq_1 + \frac{9}{4}R^2 + q_2^2 + q_3^2} - R = 0.$$

Since

$$\begin{vmatrix} \frac{\partial f_1}{\partial q_2} & \frac{\partial f_1}{\partial q_3} \\ \frac{\partial f_2}{\partial q_2} & \frac{\partial f_2}{\partial q_3} \end{vmatrix} = 0,$$

every choice of values for $q_1$, $q_2$, and $q_3$ that satisfies the constraint equations is singular relative to $q_2$ and $q_3$. In contrast, the choice

$$q_1 = \frac{3R}{4}, q_2 = \frac{\sqrt{7}R}{4}, q_3 = 0$$

is regular relative to $q_1$ and $q_2$, since

$$\begin{vmatrix} \frac{\partial f_1}{\partial q_1} & \frac{\partial f_1}{\partial q_2} \\ \frac{\partial f_2}{\partial q_1} & \frac{\partial f_2}{\partial q_2} \end{vmatrix} \left(\frac{3R}{4}, \frac{\sqrt{7}R}{4}, 0\right) = \frac{3\sqrt{7}}{8} \neq 0.$$

Exercise 4.11 shows that every solution to the constraint equations is regular relative to $q_1$ and $q_2$ or $q_1$ and $q_3$. The constrained bead thus only has a single geometric degree of freedom.

### 4.2.5 Notation and Terminology

A variable that is used to describe the configuration of a rigid body is called a *configuration coordinate*. I consistently use the symbol $q$ to denote configuration coordinates. To distinguish between different configuration coordinates, I use various subscripts and embellishments, such as $\tilde{q}_1$ or $q_{\text{ball}}$. When modeling your mechanism of choice, you may prefer to pick symbols that better reflect the physical or geometric meaning of a configuration coordinate. Instead of $q_3$, *elbowangle* may be a preferred name for a variable that controls the angle between the upper and lower arm of a human. It is never an easy task to name variables. The $q$ notation offers simplicity and lucidity.

The smallest number of configuration coordinates needed to describe the configuration of a rigid body is the number of *geometric degrees of freedom* of the rigid body. When the number of configuration coordinates equals the number of geometric degrees of freedom, the configuration coordinates are commonly called *generalized coordinates*. The latter terminology fails to convey information about what the coordinates are used to describe. Moreover, it is often advantageous to retain more configuration coordinates than the number of degrees of freedom, making the term "generalized coordinates" describe the exception rather than the rule. For these reasons, I consistently refer to variables used to describe the configuration of a rigid body as configuration coordinates.

Any condition on the configuration coordinates that can be formulated as an equation is an example of a *configuration constraint*. Thus,

$$q_1 \sin q_2 - q_3 = 0$$

is a configuration constraint, while

$$q_1 q_2 \geq 0$$

is not. Values of the configuration coordinates that satisfy all configuration constraints are said to correspond to *allowable configurations*.

For most naturally occurring configuration constraints, it is theoretically possible to solve the configuration constraints for some of the configuration coordinates in terms of the others. It follows that, typically, the imposition of configuration constraints reduces the number of required configuration coordinates, i.e., the number of geometric degrees of freedom. That it is theoretically possible to solve the configuration constraints does not imply that it is always practical, desirable, or necessary.

The notation

$${}^{\mathcal{W}}A$$

was introduced in the previous section for the coordinate representation of the point $A$ relative to the observer $\mathcal{W}$. Specifically,

$${}^{\mathcal{W}}A \stackrel{def}{=} {}^{w}\left(\mathbf{r}^{WA}\right),$$

where $W$ and $w$ are the reference point and right-handed, orthonormal basis associated with $\mathcal{W}$.

Now, let $A$ and $B$ be two points on a rigid body whose position is free to change but whose orientation is constant relative to $\mathcal{W}$. It follows that the position vector $\mathbf{r}^{AB}$ is constant while $\mathbf{r}^{WA}$ and $\mathbf{r}^{WB}$ both change with time. In fact,

$$\mathbf{r}^{WB} = \mathbf{r}^{WA} + \mathbf{r}^{AB},$$

i.e.,

$${}^{\mathcal{W}}B = {}^{\mathcal{W}}A + {}^{w}\left(\mathbf{r}^{AB}\right).$$

It follows that if three configuration coordinates suffice to describe the position of a point $A$ on the rigid body relative to $\mathcal{W}$, then they also suffice to describe the position of any arbitrary point $B$ on the rigid body relative to $\mathcal{W}$.

### 4.2.6 Multiple Moving Rigid Bodies

In the absence of any rotations, a mechanism consisting of two freely moving rigid bodies has six geometric degrees of freedom, since each of the rigid bodies has three geometric degrees of freedom. If fewer than six configuration coordinates suffice to describe the positions of the two rigid bodies, then the mechanism's configuration is constrained.

Suppose you want to describe the motion of a double pendulum – two small beads connected through an inextensible string and with one of the beads suspended from a stationary supporting plane through another inextensible string.

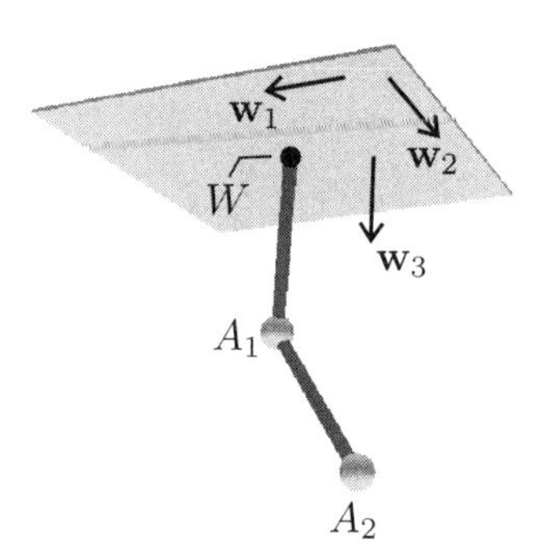


Here, the supporting plane is parallel to the $\mathbf{w}_1$ and $\mathbf{w}_2$ vectors, with $\mathbf{w}_3$ pointing away from the plane in the direction of the beads. Introduce a main observer $\mathcal{W}$, relative to which the support is stationary and with reference point $W$ at the point where the pendulum is attached. Since the positions of the beads relative to $\mathcal{W}$ change with time, introduce two auxiliary observers $\mathcal{A}_1$ and $\mathcal{A}_2$, relative to which the upper and lower beads, respectively, are stationary, and such that the reference points $A_1$ and $A_2$ coincide with the centers of the upper and lower beads, respectively.

We shall disregard any changes in orientation of the beads relative to $\mathcal{W}$ during the motion. The configuration of the observer $\mathcal{A}_1$ relative to $\mathcal{W}$ is then described by a pure translation $\mathbf{T}_{\mathcal{W}\to\mathcal{A}_1}$ corresponding to the position vector

$$\mathbf{r}^{WA_1}.$$

Since the strings are inextensible, the upper bead's position is constrained to the surface of a sphere centered on $W$ with radius equal to the length $l_1$ of the upper string. Using spherical coordinates, it follows that

$$\mathbf{r}^{WA_1} = w \begin{pmatrix} l_1 \sin q_1 \cos q_2 \\ l_1 \sin q_1 \sin q_2 \\ l_1 \cos q_1 \end{pmatrix}.$$

The configuration of the observer $\mathcal{A}_2$ relative to $\mathcal{A}_1$ is described by a pure translation $\mathbf{T}_{\mathcal{A}_1 \to \mathcal{A}_2}$ corresponding to the position vector

$$\mathbf{r}^{A_1 A_2}.$$

Since the strings are inextensible, the lower bead's position is constrained to the surface of a sphere centered on $A_1$, with radius equal to the length $l_2$ of the lower string. Using spherical coordinates, we may write

$$\mathbf{r}^{A_1 A_2} = w \begin{pmatrix} l_2 \sin q_3 \cos q_4 \\ l_2 \sin q_3 \sin q_4 \\ l_2 \cos q_3 \end{pmatrix}.$$

It follows that the configuration of the observer $\mathcal{A}_2$ relative to $\mathcal{W}$ is described by a pure translation $\mathbf{T}_{\mathcal{W} \to \mathcal{A}_2}$ corresponding to the position vector

$$\begin{aligned} \mathbf{r}^{W A_2} &= \mathbf{r}^{W A_1} + \mathbf{r}^{A_1 A_2} \\ &= w \begin{pmatrix} l_1 \sin q_1 \cos q_2 + l_2 \sin q_3 \cos q_4 \\ l_1 \sin q_1 \sin q_2 + l_2 \sin q_3 \sin q_4 \\ l_1 \cos q_1 + l_2 \cos q_3 \end{pmatrix}. \end{aligned}$$

We conclude that, in the absence of rotation, the double pendulum has four geometric degrees of freedom.

**Illustration 4.9**

In terms of Cartesian coordinates, the position vectors $\mathbf{r}^{W A_1}$ and $\mathbf{r}^{W A_2}$ are

$$\mathbf{r}^{W A_1} = w \begin{pmatrix} \tilde{q}_1 \\ \tilde{q}_2 \\ \tilde{q}_3 \end{pmatrix} \text{ and } \mathbf{r}^{W A_2} = w \begin{pmatrix} \tilde{q}_4 \\ \tilde{q}_5 \\ \tilde{q}_6 \end{pmatrix}.$$

The constraint on the upper bead implies that

$$\left\| \mathbf{r}^{W A_1} \right\| = l_1,$$

i.e.,

$$f_1 \left( \tilde{q}_1, \tilde{q}_2, \tilde{q}_3, \tilde{q}_4, \tilde{q}_5, \tilde{q}_6 \right) = \sqrt{\tilde{q}_1^2 + \tilde{q}_2^2 + \tilde{q}_3^2} - l_1 = 0.$$

Similarly, the constraint on the lower bead implies that

$$\left\| \mathbf{r}^{A_1 A_2} \right\| = l_2,$$

i.e.,

$$f_2 \left( \tilde{q}_1, \tilde{q}_2, \tilde{q}_3, \tilde{q}_4, \tilde{q}_5, \tilde{q}_6 \right) = \sqrt{\left( \tilde{q}_4 - \tilde{q}_1 \right)^2 + \left( \tilde{q}_5 - \tilde{q}_2 \right)^2 + \left( \tilde{q}_6 - \tilde{q}_3 \right)^2} - l_2 = 0,$$

2

since

$$\mathbf{r}^{A_1 A_2} = \mathbf{r}^{A_1 W} + \mathbf{r}^{W A_2} = -\mathbf{r}^{W A_1} + \mathbf{r}^{W A_2}.$$

Since

$$\begin{vmatrix} \frac{\partial f_1}{\partial \tilde{q}_4} & \frac{\partial f_1}{\partial \tilde{q}_6} \\ \frac{\partial f_2}{\partial \tilde{q}_4} & \frac{\partial f_2}{\partial \tilde{q}_6} \end{vmatrix} = 0$$

for all choices of values of $\tilde{q}_1, \ldots, \tilde{q}_6$, every solution to the constraint equations is singular relative to $\tilde{q}_4$ and $\tilde{q}_6$. On the other hand,

$$\begin{vmatrix} \frac{\partial f_1}{\partial \tilde{q}_1} & \frac{\partial f_1}{\partial \tilde{q}_2} \\ \frac{\partial f_2}{\partial \tilde{q}_1} & \frac{\partial f_2}{\partial \tilde{q}_2} \end{vmatrix} = \frac{\tilde{q}_2 \tilde{q}_4 - \tilde{q}_1 \tilde{q}_5}{l_1 l_2},$$

which differs from 0 as long as $\tilde{q}_2 \tilde{q}_4 - \tilde{q}_1 \tilde{q}_5 \neq 0$. When this is the case, the configuration coordinates $\tilde{q}_1$ and $\tilde{q}_2$ may be eliminated from the description of the configuration of the double pendulum and replaced by a function of the remaining configuration coordinates. Again, we conclude that, in the absence of rotation, the double pendulum has four geometric degrees of freedom.

## 4.3 MAMBO

The relative position of two observers can be uniquely described through a pure translation, given the selection of a reference point for each of the observers. If the observer $\mathcal{A}$ has the reference point $A$ and the observer $\mathcal{B}$ has the reference point $B$, then the position vector $\mathbf{r}^{AB}$ uniquely describes the pure translation $\mathbf{T}_{\mathcal{A} \to \mathcal{B}}$ between $\mathcal{A}$ and $\mathcal{B}$. Given a right-handed, orthonormal basis $w$, the position vector $\mathbf{r}^{AB}$ may be uniquely represented by its matrix representation relative to $w$,

$$^{w}\left(\mathbf{r}^{AB}\right).$$

### 4.3.1 A Still Life

In a MAMBO geometry description, the specification of a position vector between the reference points of successive observers is given through a **POINT** statement containing the matrix representation of the vector relative to some right-handed, orthonormal basis. In the absence of additional information, MAMBO assumes that all references to the matrix representation of a vector are made relative to a common right-handed, orthonormal basis, say $w$.

**Illustration 4.10**

The following extract from a MAMBO .geo file shows the use of the **POINT** statement to describe the relative position of successive observers:

```
MODULE W {
   BODY A {
      POINT {1,2,3}
      BODY B {
         POINT {0,0,1}
      }
   }
}
```

Here, the position of the observer $\mathcal{A}$ relative to the observer $\mathcal{W}$ is given by a pure translation $\mathbf{T}_{\mathcal{W}\to\mathcal{A}}$ corresponding to the position vector

$$\mathbf{r}^{WA} =_w \begin{pmatrix} 1 \\ 2 \\ 3 \end{pmatrix},$$

where $A$ and $W$ are the reference points of $\mathcal{A}$ and $\mathcal{W}$, respectively.

Similarly, the position of the observer $\mathcal{B}$ relative to the observer $\mathcal{A}$ is given by a pure translation $\mathbf{T}_{\mathcal{A}\to\mathcal{B}}$ corresponding to the position vector

$$\mathbf{r}^{AB} =_w \begin{pmatrix} 0 \\ 0 \\ 1 \end{pmatrix},$$

where $B$ is the reference point of $\mathcal{B}$.

By default, MAMBO interprets the absence of a **POINT** statement to be equivalent to the specification

```
POINT {0,0,0}
```

i.e., that the reference point of the current observer coincides with that of the parent observer.

The position of a rigid body relative to some observer can be uniquely described through a pure translation, given the selection of a reference point for the rigid body and a reference point for the observer. If the observer $\mathcal{A}$ has the reference point $A$ and the point $B$ is the reference point of the rigid body, then the position vector $\mathbf{r}^{AB}$ uniquely describes the pure translation $\mathbf{T}_{\mathcal{A}}$ between $\mathcal{A}$ and the rigid body. Given a right-handed, orthonormal basis $w$, the position vector $\mathbf{r}^{AB}$ may be uniquely represented by its matrix representation relative to $w$,

$$^{w}\left(\mathbf{r}^{AB}\right).$$

**Illustration 4.11**

We may use the **POINT** statement to describe the position vector between the reference point of an observer and the reference point of a rigid body.

```
MODULE W {
   BODY A {
      POINT {1,2,3}
      BODY B {
         POINT {0,0,1}
         CYLINDER {
            POINT {0,1,0}
         }
      }
      SPHERE {
         POINT {-1,0,2}
      }
   }
}
```

Here, the position of the sphere relative to the observer $\mathcal{A}$ is given by a pure translation $\mathbf{T}_{\mathcal{A}}$ corresponding to the position vector

$$\mathbf{r}^{AS} = w\begin{pmatrix} -1 \\ 0 \\ 2 \end{pmatrix},$$

where $S$ is the reference point of the sphere (assumed by MAMBO to be at the center of the sphere).

Similarly, the position of the cylinder relative to the observer $\mathcal{B}$ is given by a pure translation $\mathbf{T}_{\mathcal{B}}$ corresponding to the position vector

$$\mathbf{r}^{BC} = w\begin{pmatrix} 0 \\ 1 \\ 0 \end{pmatrix},$$

where $C$ is the reference point of the cylinder (assumed by MAMBO to be at the center of the cylinder).

The tree structure corresponding to the geometry description in the last illustration has the following form:

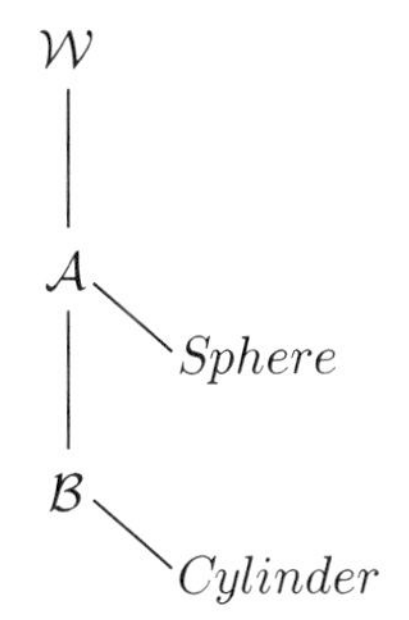


We could represent the same arrangement of rigid bodies relative to the $\mathcal{W}$ observer by relating the configuration of the sphere to the $\mathcal{B}$ observer.

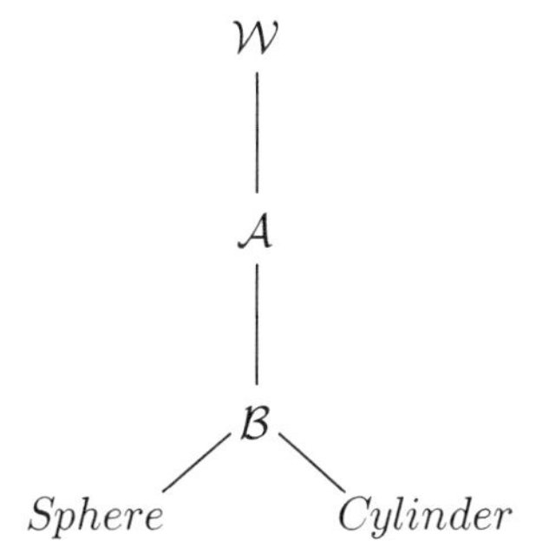


The corresponding MAMBO geometry description becomes

```
MODULE W {
   BODY A {
      POINT {1,2,3}
      BODY B {
         POINT {0,0,1}
         CYLINDER {
            POINT {0,1,0}
         }
         SPHERE {
            POINT {-1,0,1}
         }
      }
   }
}
```

Here, the **POINT** statement relating the pure translation between the $\mathcal{B}$ observer and the sphere is obtained from the following computation:

$$\begin{aligned} \mathbf{r}^{BS} &= \mathbf{r}^{BA} + \mathbf{r}^{AS} = -\mathbf{r}^{AB} + \mathbf{r}^{AS} \\ &= -w \begin{pmatrix} 0 \\ 0 \\ 1 \end{pmatrix} + w \begin{pmatrix} -1 \\ 0 \\ 2 \end{pmatrix} = w \begin{pmatrix} -1 \\ 0 \\ 1 \end{pmatrix}. \end{aligned}$$

The $\mathcal{B}$ observer may be entirely eliminated from the observer tree structure.

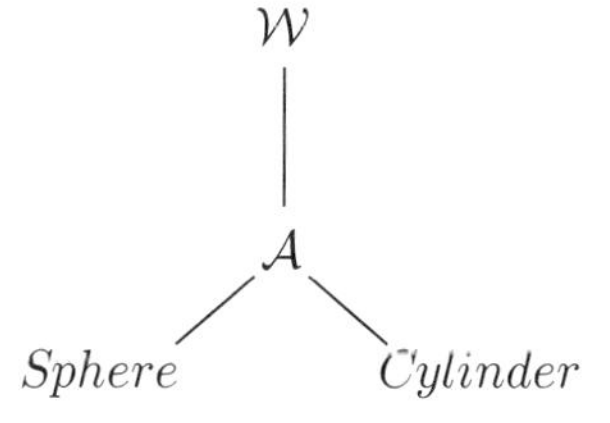


The corresponding MAMBO geometry description becomes

**MODULE** W {
  **BODY** A {
    **POINT** {1,2,3}
    **CYLINDER** {
      **POINT** {0,1,1}
    }
    **SPHERE** {
      **POINT** {-1,0,2}
    }
  }
}

Here, the **POINT** statement relating the pure translation between the $\mathcal{A}$ observer and the cylinder is obtained from the following computation:

$$\begin{aligned} \mathbf{r}^{AC} &= \mathbf{r}^{AB} + \mathbf{r}^{BC} \\ &= w \begin{pmatrix} 0 \\ 0 \\ 1 \end{pmatrix} + w \begin{pmatrix} 0 \\ 1 \\ 0 \end{pmatrix} = w \begin{pmatrix} 0 \\ 1 \\ 1 \end{pmatrix}. \end{aligned}$$

Any real number in the MAMBO geometry description may be replaced by a string of characters. For example, the **POINT** statement

**POINT** {p1,p2,p3}

uses the three labels `p1`, `p2`, and `p3` in place of numbers. The statement

```
parameters p1,p2,p3;
```

in a MAMBO motion description (a MAMBO .dyn file) establishes `p1`, `p2`, and `p3` as MAMBO *parameters*, quantities that can be changed interactively within the MAMBO application, but that do not change during an animation.

## 4.3.2 Dabbling with Motion

The **POINT** statements in the examples above all contained real numbers. Every such real number could be replaced by a mathematical expression that would evaluate to a real number. Thus, for example,

**POINT** {cos(.5)*cos(.5),sin(.5)*sin(.5),0}

is a syntactically correct statement. In addition to using real numbers or parameters inside such mathematical expressions, it is possible to include a variable corresponding to time in a MAMBO animation. The label selected to represent time is specified in a MAMBO .dyn file through the statement

```
time NAME;
```

where `NAME` represents the label. Thus, including the statement

```
time t;
```

at the top of the MAMBO .dyn file allows you to include the variable name `t` anywhere in the geometry description.

**Illustration 4.12**

In the MAMBO geometry description below, the variable `t` is used to represent the internal time variable of a MAMBO animation.

```
MODULE W {
   BODY A {
      POINT {cos(t),2-t,3}
      BODY B {
         POINT {0,t+3,1}
         CYLINDER {
            POINT {0,1,0}
         }
      }
      SPHERE {
         POINT {exp(-t),0,2}
      }
   }
}
```

From this geometry description, it follows that the position vector corresponding to the pure translation from the $\mathcal{W}$ observer to the sphere is given by

$$\begin{aligned}\mathbf{r}^{WS} &= \mathbf{r}^{WA} + \mathbf{r}^{AS} \\ &= w\begin{pmatrix}\cos t \\ 2-t \\ 3\end{pmatrix} + w\begin{pmatrix}e^{-t} \\ 0 \\ 2\end{pmatrix} = w\begin{pmatrix}\cos t + e^{-t} \\ 2-t \\ 5\end{pmatrix}.\end{aligned}$$

Similarly, the position vector corresponding to the pure translation from the $\mathcal{W}$ observer to the cylinder is given by

$$\begin{aligned}\mathbf{r}^{WC} &= \mathbf{r}^{WA} + \mathbf{r}^{AB} + \mathbf{r}^{BC} \\ &= w\begin{pmatrix}\cos t \\ 2-t \\ 3\end{pmatrix} + w\begin{pmatrix}0 \\ t+3 \\ 1\end{pmatrix} + w\begin{pmatrix}0 \\ 1 \\ 0\end{pmatrix} \\ &= w\begin{pmatrix}\cos t \\ 6 \\ 4\end{pmatrix}.\end{aligned}$$

Importing the corresponding .dyn and .geo files into the MAMBO application, running a simulation, and subsequently animating the computed dataset shows the motions of the sphere and the cylinder relative to the $\mathcal{W}$ observer as functions of time.

MAMBO's internal time variable may also be used to generate the impression of a panning motion of a camera viewing a static (or changing) scene. This is achieved by the introduction of an auxiliary observer representing the camera immediately under the world observer and containing the rest of the scene within its corresponding **BODY** block.

In the following geometry description, the camera observer has been inserted between the $\mathcal{W}$ observer and the rest of the tree structure.

```
MODULE W {
  BODY Camera {
    POINT {cos(t),sin(t),0}
    BODY A {
      POINT {1,2,3}
      BODY B {
        POINT {0,0,1}
        CYLINDER {
          POINT {0,1,0}
        }
```

```
            }
            SPHERE {
                POINT {-1,0,1}
            }
        }
    }
}
```

The explicit time-dependent position vector corresponding to the pure translation from the $\mathcal{W}$ observer to the camera observer results in a motion of the entire scene relative to the $\mathcal{W}$ observer. To the viewer, the motion of the scene relative to the screen gives the appearance of a camera motion about an unchanging scene.

MAMBO parameters were introduced to replace real numbers in a MAMBO geometry description. These parameters could be changed interactively, but remained unchanged during an animation. To replace functional expressions involving MAMBO parameters, real numbers, and the MAMBO time variable, we can introduce MAMBO *animated variables*. For example, we may replace the statement

```
POINT {cos(t)*p1,p2*t,0}
```

in a MAMBO geometry description by the statement

```
POINT {a1,a2,0}
```

where `a1` and `a2` have been introduced to replace the expressions `cos(t)*p1` and `p2*t`, respectively. The explicit nature of the animated variables `a1` and `a2` is specified in the corresponding MAMBO motion description

```
time t;
parameters p1,p2;
anims {
      a1 = cos(t)*p1;
      a2 = p2*t;
}
```

Animated variables cannot be changed interactively and are only meant to simplify the textual complexity of the MAMBO .geo file.

### 4.3.3 Multibody Mechanisms

MAMBO parameters represent quantities that can be changed interactively in the application but that do not change with time. It is also possible to introduce quantities that can be changed interactively in the application and that may change with time. These are called MAMBO *state variables* and are introduced in a MAMBO .dyn file through a statement

```
states name1,name2;
```

where `name1` and `name2` are labels for the state variables.

**Illustration 4.13**

The MAMBO geometry description

**MODULE** W {
  **BODY** A {
    **POINT** {q1,q2,q3}
    **BODY** B {
      **POINT** {0,0,1}
      **CYLINDER** {
        **POINT** {0,1,0}
      }
    }
    **SPHERE** {
      **POINT** {-1,0,1}
    }
  }
}

together with the statement

```
states q1,q2,q3;
```

in the corresponding MAMBO .dyn file specifies that the pure translation relating the configuration of the $\mathcal{A}$ observer to the $\mathcal{W}$ observer corresponds to the position vector

$$\mathbf{r}^{WA} = w \begin{pmatrix} q_1 \\ q_2 \\ q_3 \end{pmatrix},$$

where the three state variables can be changed interactively within the MAMBO application and can change with time during an animation.

MAMBO state variables can be used at any level in a MAMBO geometry description. We recommend that the reader avoid MAMBO state variables in the **POINT** statements within object blocks. This agrees with the suggestion that all motion of objects relative to the world observer be contained in the motion of auxiliary observers relative to the world observer. The **POINT** statement of a rigid body relative to its parent observer will therefore only contain real numbers, MAMBO parameters, or MAMBO animated variables that are constant.

The time history of a MAMBO state variable can be specified in two ways. On the one hand, a MAMBO dataset (a MAMBO .sds file) may be generated containing the values of any MAMBO state variables at discrete moments in time. For example, the MAMBO .geo and .dyn files corresponding to the above illustration, together with a MAMBO .sds file with the content shown in the table below, can be used to generate a motion of the sphere and cylinder objects.

| t | q1 | q2 | q3 |
|---|---|---|---|
| 0.0 | 1.00 | 0.00 | 1.00 |
| 0.10 | 0.90 | 0.05 | 1.10 |
| 0.20 | 0.80 | 0.15 | 1.00 |
| 0.30 | 0.70 | 0.40 | 0.90 |
| 0.40 | 0.60 | 0.70 | 1.00 |

More useful for purposes that will become clear in later chapters is to let the MAMBO state variables change with time according to a set of differential equations. In particular, we require that there be as many differential equations as the number of MAMBO state variables. When this condition is satisfied, the differential equations constitute a set of *kinematic differential equations.*

Suppose, for example, that the MAMBO state variables $q_1$, $q_2$, and $q_3$ satisfy the following set of kinematic differential equations:

$$\begin{aligned} \dot{q}_1 + \dot{q}_2 &= u_1(t), \\ q_2\dot{q}_1 + \dot{q}_3 &= u_2(t), \\ \dot{q}_2 q_1^2 &= u_3(t), \end{aligned}$$

where $u_1(t)$, $u_2(t)$, and $u_3(t)$ are some yet-to-be-specified functions of time. We can rewrite these equations in matrix form:

$$\begin{pmatrix} 1 & 1 & 0 \\ q_2 & 0 & 1 \\ 0 & q_1^2 & 0 \end{pmatrix} \begin{pmatrix} \dot{q}_1 \\ \dot{q}_2 \\ \dot{q}_3 \end{pmatrix} = \begin{pmatrix} u_1(t) \\ u_2(t) \\ u_3(t) \end{pmatrix}.$$

The corresponding MAMBO motion description would then include the statements

```
states q1,q2,q3;
time t;
insignals {
    u1 = 1;
    u2 = cos(t);
    u3 = 0;
}
ode {
    rhs[q1] = u1;
    rhs[q2] = u2;
    rhs[q3] = u3;
```

```
    mass[q1][q1] = 1;
    mass[q1][q2] = 1;
    mass[q1][q3] = 0;
    mass[q2][q1] = q2;
    mass[q2][q2] = 0;
    mass[q2][q3] = 1;
    mass[q3][q1] = 0;
    mass[q3][q2] = q1^2;
    mass[q3][q3] = 0;
}
```

Here, $u_1(t) = 1$, $u_2(t) = \cos t$, and $u_3(t) = 0$ as specified in the `insignals` block. The `ode` block contains information about the coefficient matrix (`mass`) and the right-hand side (`rhs`) of the kinematic differential equations. The MAMBO state variables are used to label the rows and columns of these matrices. Since the order of the equations is irrelevant, the row indices can be permuted arbitrarily. Moreover, by default, there is no need to include matrix entries that equal zero. Thus, an equivalent `ode` block could read

```
ode {
    rhs[q2] = u1;
    rhs[q1] = u2;
    rhs[q3] = u3;
    mass[q2][q1] = 1;
    mass[q2][q2] = 1;
    mass[q1][q1] = q2;
    mass[q1][q3] = 1;
    mass[q3][q2] = q1^2;
}
```

The kinematic differential equations above may be solved for the rates of change of the MAMBO state variables

$$\begin{pmatrix} \dot{q}_1 \\ \dot{q}_2 \\ \dot{q}_3 \end{pmatrix} = \begin{pmatrix} 1 & 1 & 0 \\ q_2 & 0 & 1 \\ 0 & q_1^2 & 0 \end{pmatrix}^{-1} \begin{pmatrix} u_1(t) \\ u_2(t) \\ u_3(t) \end{pmatrix},$$

provided that the inverse

$$\begin{pmatrix} 1 & 1 & 0 \\ q_2 & 0 & 1 \\ 0 & q_1^2 & 0 \end{pmatrix}^{-1}$$

exists, i.e., provided that the matrix

$$\begin{pmatrix} 1 & 1 & 0 \\ q_2 & 0 & 1 \\ 0 & q_1^2 & 0 \end{pmatrix}$$

is *non-singular*. Indeed, the matrix is non-singular as long as its determinant is non-zero. Here,

$$\begin{vmatrix} 1 & 1 & 0 \\ q_2 & 0 & 1 \\ 0 & q_1^2 & 0 \end{vmatrix} = -q_1^2,$$

which is non-zero as long as $q_1 \neq 0$. We formalize this observation by stating that the kinematic differential equations are non-singular as long as $q_1 \neq 0$. Indeed, as long as $q_1(t_0) \neq 0$, for some moment in time $t_0$, the kinematic differential equations may be solved for $q_1(t)$, $q_2(t)$, and $q_3(t)$ on some interval in time containing $t_0$. This is achieved by the MAMBO application through the use of advanced numerical methods for the approximate solution of the kinematic differential equations.

## 4.4 The MAMBO Toolbox

(Ex. 4.15 – Ex. 4.16)

### 4.4.1 Points

Points are represented within the MAMBO toolbox by entries in the global variables `GlobalPointDeclarations` and `GlobalPointDefinitions`. Changes to these variables initiated by the user are made possible through the procedures `DeclarePoints` and `DefinePoints`.

**Illustration 4.14**

In the following MAMBO toolbox session, the points $A$, $B$, and $C$ and the right-handed, orthonormal basis $a$ are declared to the program.

```
> Restart():
> DeclarePoints(A,B,C):
> DeclareTriads(a):
```

The statement

```
> DefinePoints([A,B,a,1,2,0],[B,C,a,0,2,1]):
```

establishes the position vectors

$$\mathbf{r}^{AB} = a \begin{pmatrix} 1 \\ 2 \\ 0 \end{pmatrix} \text{ and } \mathbf{r}^{BC} = a \begin{pmatrix} 0 \\ 2 \\ 1 \end{pmatrix}.$$

The effect of these statements on the global variables `GlobalPointDeclarations` and `GlobalPointDefinitions` is made clear by the following statements:

```
> print(GlobalPointDeclarations);
> print(GlobalPointDefinitions);
```

table([
$A = \{B\}$
$B = \{A, C\}$
$C = \{B\}$
])

table([
$(B, A) =$ table([
$1 =$ table([
"Coordinates" $= [-1, -2, 0]$
"Triad" $= a$
])
"Size" $- 1$
"Type" = "Vector"
])
$(A, B) =$ table([
$1 =$ table([
"Coordinates" $= [1, 2, 0]$
"Triad" $= a$
])
"Size" $= 1$
"Type" = "Vector"
])
$(C, B) =$ table([
$1 =$ table([
"Coordinates" $= [0, -2, -1]$
"Triad" $= a$
])
"Size" $= 1$
"Type" = "Vector"
])
$(B, C) =$ table([
$1 =$ table([
"Coordinates" $= [0, 2, 1]$
"Triad" $= a$
])
"Size" $= 1$
"Type" = "Vector"
])
])

The contents of the global variables `GlobalPointDeclarations` and `GlobalPointDefinitions` reflect the fact that the relative positions of the points $A$ and $B$ and the points $B$ and $C$, respectively, are now known. While `GlobalPointDeclarations` tracks all **direct** relations between points, `GlobalPointDefinitions` stores any position vectors between points in `GlobalPointDeclarations` that have been computed during a MAMBO toolbox session.

The MAMBO toolbox procedure `FindTranslation` can be invoked to compute the position vector between any two points that are declared and are (at worst, indirectly) related. Continuing with the MAMBO toolbox session in the illustration, we find

```
> FindTranslation(A,C);
```

table([
"Size" = 1
"Type" = "Vector"
1 = table([
"Triad" = $a$
"Coordinates" = [1, 4, 1]
])
])

as follows from

$$\begin{aligned}\mathbf{r}^{AC} &= \mathbf{r}^{AB} + \mathbf{r}^{BC} \\ &= a\begin{pmatrix}1\\2\\0\end{pmatrix} + a\begin{pmatrix}0\\2\\1\end{pmatrix} = a\begin{pmatrix}1\\4\\1\end{pmatrix}.\end{aligned}$$

The global variable `GlobalPointDefinitions` is automatically appended with the position vectors $\mathbf{r}^{AC}$ and $\mathbf{r}^{CA} = -\mathbf{r}^{AC}$.

### 4.4.2 Observers

To associate a point and a right-handed, orthonormal basis with an observer, the MAMBO toolbox employs the `DefineObservers` procedure. For example, if $A$, $B$, and $W$ are the reference points of three observers $\mathcal{A}$, $\mathcal{B}$, and $\mathcal{W}$, such that

$$\mathbf{r}^{WA} = a\begin{pmatrix}1\\2\\3\end{pmatrix} \text{ and } \mathbf{r}^{AB} = a\begin{pmatrix}0\\0\\1\end{pmatrix},$$

where $a$ is a right-handed, orthonormal basis associated with all three observers, then the following statements provide the necessary information to the MAMBO toolbox.

```
> Restart():
> DeclareObservers(W,A,B):
> DeclarePoints(W,A,B):
> DeclareTriads(a):
> DefinePoints([W,A,a,1,2,3],[A,B,a,0,0,1]):
> DefineObservers([W,W,a],[A,A,a],[B,B,a]):
> print(GlobalObserverDefinitions);
```

table([
$B$ = table([
"Point" = $B$
"Triad" = $a$
])
W = table([
"Point" = W
"Triad" = $a$
])
$A$ = table([
"Point" = $A$
"Triad" = $a$
])
])

The MAMBO toolbox procedure `FindPosition` can be invoked to compute the position vector between the reference points of two observers. Similarly, the `FindCoordinates` procedure computes the coordinates of a point relative to an observer. These commands are illustrated in the following statements:

```
> FindPosition(W,B);
```

table([
"Size" = 1
1 = table([
"Coordinates" = [1, 2, 4]
"Triad" = $a$
])
"Type" = "Vector"
])

```
> FindCoordinates(W,B);
```

$$[1, 2, 4]$$

We may again use the DefineNeighbors and GeometryOutput commands to generate a MAMBO geometry description using the observers introduced above. For example, the geometry description obtained from the commands

```
> DefineNeighbors([W,A],[A,B]):
> GeometryOutput(main=W);

MODULE W {
   BODY A {
      POINT {1,2,3}
      ORIENT {1,0,0,0,1,0,0,0,1}
      BODY B {
         POINT {0,0,1}
         ORIENT {1,0,0,0,1,0,0,0,1}
      }
   }
}
```

is identical to that discussed in Illustration 4.10[2].

**Illustration 4.15**

As in the previous chapter, we may reorganize the observers so as to promote $\mathcal{A}$ to be the main observer:

```
> GeometryOutput(main=A);

MODULE A {
   BODY B {
      POINT {0,0,1}
      ORIENT {1,0,0,0,1,0,0,0,1}
   }
   BODY W {
      POINT {-1,-2,-3}
      ORIENT {1,0,0,0,1,0,0,0,1}
   }
}
```

or

```
> Undo():
> DefineNeighbors([W,B],[A,B]):
> GeometryOutput(main=A);

MODULE A {
   BODY B {
      POINT {0,0,1}
      ORIENT {1,0,0,0,1,0,0,0,1}
```

[2]We will discuss the **ORIENT** statement in detail in Chapter 6.

```
        BODY W {
           POINT {-1,-2,-4}
           ORIENT {1,0,0,0,1,0,0,0,1}
        }
     }
  }
```

where we used the `Undo` utility to undo the latest change in any of the global variables.

### 4.4.3 A Sample Project

Suppose you want to visualize the motion of a small spherical bead along the edges of a wireframe representation of a stationary rectangular block, as depicted below.

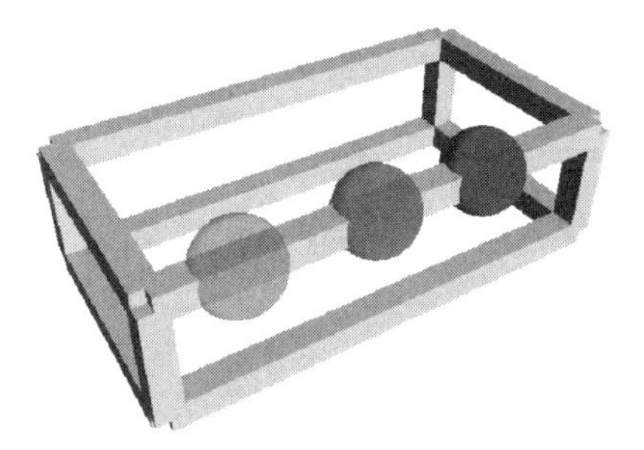

As in an earlier section, we decompose the wireframe structure into 12 rigid edges, four of which are parallel to the $\mathbf{w}_1$ basis vector, four of which are parallel to the $\mathbf{w}_2$ basis vector, and four of which are parallel to the $\mathbf{w}_3$ basis vector. The edges are labeled by pairs of integers $[i, j]$, corresponding to the $j$-th edge parallel to the $i$-th basis vector of $w$ as indicated in the figure.

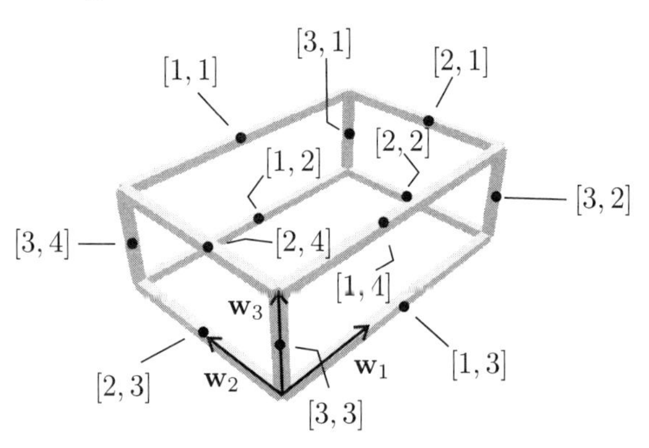

We introduce a main observer $\mathcal{W}$ with reference point $W$ at the center of the block and let the reference point $E_{[i,j]}$ of the $[i,j]$-th edge correspond to the geometric center of the edge. The following MAMBO toolbox statements declare the corresponding labels and define the observer $\mathcal{W}$:

```
> Restart():
> DeclareObservers(W):
> DeclarePoints(W,seq(seq(cat(E,i,j),i=1..3),j=1..4)):
> DeclareTriads(w):
> DefineObservers([W,W,w]):
```

The configuration of the $[i,j]$-th edge relative to $\mathcal{W}$ is then given by a pure translation $\mathbf{T}_{[i,j]}$ corresponding to the position vector

$$\mathbf{r}^{WE_{[i,j]}}.$$

Specifically, we find

$$\mathbf{r}^{WE_{[1,1]}} = w\begin{pmatrix} 0 \\ l_2/2 \\ l_3/2 \end{pmatrix}, \quad \mathbf{r}^{WE_{[1,2]}} = w\begin{pmatrix} 0 \\ l_2/2 \\ -l_3/2 \end{pmatrix},$$

$$\mathbf{r}^{WE_{[1,3]}} = w\begin{pmatrix} 0 \\ -l_2/2 \\ -l_3/2 \end{pmatrix}, \quad \mathbf{r}^{WE_{[1,4]}} = w\begin{pmatrix} 0 \\ -l_2/2 \\ l_3/2 \end{pmatrix},$$

$$\mathbf{r}^{WE_{[2,1]}} = w\begin{pmatrix} l_1/2 \\ 0 \\ l_3/2 \end{pmatrix}, \quad \mathbf{r}^{WE_{[2,2]}} = w\begin{pmatrix} l_1/2 \\ 0 \\ -l_3/2 \end{pmatrix},$$

$$\mathbf{r}^{WE_{[2,3]}} = w\begin{pmatrix} -l_1/2 \\ 0 \\ -l_3/2 \end{pmatrix}, \quad \mathbf{r}^{WE_{[2,4]}} = w\begin{pmatrix} -l_1/2 \\ 0 \\ l_3/2 \end{pmatrix},$$

$$\mathbf{r}^{WE_{[3,1]}} = w\begin{pmatrix} l_1/2 \\ l_2/2 \\ 0 \end{pmatrix}, \quad \mathbf{r}^{WE_{[3,2]}} = w\begin{pmatrix} l_1/2 \\ -l_2/2 \\ 0 \end{pmatrix},$$

$$\mathbf{r}^{WE_{[3,3]}} = w\begin{pmatrix} -l_1/2 \\ -l_2/2 \\ 0 \end{pmatrix}, \quad \text{and } \mathbf{r}^{WE_{[3,4]}} = w\begin{pmatrix} -l_1/2 \\ l_2/2 \\ 0 \end{pmatrix}.$$

Continuing with the same MAMBO toolbox session, the MAMBO toolbox procedure `DefinePoints` establishes the corresponding position vectors.

```
> DefinePoints(
> [W,E11,w,0,l2/2,l3/2],[W,E12,w,0,l2/2,-l3/2],
> [W,E13,w,0,-l2/2,-l3/2],[W,E14,w,0,-l2/2,l3/2],
> [W,E21,w,l1/2,0,l3/2],[W,E22,w,l1/2,0,-l3/2],
> [W,E23,w,-l1/2,0,-l3/2],[W,E24,w,-l1/2,0,l3/2],
> [W,E31,w,l1/2,l2/2,0],[W,E32,w,l1/2,-l2/2,0],
> [W,E33,w,-l1/2,-l2/2,0],[W,E34,w,-l1/2,l2/2,0]
> ):
```

To visualize the wireframe representation of the rectangular block, we need to add MAMBO objects to the geometry description. The MAMBO toolbox procedure `DefineObjects` associates the desired objects with the appropriate observer. Here, 12 MAMBO blocks are directly related to the $\mathcal{W}$ observer with reference points given by the $E_{[i,j]}$'s.

```
> DefineObjects(
> seq([W,'Block',point=cat(E,1,j),xlength=l1,
> ylength=(l1+l2+l3)/30,zlength=(l1+l2+l3)/30],j=1..4),
> seq([W,'Block',point=cat(E,2,j),xlength=(l1+l2+l3)/30,
> ylength=l2,zlength=(l1+l2+l3)/30],j=1..4),
> seq([W,'Block',point=cat(E,3,j),xlength=(l1+l2+l3)/30,
> ylength=(l1+l2+l3)/30,zlength=l3],j=1..4)):
```

Since the configuration of the spherical bead is taken to be time-dependent relative to $\mathcal{W}$, we introduce an auxiliary observer $\mathcal{A}$, relative to which the bead is stationary. In particular, we assume that the reference point $A$ of $\mathcal{A}$ coincides with the center of the bead, and that the orientation of the bead relative to $\mathcal{W}$ is described by the identity rotation. Then, the configuration of the observer $\mathcal{A}$ relative to $\mathcal{W}$ is given by the pure translation $\mathbf{T}_{\mathcal{W}\to\mathcal{A}}$ corresponding to the position vector

$$\mathbf{r}^{WA} = w \begin{pmatrix} q_1 \\ q_2 \\ q_3 \end{pmatrix},$$

where $q_1$, $q_2$, and $q_3$ are the configuration coordinates describing the bead's position relative to $\mathcal{W}$ as a function of time. The MAMBO toolbox statements

```
> DeclareObservers(A):
> DeclarePoints(A):
> DefineObservers([A,A,w]):
> DefineNeighbors([W,A]):
> DefinePoints([W,A,w,q1,q2,q3]):
> DefineObjects([A,'Sphere',radius=(l1+l2+l3)/20,
> color=red]):
```

append these definitions to the current geometry hierarchy. Finally, the statement

```
> GeometryOutput(main=W,filename="beadonblock.geo"):
```

exports the MAMBO geometry description to the file `beadonblock.geo`.

Every motion of the bead along the edges of the stationary rectangular block relative to $\mathcal{W}$ is equivalent to some time-dependence of the configuration coordinates. As suggested in the previous section, there are three distinct ways to implement this correspondence in the MAMBO application.

### Using MAMBO animated variables

MAMBO animated variables cannot be changed interactively by the user during a MAMBO session, but may change with time during a simulation. Thus, to achieve a desired motion, the configuration coordinates could be declared as animated variables with appropriate definitions for their time-dependence given within the `anims` block of the MAMBO motion description (the MAMBO .dyn file).

Suppose we want to visualize the time-dependence of the configuration coordinates shown in the figure below.

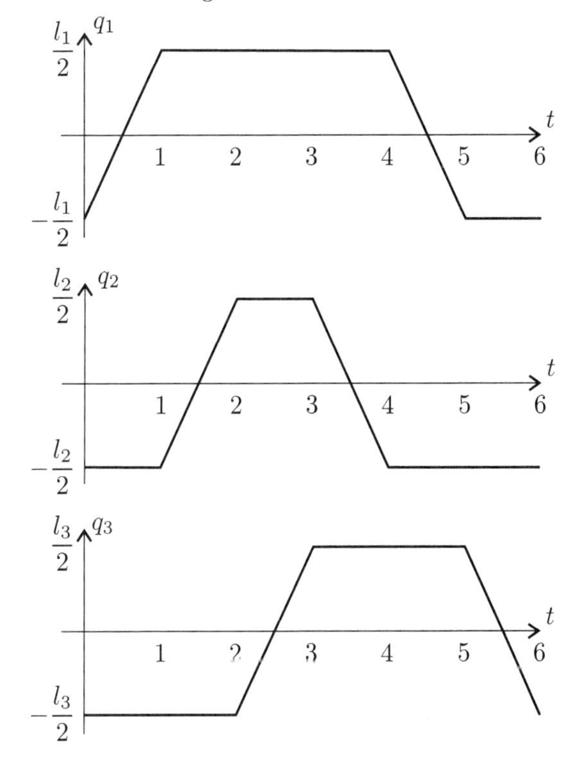


Specifically, the configuration coordinates are defined piecewise by

$$q_1(t) = \begin{cases} -\frac{l_1}{2} + l_1 t & 0 \leq t < 1 \\ \frac{l_1}{2} & 1 \leq t < 4 \\ \frac{l_1}{2} - l_1(t-4) & 4 \leq t < 5 \\ -\frac{l_1}{2} & 5 \leq t \end{cases},$$

$$q_2(t) = \begin{cases} -\frac{l_2}{2} & 0 \le t < 1 \\ -\frac{l_2}{2} + l_2(t-1) & 1 \le t < 2 \\ \frac{l_2}{2} & 2 \le t < 3 \\ \frac{l_2}{2} - l_2(t-3) & 3 \le t < 4 \\ -\frac{l_2}{2} & 4 \le t \end{cases},$$

and

$$q_3(t) = \begin{cases} -\frac{l_3}{2} & 0 \le t < 2 \\ -\frac{l_3}{2} + l_3(t-2) & 2 \le t < 3 \\ \frac{l_3}{2} & 3 \le t < 5 \\ \frac{l_3}{2} - l_3(t-5) & 5 \le t < 6 \\ -\frac{l_3}{2} & 6 \le t \end{cases}.$$

A corresponding MAMBO motion description would be obtained from the MAMBO toolbox statement

```
> MotionOutput(anims=[q1=(-l1/2+l1*t)*(t&>=0)*(t&<1)
> +(l1/2)*(t&>=1)*(t&<4)+(l1/2-l1*(t-4))*(t&>=4)*(t&<5)
> +(-l1/2)*(t&>=5),q2=(-l2/2)*((t&>=0)*(t&<1)+(t&>=4))
> +(-l2/2+l2*(t-1))*(t&>=1)*(t&<2)+(l2/2)*(t&>=2)*(t&<3)
> +(l2/2-l2*(t-3))*(t&>=3)*(t&<4),q3=(-l3/2)*((t&<2)
> +(t&>=6))+(-l3/2+l3*(t-2))*(t&>=2)*(t&<3)+
> (l3/2)*(t&>=3)*(t&<5)+(l3/2-l3*(t-5))*(t&>=5)*(t&<6)],
> parameters=[l1=1,l2=.5,l3=.25],
> filename="beadonblock.dyn"):
```

where the output from the `MotionOutput` procedure has been spooled directly to the file named `beadonblock.dyn`. Note the use of the `&<` and `&>=` operators (corresponding to $<$ and $\geq$) to generate Boolean expressions whose value is either 1 or 0 depending on whether the condition within the enclosing parentheses is satisfied or not. A visualization of the desired motion would now result from loading the geometry and motion descriptions into MAMBO and running a simulation.

### Using MAMBO state variables and a MAMBO dataset

To enable interactive changes to the configuration coordinates within a MAMBO session, these must be declared as MAMBO state variables. A corresponding MAMBO motion description would be obtained from the MAMBO toolbox statement

```
> MotionOutput(states=[q1=-.5,q2=-.25,q3=-.125],
> parameters=[l1=1,l2=.5,l3=.25],
> filename="beadonblock.dyn"):
```

where the initial values for the configuration coordinates have been specified to ensure that the bead is initially found on one corner of the block. A visualization of the desired motion could now be obtained by loading a MAMBO dataset (a MAMBO .sds file) with the following content:

```
l1      l2      l3
1.00    0.50    0.25
t       q1      q2       q3
0.00    -0.50   -0.25    -0.125
0.05    -0.45   -0.25    -0.125
0.10    -0.40   -0.25    -0.125
 ⋮       ⋮       ⋮        ⋮
0.95    0.45    -0.25    -0.125
1.00    0.50    -0.25    -0.125
1.05    0.50    -0.225   -0.125
1.10    0.50    -0.20    -0.125
 ⋮       ⋮       ⋮        ⋮
1.95    0.50    0.225    -0.125
2.00    0.50    0.25     -0.125
2.05    0.50    0.25     -0.1125
2.10    0.50    0.25     -0.10
 ⋮       ⋮       ⋮        ⋮
```

### Using MAMBO state variables and kinematic differential equations

We may retain the formulation of the configuration coordinates as MAMBO state variables while avoiding the need to generate a separate MAMBO dataset by noting that the derivatives of the configuration coordinates can be written as $\dot{q}_i = u_i$, where

$$u_1(t) = \begin{cases} l_1 & 0 \leq t < 1 \\ 0 & 1 \leq t < 4 \\ -l_1 & 4 \leq t < 5 \\ 0 & 5 \leq t \end{cases},$$

$$u_2(t) = \begin{cases} 0 & 0 \leq t < 1 \\ l_2 & 1 \leq t < 2 \\ 0 & 2 \leq t < 3 \\ -l_2 & 3 \leq t < 4 \\ 0 & 4 \leq t \end{cases},$$

and

$$u_3(t) = \begin{cases} 0 & 0 \leq t < 2 \\ l_3 & 2 \leq t < 3 \\ 0 & 3 \leq t < 5 \\ -l_3 & 5 \leq t < 6 \\ 0 & 6 \leq t \end{cases}.$$

In this formulation, the $u_i(t)$'s uniquely specify the rate of change of the configuration coordinates through the nowhere singular kinematic differential equations

$$\dot{q}_1 = u_1(t),$$
$$\dot{q}_2 = u_2(t),$$
$$\dot{q}_3 = u_3(t).$$

Taken together with specific values for the configuration coordinates at some initial time, say $q_1(0) = -\frac{l_1}{2}$, $q_2(0) = -\frac{l_2}{2}$, and $q_3(0) = -\frac{l_3}{2}$, the three functions $u_1(t)$, $u_2(t)$, and $u_3(t)$ uniquely specify the values of the configuration coordinates at subsequent times. The functions $u_1(t)$, $u_2(t)$, and $u_3(t)$ serve as input signals to the visualized motion, while the kinematic differential equations provide the connection between the inputs and the actual time evolution of the configuration coordinates.

The formulation in terms of a set of kinematic differential equations in the configuration coordinates and a set of input signals is accommodated within MAMBO through the inclusion of an `insignals` and an `ode` block in the corresponding motion description, as shown in a previous section. In particular, the MAMBO toolbox statements

```
> MotionOutput(
> ode={q1t=u1,q2t=u2,q3t=u3},
> states=[q1=-.5,q2=-.25,q3=-.125],
> parameters=[l1=1,l2=.5,l3=.25],
> insignals=[u1=l1*((t&>=0)*(t&<1)-(t&>=4)*(t&<5)),
> u2=l2*((t&>=1)*(t&<2)-(t&>=3)*(t&<4)),
> u3=l3*((t&>=2)*(t&<3)-(t&>=5)*(t&<6))],
> filename="beadonblock.dyn"):
```

create a file `beadonblock.dyn` with the following content:

```
states q1 = -.5,q2 = -.25,q3 = -.125;
parameters l1 = 1,l2 = .5,l3 = .25;
time t;
insignals {
    u1 = l1*((t>=0)*(t<1)-(t>=4)*(t<5));
    u2 = l2*((t>=1)*(t<2)-(t>=3)*(t<4));
    u3 = l3*((t>=2)*(t<3)-(t>=5)*(t<6));
}
ode {
    rhs[q1] = u1;
    rhs[q2] = u2;
    rhs[q3] = u3;
    mass[q1][q1] = 1.0;
    mass[q2][q2] = 1.0;
    mass[q3][q3] = 1.0;
}
```

We may visualize the resultant motion by loading the MAMBO geometry and motion descriptions and running a simulation.

## 4.5 Exercises

**Exercise 4.1** The position of a bead is constrained to the surface of a cylinder of radius $R$ with symmetry axis through the reference point $W$ of an observer $\mathcal{W}$ and parallel to the $\mathbf{w}_3$ basis vector of a right-handed, orthonormal basis $w$. Suppose that the coordinate representation of the reference point of the bead relative to $\mathcal{W}$ is expressed using spherical coordinates. Find the corresponding configuration constraint.

**Solution.** Let $A$ denote the reference point of the bead. Then,

$$\mathbf{r}^{WA} = w \ {}^{\mathcal{W}}A = w \begin{pmatrix} q_1 \sin q_2 \cos q_3 \\ q_1 \sin q_2 \sin q_3 \\ q_1 \cos q_2 \end{pmatrix},$$

where $q_1$, $q_2$, and $q_3$ are the spherical coordinates of the point $A$ in the coordinate system with origin at $W$ and axes parallel to the basis vectors of $w$.

Recall from Exercise 3.24 that

$$\mathbf{r}^{WA} = \left(\mathbf{r}^{WA} \bullet \mathbf{w}_3\right) \mathbf{w}_3 + \mathbf{w}_3 \times \left(\mathbf{r}^{WA} \times \mathbf{w}_3\right),$$

where the first term is parallel to $\mathbf{w}_3$ and the second term is perpendicular to $\mathbf{w}_3$. The constraint on the position of the bead then implies that

$$\left\|\mathbf{w}_3 \times \left(\mathbf{r}^{WA} \times \mathbf{w}_3\right)\right\| = R.$$

The MAMBO toolbox statements

```
> Restart():
> DeclareTriads(w):
> w3:=MakeTranslations(w,3):
> rWA:=MakeTranslations(w,
> q1*sin(q2)*cos(q3),
> q1*sin(q2)*sin(q3),q1*cos(q2)):
> simplify(VectorLength(w3 &xx
> (rWA &xx w3))=R);
```

show that the corresponding condition on the coordinates is

$$q_1 \sin q_2 = R.$$

**Exercise 4.2** The position of a bead is constrained to the surface of a cone with opening angle $\theta$, apex at the reference point $W$ of an observer $\mathcal{W}$, and with symmetry axis parallel to the basis vector $\mathbf{w}_3$ of a right-handed, orthonormal basis $w$. Suppose that the coordinate representation of the reference point of the bead relative to $\mathcal{W}$ is expressed using Cartesian coordinates. Find the corresponding configuration constraint.

**Solution.** Let $A$ denote the reference point of the bead. Then,

$$\mathbf{r}^{WA} = w \ {}^{\mathcal{W}}A = w \begin{pmatrix} q_1 \\ q_2 \\ q_3 \end{pmatrix},$$

where $q_1$, $q_2$, and $q_3$ are the Cartesian coordinates of the point $A$ in the coordinate system with origin at $W$ and axes parallel to the basis vectors of $\mathbf{w}_1$, $\mathbf{w}_2$, and $\mathbf{w}_3$.

The angle between $\mathbf{r}^{WA}$ and $\mathbf{w}_3$ is half the opening angle. It follows that

$$\cos\frac{\theta}{2} = \frac{\mathbf{r}^{WA} \bullet \mathbf{w}_3}{\left\|\mathbf{r}^{WA}\right\| \left\|\mathbf{w}_3\right\|}.$$

The MAMBO toolbox statements

```
> Restart():
> DeclareTriads(w):
> w3:=MakeTranslations(w,3):
> rWA:=MakeTranslations(w,q1,q2,q3):
> simplify(cos(theta/2)=
> (rWA &oo w3)/VectorLength(rWA)
> /VectorLength(w3));
```

show that the corresponding condition on the coordinates is

$$\cos\frac{\theta}{2} = \frac{q_3}{\sqrt{q_1^2 + q_2^2 + q_3^2}}.$$

**Exercise 4.3** The position of a bead is constrained to the surface of a sphere of radius $R$. Let $W$ be the reference point of an observer $\mathcal{W}$, such that

$$\mathbf{r}^{WS} = w\begin{pmatrix} R \\ 0 \\ 0 \end{pmatrix},$$

where $S$ denotes the center of the sphere. Find the corresponding configuration constraint if the coordinate representation of the reference point of the bead relative to $\mathcal{W}$ is formulated using a) Cartesian, b) polar, or c) spherical coordinates.

**Exercise 4.4** A small bead is attached to the end of a thin rod of length $l$, which is suspended from a spherical joint. Show that, in the absence of rotation, the bead has two geometric degrees of freedom.

**Solution.** Let $\mathcal{W}$ be an observer with reference point $W$ at the spherical joint. Denote by $A$ the reference point of the bead. Since the rod has constant length, it follows that

$$\left\|\mathbf{r}^{WA}\right\| = l.$$

The position of the bead is constrained to the surface of a sphere of radius $l$. From the text, we recall that the coordinate representation of $A$ relative to $\mathcal{W}$ can be written as

$$^{\mathcal{W}}A = \begin{pmatrix} l\sin q_1 \cos q_2 \\ l\sin q_1 \sin q_2 \\ l\cos q_1 \end{pmatrix}.$$

It follows that no more than two configuration coordinates are necessary to describe the position of the bead, i.e., the number of geometric degrees of freedom is $\leq 2$. It is not possible to describe the position using fewer than two coordinates, i.e., the number of geometric degrees of freedom is $\geq 2$ and the claim follows.

**Exercise 4.5** Suppose that the second bead on a double pendulum is constrained to the surface of a sphere of radius $R < l_1 + l_2$, where $l_1$ and $l_2$ are the lengths of the two pendulum segments. Find the corresponding configuration constraint and the number of geometric degrees of freedom of the constrained mechanism.

**Exercise 4.6** Suppose that the second bead on a double pendulum is constrained to a specific point on the surface of the sphere introduced in the previous exercise. Find the corresponding configuration constraint and the number of geometric degrees of freedom of the constrained mechanism.

**Exercise 4.7** Suppose that the first bead on the double pendulum from the previous exercise is constrained to a plane through the supporting point and the second bead. Find the corresponding configuration constraint and the number of geometric degrees of freedom of the constrained mechanism.

**Exercise 4.8** Suppose that the beads of a triple pendulum are constrained to a plane through the supporting point. Find the corresponding configuration constraint and the number of geometric degrees of freedom of the constrained mechanism.

**Exercise 4.9** Suppose that the third bead on the triple pendulum from the previous exercise is constrained to a specific point on the plane. Find the corresponding configuration constraint and the number of geometric degrees of freedom.

**Exercise 4.10** Consider the configuration constraint

$$q_1^2 + q_2^2 + q_3^2 = R^2.$$

Show that any choice of values that satisfies this constraint is regular relative to at least one of the three configuration coordinates $q_1$, $q_2$, and $q_3$.

**Solution.** From

$$f(q_1, q_2, q_3) = q_1^2 + q_2^2 + q_3^2 - R^2$$

we find

$$\frac{\partial f}{\partial q_1} = 2q_1, \frac{\partial f}{\partial q_2} = 2q_2, \text{ and } \frac{\partial f}{\partial q_3} = 2q_3.$$

Since

$$f(0,0,0) \neq 0,$$

it follows that at least one of the partial derivatives must be non-zero for a choice of values that satisfies the configuration constraint, thus confirming the claim.

**Exercise 4.11** Consider the configuration constraints

$$q_1^2 + q_2^2 + q_3^2 = R^2,$$
$$\left(q_1 - \frac{3}{2}R\right)^2 + q_2^2 + q_3^2 = R^2.$$

Show that any choice of values that satisfies these constraints is regular relative to the pairs of configuration coordinates $q_1$ and $q_2$ or $q_1$ and $q_3$.

**Solution.** The following Maple statements prove the statement (show this!):

```
> f1:=q1^2+q2^2+q3^2-R^2:
> f2:=(q1-3*R/2)^2+q2^2+q3^2-R^2:
> det(matrix(2,2,[[diff(f1,q1),
> diff(f1,q2)],[diff(f2,q1),
> diff(f2,q2)]]));
```

$$6\, q2\, R$$

```
> det(matrix(2,2,[[diff(f1,q1),
> diff(f1,q3)],[diff(f2,q1),
> diff(f2,q3)]]));
```

$$6\, q3\, R$$

**Exercise 4.12** Consider the configuration constraints

$$l_1 \cos q_1 + l_2 \cos q_2 + l_3 \cos q_3 = L,$$
$$l_1 \sin q_1 + l_2 \sin q_2 + l_3 \sin q_3 = 0.$$

Are all choices of values for $q_1$, $q_2$, and $q_3$ that satisfy these constraints regular relative to i) $q_1$, ii) $q_2$, iii) $q_3$?

**Exercise 4.13** Consider the configuration constraints derived in Exercises 4.1–4.9. For what choices of values of the configuration coordinates does the implicit function theorem apply?

**Exercise 4.14** That a configuration constraint is regular relative to one of the configuration coordinates for some choice of values $q_1 = q_{1,0}, \ldots, q_n = q_{n,0}$ implies that this coordinate may be thought of as a function of the remaining $n-1$ configuration coordinates for $q_1 \approx q_{1,0}, \ldots, q_n \approx q_{n,0}$, thus reducing the number of geometric degrees of freedom by one. Use this observation to contrast the constraint equation

$$q_1^2 + q_2^2 = 0$$

to the equivalent pair of constraint equations

$$q_1 = 0,$$
$$q_2 = 0.$$

**Solution.** The pair of constraint equations corresponds to the functions

$$f_1(q_1, \dots, q_n) = q_1,$$
$$f_2(q_1, \dots, q_n) = q_2.$$

Here, any choice of values of the configuration coordinates that satisfies $f_1 = 0$ is regular relative to $q_1$. This implies that $q_1$ may be thought of as a function of the $n-1$ remaining configuration coordinates near any such choice of values. Similarly, after substitution of $q_1$ in terms of the other $n-1$ coordinates into $f_2$, we find that any choice of values of the remaining configuration coordinates that satisfies $f_2 = 0$ is regular relative to $q_2$. This implies that $q_2$ may be thought of as a function of the $n-2$ remaining configuration coordinates. In conclusion, the pair of constraint equations reduces the number of geometric degrees of freedom by two.

In contrast, let the single constraint equation correspond to the function

$$f(q_1, \dots, q_n) = q_1^2 + q_2^2.$$

Here,

$$\frac{\partial f}{\partial q_1} = 2q_1$$

and

$$\frac{\partial f}{\partial q_2} = 2q_2.$$

In this case, the configuration constraint is only satisfied at $q_1 = q_2 = 0$. But this choice of values is singular relative to $q_1$ and relative to $q_2$. In contrast to the regular situation above, although $f$ corresponds to a single constraint equation, the number of geometric degrees of freedom is reduced by two.

**Exercise 4.15** For each of the collections of rigid bodies below, use the MAMBO toolbox to formulate a MAMBO geometry description and visualize it using MAMBO. You may find the information in the MAMBO reference manual regarding the geometric properties of MAMBO spheres, blocks, and cylinders helpful.

a) Tetrahedral arrangement of spheres
b) Icosahedral arrangement of spheres
c) Brick wall
d) Pile of parallel logs
e) Books in book shelf
f) Tiled bathroom floor
g) Hardwood floor
h) Rectangular bird cage
i) Multiple rows of rectangular chairs
j) Parallel rows of marble columns

**Exercise 4.16** For each of the scenes below, use the MAMBO toolbox to formulate a MAMBO geometry description and implement different animation sequences in MAMBO.

a) A game of checkers
b) Fitting a table with an extension
c) Dialing on a digital dialing pad
d) Typing on a computer keyboard
e) Packets traveling through a network
f) Road work with a pneumatic drill

### Summary of notation

Upper-case, italicized letters with calligraphy-style left superscripts, such as ${}^{\mathcal{W}}A$ and ${}^{\mathcal{A}}R$, were used in this chapter to denote the coordinate representation of a point relative to an observer.

Lower-case $q$'s with various subscripts and decorations, such as $q_1$ and $\tilde{q}_3$, were used in this chapter to denote configuration coordinates.

### Summary of terminology

The *coordinate representation* of a point relative to an observer is the matrix representation of the position vector from the reference point of the observer to the point relative to the right-handed, orthonormal basis associated with the observer. (Page 159)

A mechanism is *constrained* if its configuration is limited. (Page 161)

A variable that is used to describe the configuration of a mechanism is called a *configuration coordinate.* (Page 168)

The number of *geometric degrees of freedom* of a mechanism is the smallest number of configuration coordinates necessary to describe the configuration of the mechanism. (Page 168)

An equation in the configuration coordinates that corresponds to a constraint on the configuration of a mechanism is called a *configuration constraint.* (Page 169)

A complete set of differential equations in the configuration coordinates is called a set of *kinematic differential equations.* (Page 181)

In Mambo, the position of an observer or a rigid body relative to an observer is given through a **POINT** statement in the .geo file. (Page 172)

In Mambo, parameters are declared through a `parameters` statement in the .dyn file. (Page 177)

In Mambo, the time variable is labeled through a `time` statement in the .dyn file. (Page 177)

In Mambo, animated variables are declared through an `anims` block in the .dyn file. (Page 179)

In Mambo, states are declared through a `states` statement in the .dyn file. (Page 179)

In Mambo, the kinematic differential equations are declared in the `ode` block in the .dyn file. (Page 182)

(Page 182) In Mambo, input signals to the kinematic differential equations are declared in the `insignals` block in the .dyn file.

(Page 183) In the Mambo toolbox, the global variable `GlobalPointDeclarations` contains the names of all declared points.

(Page 183) In the Mambo toolbox, the global variable `GlobalPointDefinitions` contains position vectors relating declared points.

(Page 183) In the Mambo toolbox, the procedure `DeclarePoints` appends `GlobalPointDeclarations` with any number of point labels.

(Page 183) In the Mambo toolbox, the procedure `DefinePoints` appends `GlobalPointDefinitions` with any number of position vectors relating declared points.

(Page 185) In the Mambo toolbox, the procedure `FindTranslation` is used to find the position vector between two defined points.

(Page 185) In the Mambo toolbox, the procedure `DefineObservers` appends `GlobalObserverDefinitions` with any number of associations between observers and pairs of declared reference points and declared right-handed, orthonormal bases.

(Page 186) In the Mambo toolbox, the procedure `FindPosition` is used to find the position vector between the reference points of two observers.

(Page 187) In the Mambo toolbox, the procedure `FindCoordinates` is used to compute the coordinate representation of a point relative to an observer.

(Page 187) In the Mambo toolbox, the procedure `Undo` is used to undo the latest change to any of the global variables.

(Page 190) In the Mambo toolbox, the procedure `DefineObjects` is used to associate objects with defined observers.

(Page 192) In the Mambo toolbox, the procedure `MotionOutput` generates a Mambo motion description.

# Chapter 5

# Rotations

*wherein the reader learns of:*

- *Using rotation matrices to describe the relation between right-handed, orthonormal bases;*
- *Computing rotation matrices for a variety of common situations.*

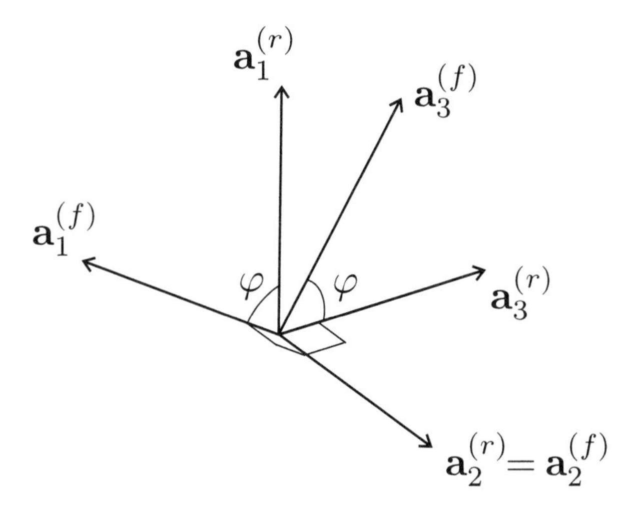

Practicum

Make it a habit to represent a right-handed, orthonormal basis with the first three fingers of your right hand. Let your index finger represent the first basis vector, the middle finger the second basis vector, and the thumb the third basis vector. This gesture will be very handy (!) when attempting to visualize the relations between multiple right-handed, orthonormal bases. It will help you concretize the idea of rotation matrices as relations between such bases.

Almost every computation in this chapter can be completed using the MAMBO toolbox routines presented at the end of the chapter. This is good practice and will make the purpose of the computer-algebra procedures more evident.

## 5.1 Triads

Right-handed, orthonormal bases of space play a central role in the analysis and visualization of multibody mechanisms. From this point on, right-handed, orthonormal bases will be referred to as *triads*, reflecting the triplet of basis vectors.

### 5.1.1 Notation

To denote triads, I consistently use lower-case, italicized letters, e.g.,

$$a,\ r,\ \text{or}\ x.$$

The triad $a$ consists of the three basis vectors $\mathbf{a}_1$, $\mathbf{a}_2$, and $\mathbf{a}_3$. Similarly, the triad $r$ consists of the three basis vectors $\mathbf{r}_1$, $\mathbf{r}_2$, and $\mathbf{r}_3$, and so on. The choice of letter is not important, unless you are trying to give the person you are communicating with additional information by a clever choice of letter. For example, if a triad is to be used as the reference triad of a specific observer, it may be advantageous to denote it by the same letter that was used to denote the observer. To distinguish between multiple triads that use the same letter, I introduce appropriate superscripts, for example, $a^{(1)}$, $r^{(\text{reference})}$, and $c^{(\text{world})}$. The parentheses are included to eliminate the risk for confusing the superscript with an exponent.

To graphically represent a triad, this text consistently uses three, mutually perpendicular separations representing the basis vectors often (but not necessarily) emanating from a common starting point. For later reference, it is a good idea to place the corresponding vector symbols adjacent to each of the basis vectors.

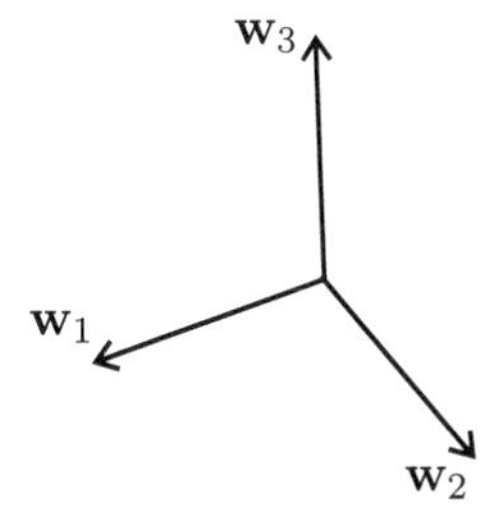


### 5.1.2 Common Misconceptions

Triads **do not** have a specific location. After all, triads consist of three, mutually perpendicular basis vectors, each of which corresponds to an infinite collection of equivalent separations. It is important not to confuse the location of the three separations representing the triad with an actual location of the triad. Any three separations representing the basis vectors can be chosen in the graphical depiction of the triad. There is, for example, no need to pick separations that are based at some specific reference point.

A common notation for the basis vectors of a right-handed, orthonormal basis is **i**, **j**, and **k**. Since it is not clear how to distinguish between different bases, we do not use this notation here.

Instead of labeling the three basis vectors by numerical subscripts, it is common to use the letters $x$, $y$, and $z$ as subscripts. With this choice, the triad $a$ would be represented by the basis vectors $\mathbf{a}_x$, $\mathbf{a}_y$, and $\mathbf{a}_z$. This is clearly motivated by the use of $x$, $y$, and $z$ to denote the coordinates of a right-handed coordinate system. There is certainly nothing wrong with this notation, but it is inconsistent with my expressed desire to represent triads with matrices

$$a = \begin{pmatrix} \mathbf{a}_1 & \mathbf{a}_2 & \mathbf{a}_3 \end{pmatrix}.$$

Here, the numerical subscripts act not only to distinguish the basis vectors from another, but also to denote placement in the matrix. The $x$, $y$, and $z$ subscripts achieve the former, but do not reflect the latter.

(Ex. 5.1 – Ex. 5.24)

## 5.2 Rotation Matrices

### 5.2.1 Fundamental Relations

**Illustration 5.1**

Let $a$ be a triad. Consider the vector

$$\mathbf{v} = \mathbf{a}_1 + 2\mathbf{a}_3 = a \begin{pmatrix} 1 \\ 0 \\ 2 \end{pmatrix}.$$

We may construct a second triad $b$, such that $\mathbf{b}_3$ is parallel to $\mathbf{v}$ by the method of Chapter 3. In particular, let

$$\mathbf{b}_3 = \frac{\mathbf{v}}{\|\mathbf{v}\|} = a \begin{pmatrix} \frac{1}{\sqrt{5}} \\ 0 \\ \frac{2}{\sqrt{5}} \end{pmatrix},$$

since

$$\|\mathbf{v}\|^2 = \mathbf{v} \bullet \mathbf{v} = \begin{pmatrix} 1 & 0 & 2 \end{pmatrix} \begin{pmatrix} 1 \\ 0 \\ 2 \end{pmatrix} = 5.$$

Moreover,

$$\mathbf{b}_1 = \frac{\mathbf{a}_1 \times \mathbf{v}}{\|\mathbf{a}_1 \times \mathbf{v}\|} = a \begin{pmatrix} 0 \\ -1 \\ 0 \end{pmatrix}$$

and

$$\mathbf{b}_2 = \mathbf{b}_3 \times \mathbf{b}_1 = a \begin{pmatrix} \frac{2}{\sqrt{5}} \\ 0 \\ -\frac{1}{\sqrt{5}} \end{pmatrix}.$$

In particular, we may write

$$\begin{pmatrix} \mathbf{b}_1 & \mathbf{b}_2 & \mathbf{b}_3 \end{pmatrix} = \begin{pmatrix} \mathbf{a}_1 & \mathbf{a}_2 & \mathbf{a}_3 \end{pmatrix} \begin{pmatrix} 0 & \frac{2}{\sqrt{5}} & \frac{1}{\sqrt{5}} \\ -1 & 0 & 0 \\ 0 & -\frac{1}{\sqrt{5}} & \frac{2}{\sqrt{5}} \end{pmatrix}.$$

The $[i, j]$-th entry in the scalar matrix can be found from the dot product $\mathbf{a}_i \bullet \mathbf{b}_j$. For example,

$$\mathbf{a}_1 \bullet \mathbf{b}_3 = \mathbf{a}_1 \bullet \left( \frac{1}{\sqrt{5}} \mathbf{a}_1 + \frac{2}{\sqrt{5}} \mathbf{a}_3 \right) = \frac{1}{\sqrt{5}}$$

and so on.

In a pure rotation, one point on the rigid body is kept fixed relative to the reference configuration. Denote this point by $A$ and let $B$ and $C$ be two other points on the rigid body, such that the separations $\overrightarrow{AB}$ and $\overrightarrow{AC}$ have unit length and are mutually perpendicular. Let a $_r$ subscript denote the corresponding points in the reference configuration and let a $_f$ subscript denote the corresponding points in the final configuration. The final orientation of the rigid body is uniquely determined by the locations of the points $B_f$ and $C_f$ relative to $B_r$ and $C_r$, respectively.

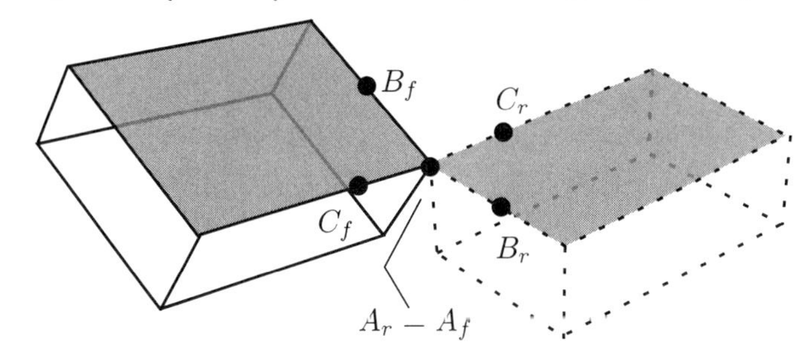


It follows that

$$\mathbf{a}_1^{(r)} = \left[\overrightarrow{A_r B_r}\right], \mathbf{a}_2^{(r)} = \left[\overrightarrow{A_r C_r}\right], \mathbf{a}_3^{(r)} = \mathbf{a}_1^{(r)} \times \mathbf{a}_2^{(r)}$$

and

$$\mathbf{a}_1^{(f)} = \left[\overrightarrow{A_f B_f}\right], \mathbf{a}_2^{(f)} = \left[\overrightarrow{A_f C_f}\right], \mathbf{a}_3^{(f)} = \mathbf{a}_1^{(f)} \times \mathbf{a}_2^{(f)}$$

are the components of two triads that uniquely describe the reference orientation and the final orientation of the rigid body.

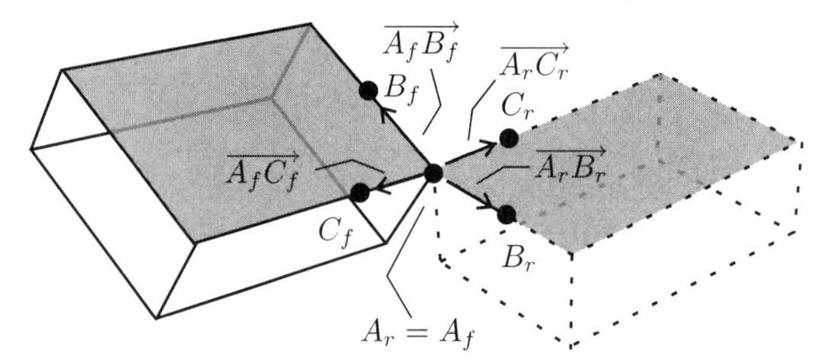


As in the illustration above, we may express the triad $a^{(f)}$ in terms of the triad $a^{(r)}$ :

$$a^{(f)} = a^{(r)} R_{a^{(r)} a^{(f)}},$$

where $R_{a^{(r)} a^{(f)}}$ is a $3 \times 3$ matrix of real numbers.

> Every pure rotation keeping the point $A$ fixed corresponds to a unique choice of entries of $R_{a^{(r)} a^{(f)}}$.

Below, we will consider the conditions that the entries of a $3 \times 3$ matrix must satisfy to ensure that the matrix corresponds to some pure rotation.

### Illustration 5.2

Let $a$ and $b$ be arbitrary triads. We recall that for an arbitrary vector

$$\mathbf{v} = v_1 \mathbf{a}_1 + v_2 \mathbf{a}_2 + v_3 \mathbf{a}_3,$$

where

$$v_i = \mathbf{a}_i \bullet \mathbf{v},$$

i.e.,

$$\mathbf{v} = \mathbf{a}_1 (\mathbf{a}_1 \bullet \mathbf{v}) + \mathbf{a}_2 (\mathbf{a}_2 \bullet \mathbf{v}) + \mathbf{a}_3 (\mathbf{a}_3 \bullet \mathbf{v}).$$

In particular, this is true for each of the basis vectors in the triad $b$:

$$\begin{aligned}
\mathbf{b}_1 &= \mathbf{a}_1 (\mathbf{a}_1 \bullet \mathbf{b}_1) + \mathbf{a}_2 (\mathbf{a}_2 \bullet \mathbf{b}_1) + \mathbf{a}_3 (\mathbf{a}_3 \bullet \mathbf{b}_1), \\
\mathbf{b}_2 &= \mathbf{a}_1 (\mathbf{a}_1 \bullet \mathbf{b}_2) + \mathbf{a}_2 (\mathbf{a}_2 \bullet \mathbf{b}_2) + \mathbf{a}_3 (\mathbf{a}_3 \bullet \mathbf{b}_2), \\
\mathbf{b}_3 &= \mathbf{a}_1 (\mathbf{a}_1 \bullet \mathbf{b}_3) + \mathbf{a}_2 (\mathbf{a}_2 \bullet \mathbf{b}_2) + \mathbf{a}_3 (\mathbf{a}_3 \bullet \mathbf{b}_3).
\end{aligned}$$

In terms of the matrices introduced in the previous chapter, this is equivalent to

$$\begin{pmatrix} \mathbf{b}_1 & \mathbf{b}_2 & \mathbf{b}_3 \end{pmatrix} = \begin{pmatrix} \mathbf{a}_1 & \mathbf{a}_2 & \mathbf{a}_3 \end{pmatrix} \begin{pmatrix} \mathbf{a}_1 \bullet \mathbf{b}_1 & \mathbf{a}_1 \bullet \mathbf{b}_2 & \mathbf{a}_1 \bullet \mathbf{b}_3 \\ \mathbf{a}_2 \bullet \mathbf{b}_1 & \mathbf{a}_2 \bullet \mathbf{b}_2 & \mathbf{a}_2 \bullet \mathbf{b}_3 \\ \mathbf{a}_3 \bullet \mathbf{b}_1 & \mathbf{a}_3 \bullet \mathbf{b}_2 & \mathbf{a}_3 \bullet \mathbf{b}_3 \end{pmatrix},$$

where the matrix

$$R_{ab} = \begin{pmatrix} \mathbf{a}_1 \bullet \mathbf{b}_1 & \mathbf{a}_1 \bullet \mathbf{b}_2 & \mathbf{a}_1 \bullet \mathbf{b}_3 \\ \mathbf{a}_2 \bullet \mathbf{b}_1 & \mathbf{a}_2 \bullet \mathbf{b}_2 & \mathbf{a}_2 \bullet \mathbf{b}_3 \\ \mathbf{a}_3 \bullet \mathbf{b}_1 & \mathbf{a}_3 \bullet \mathbf{b}_2 & \mathbf{a}_3 \bullet \mathbf{b}_3 \end{pmatrix}$$

is called *the rotation matrix from $a$ to $b$.*

This definition allows us to write

$$b = aR_{ab},$$

where

$$R_{ab} = a^T \bullet b.$$

Let $\mathbf{v}$ and $\mathbf{w}$ be two arbitrary vectors. The dot product

$$\mathbf{v} \bullet \mathbf{w}$$

was defined in the previous chapter as the product of the lengths of the two vectors and cosine of the angle between the vectors:

$$\|\mathbf{v}\| \, \|\mathbf{w}\| \cos\theta \left(\mathbf{v}, \mathbf{w}\right).$$

This definition did not rely on the introduction of any particular basis relative to which either vector was expressed. The dot product was independent of the choice of basis.

While the geometric definition of the dot product is useful to show its independence of basis, it is not a very convenient method for computing the value of the dot product. Instead, the use of a basis greatly simplifies the computation. Here, given a triad $a$, the dot product between the two vectors equals

$$\begin{aligned} \mathbf{v} \bullet \mathbf{w} &= \left({}^a v_1 \mathbf{a}_1 + {}^a v_2 \mathbf{a}_2 + {}^a v_3 \mathbf{a}_3\right) \bullet \left({}^a w_1 \mathbf{a}_1 + {}^a w_2 \mathbf{a}_2 + {}^a w_3 \mathbf{a}_3\right) \\ &= \begin{pmatrix} {}^a v_1 & {}^a v_2 & {}^a v_3 \end{pmatrix} \begin{pmatrix} \mathbf{a}_1 \bullet \mathbf{a}_1 & \mathbf{a}_1 \bullet \mathbf{a}_2 & \mathbf{a}_1 \bullet \mathbf{a}_3 \\ \mathbf{a}_2 \bullet \mathbf{a}_1 & \mathbf{a}_2 \bullet \mathbf{a}_2 & \mathbf{a}_2 \bullet \mathbf{a}_3 \\ \mathbf{a}_3 \bullet \mathbf{a}_1 & \mathbf{a}_3 \bullet \mathbf{a}_2 & \mathbf{a}_3 \bullet \mathbf{a}_3 \end{pmatrix} \begin{pmatrix} {}^a w_1 \\ {}^a w_2 \\ {}^a w_3 \end{pmatrix} \\ &= \left({}^a v\right)^T {}^a w, \end{aligned}$$

since

$$a^T \bullet a = \begin{pmatrix} \mathbf{a}_1 \bullet \mathbf{a}_1 & \mathbf{a}_1 \bullet \mathbf{a}_2 & \mathbf{a}_1 \bullet \mathbf{a}_3 \\ \mathbf{a}_2 \bullet \mathbf{a}_1 & \mathbf{a}_2 \bullet \mathbf{a}_2 & \mathbf{a}_2 \bullet \mathbf{a}_3 \\ \mathbf{a}_3 \bullet \mathbf{a}_1 & \mathbf{a}_3 \bullet \mathbf{a}_2 & \mathbf{a}_3 \bullet \mathbf{a}_3 \end{pmatrix}$$

equals the identity matrix for an orthonormal basis. A similar expression would result from replacing the triad $a$ with a different triad $b$:

$$\mathbf{v} \bullet \mathbf{w} = \left({}^b v\right)^T {}^b w.$$

If, instead, we express $\mathbf{v}$ relative to the $a$ triad and $\mathbf{w}$ relative to the $b$ triad, we find

$$\begin{aligned}
\mathbf{v} \bullet \mathbf{w} &= \left({}^a v_1 \mathbf{a}_1 + {}^a v_2 \mathbf{a}_2 + {}^a v_3 \mathbf{a}_3\right) \bullet \left({}^b w_1 \mathbf{b}_1 + {}^b w_2 \mathbf{b}_2 + {}^b w_3 \mathbf{b}_3\right) \\
&= \begin{pmatrix} {}^a v_1 & {}^a v_2 & {}^a v_3 \end{pmatrix} \begin{pmatrix} \mathbf{a}_1 \bullet \mathbf{b}_1 & \mathbf{a}_1 \bullet \mathbf{b}_2 & \mathbf{a}_1 \bullet \mathbf{b}_3 \\ \mathbf{a}_2 \bullet \mathbf{b}_1 & \mathbf{a}_2 \bullet \mathbf{b}_2 & \mathbf{a}_2 \bullet \mathbf{b}_3 \\ \mathbf{a}_3 \bullet \mathbf{b}_1 & \mathbf{a}_3 \bullet \mathbf{b}_2 & \mathbf{a}_3 \bullet \mathbf{b}_3 \end{pmatrix} \begin{pmatrix} {}^b w_1 \\ {}^b w_2 \\ {}^b w_3 \end{pmatrix} \\
&= \left({}^a v\right)^T R_{ab} {}^b w.
\end{aligned}$$

But the dot product is independent of the triad(s) used to express the vectors. It follows that

$$\begin{aligned}
\mathbf{v} \bullet \mathbf{w} &= \left({}^a v\right)^T {}^a w \\
&= \left({}^a v\right)^T R_{ab} {}^b w.
\end{aligned}$$

Since the vectors $\mathbf{v}$ and $\mathbf{w}$ were arbitrary, we conclude that

$${}^a w = R_{ab} {}^b w.$$

We have shown that the rotation matrix $R_{ab}$ between the triads $a$ and $b$ satisfies the following relations:

$$\begin{aligned}
R_{ab} &= a^T \bullet b, \\
b &= a R_{ab},
\end{aligned}$$

and

$${}^a v = R_{ab} {}^b v$$

for an arbitrary vector $\mathbf{v}$.

From the symmetry of the dot product, it follows that

$$\begin{aligned}
R_{ba} &= \begin{pmatrix} \mathbf{b}_1 \bullet \mathbf{a}_1 & \mathbf{b}_1 \bullet \mathbf{a}_2 & \mathbf{b}_1 \bullet \mathbf{a}_3 \\ \mathbf{b}_2 \bullet \mathbf{a}_1 & \mathbf{b}_2 \bullet \mathbf{a}_2 & \mathbf{b}_2 \bullet \mathbf{a}_3 \\ \mathbf{b}_3 \bullet \mathbf{a}_1 & \mathbf{b}_3 \bullet \mathbf{a}_2 & \mathbf{b}_3 \bullet \mathbf{a}_3 \end{pmatrix} \\
&= \begin{pmatrix} \mathbf{a}_1 \bullet \mathbf{b}_1 & \mathbf{a}_2 \bullet \mathbf{b}_1 & \mathbf{a}_3 \bullet \mathbf{b}_1 \\ \mathbf{a}_1 \bullet \mathbf{b}_2 & \mathbf{a}_2 \bullet \mathbf{b}_2 & \mathbf{a}_3 \bullet \mathbf{b}_2 \\ \mathbf{a}_1 \bullet \mathbf{b}_3 & \mathbf{a}_2 \bullet \mathbf{b}_3 & \mathbf{a}_3 \bullet \mathbf{b}_3 \end{pmatrix} \\
&= \begin{pmatrix} \mathbf{a}_1 \bullet \mathbf{b}_1 & \mathbf{a}_1 \bullet \mathbf{b}_2 & \mathbf{a}_1 \bullet \mathbf{b}_3 \\ \mathbf{a}_2 \bullet \mathbf{b}_1 & \mathbf{a}_2 \bullet \mathbf{b}_2 & \mathbf{a}_2 \bullet \mathbf{b}_3 \\ \mathbf{a}_3 \bullet \mathbf{b}_1 & \mathbf{a}_3 \bullet \mathbf{b}_2 & \mathbf{a}_3 \bullet \mathbf{b}_3 \end{pmatrix}^T \\
&= \left(R_{ab}\right)^T.
\end{aligned}$$

In other words, the rotation matrix from $b$ to $a$ is the transpose of the rotation matrix from $a$ to $b$.

$$R_{ab} = a^T \bullet b = \left(b^T \bullet a\right)^T = \left(R_{ba}\right)^T .$$

Any two triads $a$ and $b$ are related through a $3 \times 3$ matrix $R_{ab}$ of real numbers. But under what conditions does a $3 \times 3$ matrix qualify as a rotation matrix? Put differently, under what conditions on the $3 \times 3$ matrix $R$ of real numbers is the result of the matrix product

$$aR$$

a triad if $a$ is a triad?

Recall that a basis $b$ is orthonormal if and only if the matrix product

$$b^T \bullet b = \begin{pmatrix} \mathbf{b}_1 \bullet \mathbf{b}_1 & \mathbf{b}_1 \bullet \mathbf{b}_2 & \mathbf{b}_1 \bullet \mathbf{b}_3 \\ \mathbf{b}_2 \bullet \mathbf{b}_1 & \mathbf{b}_2 \bullet \mathbf{b}_2 & \mathbf{b}_2 \bullet \mathbf{b}_3 \\ \mathbf{b}_3 \bullet \mathbf{b}_1 & \mathbf{b}_3 \bullet \mathbf{b}_2 & \mathbf{b}_3 \bullet \mathbf{b}_3 \end{pmatrix}$$

equals the identity matrix. Thus, $aR$ is an orthonormal basis if and only if the matrix product

$$(aR)^T \bullet (aR) = R^T \left(a^T \bullet a\right) R = R^T R$$

equals the identity matrix.

To find the condition that ensures that $aR$ is right-handed, we proceed as follows. Let the $[i, j]$-th entry of $R$ be denoted by $r_{ij}$, i.e.,

$$R = \begin{pmatrix} r_{11} & r_{12} & r_{13} \\ r_{21} & r_{22} & r_{23} \\ r_{31} & r_{32} & r_{33} \end{pmatrix} .$$

Then, from

$$\begin{pmatrix} \mathbf{v}_1 & \mathbf{v}_2 & \mathbf{v}_3 \end{pmatrix} = aR,$$

we find

$$\begin{aligned} \mathbf{v}_1 &= r_{11}\mathbf{a}_1 + r_{21}\mathbf{a}_2 + r_{31}\mathbf{a}_3, \\ \mathbf{v}_2 &= r_{12}\mathbf{a}_1 + r_{22}\mathbf{a}_2 + r_{32}\mathbf{a}_3, \\ \mathbf{v}_3 &= r_{13}\mathbf{a}_1 + r_{23}\mathbf{a}_2 + r_{33}\mathbf{a}_3. \end{aligned}$$

The orthonormal basis $\{\mathbf{v}_1, \mathbf{v}_2, \mathbf{v}_3\}$ is right-handed if and only if

$$\mathbf{v}_1 \bullet (\mathbf{v}_2 \times \mathbf{v}_3) = 1.$$

1

But

$$\begin{aligned}
\mathbf{v}_1 \bullet (\mathbf{v}_2 \times \mathbf{v}_3) &= (r_{11}\mathbf{a}_1 + r_{21}\mathbf{a}_2 + r_{31}\mathbf{a}_3) \bullet \\
&\quad [(r_{12}\mathbf{a}_1 + r_{22}\mathbf{a}_2 + r_{32}\mathbf{a}_3) \times (r_{13}\mathbf{a}_1 + r_{23}\mathbf{a}_2 + r_{33}\mathbf{a}_3)] \\
&= (r_{11}\mathbf{a}_1 + r_{21}\mathbf{a}_2 + r_{31}\mathbf{a}_3) \bullet \\
&\quad \left( \begin{array}{c} r_{12}r_{13}\mathbf{a}_1 \times \mathbf{a}_1 + r_{12}r_{23}\mathbf{a}_1 \times \mathbf{a}_2 + r_{12}r_{33}\mathbf{a}_1 \times \mathbf{a}_3 \\ +r_{22}r_{13}\mathbf{a}_2 \times \mathbf{a}_1 + r_{22}r_{23}\mathbf{a}_2 \times \mathbf{a}_2 + r_{22}r_{33}\mathbf{a}_2 \times \mathbf{a}_3 \\ +r_{32}r_{13}\mathbf{a}_3 \times \mathbf{a}_1 + r_{32}r_{23}\mathbf{a}_3 \times \mathbf{a}_2 + r_{32}r_{33}\mathbf{a}_3 \times \mathbf{a}_3 \end{array} \right) \\
&= r_{11}(r_{22}r_{33} - r_{23}r_{32}) - r_{21}(r_{12}r_{33} - r_{32}r_{13}) \\
&\quad + r_{31}(r_{12}r_{23} - r_{22}r_{13}) \\
&= \begin{vmatrix} r_{11} & r_{12} & r_{13} \\ r_{21} & r_{22} & r_{23} \\ r_{31} & r_{32} & r_{33} \end{vmatrix} \\
&= \det R,
\end{aligned}$$

where we used the fact that the basis $a$ is right-handed, i.e., that

$$\mathbf{a}_1 \times \mathbf{a}_2 = \mathbf{a}_3, \ \mathbf{a}_2 \times \mathbf{a}_3 = \mathbf{a}_1, \text{ and } \mathbf{a}_3 \times \mathbf{a}_1 = \mathbf{a}_2.$$

It follows that the basis $aR$ is right-handed if and only if

$$\det R = 1.$$

### 5.2.2 Mnemonics

Rotation matrices are denoted by the letter $R$ and the names of a pair of triads as subscripts, e.g.,

$$R_{ab}, \ R_{c^{(1)}c^{(2)}}, \text{ or } R_{b^{(\text{lab})}b^{(\text{ball})}}.$$

Rotation matrices are $3 \times 3$ matrices of real numbers. Recall that a triad $a$ is represented by the $1 \times 3$ matrix of vectors

$$a = \begin{pmatrix} \mathbf{a}_1 & \mathbf{a}_2 & \mathbf{a}_3 \end{pmatrix}.$$

By the standard rules of matrix multiplication, the matrix product

$$aR_{ac}$$

is defined and evaluates to a $1 \times 3$ matrix of vectors, namely the triad $c$. On the other hand, the matrix product

$$R_{ac}a$$

is not defined, since the number of columns of $R_{ac}$ does not agree with the number of rows of $a$.

The rotation matrix $R_{cd}$ can be computed from the formula

$$R_{cd} = c^T \bullet d,$$

where $c$ and $d$ are two triads. The right-hand side is the dot product of a $3 \times 1$ matrix of vectors with a $1 \times 3$ matrix of vectors. By the rules of matrix multiplication, the result is therefore a $3 \times 3$ matrix of numbers. In contrast, the expression

$$c \bullet d^T$$

is the dot product of a $1 \times 3$ matrix of vectors with a $3 \times 1$ matrix of vectors. By the rules of matrix multiplication, the result is therefore a $1 \times 1$ matrix of numbers, or simply a single real number.

For an arbitrary vector $\mathbf{v}$, the matrix representations relative to two triads $a$ and $b$ are related through

$${}^a v = R_{ab} \, {}^b v.$$

The right-hand side is a product of a $3 \times 3$ matrix of real numbers with a $3 \times 1$ matrix of real numbers. By the rules of matrix multiplication, the result is therefore a $3 \times 1$ matrix of real numbers. In contrast, the expression

$${}^b v R_{ab}$$

is undefined, since the number of columns of ${}^b v$ does not equal the number of rows of $R_{ab}$.

### 5.2.3 Rotating Vectors

Let $P$ and $Q$ be two arbitrary points on a rigid body. Then the vector

$$\mathbf{v}_r = \left[\overrightarrow{P_r Q_r}\right]$$

is the collection of all separations that are equivalent to $\overrightarrow{P_r Q_r}$, where the $_r$ subscript refers to points in the reference configuration. Similarly, the vector

$$\mathbf{v}_f = \left[\overrightarrow{P_f Q_f}\right]$$

is the collection of all separations that are equivalent to $\overrightarrow{P_f Q_f}$, where the $_f$ subscript refers to points in the final configuration.

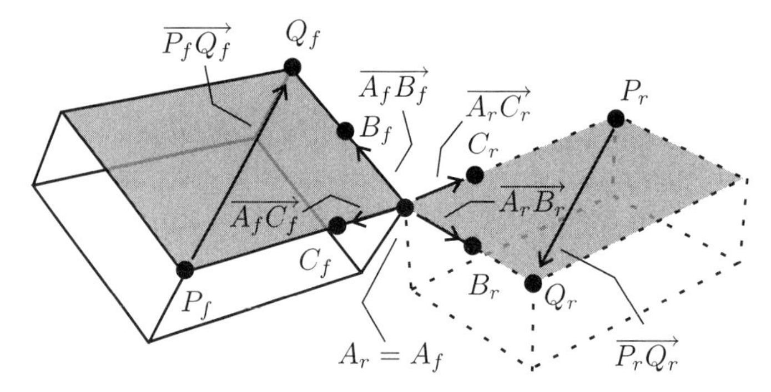


Since the positions of the points $P$ and $Q$ are unchanged relative to the triplet $A$, $B$, and $C$ introduced above, it follows that

$$^{a^{(f)}}v_f = {}^{a^{(r)}}v_r$$

or

$$\begin{aligned} ^{a^{(r)}}v_f &= R_{a^{(r)}a^{(f)}} {}^{a^{(f)}}v_f \\ &= R_{a^{(r)}a^{(f)}} {}^{a^{(r)}}v_r. \end{aligned}$$

**Illustration 5.3**

The pure rotation rotates the straight line spanned by $\mathbf{v}_r$ until it coincides with the straight line spanned by $\mathbf{v}_f$. The equality

$$^{a^{(f)}}v_f = {}^{a^{(r)}}v_r$$

shows that $\|\mathbf{v}_f\| = \|\mathbf{v}_r\|$.

From the definition of the dot product, we recall that the angle $\theta(\mathbf{v}_r, \mathbf{v}_f)$ between the vectors satisfies the relation

$$\begin{aligned} \cos\theta(\mathbf{v}_r, \mathbf{v}_f) &= \frac{\mathbf{v}_r \bullet \mathbf{v}_f}{\|\mathbf{v}_r\| \cdot \|\mathbf{v}_f\|} \\ &= \frac{1}{\|\mathbf{v}_r\|^2} \left({}^{a^{(r)}}v_r\right)^T {}^{a^{(r)}}v_f \\ &= \frac{1}{\|\mathbf{v}_r\|^2} \left({}^{a^{(r)}}v_r\right)^T R_{a^{(r)}a^{(f)}} {}^{a^{(r)}}v_r. \end{aligned}$$

2

### 5.2.4 Orthogonal Matrices

From Exercise 5.11, we conclude that

$$(R_{ab})^T R_{ab}$$

equals the identity matrix, i.e., the rotation matrix $R_{ab}$ from the triad $a$ to the triad $b$ is an example of an *orthogonal matrix*.

---

**Definition 5.1** An $n$ by $n$ matrix $A$ such that $A^T A = I$, where $I$ is the $n$ by $n$ unit matrix, is called an orthogonal matrix. The following is true for general orthogonal matrices:

- *Orientation preserving or reversing:* $\det(A) = \pm 1$, since $\det\left(A^T\right) = \det(A)$, and $\det(AB) = \det(A)\det(B)$;
- *Preserves the inner product:* $(Av)^T (Aw) = v^T A^T A w = v^T w$;
- *Degrees of freedom:* $A$ is determined by $\frac{n(n-1)}{2}$ independent quantities, since the orthogonality condition imposes $\frac{n(n+1)}{2}$ conditions on the $n^2$ matrix components.

---

It follows that rotation matrices are orientation-preserving.

**Illustration 5.4**
That the rotation matrix $R_{ab}$ preserves the vector dot product follows from the following argument. Let **v** and **w** be two arbitrary vectors. Then

$${}^a v = R_{ab}{}^b v$$

and

$${}^a w = R_{ab}{}^b w$$

imply that

$$\begin{aligned} \left({}^a v\right)^T {}^a w &= \left(R_{ab}{}^b v\right)^T R_{ab}{}^b w \\ &= \left({}^b v\right)^T (R_{ab})^T R_{ab}{}^b w \\ &= \left({}^b v\right)^T {}^b w. \end{aligned}$$

But this is just a restatement of the observation that the dot product $\mathbf{v} \bullet \mathbf{w}$ is independent of the triad relative to which the vectors are expressed.

2

The result in Exercise 5.15 shows that for every rotation matrix $R_{ab}$ there exists an **orthogonal** matrix $V$, such that

$$V^T R_{ab} V = \begin{pmatrix} 1 & 0 & 0 \\ 0 & t_{11} & t_{12} \\ 0 & t_{21} & t_{22} \end{pmatrix} \Leftrightarrow R_{ab} = V \begin{pmatrix} 1 & 0 & 0 \\ 0 & t_{11} & t_{12} \\ 0 & t_{21} & t_{22} \end{pmatrix} V^T,$$

where $t_{11}$, $t_{12}$, $t_{21}$, and $t_{22}$ are some constants. But since

$$\begin{aligned} \left(V^T R_{ab} V\right)^T \left(V^T R_{ab} V\right) &= V^T R_{ba} V V^T R_{ab} V \\ &= V^T R_{ba} R_{ab} V \\ &= V^T V \end{aligned}$$

equals the identity matrix, it follows that

$$\begin{pmatrix} 1 & 0 & 0 \\ 0 & t_{11} & t_{12} \\ 0 & t_{21} & t_{22} \end{pmatrix}$$

must be orthogonal, i.e.,

$$\begin{aligned} \begin{pmatrix} 1 & 0 & 0 \\ 0 & 1 & 0 \\ 0 & 0 & 1 \end{pmatrix} &= \begin{pmatrix} 1 & 0 & 0 \\ 0 & t_{11} & t_{12} \\ 0 & t_{21} & t_{22} \end{pmatrix}^T \begin{pmatrix} 1 & 0 & 0 \\ 0 & t_{11} & t_{12} \\ 0 & t_{21} & t_{22} \end{pmatrix} \\ &= \begin{pmatrix} 1 & 0 & 0 \\ 0 & t_{11}^2 + t_{21}^2 & t_{11}t_{12} + t_{21}t_{22} \\ 0 & t_{11}t_{12} + t_{21}t_{22} & t_{12}^2 + t_{22}^2 \end{pmatrix}. \end{aligned}$$

Moreover,

$$\begin{aligned} \det\left(V^T R_{ab} V\right) &= \det V^T \det R_{ab} \det V \\ &= \det R_{ab} \left(\det V\right)^2 \\ &= 1 \end{aligned}$$

implies that

$$1 = \det \begin{pmatrix} 1 & 0 & 0 \\ 0 & t_{11} & t_{12} \\ 0 & t_{21} & t_{22} \end{pmatrix} = t_{11}t_{22} - t_{12}t_{21}.$$

Solving these equations for $t_{11}$, $t_{12}$, $t_{21}$, and $t_{22}$ yields

$$t_{11} = \cos\varphi,\ t_{12} = -\sin\varphi,\ t_{21} = \sin\varphi, \text{ and } t_{22} = \cos\varphi$$

for some real number $\varphi$.

Using the explicit expression for $V$ from Exercise 5.15, the $[i, j]$-th entry of $R_{ab}$ is then found to equal

$$\left[R_{ab}\right]_{ij} = \delta_{ij}\cos\varphi + (1 - \cos\varphi)\, v_i v_j - \sin\varphi \sum_k \varepsilon_{ijk} v_k,$$

where

$$\delta_{11} = \delta_{22} = \delta_{33} = \varepsilon_{123} = \varepsilon_{231} = \varepsilon_{312} = -\varepsilon_{321} = -\varepsilon_{213} = -\varepsilon_{132} = 1$$

and zero otherwise,

$$\begin{pmatrix} v_1 \\ v_2 \\ v_3 \end{pmatrix}$$

is the eigenvector of $R_{ab}$ corresponding to the eigenvalue 1, and

$$v_1^2 + v_2^2 + v_3^2 = 1.$$

By restricting attention to $v_3 \geq 0$, it follows that the rotation matrix is uniquely determined by the three independent quantities $\varphi$, $v_1$, and $v_2$. This agrees with the contention in the definition above regarding the number of degrees of freedom of an orthogonal matrix.

### 5.2.5 Algebra of Rotation Matrices

#### The identity rotation

In the absence of any rotation, the two triads $a^{(r)}$ and $a^{(f)}$ must be equal, i.e.,

$$a^{(f)} = a^{(r)} R_{a^{(r)}a^{(f)}} = a^{(r)}.$$

This implies that the rotation matrix

$$R_{a^{(r)}a^{(f)}}$$

must equal the identity matrix.

#### Scaling of rotations

Recall from the above discussion that the rotation matrix $R_{a^{(r)}a^{(f)}}$ is given by the expression $R(\varphi, v_1, v_2, v_3) =$

$$\begin{pmatrix} v_1^2 + \left(1 - v_1^2\right)\cos\varphi & (1-\cos\varphi)\, v_1 v_2 - v_3 \sin\varphi & (1-\cos\varphi)\, v_1 v_3 + v_2 \sin\varphi \\ (1-\cos\varphi)\, v_1 v_2 + v_3 \sin\varphi & v_2^2 + \left(1 - v_2^2\right)\cos\varphi & (1-\cos\varphi)\, v_2 v_3 - v_1 \sin\varphi \\ (1-\cos\varphi)\, v_1 v_3 - v_2 \sin\varphi & (1-\cos\varphi)\, v_2 v_3 + v_1 \sin\varphi & v_3^2 + \left(1 - v_3^2\right)\cos\varphi \end{pmatrix},$$

where

$$v_1^2 + v_2^2 + v_3^2 = 1.$$

1

In particular,

$$R_{a^{(r)}a^{(f)}}\begin{pmatrix} v_1 \\ v_2 \\ v_3 \end{pmatrix} = \begin{pmatrix} v_1 \\ v_2 \\ v_3 \end{pmatrix}$$

$$\Leftrightarrow$$

$$\begin{pmatrix} v_1 \\ v_2 \\ v_3 \end{pmatrix} = R_{a^{(f)}a^{(r)}} R_{a^{(r)}a^{(f)}} \begin{pmatrix} v_1 \\ v_2 \\ v_3 \end{pmatrix} = R_{a^{(f)}a^{(r)}} \begin{pmatrix} v_1 \\ v_2 \\ v_3 \end{pmatrix},$$

i.e.,

$$\begin{pmatrix} v_1 \\ v_2 \\ v_3 \end{pmatrix}$$

is the eigenvector of $R_{a^{(r)}a^{(f)}}$ and $R_{a^{(f)}a^{(r)}}$ corresponding to the eigenvalue that equals 1.

If

$$\mathbf{v}_r = a^{(r)} \begin{pmatrix} v_1 \\ v_2 \\ v_3 \end{pmatrix}$$

it follows that

$$\begin{aligned} {}^{a^{(r)}}v_f &= R_{a^{(r)}a^{(f)}} {}^{a^{(f)}}v_f \\ &= R_{a^{(r)}a^{(f)}} {}^{a^{(r)}}v_r \\ &= R_{a^{(r)}a^{(f)}} \begin{pmatrix} v_1 \\ v_2 \\ v_3 \end{pmatrix} \\ &= \begin{pmatrix} v_1 \\ v_2 \\ v_3 \end{pmatrix}, \end{aligned}$$

i.e.,

$$\mathbf{v}_f = \mathbf{v}_r.$$

If we let $A$ denote the point kept fixed by the pure rotation, then any point $B$ for which

$$A \overset{\alpha \mathbf{v}_r}{\curvearrowright} B$$

is also kept fixed by the pure rotation. It follows that $\mathbf{v}_r$ spans the axis held fixed by the pure rotation.

Now, let

$$\mathbf{w}_r = a^{(r)} \begin{pmatrix} w_1 \\ w_2 \\ w_3 \end{pmatrix}$$

be some arbitrary vector that is perpendicular to $\mathbf{v}_r$. It follows that the angle between $\mathbf{v}_f$ and $\mathbf{w}_f$, $\theta(\mathbf{v}_f, \mathbf{w}_f)$, satisfies the equality

$$\begin{aligned}
\cos\theta(\mathbf{v}_f, \mathbf{w}_f) &= \frac{\mathbf{v}_f \bullet \mathbf{w}_f}{\|\mathbf{v}_f\| \cdot \|\mathbf{w}_f\|} \\
&= \frac{1}{\|\mathbf{v}_r\| \cdot \|\mathbf{w}_r\|} \left({}^{a^{(r)}} v_f\right)^T {}^{a^{(r)}} w_f \\
&= \frac{1}{\|\mathbf{v}_r\| \cdot \|\mathbf{w}_r\|} \left(R_{a^{(r)}a^{(f)}} {}^{a^{(f)}} v_f\right)^T R_{a^{(r)}a^{(f)}} {}^{a^{(f)}} w_f \\
&= \frac{1}{\|\mathbf{v}_r\| \cdot \|\mathbf{w}_r\|} \left({}^{a^{(r)}} v_r\right)^T \left(R_{a^{(r)}a^{(f)}}\right)^T R_{a^{(r)}a^{(f)}} {}^{a^{(r)}} w_r \\
&= \frac{\mathbf{v}_r \bullet \mathbf{w}_r}{\|\mathbf{v}_r\| \cdot \|\mathbf{w}_r\|} \\
&= \cos\theta(\mathbf{v}_r, \mathbf{w}_r) \\
&= 0,
\end{aligned}$$

where the last equality follows, since $\mathbf{v}_r$ is perpendicular to $\mathbf{w}_r$. It follows that $\mathbf{w}_f$ is also perpendicular to $\mathbf{v}_r = \mathbf{v}_f$.

Moreover, the angle between $\mathbf{w}_r$ and $\mathbf{w}_f$, $\theta(\mathbf{w}_r, \mathbf{w}_f)$, satisfies the equality

$$\begin{aligned}
\cos\theta(\mathbf{w}_r, \mathbf{w}_f) &= \frac{\mathbf{w}_r \bullet \mathbf{w}_f}{\|\mathbf{w}_r\| \cdot \|\mathbf{w}_f\|} \\
&= \frac{1}{w_1^2 + w_2^2 + w_3^2} \left({}^{a^{(r)}} w_r\right)^T R_{a^{(r)}a^{(f)}} {}^{a^{(r)}} w_r,
\end{aligned}$$

which, using the expression for $R_{a^{(r)}a^{(f)}}$ shown above, is found to equal $\cos\varphi$. It follows that the quantity $\varphi$ corresponds to the angle of rotation about the axis spanned by $\mathbf{v}_r$.

- Every pure rotation corresponds to a rotation about some fixed axis by some angle;
- The axis is spanned by the vector whose matrix representation relative to both $a^{(r)}$ and $a^{(f)}$ is the eigenvector of the corresponding rotation matrix corresponding to the eigenvalue 1;
- The angle of rotation is the quantity $\varphi$ that appeared in the expression for the rotation matrix.

It follows that to scale the pure rotation by the real number $\alpha$ corresponds to multiplying $\varphi$ by $\alpha$:

$$R(\alpha\varphi, v_1, v_2, v_3).$$

### Composition

If $a$, $b$, and $c$ are three arbitrary triads and $\mathbf{v}$ is an arbitrary vector, then

$$^{a}v = R_{ab}{}^{b}v,$$
$$^{b}v = R_{bc}{}^{c}v,$$

and

$$^{a}v = R_{ac}{}^{c}v.$$

Substitution of the second expression into the first yields

$$^{a}v = R_{ab}R_{bc}{}^{c}v.$$

Since $\mathbf{v}$ was arbitrary, comparison with the third expression implies that

$$R_{ac} = R_{ab}R_{bc}.$$

**Illustration 5.5**

Let $A$ be the reference point of a rigid body and introduce a body-fixed triad $a$. Let the $^{(r)}$ superscript refer to the corresponding triad in the reference configuration. Suppose that $\mathbf{R}_1$ is a pure rotation of the rigid body by a quarter of a turn about a direction parallel to the $\mathbf{a}_1^{(r)}$ basis vector and that $\mathbf{R}_2$ is a pure rotation of the rigid body by a quarter of a turn about a direction parallel to the vector $\mathbf{a}_3^{(r)}$.

Consider the composition $\mathbf{R}_2 \circ \mathbf{R}_1$ and denote by $a^{(i_1)}$ the corresponding triad after applying the rotation $\mathbf{R}_1$ and by $a^{(f_1)}$ the corresponding triad after applying $\mathbf{R}_2$. Then, the pure rotation $\mathbf{R}_1$ corresponds to the rotation matrix

$$R_{a^{(r)}a^{(i_1)}} = R\left(\frac{\pi}{2}, 1, 0, 0\right).$$

Moreover, since

$$\begin{aligned}\mathbf{a}_3^{(r)} &= a^{(r)} \begin{pmatrix} 0 \\ 0 \\ 1 \end{pmatrix} \\ &= a^{(i_1)} R_{a^{(i_1)}a^{(r)}} \begin{pmatrix} 0 \\ 0 \\ 1 \end{pmatrix} \\ &= a^{(i_1)} \begin{pmatrix} 0 \\ 1 \\ 0 \end{pmatrix},\end{aligned}$$

the subsequent application of $\mathbf{R}_2$ corresponds to the rotation matrix

$$R_{a^{(i_1)}a^{(f_1)}} = R\left(\frac{\pi}{2}, 0, 1, 0\right).$$

The rotation matrix corresponding to the composite rotation $\mathbf{R}_2 \circ \mathbf{R}_1$ is then given by

$$\begin{aligned} R_{a^{(r)}a^{(f_1)}} &= R_{a^{(r)}a^{(i_1)}} R_{a^{(i_1)}a^{(f_1)}} \\ &= \begin{pmatrix} 0 & 0 & 1 \\ 1 & 0 & 0 \\ 0 & 1 & 0 \end{pmatrix}. \end{aligned}$$

Consider instead the composition $\mathbf{R}_1 \circ \mathbf{R}_2$ and denote by $a^{(i_2)}$ the corresponding triad after applying the rotation $\mathbf{R}_2$ and by $a^{(f_2)}$ the corresponding triad after applying $\mathbf{R}_1$. Then, the pure rotation $\mathbf{R}_2$ corresponds to the rotation matrix

$$R_{a^{(r)}a^{(i_2)}} = R\left(\frac{\pi}{2}, 0, 0, 1\right).$$

Moreover, since

$$\begin{aligned} \mathbf{a}_1^{(r)} &= a^{(r)} \begin{pmatrix} 1 \\ 0 \\ 0 \end{pmatrix} \\ &= a^{(i_2)} R_{a^{(i_2)}a^{(r)}} \begin{pmatrix} 1 \\ 0 \\ 0 \end{pmatrix} \\ &= a^{(i_2)} \begin{pmatrix} 0 \\ -1 \\ 0 \end{pmatrix}, \end{aligned}$$

the subsequent application of $\mathbf{R}_1$ corresponds to the rotation matrix

$$R_{a^{(i_2)}a^{(f_2)}} = R\left(\frac{\pi}{2}, 0, -1, 0\right).$$

The rotation matrix corresponding to the composite rotation $\mathbf{R}_1 \circ \mathbf{R}_2$ is then given by

$$\begin{aligned} R_{a^{(r)}a^{(f_2)}} &= R_{a^{(r)}a^{(i_2)}} R_{a^{(i_2)}a^{(f_2)}} \\ &= \begin{pmatrix} 0 & -1 & 0 \\ 0 & 0 & -1 \\ 1 & 0 & 0 \end{pmatrix}. \end{aligned}$$

Clearly,

$$\mathbf{R}_1 \circ \mathbf{R}_2 \neq \mathbf{R}_2 \circ \mathbf{R}_1$$

as first argued in Chapter 1. Note also that the rotation matrices corresponding to the pure rotation $\mathbf{R}_1$ depend on the triad that is used to compute the rotation matrix.

If the pure rotations correspond to rotations about the same direction, then the corresponding rotation matrices are

$$R_{a^{(r)}a^{(i_1)}} = R_{a^{(i_2)}a^{(f_2)}} = R\left(\varphi_1, v_1, v_2, v_3\right)$$

and

$$R_{a^{(i_1)}a^{(f_1)}} = R_{a^{(r)}a^{(i_2)}} = R\left(\varphi_2, v_1, v_2, v_3\right).$$

Moreover, since (show this!)

$$R\left(\varphi_1, v_1, v_2, v_3\right) R\left(\varphi_2, v_1, v_2, v_3\right) = R\left(\varphi_1 + \varphi_2, v_1, v_2, v_3\right),$$

it follows that

$$\mathbf{R}_1 \circ \mathbf{R}_2 = \mathbf{R}_2 \circ \mathbf{R}_1$$

in this case.

### Inverses

The result of Exercise 5.11 can be expressed as

$$\left(R_{ab}\right)^{-1} = R_{ba}.$$

But $R_{ab}$ is the rotation matrix corresponding to the pure rotation $\mathbf{R}$ that brings the triad $a$ to coincide with $b$. This observation thus implies that the inverse $\mathbf{R}^{-1}$ corresponds to the rotation matrix $R_{ba}$.

If

$$R\left(\varphi, v_1, v_2, v_3\right)$$

is the rotation matrix corresponding to the pure rotation $\mathbf{R}$, then the inverse $\mathbf{R}^{-1}$ corresponds to a rotation about the same axis and by the same amount as $\mathbf{R}$ but in the opposite direction. The corresponding rotation matrix is then given by

$$R\left(-\varphi, v_1, v_2, v_3\right).$$

From the illustration in the previous section, we recall that

$$R\left(\varphi, v_1, v_2, v_3\right) R\left(-\varphi, v_1, v_2, v_3\right) = R\left(0, v_1, v_2, v_3\right),$$

i.e., the identity matrix.

## 5.3 Special Cases

(Ex. 5.25 – Ex. 5.31)

Let $A$ be the point on a rigid body kept fixed by the pure rotation $\mathbf{R}$. Let $B$ and $C$ be two other points on the rigid body, such that the separations $\overrightarrow{AB}$ and $\overrightarrow{AC}$ have unit length and are mutually perpendicular. Then

$$\mathbf{a}_1 = \left[\overrightarrow{AB}\right], \mathbf{a}_2 = \left[\overrightarrow{AC}\right], \text{ and } \mathbf{a}_3 = \mathbf{a}_1 \times \mathbf{a}_2$$

are the components of a triad that is fixed relative to the rigid body. Let a $^{(r)}$ superscript denote the corresponding triad in the reference configuration and let a $^{(f)}$ superscript denote the corresponding triad in the final configuration. The rotation matrix

$$R_{a^{(r)}a^{(f)}} = \left(a^{(r)}\right)^T \bullet a^{(f)}$$

describes the relationship between the two triads. In this section, we consider some special cases of rotation matrices and the corresponding pure rotations.

### 5.3.1 Rotations about $\mathbf{a}_1$

Suppose that the pure rotation $\mathbf{R}$ corresponds to a rotation of the rigid body about the axis through $A$ that is parallel to the vector $\mathbf{a}_1^{(r)}$ by an angle $\varphi$.

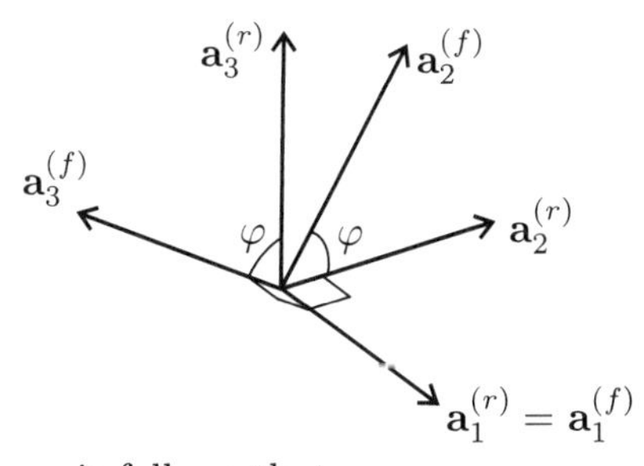


From the figure, it follows that

$$\begin{aligned}
\mathbf{a}_1^{(f)} &= \mathbf{a}_1^{(r)}, \\
\mathbf{a}_2^{(f)} &= \cos\varphi \mathbf{a}_2^{(r)} + \sin\varphi \mathbf{a}_3^{(r)}, \\
\mathbf{a}_3^{(f)} &= -\sin\varphi \mathbf{a}_2^{(r)} + \cos\varphi \mathbf{a}_3^{(r)}.
\end{aligned}$$

The corresponding rotation matrix then becomes

$$\begin{aligned}
R_{a^{(r)}a^{(f)}} &= \left(a^{(r)}\right)^T \bullet a^{(f)} \\
&= \begin{pmatrix} \mathbf{a}_1^{(r)} \bullet \mathbf{a}_1^{(f)} & \mathbf{a}_1^{(r)} \bullet \mathbf{a}_2^{(f)} & \mathbf{a}_1^{(r)} \bullet \mathbf{a}_3^{(f)} \\ \mathbf{a}_2^{(r)} \bullet \mathbf{a}_1^{(f)} & \mathbf{a}_2^{(r)} \bullet \mathbf{a}_2^{(f)} & \mathbf{a}_2^{(r)} \bullet \mathbf{a}_3^{(f)} \\ \mathbf{a}_3^{(r)} \bullet \mathbf{a}_1^{(f)} & \mathbf{a}_3^{(r)} \bullet \mathbf{a}_2^{(f)} & \mathbf{a}_3^{(r)} \bullet \mathbf{a}_3^{(f)} \end{pmatrix} \\
&= \begin{pmatrix} 1 & 0 & 0 \\ 0 & \cos\varphi & -\sin\varphi \\ 0 & \sin\varphi & \cos\varphi \end{pmatrix}.
\end{aligned}$$

**Illustration 5.6**

Let $D$ be a fourth point on the rigid body, such that

$$\mathbf{r}^{AD_r} = a^{(r)} \begin{pmatrix} 0 \\ y \\ z \end{pmatrix}.$$

It follows that

$$\begin{aligned}
\mathbf{r}^{AD_f} &= a^{(f)} \begin{pmatrix} 0 \\ y \\ z \end{pmatrix} \\
&= a^{(r)} R_{a^{(r)}a^{(f)}} \begin{pmatrix} 0 \\ y \\ z \end{pmatrix} \\
&= a^{(r)} \begin{pmatrix} 0 \\ y\cos\varphi - z\sin\varphi \\ y\sin\varphi + z\cos\varphi \end{pmatrix}
\end{aligned}$$

and consequently,

$$\begin{aligned}
\cos\theta\left(\mathbf{r}^{AD_r}, \mathbf{r}^{AD_f}\right) &= \frac{\mathbf{r}^{AD_r} \bullet \mathbf{r}^{AD_f}}{\left\|\mathbf{r}^{AD_r}\right\| \left\|\mathbf{r}^{AD_f}\right\|} \\
&= \cos\varphi.
\end{aligned}$$

It is straightforward to see that

$$R_{a^{(r)}a^{(f)}} = R(\varphi, 1, 0, 0).$$

### 5.3.2 Rotations about $\mathbf{a}_2$

Suppose that the pure rotation $\mathbf{R}$ corresponds to a rotation of the rigid body about the axis through $A$ that is parallel to the vector $\mathbf{a}_2^{(r)}$ by an angle $\varphi$.

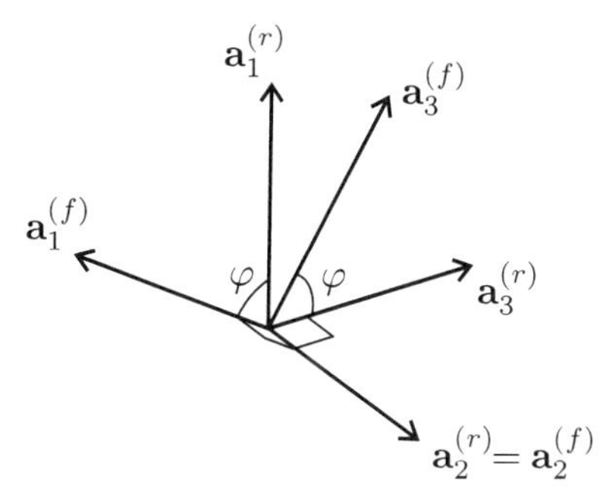


From the figure, it follows that

$$\begin{aligned}
\mathbf{a}_1^{(f)} &= \cos\varphi \mathbf{a}_1^{(r)} - \sin\varphi \mathbf{a}_3^{(r)}, \\
\mathbf{a}_2^{(f)} &= \mathbf{a}_2^{(r)}, \\
\mathbf{a}_3^{(f)} &= \sin\varphi \mathbf{a}_1^{(r)} + \cos\varphi \mathbf{a}_3^{(r)}.
\end{aligned}$$

The corresponding rotation matrix then becomes

$$\begin{aligned}
R_{a^{(r)}a^{(f)}} &= \left(a^{(r)}\right)^T \bullet a^{(f)} \\
&= \begin{pmatrix} \mathbf{a}_1^{(r)} \bullet \mathbf{a}_1^{(f)} & \mathbf{a}_1^{(r)} \bullet \mathbf{a}_2^{(f)} & \mathbf{a}_1^{(r)} \bullet \mathbf{a}_3^{(f)} \\ \mathbf{a}_2^{(r)} \bullet \mathbf{a}_1^{(f)} & \mathbf{a}_2^{(r)} \bullet \mathbf{a}_2^{(f)} & \mathbf{a}_2^{(r)} \bullet \mathbf{a}_3^{(f)} \\ \mathbf{a}_3^{(r)} \bullet \mathbf{a}_1^{(f)} & \mathbf{a}_3^{(r)} \bullet \mathbf{a}_2^{(f)} & \mathbf{a}_3^{(r)} \bullet \mathbf{a}_3^{(f)} \end{pmatrix} \\
&= \begin{pmatrix} \cos\varphi & 0 & \sin\varphi \\ 0 & 1 & 0 \\ -\sin\varphi & 0 & \cos\varphi \end{pmatrix}.
\end{aligned}$$

**Illustration 5.7**

Let $D$ be a fourth point on the rigid body, such that

$$\mathbf{r}^{AD_r} = a^{(r)} \begin{pmatrix} x \\ 0 \\ z \end{pmatrix}.$$

It follows that

$$\begin{aligned}
\mathbf{r}^{AD_f} &= a^{(f)} \begin{pmatrix} x \\ 0 \\ z \end{pmatrix} \\
&= a^{(r)} R_{a^{(r)}a^{(f)}} \begin{pmatrix} x \\ 0 \\ z \end{pmatrix} \\
&= a^{(r)} \begin{pmatrix} x\cos\varphi + z\sin\varphi \\ 0 \\ -x\sin\varphi + z\cos\varphi \end{pmatrix}
\end{aligned}$$

and consequently,

$$\begin{aligned}\cos\theta\left(\mathbf{r}^{AD_r},\mathbf{r}^{AD_f}\right) &= \frac{\mathbf{r}^{AD_r}\bullet\mathbf{r}^{AD_f}}{\left\|\mathbf{r}^{AD_r}\right\|\left\|\mathbf{r}^{AD_f}\right\|} \\ &= \cos\varphi.\end{aligned}$$

It is straightforward to see that

$$R_{a^{(r)}a^{(f)}} = R(\varphi,0,1,0).$$

### 5.3.3 Rotations about $\mathbf{a}_3$

Suppose that the pure rotation $\mathbf{R}$ corresponds to a rotation of the rigid body about the axis through $A$ that is parallel to the vector $\mathbf{a}_3^{(r)}$ by an angle $\varphi$.

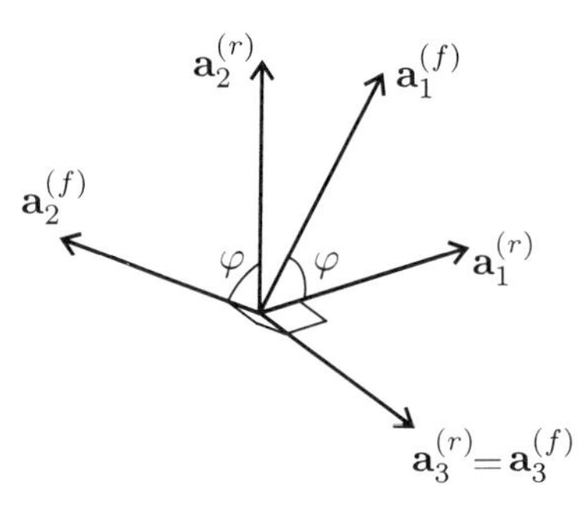


From the figure, it follows that

$$\begin{aligned}\mathbf{a}_1^{(f)} &= \cos\varphi\mathbf{a}_1^{(r)} + \sin\varphi\mathbf{a}_2^{(r)}, \\ \mathbf{a}_2^{(f)} &= -\sin\varphi\mathbf{a}_1^{(r)} + \cos\varphi\mathbf{a}_2^{(r)}, \\ \mathbf{a}_3^{(f)} &= \mathbf{a}_3^{(r)}.\end{aligned}$$

The corresponding rotation matrix then becomes

$$\begin{aligned}R_{a^{(r)}a^{(f)}} &= \left(a^{(r)}\right)^T \bullet a^{(f)} \\ &= \begin{pmatrix} \mathbf{a}_1^{(r)}\bullet\mathbf{a}_1^{(f)} & \mathbf{a}_1^{(r)}\bullet\mathbf{a}_2^{(f)} & \mathbf{a}_1^{(r)}\bullet\mathbf{a}_3^{(f)} \\ \mathbf{a}_2^{(r)}\bullet\mathbf{a}_1^{(f)} & \mathbf{a}_2^{(r)}\bullet\mathbf{a}_2^{(f)} & \mathbf{a}_2^{(r)}\bullet\mathbf{a}_3^{(f)} \\ \mathbf{a}_3^{(r)}\bullet\mathbf{a}_1^{(f)} & \mathbf{a}_3^{(r)}\bullet\mathbf{a}_2^{(f)} & \mathbf{a}_3^{(r)}\bullet\mathbf{a}_3^{(f)} \end{pmatrix} \\ &= \begin{pmatrix} \cos\varphi & -\sin\varphi & 0 \\ \sin\varphi & \cos\varphi & 0 \\ 0 & 0 & 1 \end{pmatrix}.\end{aligned}$$

**Illustration 5.8**

Let $D$ be a fourth point on the rigid body, such that

$$\mathbf{r}^{AD_r} = a^{(r)} \begin{pmatrix} x \\ y \\ 0 \end{pmatrix}.$$

It follows that

$$\begin{aligned}
\mathbf{r}^{AD_f} &= a^{(f)} \begin{pmatrix} x \\ y \\ 0 \end{pmatrix} \\
&= a^{(r)} R_{a^{(r)}a^{(f)}} \begin{pmatrix} x \\ y \\ 0 \end{pmatrix} \\
&= a^{(r)} \begin{pmatrix} x\cos\varphi - y\sin\varphi \\ x\sin\varphi + y\cos\varphi \\ 0 \end{pmatrix}
\end{aligned}$$

and consequently,

$$\begin{aligned}
\cos\theta\left(\mathbf{r}^{AD_r}, \mathbf{r}^{AD_f}\right) &= \frac{\mathbf{r}^{AD_r} \bullet \mathbf{r}^{AD_f}}{\left\|\mathbf{r}^{AD_r}\right\| \left\|\mathbf{r}^{AD_f}\right\|} \\
&= \cos\varphi.
\end{aligned}$$

It is straightforward to see that

$$R_{a^{(r)}a^{(f)}} = R(\varphi, 0, 0, 1).$$

### 5.3.4 Euler Angles

**Illustration 5.9**

Let the pure rotation $\mathbf{R}_1$ correspond to a rotation about an axis through the point $A$ parallel to the vector $\mathbf{a}_3^{(r)}$ by an angle $\varphi_1$. The corresponding rotation matrix is

$$R_{a^{(r)}a^{(i)}} = \begin{pmatrix} c_1 & -s_1 & 0 \\ s_1 & c_1 & 0 \\ 0 & 0 & 1 \end{pmatrix},$$

where $c_1 = \cos\varphi_1$ and $s_1 = \sin\varphi_1$. Here, $a^{(i)}$ corresponds to the right-handed, orthonormal basis

$$\mathbf{a}_1^{(i)} = \left[\overrightarrow{AB_i}\right], \; \mathbf{a}_2^{(i)} = \left[\overrightarrow{AC_i}\right], \text{ and } \mathbf{a}_3^{(i)} = \mathbf{a}_1^{(i)} \times \mathbf{a}_2^{(i)}$$

and the $^{(i)}$ superscript refers to an intermediate configuration.

Let the pure rotation $\mathbf{R}_2$ correspond to a rotation about an axis through the point $A$ parallel to the vector $\mathbf{a}_2^{(i)}$ by an angle $\varphi_2$. The corresponding rotation matrix is

$$R_{a^{(i)}a^{(f)}} = \begin{pmatrix} c_2 & 0 & s_2 \\ 0 & 1 & 0 \\ -s_2 & 0 & c_2 \end{pmatrix}.$$

From the previous section, we conclude that the rotation matrix corresponding to the composition $\mathbf{R}_2 \circ \mathbf{R}_1$ equals

$$\begin{aligned} R_{a^{(r)}a^{(f)}} &= R_{a^{(r)}a^{(i)}} R_{a^{(i)}a^{(f)}} \\ &= \begin{pmatrix} c_1 & -s_1 & 0 \\ s_1 & c_1 & 0 \\ 0 & 0 & 1 \end{pmatrix} \begin{pmatrix} c_2 & 0 & s_2 \\ 0 & 1 & 0 \\ -s_2 & 0 & c_2 \end{pmatrix} \\ &= \begin{pmatrix} c_1 c_2 & -s_1 & c_1 s_2 \\ s_1 c_2 & c_1 & s_1 s_2 \\ -s_2 & 0 & c_2 \end{pmatrix}. \end{aligned}$$

Suppose $\mathbf{R}$ is an arbitrary pure rotation or, equivalently, that $a^{(f)}$ has some arbitrary orientation relative to $a^{(r)}$. Recall that

$$\mathbf{a}_1 = \left[\overrightarrow{AB}\right]$$

and that the point $B_f$ lies on a sphere of radius $\|\mathbf{a}_1\| = 1$ centered at $A$.

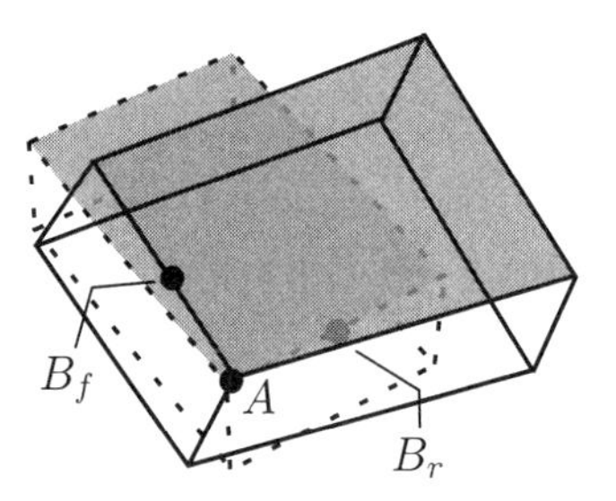


We can align the point $B$ on the rigid body with the point $B_f$ in the final configuration by a sequence of two pure rotations $\mathbf{R}_1$ and $\mathbf{R}_2$. In particular, let $\mathbf{R}_1$ correspond to a rotation about the axis through $A$ parallel to the vector $\mathbf{a}_3^{(r)}$ by an angle $\varphi_1$, such that the body-fixed triad $a = a^{(i)}$, and let $\mathbf{R}_2$ correspond to a rotation about the axis through $A$ parallel to the vector $\mathbf{a}_2^{(i)}$ by an angle $\varphi_2$, such that $\mathbf{a}_1 = \mathbf{a}_1^{(f)}$. Naturally, the angles $\varphi_1$ and $\varphi_2$ depend on the location of $B_f$ relative to $A$.

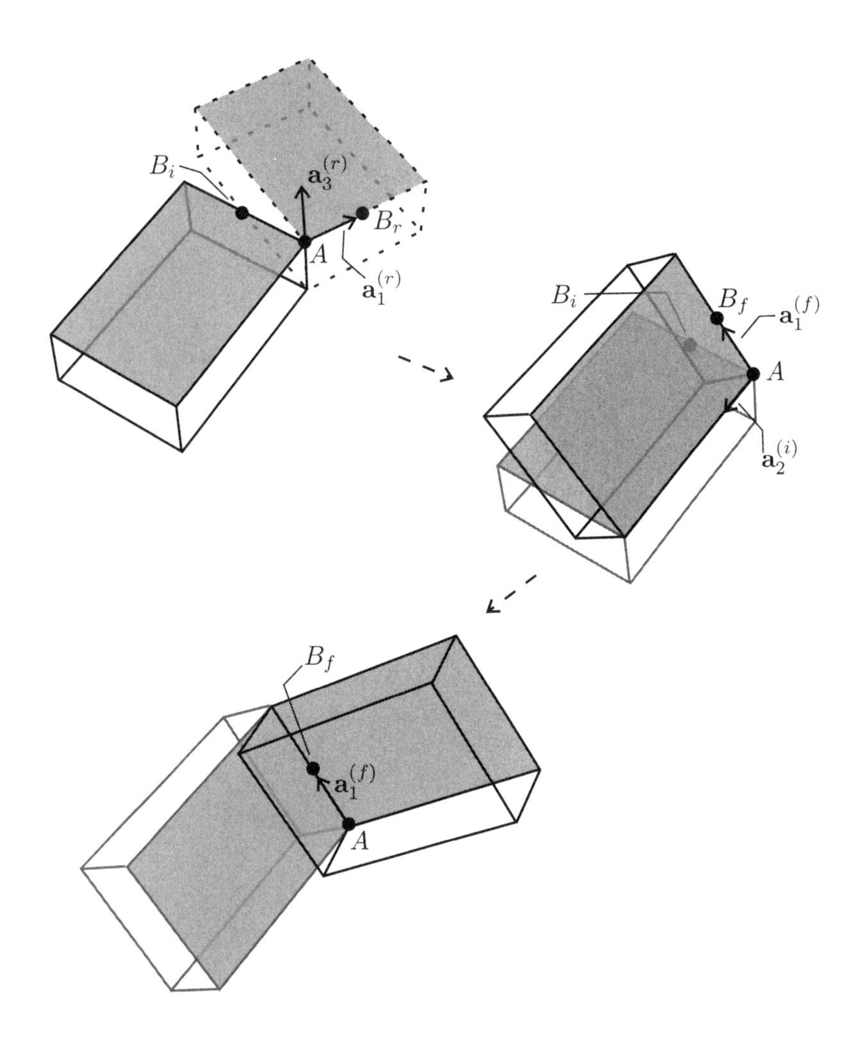


The composition $\mathbf{R}_2 \circ \mathbf{R}_1$ aligns the vector $\mathbf{a}_1$ with the corresponding vector in the final configuration $\mathbf{a}_1^{(f)}$. From the illustration, it follows that the corresponding rotation matrix is

$$\begin{pmatrix} c_1 c_2 & -s_1 & c_1 s_2 \\ s_1 c_2 & c_1 & s_1 s_2 \\ -s_2 & 0 & c_2 \end{pmatrix},$$

where $c_i = \cos \varphi_i$ and $s_i = \sin \varphi_i$. The remaining basis vectors can now be aligned with the corresponding vectors in the final configuration through a pure rotation $\mathbf{R}_3$ about an axis through $A$ parallel to $\mathbf{a}_1^{(f)}$ by an angle $\varphi_3$. It follows that the pure rotation $\mathbf{R}$ is given by the composition

$$\mathbf{R} = \mathbf{R}_3 \circ \mathbf{R}_2 \circ \mathbf{R}_1.$$

The corresponding rotation matrix becomes

$$
\begin{aligned}
R_{a^{(r)}a^{(f)}} &= \begin{pmatrix} c_1c_2 & -s_1 & c_1s_2 \\ s_1c_2 & c_1 & s_1s_2 \\ -s_2 & 0 & c_2 \end{pmatrix} \begin{pmatrix} 1 & 0 & 0 \\ 0 & c_3 & -s_3 \\ 0 & s_3 & c_3 \end{pmatrix} \\
&= \begin{pmatrix} c_1c_2 & -s_1c_3 + c_1s_2s_3 & s_1s_3 + c_1s_2c_3 \\ s_1c_2 & c_1c_3 + s_1s_2s_3 & -c_1s_3 + s_1s_2c_3 \\ -s_2 & c_2s_3 & c_2c_3 \end{pmatrix},
\end{aligned}
$$

where $c_3 = \cos\varphi_3$ and $s_3 = \sin\varphi_3$.

It is always possible to find values of $\varphi_1$, $\varphi_2$, and $\varphi_3$ such that the rotation matrix $R_{a^{(r)}a^{(f)}}$ is given by the expression above. Equivalently, every pure rotation can be decomposed into a sequence of three pure rotations, each of which is about an axis parallel to a basis vector. Conversely, every choice of values for $\varphi_1$, $\varphi_2$, and $\varphi_3$ corresponds to some pure rotation $\mathbf{R}$ of the rigid body.

The angles $\varphi_1$, $\varphi_2$, and $\varphi_3$ are called *Euler angles*. In this case, they correspond to the rotation sequence $3-2-1$, i.e., a pure rotation about the axis parallel to the third basis vector, followed by a pure rotation about the axis parallel to the second basis vector, followed by a pure rotation about the axis parallel to the first basis vector. While the first pure rotation is about a basis vector in the original triad, the second and third rotations are about basis vectors in the two intermediate triads.

There are many other possible choices for Euler angles. For example, the rotation sequence $1-3-1$ could also be used to represent an arbitrary pure rotation, as could the sequence $2-3-1$. Naturally, the form of the rotation matrix would differ between these cases.

### 5.3.5 Alignment

Suppose we seek a pure rotation that will align the $\mathbf{a}_1$ vector with an axis through $A$ and an additional point $B$, such that

$$
\mathbf{r}^{AB} = a^{(r)} \begin{pmatrix} 2 \\ 3 \\ 1 \end{pmatrix},
$$

i.e., such that $\mathbf{a}_1^{(f)}$ is parallel to $\mathbf{r}^{AB}$. Following the methodology presented in Chapter 3, we let

$$
\mathbf{a}_1^{(f)} = \frac{\mathbf{r}^{AB}}{\|\mathbf{r}^{AB}\|} = a^{(r)} \begin{pmatrix} \frac{2}{\sqrt{14}} \\ \frac{3}{\sqrt{14}} \\ \frac{1}{\sqrt{14}} \end{pmatrix},
$$

$$
\mathbf{a}_2^{(f)} = \frac{\mathbf{a}_1^{(f)} \times \mathbf{a}_1^{(r)}}{\left\|\mathbf{a}_1^{(f)} \times \mathbf{a}_1^{(r)}\right\|} = a^{(r)} \begin{pmatrix} 0 \\ \frac{1}{\sqrt{10}} \\ -\frac{3}{\sqrt{10}} \end{pmatrix},
$$

and

$$\mathbf{a}_3^{(f)} = \mathbf{a}_1^{(f)} \times \mathbf{a}_2^{(f)} = a^{(r)} \begin{pmatrix} -\frac{5}{\sqrt{35}} \\ \frac{3}{\sqrt{35}} \\ \frac{1}{\sqrt{35}} \end{pmatrix}.$$

It follows that

$$a^{(f)} = a^{(r)} \begin{pmatrix} \frac{2}{\sqrt{14}} & 0 & -\frac{5}{\sqrt{35}} \\ \frac{3}{\sqrt{14}} & \frac{1}{\sqrt{10}} & \frac{3}{\sqrt{35}} \\ \frac{1}{\sqrt{14}} & -\frac{3}{\sqrt{10}} & \frac{1}{\sqrt{35}} \end{pmatrix} = a^{(r)} R_{a^{(r)}a^{(f)}}.$$

There is a unique pure rotation corresponding to the rotation matrix $R_{a^{(r)}a^{(f)}}$ found above. This pure rotation is not the only solution to the problem of aligning $\mathbf{a}_1$ with $\mathbf{r}^{AB}$, however. That there are infinitely many solutions follows from the possibility of using a different vector than $\mathbf{a}_1^{(r)}$ in the computation for $\mathbf{a}_2^{(f)}$. Which of these is most appropriate depends on the application.

## 5.4 The MAMBO Toolbox

(Ex. 5.32)

Rotation matrices are computed in the MAMBO toolbox using the `MakeRotations` procedure.

**Illustration 5.10**

In the following MAMBO toolbox session, the rotation matrices $R_{a^{(r)}a^{(f)}}$ corresponding to pure rotations about axes parallel to the basis vectors $\mathbf{a}_1^{(r)}$, $\mathbf{a}_2^{(r)}$, and $\mathbf{a}_3^{(r)}$, respectively, are computed using the `MakeRotations` procedure.

```
> MakeRotations(theta,1);
```

$$\begin{bmatrix} 1 & 0 & 0 \\ 0 & \cos(\theta) & -\sin(\theta) \\ 0 & \sin(\theta) & \cos(\theta) \end{bmatrix}$$

```
> MakeRotations(phi,2);
```

$$\begin{bmatrix} \cos(\phi) & 0 & \sin(\phi) \\ 0 & 1 & 0 \\ -\sin(\phi) & 0 & \cos(\phi) \end{bmatrix}$$

```
> MakeRotations(psi,3);
```

$$\begin{bmatrix} \cos(\psi) & -\sin(\psi) & 0 \\ \sin(\psi) & \cos(\psi) & 0 \\ 0 & 0 & 1 \end{bmatrix}$$

By combining these into a single call to `MakeRotations`, we obtain the rotation matrix corresponding to the $1-2-3$ sequence of pure rotations introduced in the context of Euler angles.

```
> MakeRotations([theta,1],[phi,2],[psi,3]);
```

$$[\cos(\phi)\cos(\psi)\,,\ -\cos(\phi)\sin(\psi)\,,\ \sin(\phi)]$$
$$[\sin(\theta)\sin(\phi)\cos(\psi)+\cos(\theta)\sin(\psi)\,,$$
$$-\sin(\theta)\sin(\phi)\sin(\psi)+\cos(\theta)\cos(\psi)\,,\ -\sin(\theta)\cos(\phi)]$$
$$[-\cos(\theta)\sin(\phi)\cos(\psi)+\sin(\theta)\sin(\psi)\,,$$
$$\cos(\theta)\sin(\phi)\sin(\psi)+\sin(\theta)\cos(\psi)\,,\ \ \cos(\theta)\cos(\phi)]$$

Note the use of brackets to separate the individual rotations.

Since the output from the `MakeRotations` procedure is a MAPLE matrix, normal MAPLE matrix operations apply. For example, the commands **transpose**, **inverse**, and **eigenvals**, respectively, will compute the transpose, inverse, and eigenvalues of a rotation matrix.

```
> rotmat:=MakeRotations(theta,1);
```

$$rotmat := \begin{bmatrix} 1 & 0 & 0 \\ 0 & \cos(\theta) & -\sin(\theta) \\ 0 & \sin(\theta) & \cos(\theta) \end{bmatrix}$$

```
> transpose(rotmat);
```

$$\begin{bmatrix} 1 & 0 & 0 \\ 0 & \cos(\theta) & \sin(\theta) \\ 0 & -\sin(\theta) & \cos(\theta) \end{bmatrix}$$

```
> inverse(rotmat);
```

$$\begin{bmatrix} 1 & 0 & 0 \\ 0 & \frac{\cos(\theta)}{\%1} & \frac{\sin(\theta)}{\%1} \\ 0 & -\frac{\sin(\theta)}{\%1} & \frac{\cos(\theta)}{\%1} \end{bmatrix}$$
$$\%1 := \cos(\theta)^2 + \sin(\theta)^2$$

```
> eigenvals(rotmat);
```

$$1,\ \cos(\theta) + \sqrt{\cos(\theta)^2 - 1},\ \cos(\theta) - \sqrt{\cos(\theta)^2 - 1}$$

The `MakeRotations` procedure may also be invoked to compute the rotation matrix $R(\theta, v_1, v_2, v_3)$ corresponding to a rotation by an angle $\theta$ about some arbitrary axis parallel to the vector

$$\mathbf{v} = a^{(r)} \begin{pmatrix} v_1 \\ v_2 \\ v_3 \end{pmatrix}.$$

The command

```
> MakeRotations(theta,v1,v2,v3);
```

$$\left[\cos(\theta) + \frac{(1-\cos(\theta))\, v1^2}{\%1},\ \frac{(1-\cos(\theta))\, v1\, v2}{\%1} - \frac{\sin(\theta)\, v3}{\sqrt{\%1}},\ \frac{(1-\cos(\theta))\, v1\, v3}{\%1} + \frac{\sin(\theta)\, v2}{\sqrt{\%1}}\right]$$
$$\left[\frac{(1-\cos(\theta))\, v1\, v2}{\%1} + \frac{\sin(\theta)\, v3}{\sqrt{\%1}},\ \cos(\theta) + \frac{(1-\cos(\theta))\, v2^2}{\%1},\ \frac{(1-\cos(\theta))\, v2\, v3}{\%1} - \frac{\sin(\theta)\, v1}{\sqrt{\%1}}\right]$$
$$\left[\frac{(1-\cos(\theta))\, v1\, v3}{\%1} - \frac{\sin(\theta)\, v2}{\sqrt{\%1}},\ \frac{(1-\cos(\theta))\, v2\, v3}{\%1} + \frac{\sin(\theta)\, v1}{\sqrt{\%1}},\ \cos(\theta) + \frac{(1-\cos(\theta))\, v3^2}{\%1}\right]$$
$$\%1 := v1^2 + v2^2 + v3^2$$

returns the matrix derived in Section 5.2. Note that `MakeRotations` relaxes the requirement that $v_1^2 + v_2^2 + v_3^2 = 1$ and introduces the appropriate normalization where necessary.

**Illustration 5.11**

We may also appeal to the definition of the rotation matrix from a triad $a$ to a triad $b$:

$$R_{ab} = a^T \bullet b.$$

Suppose that we seek to introduce a triad $b$, such that $\mathbf{b}_1$ is parallel to the vector

$$a \begin{pmatrix} 2 \\ 3 \\ 1 \end{pmatrix}.$$

Then, the following sequence of MAMBO toolbox statements generates the triad $b$ and computes the associated rotation matrix $R_{ab}$.

```
> DeclareTriads(a):
> a1:=MakeTranslations(a,1):
> a2:=MakeTranslations(a,2):
> a3:=MakeTranslations(a,3):
> v:=MakeTranslations(a,2,3,1):
> b1:=(1/VectorLength(v)) &** v:
> b2:=(1/VectorLength(b1 &xx a1)) &** (b1 &xx a1):
> b3:=b1 &xx b2:
> matrix(3,3,(i,j)->cat(a,i) &oo cat(b,j));
```

$$\begin{bmatrix} \frac{1}{7}\sqrt{14} & 0 & -\frac{1}{7}\sqrt{35} \\ \frac{3}{14}\sqrt{14} & \frac{1}{70}\sqrt{35}\sqrt{14} & \frac{3}{35}\sqrt{35} \\ \frac{1}{14}\sqrt{14} & -\frac{3}{70}\sqrt{35}\sqrt{14} & \frac{1}{35}\sqrt{35} \end{bmatrix}$$

## 5.5 Exercises

**Exercise 5.1** Let $a$ and $b$ be two triads, such that

$$R_{ab} = \begin{pmatrix} \frac{1}{2} & -\frac{3}{4} & \frac{\sqrt{3}}{4} \\ \frac{\sqrt{3}}{2} & \frac{\sqrt{3}}{4} & -\frac{1}{4} \\ 0 & \frac{1}{2} & \frac{\sqrt{3}}{2} \end{pmatrix}.$$

Find the matrix representation of the vector $\mathbf{b}_2$ relative to the triad $a$.

**Solution.** Expanding the general formula

$$b = aR_{ab}$$

we find

$$\begin{pmatrix} \mathbf{b}_1 & \mathbf{b}_2 & \mathbf{b}_3 \end{pmatrix} = \begin{pmatrix} \mathbf{a}_1 & \mathbf{a}_2 & \mathbf{a}_3 \end{pmatrix} \begin{pmatrix} \frac{1}{2} & -\frac{3}{4} & \frac{\sqrt{3}}{4} \\ \frac{\sqrt{3}}{2} & \frac{\sqrt{3}}{4} & -\frac{1}{4} \\ 0 & \frac{1}{2} & \frac{\sqrt{3}}{2} \end{pmatrix},$$

which implies that

$$\begin{aligned} \mathbf{b}_2 &= -\frac{3}{4}\mathbf{a}_1 + \frac{\sqrt{3}}{4}\mathbf{a}_2 + \frac{1}{2}\mathbf{a}_3 \\ &= a \begin{pmatrix} -3/4 \\ \sqrt{3}/4 \\ 1/2 \end{pmatrix} = a\ {}^{a}(\mathbf{b}_2)\,. \end{aligned}$$

**Exercise 5.2** Let $a$ and $b$ be two triads, such that

$$R_{ab} = \begin{pmatrix} \frac{1}{2} & -\frac{3}{4} & \frac{\sqrt{3}}{4} \\ \frac{\sqrt{3}}{2} & \frac{\sqrt{3}}{4} & -\frac{1}{4} \\ 0 & \frac{1}{2} & \frac{\sqrt{3}}{2} \end{pmatrix}.$$

Find the matrix representation of the vector $\mathbf{a}_3$ relative to the triad $b$.

**Solution.** Since $R_{ba} = (R_{ab})^T$, we have

$$a = bR_{ba} = b\,(R_{ab})^T$$

or

$$\begin{pmatrix} \mathbf{a}_1 & \mathbf{a}_2 & \mathbf{a}_3 \end{pmatrix} = \begin{pmatrix} \mathbf{b}_1 & \mathbf{b}_2 & \mathbf{b}_3 \end{pmatrix} \begin{pmatrix} \frac{1}{2} & \frac{\sqrt{3}}{2} & 0 \\ -\frac{3}{4} & \frac{\sqrt{3}}{4} & \frac{1}{2} \\ \frac{\sqrt{3}}{4} & -\frac{1}{4} & \frac{\sqrt{3}}{2} \end{pmatrix},$$

i.e.,

$$\begin{aligned} \mathbf{a}_3 &= \frac{1}{2}\mathbf{b}_2 + \frac{\sqrt{3}}{2}\mathbf{b}_3 \\ &= b \begin{pmatrix} 0 \\ 1/2 \\ \sqrt{3}/2 \end{pmatrix} = b\ {}^{b}(\mathbf{a}_3)\,. \end{aligned}$$

**Exercise 5.3** Let $a$ and $b$ be two arbitrary triads. Find the matrix representations of the basis vectors $\mathbf{a}_1$, $\mathbf{a}_2$, and $\mathbf{a}_3$ relative to the $b$ triad and of the basis vectors $\mathbf{b}_1$, $\mathbf{b}_2$, and $\mathbf{b}_3$ relative to the $a$ triad if $R_{ab} =$

a) $\begin{pmatrix} \frac{3}{4} & -\frac{\sqrt{6}}{4} & \frac{1}{4} \\ \frac{\sqrt{6}}{4} & \frac{1}{2} & -\frac{\sqrt{6}}{4} \\ \frac{1}{4} & \frac{\sqrt{6}}{4} & \frac{3}{4} \end{pmatrix}$

b) $\begin{pmatrix} 0 & -\frac{1}{\sqrt{2}} & \frac{1}{\sqrt{2}} \\ \frac{1}{\sqrt{2}} & \frac{1}{2} & \frac{1}{2} \\ -\frac{1}{\sqrt{2}} & \frac{1}{2} & \frac{1}{2} \end{pmatrix}$

c) $\begin{pmatrix} \frac{4}{5} & \frac{2}{5} & \frac{1}{\sqrt{5}} \\ \frac{2}{5} & \frac{1}{5} & -\frac{2}{\sqrt{5}} \\ -\frac{1}{\sqrt{5}} & \frac{2}{\sqrt{5}} & 0 \end{pmatrix}$

d) $\begin{pmatrix} \frac{1}{\sqrt{2}} & -\frac{1}{\sqrt{2}} & 0 \\ \frac{1}{\sqrt{2}} & \frac{1}{\sqrt{2}} & 0 \\ 0 & 0 & 1 \end{pmatrix}$

e) $\begin{pmatrix} -\frac{1}{\sqrt{2}} & 0 & \frac{1}{\sqrt{2}} \\ 0 & -1 & 0 \\ \frac{1}{\sqrt{2}} & 0 & \frac{1}{\sqrt{2}} \end{pmatrix}$

f) $\begin{pmatrix} \frac{1}{3} - \frac{1}{\sqrt{3}} & \frac{1}{3} + \frac{1}{\sqrt{3}} & \frac{1}{3} \\ \frac{1}{\sqrt{3}} + \frac{1}{6} & \frac{1}{2\sqrt{3}} - \frac{1}{3} & \frac{2}{3} \\ \frac{1}{3} + \frac{1}{2\sqrt{3}} & -\frac{1}{6} + \frac{1}{\sqrt{3}} & -\frac{2}{3} \end{pmatrix}$

**Exercise 5.4** Let $a$ and $b$ be two arbitrary triads. Find $R_{ab}$ and $R_{ba}$ when

a) $\mathbf{a}_1 = \mathbf{b}_2$, $\mathbf{a}_2 = \mathbf{b}_3$, $\mathbf{a}_3 = \mathbf{b}_1$

b) $\mathbf{b}_1 = -\frac{1}{\sqrt{2}}\mathbf{a}_1 + \frac{1}{\sqrt{2}}\mathbf{a}_3$, $\mathbf{b}_2 = -\mathbf{a}_2$, $\mathbf{b}_3 = \frac{1}{\sqrt{2}}\mathbf{a}_1 + \frac{1}{\sqrt{2}}\mathbf{a}_3$

c) $\mathbf{b}_1 = \mathbf{a}_2$, $\mathbf{b}_2 = -\mathbf{a}_1$, $\mathbf{b}_3 = \mathbf{a}_3$

d) $\mathbf{a}_1 = \frac{1}{\sqrt{2}}\mathbf{b}_1 - \frac{1}{\sqrt{2}}\mathbf{b}_2$, $\mathbf{a}_2 = \frac{1}{\sqrt{2}}\mathbf{b}_1 + \frac{1}{\sqrt{2}}\mathbf{b}_2$, $\mathbf{a}_3 = \mathbf{b}_3$

e) $\mathbf{a}_1 = -\mathbf{b}_2$, $\mathbf{a}_2 = -\mathbf{b}_3$, $\mathbf{a}_3 = \mathbf{b}_1$

$$\text{f)}\quad \begin{aligned} \mathbf{b}_1 &= \tfrac{1}{2}\mathbf{a}_1 - \tfrac{3}{4}\mathbf{a}_2 + \tfrac{\sqrt{3}}{4}\mathbf{a}_3 \\ \mathbf{b}_2 &= \tfrac{\sqrt{3}}{2}\mathbf{a}_1 + \tfrac{\sqrt{3}}{4}\mathbf{a}_2 - \tfrac{1}{4}\mathbf{a}_3 \\ \mathbf{b}_3 &= \tfrac{1}{2}\mathbf{a}_2 + \tfrac{\sqrt{3}}{2}\mathbf{a}_3 \end{aligned}$$

**Exercise 5.5** Let $a$ and $b$ be two triads, such that

$$R_{ab} = \begin{pmatrix} \frac{1}{2} & -\frac{3}{4} & \frac{\sqrt{3}}{4} \\ \frac{\sqrt{3}}{2} & \frac{\sqrt{3}}{4} & -\frac{1}{4} \\ 0 & \frac{1}{2} & \frac{\sqrt{3}}{2} \end{pmatrix}$$

and consider a vector $\mathbf{v}$ whose matrix representation relative to $a$ is given by

$${}^{a}v = \begin{pmatrix} 1 \\ 0 \\ -1 \end{pmatrix}.$$

Find the matrix representation of the vector $\mathbf{v}$ relative to the triad $b$.

**Solution.** We have

$$\begin{aligned} {}^{b}v &= R_{ba}\,{}^{a}v \\ &= (R_{ab})^{T}\,{}^{a}v \\ &= \begin{pmatrix} \frac{1}{2} & \frac{\sqrt{3}}{2} & 0 \\ -\frac{3}{4} & \frac{\sqrt{3}}{4} & \frac{1}{2} \\ \frac{\sqrt{3}}{4} & -\frac{1}{4} & \frac{\sqrt{3}}{2} \end{pmatrix} \begin{pmatrix} 1 \\ 0 \\ -1 \end{pmatrix} \\ &= \begin{pmatrix} \frac{1}{2} \\ -\frac{5}{4} \\ -\frac{\sqrt{3}}{4} \end{pmatrix}. \end{aligned}$$

**Exercise 5.6** Let $a$ and $b$ be two arbitrary triads and consider a vector $\mathbf{v}$ whose matrix representation relative to $b$ is given by

$${}^{b}v = \begin{pmatrix} 1 \\ 0 \\ -1 \end{pmatrix}.$$

Find the matrix representation of the vector $\mathbf{v}$ relative to the triad $a$ when $R_{ab} =$

$$\text{a)}\ \begin{pmatrix} \frac{3}{4} & -\frac{\sqrt{6}}{4} & \frac{1}{4} \\ \frac{\sqrt{6}}{4} & \frac{1}{2} & -\frac{\sqrt{6}}{4} \\ \frac{1}{4} & \frac{\sqrt{6}}{4} & \frac{3}{4} \end{pmatrix}$$

$$\text{b)}\ \begin{pmatrix} 0 & -\frac{1}{\sqrt{2}} & \frac{1}{\sqrt{2}} \\ \frac{1}{\sqrt{2}} & \frac{1}{2} & \frac{1}{2} \\ -\frac{1}{\sqrt{2}} & \frac{1}{2} & \frac{1}{2} \end{pmatrix}$$

$$\text{c)}\ \begin{pmatrix} \frac{4}{5} & \frac{2}{5} & \frac{1}{\sqrt{5}} \\ \frac{2}{5} & \frac{1}{5} & -\frac{2}{\sqrt{5}} \\ -\frac{1}{\sqrt{5}} & \frac{2}{\sqrt{5}} & 0 \end{pmatrix}$$

$$\text{d)}\ \begin{pmatrix} \frac{1}{\sqrt{2}} & -\frac{1}{\sqrt{2}} & 0 \\ \frac{1}{\sqrt{2}} & \frac{1}{\sqrt{2}} & 0 \\ 0 & 0 & 1 \end{pmatrix}$$

$$\text{e)}\ \begin{pmatrix} -\frac{1}{\sqrt{2}} & 0 & \frac{1}{\sqrt{2}} \\ 0 & -1 & 0 \\ \frac{1}{\sqrt{2}} & 0 & \frac{1}{\sqrt{2}} \end{pmatrix}$$

$$\text{f)}\ \begin{pmatrix} \frac{1}{3} - \frac{1}{\sqrt{3}} & \frac{1}{3} + \frac{1}{\sqrt{3}} & \frac{1}{3} \\ \frac{1}{\sqrt{3}} + \frac{1}{6} & \frac{1}{2\sqrt{3}} - \frac{1}{3} & \frac{2}{3} \\ \frac{1}{3} + \frac{1}{2\sqrt{3}} & -\frac{1}{6} + \frac{1}{\sqrt{3}} & -\frac{2}{3} \end{pmatrix}$$

**Exercise 5.7** Recall the geometric definition of the dot product between the two vectors $\mathbf{v}$ and $\mathbf{w}$:

$$\mathbf{v} \bullet \mathbf{w} = \|\mathbf{v}\|\,\|\mathbf{w}\| \cos\theta\,(\mathbf{v}, \mathbf{w}),$$

where $\theta\,(\mathbf{v}, \mathbf{w})$ is the angle between the vectors. Let $a$ and $b$ be two arbitrary triads and show that

$${}^{a}v^{T}\,{}^{a}w = {}^{b}v^{T}\,{}^{b}w.$$

Use this fact to prove that

$$\mathbf{v} \bullet \mathbf{w} = {}^{b}v^{T}\,{}^{b}w$$

for an arbitrary triad $b$.

**Solution.** We have

$$\begin{aligned} {}^{a}v^{T}\,{}^{a}w &= \left(R_{ab}\,{}^{b}v\right)^{T}\left(R_{ab}\,{}^{b}w\right) \\ &= {}^{b}v^{T}R_{ba}R_{ab}\,{}^{b}w \\ &= {}^{b}v^{T}\,{}^{b}w \end{aligned}$$

proving the first half of the claim. Now, consider, without any additional assumptions on $\mathbf{v}$ and $\mathbf{w}$, the triad $a$, such that

$$\mathbf{v} = a \begin{pmatrix} \|\mathbf{v}\| \\ 0 \\ 0 \end{pmatrix}$$

and

$$\mathbf{w} = a \begin{pmatrix} \|\mathbf{w}\| \cos\theta(\mathbf{v},\mathbf{w}) \\ \|\mathbf{w}\| \sin\theta(\mathbf{v},\mathbf{w}) \\ 0 \end{pmatrix}.$$

It follows that

$${}^a v^T \, {}^a w = \|\mathbf{v}\| \, \|\mathbf{w}\| \cos\theta(\mathbf{v},\mathbf{w}) = \mathbf{v} \bullet \mathbf{w}$$

and the second claim follows.

**Exercise 5.8** Recall the geometric definition of the cross product between the two vectors $\mathbf{v}$ and $\mathbf{w}$ as the vector whose direction is given by the right-hand rule and whose length is

$$\|\mathbf{v} \times \mathbf{w}\| = \|\mathbf{v}\| \, \|\mathbf{w}\| \sin\theta(\mathbf{v},\mathbf{w}),$$

where $\theta(\mathbf{v},\mathbf{w})$ is the angle between the vectors. Let $a$ and $b$ be two arbitrary triads and show that

$$\begin{vmatrix} \mathbf{a}_1 & \mathbf{a}_2 & \mathbf{a}_3 \\ {}^a v_1 & {}^a v_2 & {}^a v_3 \\ {}^a w_1 & {}^a w_2 & {}^a w_3 \end{vmatrix} = \begin{vmatrix} \mathbf{b}_1 & \mathbf{b}_2 & \mathbf{b}_3 \\ {}^b v_1 & {}^b v_2 & {}^b v_3 \\ {}^b w_1 & {}^b w_2 & {}^b w_3 \end{vmatrix}.$$

Use this fact to prove that

$$\mathbf{v} \times \mathbf{w} = \begin{vmatrix} \mathbf{b}_1 & \mathbf{b}_2 & \mathbf{b}_3 \\ {}^b v_1 & {}^b v_2 & {}^b v_3 \\ {}^b w_1 & {}^b w_2 & {}^b w_3 \end{vmatrix}$$

for an arbitrary triad $b$.

[Hint: Let $R_{ab}$ be the matrix whose $[i,j]$-th component is $r_{ij}$. Replace all components in the leftmost determinant using the formulae

$$a = b\,(R_{ab})^T,$$
$${}^a v = R_{ab} \, {}^b v,$$

and

$${}^a w = R_{ab} \, {}^b w$$

and use the fact that $\det R_{ab} = 1$ to establish the first claim. Next, consider, without any further assumptions on $\mathbf{v}$ and $\mathbf{w}$, the triad $a$ for which $\mathbf{v}$ and $\mathbf{w}$ take the form in the solution to the previous exercise and proceed from there to show the equivalence between the geometric and the algebraic formulae for the cross product.]

**Exercise 5.9** For each of the following matrices, determine if it qualifies as a rotation matrix.

a) $\begin{pmatrix} 0 & 1 & 0 \\ 0 & 0 & 1 \\ 1 & 0 & 0 \end{pmatrix}$

b) $\begin{pmatrix} 0 & -1 & 0 \\ 0 & 0 & 1 \\ 1 & 0 & 0 \end{pmatrix}$

c) $\begin{pmatrix} 0 & -1 & 0 \\ 0 & 0 & -1 \\ 1 & 0 & 0 \end{pmatrix}$

d) $\begin{pmatrix} \frac{1}{\sqrt{2}} & -\frac{1}{\sqrt{2}} & 0 \\ \frac{1}{\sqrt{2}} & \frac{1}{\sqrt{2}} & 0 \\ 0 & 0 & 1 \end{pmatrix}$

e) $\begin{pmatrix} \frac{1}{\sqrt{2}} & -\frac{1}{\sqrt{2}} & 0 \\ \frac{1}{\sqrt{2}} & -\frac{1}{\sqrt{2}} & 0 \\ 0 & 0 & 1 \end{pmatrix}$

f) $\begin{pmatrix} \frac{1}{\sqrt{2}} & -\frac{1}{\sqrt{2}} & 0 \\ \frac{1}{\sqrt{2}} & \frac{1}{\sqrt{2}} & 0 \\ 0 & 0 & -1 \end{pmatrix}$

**Exercise 5.10** Let $a$ denote a triad that is fixed relative to a rigid body and denote by $a^{(r)}$ and $a^{(f)}$ the corresponding triads when the body is in the reference and final configurations, respectively, such that

$$R_{a^{(r)}a^{(f)}} = \begin{pmatrix} 0 & -\frac{1}{\sqrt{2}} & \frac{1}{\sqrt{2}} \\ \frac{1}{\sqrt{2}} & \frac{1}{2} & \frac{1}{2} \\ -\frac{1}{\sqrt{2}} & \frac{1}{2} & \frac{1}{2} \end{pmatrix}.$$

Let $\mathbf{v}_r$ and $\mathbf{v}_f$ denote two vectors that contain a separation between two fixed points on the rigid body in the reference and final configurations, respectively. Compute ${}^{a^{(f)}}v_f$, ${}^{a^{(r)}}v_f$ and the angle between $\mathbf{v}_r$ and $\mathbf{v}_f$, when $\mathbf{v}_r =$

a) $a^{(r)} \begin{pmatrix} 1 \\ 0 \\ 1 \end{pmatrix}$ b) $a^{(r)} \begin{pmatrix} 0 \\ 1 \\ 2 \end{pmatrix}$

c) $a^{(r)} \begin{pmatrix} 0 \\ 1 \\ 1 \end{pmatrix}$ d) $a^{(r)} \begin{pmatrix} 1 \\ 0 \\ 0 \end{pmatrix}$

e) $a^{(r)} \begin{pmatrix} 0 \\ 2 \\ 2 \end{pmatrix}$ f) $a^{(r)} \begin{pmatrix} 1 \\ 1 \\ 1 \end{pmatrix}$

**Exercise 5.11** Let $a$ and $b$ be two arbitrary triads. Show that $R_{ba} = (R_{ab})^{-1}$, i.e., that $R_{ba}R_{ab} = (R_{ab})^T R_{ab}$ equals the identity matrix.

**Solution.** Let $\mathbf{v}$ be an arbitrary vector. Then,

$${}^b v = R_{ba}\, {}^a v$$

and

$${}^a v = R_{ab}\, {}^b v.$$

Substitution of the latter expression into the former then yields

$${}^b v = R_{ba} R_{ab}\, {}^b v.$$

Since this is true for an arbitrary vector, it follows that

$$R_{ba} R_{ab} = \begin{pmatrix} 1 & 0 & 0 \\ 0 & 1 & 0 \\ 0 & 0 & 1 \end{pmatrix}$$

and the claim follows.

**Exercise 5.12** Let $a$ and $b$ be two arbitrary triads. Find the eigenvalues and eigenvectors of the rotation matrix $R_{ab}$ when $R_{ab} =$

a) $\begin{pmatrix} \frac{3}{4} & -\frac{\sqrt{6}}{4} & \frac{1}{4} \\ \frac{\sqrt{6}}{4} & \frac{1}{2} & -\frac{\sqrt{6}}{4} \\ \frac{1}{4} & \frac{\sqrt{6}}{4} & \frac{3}{4} \end{pmatrix}$

b) $\begin{pmatrix} 0 & -\frac{1}{\sqrt{2}} & \frac{1}{\sqrt{2}} \\ \frac{1}{\sqrt{2}} & \frac{1}{2} & \frac{1}{2} \\ -\frac{1}{\sqrt{2}} & \frac{1}{2} & \frac{1}{2} \end{pmatrix}$

c) $\begin{pmatrix} \frac{4}{5} & \frac{2}{5} & \frac{1}{\sqrt{5}} \\ \frac{2}{5} & \frac{1}{5} & -\frac{2}{\sqrt{5}} \\ -\frac{1}{\sqrt{5}} & \frac{2}{\sqrt{5}} & 0 \end{pmatrix}$

d) $\begin{pmatrix} \frac{1}{\sqrt{2}} & -\frac{1}{\sqrt{2}} & 0 \\ \frac{1}{\sqrt{2}} & \frac{1}{\sqrt{2}} & 0 \\ 0 & 0 & 1 \end{pmatrix}$

e) $\begin{pmatrix} -\frac{1}{\sqrt{2}} & 0 & \frac{1}{\sqrt{2}} \\ 0 & -1 & 0 \\ \frac{1}{\sqrt{2}} & 0 & \frac{1}{\sqrt{2}} \end{pmatrix}$

f) $\begin{pmatrix} \frac{1}{3} - \frac{1}{\sqrt{3}} & \frac{1}{3} + \frac{1}{\sqrt{3}} & \frac{1}{3} \\ \frac{1}{\sqrt{3}} + \frac{1}{6} & \frac{1}{2\sqrt{3}} - \frac{1}{3} & \frac{2}{3} \\ \frac{1}{3} + \frac{1}{2\sqrt{3}} & -\frac{1}{6} + \frac{1}{\sqrt{3}} & -\frac{2}{3} \end{pmatrix}$

**Exercise 5.13** Show that the eigenvalues of $R_{ab}$ lie on the unit circle in the complex plane.

**Solution.** Let $\lambda$ denote a (possibly complex) eigenvalue of $R_{ab}$ corresponding to the (possibly complex) eigenvector $v$. Then

$$R_{ab} v = \lambda v$$

Taking complex conjugates on both sides, we obtain

$$R_{ab} v^* = \lambda^* v^*,$$

since $R_{ab}$ is a matrix of real numbers[1]. Then

$$\begin{aligned} (v^*)^T v &= (v^*)^T (R_{ab})^T R_{ab} v \\ &= (R_{ab} v^*)^T R_{ab} v \\ &= \lambda^* \lambda (v^*)^T v, \end{aligned}$$

where the first equality follows from the fact that

$$(R_{ab})^T R_{ab} = R_{ba} R_{ab}$$

equals the identity. But,

$$(v^*)^T v$$

is positive for all $v \neq \begin{pmatrix} 0 & 0 & 0 \end{pmatrix}^T$. It follows that

$$\lambda^* \lambda = |\lambda|^2$$

must equal 1 and the claim follows.

**Exercise 5.14** Let $a$ and $b$ be two arbitrary triads. Show that one of the eigenvalues of $R_{ab}$ equals 1.

**Solution.** Since $R_{ab}$ is a matrix of real numbers, every complex eigenvalue is a member of a complex conjugate pair of eigenvalues. Since $R_{ab}$ is a $3 \times 3$ matrix, it has at most three separate eigenvalues. Thus, either all eigenvalues of $R_{ab}$ are real, or one is real and the other two are complex conjugates. Since all eigenvalues must lie on the unit circle, there are only six possibilities:

a) $\lambda_1 = \lambda_2 = \lambda_3 = 1$,
b) $\lambda_1 = \lambda_2 = 1$, $\lambda_3 = -1$,
c) $\lambda_1 = 1$, $\lambda_2 = \lambda_3 = -1$,
d) $\lambda_1 = \lambda_2 = \lambda_3 = -1$,
e) $\lambda_1 = 1$, $\lambda_{2,3} = e^{\pm i\theta}$, or
f) $\lambda_1 = -1, \lambda_{2,3} = e^{\pm i\theta}$,

where $\theta \in (0, \pi)$. Finally, recall that the determinant of a matrix is the product of its eigenvalues. Since $R_{ab}$ is orientation-preserving, it follows that

$$1 = \det R_{ab} = \lambda_1 \lambda_2 \lambda_3.$$

Of the six cases, only a), c), and e) satisfy this condition, confirming the claim.

**Exercise 5.15** Let $a$ and $b$ be two arbitrary triads. Let

$$e_1 = \begin{pmatrix} v_1 \\ v_2 \\ v_3 \end{pmatrix}$$

be the eigenvector of $R_{ab}$ that corresponds to the eigenvalue 1, such that

$$(e_1)^T e_1 = v_1^2 + v_2^2 + v_3^2 = 1.$$

Show that there exists an orthogonal $3 \times 3$ matrix $V$, such that

$$R_{ab} = V \begin{pmatrix} 1 & 0 & 0 \\ 0 & t_{11} & t_{12} \\ 0 & t_{21} & t_{22} \end{pmatrix} V^T,$$

for some constants $t_{11}$, $t_{12}$, $t_{21}$, and $t_{22}$.

**Solution.** By assumption,

$$R_{ab} e_1 = e_1$$

and thus

$$e_1 = R_{ba} R_{ab} e_1 = R_{ba} e_1.$$

If $w$ is an arbitrary $3 \times 1$ column matrix, such that

$$w^T e_1 = 0,$$

then it follows that

$$(R_{ab} w)^T e_1 = w^T R_{ba} e_1 = w^T e_1 = 0.$$

Now, define

$$e_2 = \begin{pmatrix} -v_2 / \sqrt{v_1^2 + v_2^2} \\ v_1 / \sqrt{v_1^2 + v_2^2} \\ 0 \end{pmatrix}$$

[1]Here, $^*$ is used to denote complex conjugation.

and

$$e_3 = \begin{pmatrix} -v_1 v_3 / \sqrt{v_1^2 + v_2^2} \\ -v_2 v_3 / \sqrt{v_1^2 + v_2^2} \\ \sqrt{v_1^2 + v_2^2} \end{pmatrix}$$

if $v_1^2 + v_2^2 > 0$ and

$$e_2 = \begin{pmatrix} 1 \\ 0 \\ 0 \end{pmatrix} \text{ and } e_3 = \begin{pmatrix} 0 \\ 1 \\ 0 \end{pmatrix}$$

if $v_1 = v_2 = 0$. In both cases, the column matrices $e_1$, $e_2$, and $e_3$ satisfy the conditions

$$\begin{aligned} (e_1)^T e_1 = (e_2)^T e_2 = (e_3)^T e_3 = 1, \\ (e_1)^T e_2 = (e_1)^T e_3 = (e_2)^T e_3 \\ = (e_2)^T e_1 = (e_3)^T e_1 \\ = (e_3)^T e_2 = 0. \end{aligned}$$

It follows that the matrix

$$V = \begin{pmatrix} | & | & | \\ e_1 & e_2 & e_3 \\ | & | & | \end{pmatrix}.$$

is orthogonal, since $V^T V =$

$$\begin{aligned} &\begin{pmatrix} - & (e_1)^T & - \\ - & (e_2)^T & - \\ - & (e_3)^T & - \end{pmatrix} \begin{pmatrix} | & | & | \\ e_1 & e_2 & e_3 \\ | & | & | \end{pmatrix} \\ &= \begin{pmatrix} (e_1)^T e_1 & (e_1)^T e_2 & (e_1)^T e_3 \\ (e_2)^T e_1 & (e_2)^T e_2 & (e_2)^T e_3 \\ (e_3)^T e_1 & (e_3)^T e_2 & (e_3)^T e_3 \end{pmatrix} \\ &= \begin{pmatrix} 1 & 0 & 0 \\ 0 & 1 & 0 \\ 0 & 0 & 1 \end{pmatrix}. \end{aligned}$$

Moreover, $V^T R_{ab} V =$

$$\begin{aligned} &\begin{pmatrix} - & (e_1)^T & - \\ - & (e_2)^T & - \\ - & (e_3)^T & - \end{pmatrix} R_{ab} \begin{pmatrix} | & | & | \\ e_1 & e_2 & e_3 \\ | & | & | \end{pmatrix} \\ &= \begin{pmatrix} (e_1)^T R_{ab} e_1 & (e_1)^T R_{ab} e_2 & (e_1)^T R_{ab} e_3 \\ (e_2)^T R_{ab} e_1 & (e_2)^T R_{ab} e_2 & (e_2)^T R_{ab} e_3 \\ (e_3)^T R_{ab} e_1 & (e_3)^T R_{ab} e_2 & (e_3)^T R_{ab} e_3 \end{pmatrix} \\ &= \begin{pmatrix} (e_1)^T e_1 & (R_{ba} e_1)^T e_2 & (R_{ba} e_1)^T e_3 \\ (e_2)^T e_1 & (e_2)^T R_{ab} e_2 & (e_2)^T R_{ab} e_3 \\ (e_3)^T e_1 & (e_3)^T R_{ab} e_2 & (e_3)^T R_{ab} e_3 \end{pmatrix} \\ &= \begin{pmatrix} (e_1)^T e_1 & (e_1)^T e_2 & (e_1)^T e_3 \\ (e_2)^T e_1 & (e_2)^T R_{ab} e_2 & (e_2)^T R_{ab} e_3 \\ (e_3)^T e_1 & (e_3)^T R_{ab} e_2 & (e_3)^T R_{ab} e_3 \end{pmatrix} \\ &= \begin{pmatrix} 1 & 0 & 0 \\ 0 & (e_2)^T R_{ab} e_2 & (e_2)^T R_{ab} e_3 \\ 0 & (e_3)^T R_{ab} e_2 & (e_3)^T R_{ab} e_3 \end{pmatrix}. \end{aligned}$$

The claim follows from the observation that

$$R_{ab} = V \left( V^T R_{ab} V \right) V^T,$$

since $V$ is orthogonal.

**Exercise 5.16** Show that the columns of an orthogonal matrix are linearly independent.

**Solution.** If the columns (or rows) were linearly dependent, then the determinant would vanish. For a rotation matrix, however, the determinant equals $\pm 1$, confirming the claim.

**Exercise 5.17** Show that the set of orthogonal matrices with the normal rule for matrix multiplication is a group.

**Exercise 5.18** Show by an example that the group of orthogonal matrices is not Abelian.

**Exercise 5.19** Show that the set of orientation-preserving orthogonal matrices with the normal rule for matrix multiplication is a group.

**Exercise 5.20** We say that two orthonormal bases $a$ and $b$ are equivalent, i.e., that $a \sim b$, if $\det R_{ab} = 1$. Show that the corresponding quotient set has only two elements, namely the collection of right-handed, orthonormal bases and the collection of left-handed, orthonormal bases.

**Exercise 5.21** For each of the rotation matrices below, find $\varphi$, $v_1$, $v_2$, and $v_3$, such that the matrix is given by $R(\varphi, v_1, v_2, v_3)$.

a) $$\begin{pmatrix} \frac{3}{4} & -\frac{\sqrt{6}}{4} & \frac{1}{4} \\ \frac{\sqrt{6}}{4} & \frac{1}{2} & -\frac{\sqrt{6}}{4} \\ \frac{1}{4} & \frac{\sqrt{6}}{4} & \frac{3}{4} \end{pmatrix}$$

b) $$\begin{pmatrix} 0 & -\frac{1}{\sqrt{2}} & \frac{1}{\sqrt{2}} \\ \frac{1}{\sqrt{2}} & \frac{1}{2} & \frac{1}{2} \\ -\frac{1}{\sqrt{2}} & \frac{1}{2} & \frac{1}{2} \end{pmatrix}$$

c) $$\begin{pmatrix} \frac{4}{5} & \frac{2}{5} & \frac{1}{\sqrt{5}} \\ \frac{2}{5} & \frac{1}{5} & -\frac{2}{\sqrt{5}} \\ -\frac{1}{\sqrt{5}} & \frac{2}{\sqrt{5}} & 0 \end{pmatrix}$$

d) $$\begin{pmatrix} \frac{1}{\sqrt{2}} & -\frac{1}{\sqrt{2}} & 0 \\ \frac{1}{\sqrt{2}} & \frac{1}{\sqrt{2}} & 0 \\ 0 & 0 & 1 \end{pmatrix}$$

e) $$\begin{pmatrix} -\frac{1}{\sqrt{2}} & 0 & \frac{1}{\sqrt{2}} \\ 0 & -1 & 0 \\ \frac{1}{\sqrt{2}} & 0 & \frac{1}{\sqrt{2}} \end{pmatrix}$$

f) $$\begin{pmatrix} \frac{1}{3} - \frac{1}{\sqrt{3}} & \frac{1}{3} + \frac{1}{\sqrt{3}} & \frac{1}{3} \\ \frac{1}{\sqrt{3}} + \frac{1}{6} & \frac{1}{2\sqrt{3}} - \frac{1}{3} & \frac{2}{3} \\ \frac{1}{3} + \frac{1}{2\sqrt{3}} & -\frac{1}{6} + \frac{1}{\sqrt{3}} & -\frac{2}{3} \end{pmatrix}$$

**Exercise 5.22** Consider applying a pure rotation to a block in its reference configuration corresponding to a half turn about an edge through a given corner on the block, followed by a pure rotation corresponding to a quarter turn about a different edge through the same corner as shown in the figure below.

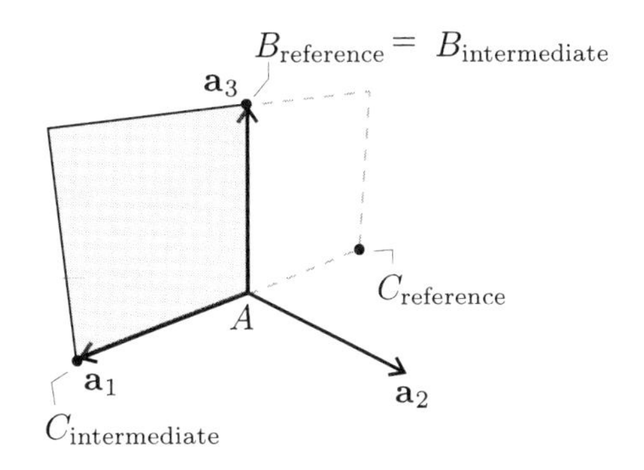


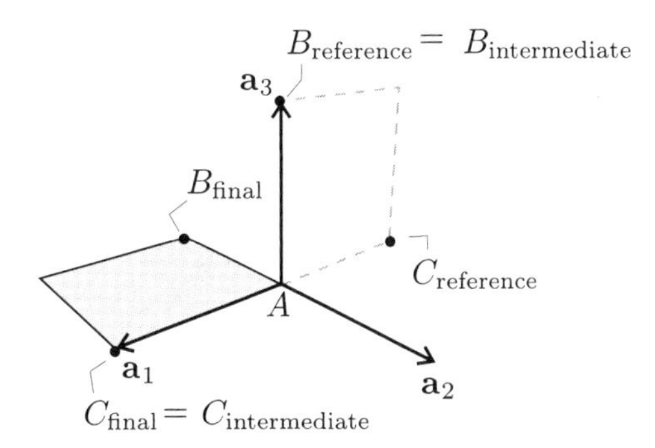


Show that the final configuration is related to the reference configuration by a single pure rotation about an axis through the corner making an angle of $\theta_1 = 45°$ with the first edge and $\theta_2 = 90°$ with the second edge.

**Solution.** Introduce a triad $a$, such that the first edge is parallel to $\mathbf{a}_3$ and the second edge is parallel to $\mathbf{a}_1$. Denote by $b$ a body-fixed triad, such that $b^{(\text{reference})} = a$. Then, the first rotation corresponds to the rotation matrix

$$\begin{aligned} R_{b^{(\text{reference})}b^{(\text{intermediate})}} &= R(\pi, 0, 0, 1) \\ &= \begin{pmatrix} -1 & 0 & 0 \\ 0 & -1 & 0 \\ 0 & 0 & 1 \end{pmatrix}. \end{aligned}$$

Furthermore,

$$\begin{aligned} & {}^{b^{(\text{intermediate})}}(\mathbf{a}_1) \\ &= R_{b^{(\text{intermediate})}b^{(\text{reference})}}\ {}^{b^{(\text{reference})}}(\mathbf{a}_1) \\ &= \begin{pmatrix} -1 & 0 & 0 \\ 0 & -1 & 0 \\ 0 & 0 & 1 \end{pmatrix} \begin{pmatrix} 1 \\ 0 \\ 0 \end{pmatrix} \\ &= \begin{pmatrix} -1 \\ 0 \\ 0 \end{pmatrix}, \end{aligned}$$

i.e., the second rotation corresponds to the rotation matrix

$$\begin{aligned} R_{b^{(\text{intermediate})}b^{(\text{final})}} &= R\left(\frac{\pi}{2}, -1, 0, 0\right) \\ &= \begin{pmatrix} 1 & 0 & 0 \\ 0 & 0 & 1 \\ 0 & -1 & 0 \end{pmatrix}. \end{aligned}$$

Since the same corner is kept fixed by the pure rotations, it follows that the final configuration is related to the reference configuration by a single pure rotation keeping the corner fixed. From Exercise 1.8, we recall that every pure rotation is equivalent to a rotation about a unique axis through the point kept fixed. Moreover, the axis of rotation is parallel to the vector whose coordinate representation relative to $b^{(\text{reference})}$ (i.e., $a$) is given by the eigenvector corresponding to the 1 eigenvalue of the rotation matrix $R_{b^{(\text{reference})}b^{(\text{final})}}$. But,

$$\begin{aligned} & R_{b^{(\text{reference})}b^{(\text{final})}} \\ &= R_{b^{(\text{reference})}b^{(\text{intermediate})}} R_{b^{(\text{intermediate})}b^{(\text{final})}} \\ &= \begin{pmatrix} -1 & 0 & 0 \\ 0 & 0 & -1 \\ 0 & -1 & 0 \end{pmatrix}, \end{aligned}$$

from which we find the corresponding eigenvector

$$v = \begin{pmatrix} 0 \\ -1 \\ 1 \end{pmatrix},$$

i.e., the axis of rotation is parallel to the vector $-\mathbf{a}_2 + \mathbf{a}_3$. Indeed,

$$\cos\theta_1 = \frac{(-\mathbf{a}_2 + \mathbf{a}_3) \bullet \mathbf{a}_1}{\|-\mathbf{a}_2 + \mathbf{a}_3\|\ \|\mathbf{a}_1\|} = 0$$

and

$$\cos\theta_1 = \frac{(-\mathbf{a}_2 + \mathbf{a}_3) \bullet \mathbf{a}_3}{\|-\mathbf{a}_2 + \mathbf{a}_3\|\ \|\mathbf{a}_3\|} = \frac{1}{\sqrt{2}},$$

from which the claim follows.

**Exercise 5.23** Consider applying a pure rotation to a block in its reference configuration corresponding to a half turn about an edge through a given corner on a block followed by a pure rotation by an angle $\theta$ about a different edge through the same corner. The final configuration is related to the reference configuration by a single pure rotation about an axis through the corner making an angle $\phi$ with the first edge and perpendicular to the second edge. Show that

$$\phi = \frac{|\theta|}{2}.$$

**Exercise 5.24** Consider applying a pure rotation to a block in its reference configuration corresponding to a half turn about some axis through a given corner on a block followed by a pure rotation corresponding to a quarter turn about a different axis through the same corner making an angle $\theta$ with the first axis. The final configuration is related to the reference configuration by a single pure rotation about an axis through the corner making an angle $\phi_1$ with the first axis and $\phi_2$ with the second axis. Show that

$$\cos\phi_1 = \frac{1}{\sqrt{1 + \sin^2\theta}}$$

and

$$\cos\phi_2 = \frac{\cos\theta}{\sqrt{1 + \sin^2\theta}}.$$

[Hint: Let the first axis be parallel to the basis vector $\mathbf{a}_3$ of a right-handed, orthonormal basis and let the second axis be parallel to the vector $\sin\theta\mathbf{a}_1 + \cos\theta\mathbf{a}_3$.]

**Exercise 5.25** Consider three reference triads $a$, $b$, and $c$, such that $b$ is rotated relative to $a$ by 60° about the common 2-direction, and $c$ is rotated relative to $b$ by 45° about the common 3-direction. Find the rotation matrix $R_{ac}$.

**Solution.** We have

$$\begin{aligned} R_{ab} &= R\left(\frac{\pi}{3}, 0, 1, 0\right) \\ &= \begin{pmatrix} \cos\frac{\pi}{3} & 0 & \sin\frac{\pi}{3} \\ 0 & 1 & 0 \\ -\sin\frac{\pi}{3} & 0 & \cos\frac{\pi}{3} \end{pmatrix} \\ &= \begin{pmatrix} \frac{1}{2} & 0 & \frac{\sqrt{3}}{2} \\ 0 & 1 & 0 \\ -\frac{\sqrt{3}}{2} & 0 & \frac{1}{2} \end{pmatrix} \end{aligned}$$

and

$$\begin{aligned} R_{bc} &= R\left(\frac{\pi}{4}, 0, 0, 1\right) \\ &= \begin{pmatrix} \cos\frac{\pi}{4} & -\sin\frac{\pi}{4} & 0 \\ \sin\frac{\pi}{4} & \cos\frac{\pi}{4} & 0 \\ 0 & 0 & 1 \end{pmatrix} \\ &= \begin{pmatrix} \frac{\sqrt{2}}{2} & -\frac{\sqrt{2}}{2} & 0 \\ \frac{\sqrt{2}}{2} & \frac{\sqrt{2}}{2} & 0 \\ 0 & 0 & 1 \end{pmatrix}. \end{aligned}$$

It follows that

$$R_{ac} = R_{ab}R_{bc} = \begin{pmatrix} \frac{\sqrt{2}}{4} & -\frac{\sqrt{2}}{4} & \frac{\sqrt{3}}{2} \\ \frac{\sqrt{2}}{2} & \frac{\sqrt{2}}{2} & 0 \\ -\frac{\sqrt{6}}{4} & \frac{\sqrt{6}}{4} & \frac{1}{2} \end{pmatrix}.$$

**Exercise 5.26** Consider the rotation matrix $R$ corresponding to a $3-1-3$ sequence of pure rotations. Find the eigenvalues of $R$.

**Solution.** We have

$$\begin{aligned} R &= R(\varphi_1, 0, 0, 1)\, R(\varphi_2, 1, 0, 0)\, R(\varphi_3, 0, 0, 1) \\ &= \begin{pmatrix} c_1c_3 - s_1c_2s_3 & -c_1s_3 - s_1c_2c_3 & s_1s_2 \\ s_1c_3 + c_1c_2s_3 & -s_1s_3 + c_1c_2c_3 & -c_1s_2 \\ s_2s_3 & s_2c_3 & c_2 \end{pmatrix}, \end{aligned}$$

where $c_i = \cos\varphi_i$ and $s_i = \sin\varphi_i$. An eigenvector of $R$ is a non-zero column matrix solution to the equation

$$Rv = \lambda v,$$

for some $\lambda$. It follows that

$$(R - \lambda I)\, v = \begin{pmatrix} 0 \\ 0 \\ 0 \end{pmatrix},$$

which has a non-trivial solution if and only if

$$\begin{aligned} 0 &= \det(R - \lambda I) \\ &= -\lambda^3 + \begin{pmatrix} c_1c_2c_3 - s_1c_2s_3 + c_2 \\ +c_1c_3 - s_1s_3 \end{pmatrix} \lambda^2 \\ &\quad + \begin{pmatrix} -c_2 + s_1s_3 - c_1c_3 \\ +s_1c_2s_3 - c_1c_2c_3 \end{pmatrix} \lambda + 1 \\ &= (1 - \lambda)\left[\lambda^2 + \mu\lambda + 1\right], \end{aligned}$$

where $-c_1c_2c_3 + s_1c_2s_3 - c_2 - c_1c_3 + s_1s_3 + 1 = \mu \in [-2, 2]$. The eigenvalues are thus given by

$$\lambda_1 = 1,\ \lambda_{2,3} = e^{\pm i\eta},$$

where

$$\eta = \arctan\left(-\frac{\sqrt{4-\mu^2}}{\mu}\right).$$

**Exercise 5.27** Consider the combined action on a rigid body of a pure translation $\mathbf{T}$ corresponding to a vector $\mathbf{u}$ and a pure rotation $\mathbf{R}$ corresponding to a rotation along a unit vector $\mathbf{n}$ by an amount $\varphi$. Recall from Exercise 1.12 that every combination of a pure translation and a pure rotation can be expressed in terms of a pure translation and a pure rotation along a common axis whose direction is fixed relative to the reference configuration. This is known as a *screw* motion. Find the axis of the screw in terms of $\mathbf{u}$, $\mathbf{n}$, and $\varphi$.

[Hint: We can identify an axis in space relative to any observer $\mathcal{A}$ by a unit vector $\mathbf{l}$ that is parallel to the axis and by the position vector $\mathbf{r}^{AP}$ from the reference point $A$ of $\mathcal{A}$ to a (non-unique) point $P$ on the line. From Exercise 3.24, we recall that

$$\mathbf{r}^{AP} = \left(\mathbf{r}^{AP} \bullet \mathbf{l}\right)\mathbf{l} + \mathbf{l} \times \left(\mathbf{r}^{AP} \times \mathbf{l}\right),$$

where the first term on the right-hand side is a vector parallel to the axis and the second term is a vector perpendicular to the axis. Indeed,

$$\mathbf{l} \times \left(\mathbf{r}^{AP} \times \mathbf{l}\right) = \mathbf{r}^{AP} - \left(\mathbf{r}^{AP} \bullet \mathbf{l}\right)\mathbf{l}$$

is a vector that points from $A$ to the point on the axis that is closest to $A$. It follows that the axis is uniquely specified by the two vectors $\mathbf{l}$ and $\mathbf{r}^{AP} \times \mathbf{l}$, where the latter vector is independent of the point $P$ (as long as $P$ lies on the line).

Let $\mathcal{A}$ and $\mathcal{B}$ be two observers that coincide with the rigid body in its initial and final configurations, respectively. Let $(A, a)$ and $(B, b)$ denote the pair of reference point and reference triad for $\mathcal{A}$ and $\mathcal{B}$, respectively. Let $L$ be an arbitrary line that is stationary relative to the rigid body. Let the initial configuration of $L$ relative to $\mathcal{A}$ be specified by the unit vector $\mathbf{l}_i$ and the vector $\mathbf{r}^{AP_i} \times \mathbf{l}_i$, where $P_i$ is an arbitrary point on $L$ in the initial configuration. Similarly, let the final configuration of $L$ relative to $\mathcal{A}$ be specified by the unit vector $\mathbf{l}_f$ and the vector $\mathbf{r}^{AP_f} \times \mathbf{l}_f$, where $P_f$ is the point corresponding to $P_i$ in the final configuration of $L$. Show that

1.
$$^{b}l_f = {}^{a}l_i$$
and thus
$$^{a}l_f = R_{ab}\, {}^{a}l_i.$$

2.
$$^{b}\left(\mathbf{r}^{BP_f} \times \mathbf{l}_f\right) = {}^{a}\left(\mathbf{r}^{AP_i} \times \mathbf{l}_i\right)$$
and thus
$$\begin{aligned} &^{a}\left(\mathbf{r}^{AP_f} \times \mathbf{l}_f\right) \\ &= {}^{a}\left(\mathbf{r}^{AB} \times \mathbf{l}_f + \mathbf{r}^{BP_f} \times \mathbf{l}_f\right) \\ &= {}^{a}\left(\mathbf{u} \times \mathbf{l}_f\right) + R_{ab}\, {}^{a}\left(\mathbf{r}^{AP_i} \times \mathbf{l}_i\right). \end{aligned}$$

3. If $L$ is the screw axis, then
$$^{a}l_i = {}^{a}l_f$$
and
$$^{a}\left(\mathbf{r}^{AP_i} \times \mathbf{l}_i\right) = {}^{a}\left(\mathbf{r}^{AP_f} \times \mathbf{l}_f\right),$$
which imply that
$$^{a}l_i = R_{ab}\, {}^{a}l_i$$
and
$$\begin{aligned} &^{a}\left(\mathbf{r}^{AP_i} \times \mathbf{l}_i\right) \\ &= {}^{a}\left(\mathbf{u} \times \mathbf{l}_f\right) + R_{ab}\, {}^{a}\left(\mathbf{r}^{AP_i} \times \mathbf{l}_i\right). \end{aligned}$$

4. It then follows that
$$\mathbf{l}_i = \mathbf{l}_f = \pm\mathbf{n}$$
and
$$\left(R_{ab} - Id\right)\, {}^{a}\left(\mathbf{r}^{AP_i} \times \mathbf{n}\right) = {}^{a}\left(\mathbf{n} \times \mathbf{u}\right).$$

5. Assuming, without loss of generality that $\mathbf{a}_3 = \mathbf{n}$, the latter equation implies that
$$\begin{aligned} &\mathbf{r}^{AP_i} \times \mathbf{n} \\ &= \left(\frac{1}{2}\mathbf{u} + \cot\frac{\varphi}{2}\mathbf{n} \times \mathbf{u}\right) \times \mathbf{n} \\ &= \left(\begin{array}{c} \frac{1}{2}\left(\mathbf{u} \bullet \mathbf{n}\right)\mathbf{n} + \frac{1}{2}\mathbf{n} \times \left(\mathbf{u} \times \mathbf{n}\right) \\ + \cot\frac{\varphi}{2}\mathbf{n} \times \mathbf{u} \end{array}\right) \times \mathbf{n} \\ &= \left(\frac{1}{2}\mathbf{n} \times \left(\mathbf{u} \times \mathbf{n}\right) + \cot\frac{\varphi}{2}\mathbf{n} \times \mathbf{u}\right) \times \mathbf{n}, \end{aligned}$$

i.e., the point on the screw axis closest to $A$ is given by the position vector

$$\mathbf{r} = \frac{1}{2}\mathbf{n} \times (\mathbf{u} \times \mathbf{n}) + \cot\frac{\varphi}{2}\mathbf{n} \times \mathbf{u}.$$

]

**Exercise 5.28** Use the fact that

$$R(\varphi_1, v_1, v_2, v_3)\, R(\varphi_2, v_1, v_2, v_3) = R(\varphi_1 + \varphi_2, v_1, v_2, v_3)$$

to derive the trigonometric addition formulae for the cosine and sine functions.

**Solution.** Consider as a special case, $v_1 = 1$, $v_2 = v_3 = 0$. Then,

$$R(\varphi_1, 1, 0, 0) = \begin{pmatrix} 1 & 0 & 0 \\ 0 & c_1 & -s_1 \\ 0 & s_1 & c_1 \end{pmatrix},$$

$$R(\varphi_2, 1, 0, 0) = \begin{pmatrix} 1 & 0 & 0 \\ 0 & c_2 & -s_2 \\ 0 & s_2 & c_2 \end{pmatrix},$$

where $c_i = \cos\varphi_1$ and $s_i = \sin\varphi_i$. Moreover, $R(\varphi_1 + \varphi_2, 1, 0, 0)$

$$= \begin{pmatrix} 1 & 0 & 0 \\ 0 & \cos(\varphi_1 + \varphi_2) & -\sin(\varphi_1 + \varphi_2) \\ 0 & \sin(\varphi_1 + \varphi_2) & \cos(\varphi_1 + \varphi_2) \end{pmatrix}.$$

But $R(\varphi_1, 1, 0, 0)\, R(\varphi_2, 1, 0, 0)$

$$= \begin{pmatrix} 1 & 0 & 0 \\ 0 & c_1 & -s_1 \\ 0 & s_1 & c_1 \end{pmatrix} \begin{pmatrix} 1 & 0 & 0 \\ 0 & c_2 & -s_2 \\ 0 & s_2 & c_2 \end{pmatrix}$$
$$= \begin{pmatrix} 1 & 0 & 0 \\ 0 & c_1c_2 - s_1s_2 & -c_1s_2 - s_1c_2 \\ 0 & s_1c_2 + c_1s_2 & c_1c_2 - s_1s_2 \end{pmatrix},$$

i.e.,

$$\cos(\varphi_1 + \varphi_2) = \cos\varphi_1 \cos\varphi_2 - \sin\varphi_1 \sin\varphi_2$$
$$\sin(\varphi_1 + \varphi_2) = \sin\varphi_1 \cos\varphi_2 + \cos\varphi_1 \sin\varphi_2$$

corresponding to the trigonometric addition formulae.

**Exercise 5.29** Use rotation matrices to find the relation between Cartesian and polar coordinates relative to a coordinate system with axes parallel to the basis vectors of a triad $a$.

**Solution.** Let $A$ denote the origin of the coordinate system and let $P$ denote an arbitrary point. The position vector

$$\mathbf{r}^{AP} = a\begin{pmatrix} x \\ y \\ z \end{pmatrix},$$

where $x$, $y$, and $z$ are the Cartesian coordinates of the point $P$. In contrast, the polar coordinates $\rho$, $\theta$, and $z$ of the point $P$ are defined by:

- The distance from $P$ to the axis parallel with the $\mathbf{a}_3$ vector;
- The angle between the position vector $\mathbf{r}^{AP'}$ to the projection of $P$ onto the plane through $A$ spanned by $\mathbf{a}_1$ and $\mathbf{a}_2$ and the axis parallel to $\mathbf{a}_1$;
- The distance from $P$ to the plane through $A$ spanned by $\mathbf{a}_1$ and $\mathbf{a}_2$,

respectively.

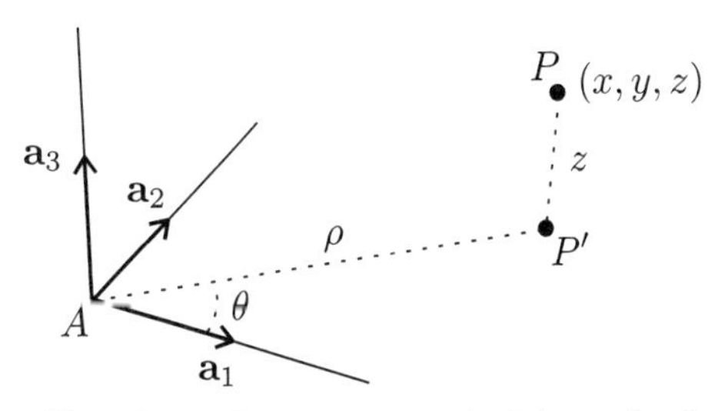


Now introduce a new triad $b$, such that $b$ is rotated relative to $a$ about the common 3-direction by an angle $\theta$, i.e.,

$$R_{ab} = \begin{pmatrix} \cos\theta & -\sin\theta & 0 \\ \sin\theta & \cos\theta & 0 \\ 0 & 0 & 1 \end{pmatrix}.$$

It follows that

$$\begin{aligned} \mathbf{r}^{AP} &= b\begin{pmatrix} \rho \\ 0 \\ z \end{pmatrix} = aR_{ab}\begin{pmatrix} \rho \\ 0 \\ z \end{pmatrix} \\ &= a\begin{pmatrix} \rho\cos\theta \\ \rho\sin\theta \\ z \end{pmatrix}, \end{aligned}$$

i.e.,

$$\begin{aligned} x &= \rho\cos\theta, \\ y &= \rho\sin\theta, \\ z &= z. \end{aligned}$$

and

$$R_{ab} = \begin{pmatrix} x/\rho & -y/\rho & 0 \\ y/\rho & x/\rho & 0 \\ 0 & 0 & 1 \end{pmatrix}.$$

Alternatively,

$$\begin{aligned} \mathbf{r}^{AP} &= a\begin{pmatrix} x \\ y \\ z \end{pmatrix} = bR_{ba}\begin{pmatrix} x \\ y \\ z \end{pmatrix} \\ &= b\begin{pmatrix} \left(x^2+y^2\right)/\rho \\ 0 \\ z \end{pmatrix}, \end{aligned}$$

which implies that

$$\begin{aligned} \rho &= \sqrt{x^2+y^2}, \\ \theta &= \arctan\frac{y}{x}, \\ z &= z. \end{aligned}$$

**Exercise 5.30** Use rotation matrices to find the relation between Cartesian and spherical coordinates relative to a coordinate system with axes parallel to the basis vectors of a triad $a$.

**Solution.** Let $A$ denote the origin of the coordinate system and let $P$ denote an arbitrary point. The position vector

$$\mathbf{r}^{AP} = a\begin{pmatrix} x \\ y \\ z \end{pmatrix},$$

where $x$, $y$, and $z$ are the Cartesian coordinates of the point $P$. In contrast, the spherical coordinates $\varrho$, $\theta$, and $\phi$ of the point $P$ are defined by:

- The distance from $P$ to $A$;
- The angle between the position vector $\mathbf{r}^{AP'}$ to the projection of $P$ onto the plane through $A$ spanned by $\mathbf{a}_1$ and $\mathbf{a}_2$ and the axis parallel to $\mathbf{a}_1$;
- The angle between the $\mathbf{r}^{AP}$ and the axis parallel to $\mathbf{a}_3$,

respectively. Now introduce new triads $b$ and $c$, such that $b$ is rotated relative to $a$ about the common 3-direction by an angle $\theta$, i.e.,

$$R_{ab} = \begin{pmatrix} \cos\theta & -\sin\theta & 0 \\ \sin\theta & \cos\theta & 0 \\ 0 & 0 & 1 \end{pmatrix}$$

and $c$ is rotated relative to $b$ about the common 2-direction by an angle $\phi$, i.e.,

$$R_{bc} = \begin{pmatrix} \cos\phi & 0 & \sin\phi \\ 0 & 1 & 0 \\ -\sin\phi & 0 & \cos\phi \end{pmatrix}.$$

It follows that

$$\begin{aligned} \mathbf{r}^{AP} &= c\begin{pmatrix} 0 \\ 0 \\ \varrho \end{pmatrix} = aR_{ab}R_{bc}\begin{pmatrix} 0 \\ 0 \\ \varrho \end{pmatrix} \\ &= a\begin{pmatrix} \varrho\cos\theta\sin\phi \\ \varrho\sin\theta\sin\phi \\ \varrho\cos\phi \end{pmatrix}, \end{aligned}$$

i.e.,

$$\begin{aligned} x &= \varrho\cos\theta\sin\phi, \\ y &= \varrho\sin\theta\sin\phi, \\ z &= \varrho\cos\phi, \end{aligned}$$

and

$$R_{ab} = \begin{pmatrix} x/\sqrt{x^2+y^2} & -y/\sqrt{x^2+y^2} & 0 \\ y/\sqrt{x^2+y^2} & x/\sqrt{x^2+y^2} & 0 \\ 0 & 0 & 1 \end{pmatrix},$$

$$R_{bc} = \begin{pmatrix} z/\varrho & 0 & \sqrt{x^2+y^2}/\varrho \\ 0 & 1 & 0 \\ -\sqrt{x^2+y^2}/\varrho & 0 & z/\varrho \end{pmatrix}.$$

Alternatively,

$$\begin{aligned} \mathbf{r}^{AP} &= a\begin{pmatrix} x \\ y \\ z \end{pmatrix} = cR_{cb}R_{ba}\begin{pmatrix} x \\ y \\ z \end{pmatrix} \\ &= c\begin{pmatrix} 0 \\ 0 \\ \left(x^2+y^2+z^2\right)/\varrho \end{pmatrix}, \end{aligned}$$

which implies that

$$\begin{aligned} \varrho &= \sqrt{x^2+y^2+z^2}, \\ \theta &= \arctan\frac{y}{x}, \\ \phi &= \arccos\left(\frac{z}{\sqrt{x^2+y^2+z^2}}\right). \end{aligned}$$

**Exercise 5.31** Use rotation matrices to find the relation between polar and spherical coordinates relative to a coordinate system with axes parallel to the basis vectors of a triad $a$.

**Exercise 5.32** Repeat Exercises 5.4, 5.12, and 5.24 using the MAMBO toolbox.

### Summary of notation

An upper-case $R$ with a pair of triad labels as subscript, such as $R_{cd}$ and $R_{a^{(r)}a^{(f)}}$, was used in this chapter to denote the rotation matrix between the two triads.

An upper-case $R$ followed by four arguments within parentheses, such as

$$R(\varphi, v_1, v_2, v_3)$$

was used in this chapter to denote the rotation matrix with eigenvector corresponding to the eigenvalue 1 given by the column matrix $\begin{pmatrix} v_1 & v_2 & v_3 \end{pmatrix}^T$ and whose other eigenvalues are $e^{\pm i\varphi}$.

### Summary of terminology

(Page 203) Right-handed, orthonormal bases are referred to as *triads.*

(Page 207) The *rotation matrix* $R_{ab}$ is the matrix whose columns are the matrix representations of the basis vectors of the triad $b$ relative to the triad $a$.

(Page 228) An arbitrary rotation matrix can be decomposed into a product of rotation matrices corresponding to rotations about basis vectors. The corresponding angles are called *Euler angles.*

(Page 229) In the MAMBO toolbox, the procedure `MakeRotations` generates a rotation matrix.

# Chapter 6

# Orientations

*wherein the reader learns of:*

- *Using rotation matrices to describe the orientation of a rigid body or observer relative to another observer;*
- *Using configuration coordinates to describe time-dependent orientations;*
- *Configuration constraints and their implications on the allowable configurations of a mechanism.*

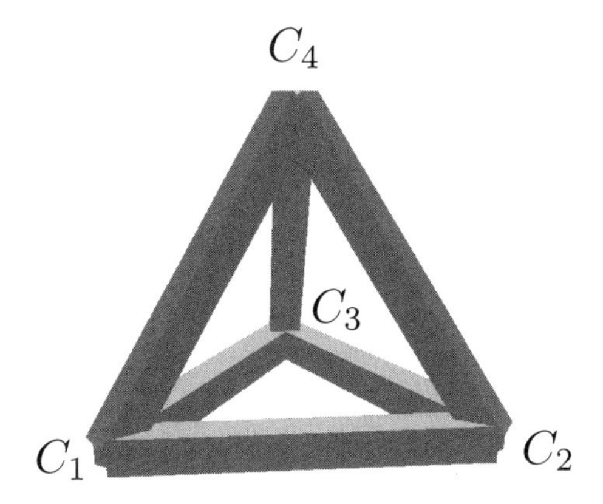

## Practicum

A fascinating array of real-life mechanisms rely on rotational motion for their function. Particularly intriguing are mechanisms that use intricate arrangements of gears to transmit rotational action about one axis into rotational or translational motion about some other axis, e.g., egg beaters, vehicle transmissions, car differentials, lawn sprinklers, and so on. Similarly, interesting uses of translational motion to generate changes in orientation are found, for example, in flight simulators.

Pick up a book with schematic outlines of some of these mechanisms and translate their action and design into MAMBO projects. Try to resolve complex arrangements of detailed parts through combinations of translations and rotations. This will hone your modeling skills, pique your curiosity, and bring the envy of your peers!

# 6.1 Review

## 6.1.1 Reference Triads

We recall the following observations from Chapter 1:

- The configuration of a rigid body relative to a reference configuration is uniquely described through a combination of a pure translation and a pure rotation, given the selection of a specific point on the body that is kept fixed by the pure rotation;
- The pure translation shifts all points on the body from the reference configuration to an intermediate configuration, such that the selected point on the rigid body coincides with its location in the final configuration;
- The subsequent pure rotation is determined by the location in the final configuration of two other points on the rigid body relative to the corresponding points in the intermediate configuration.

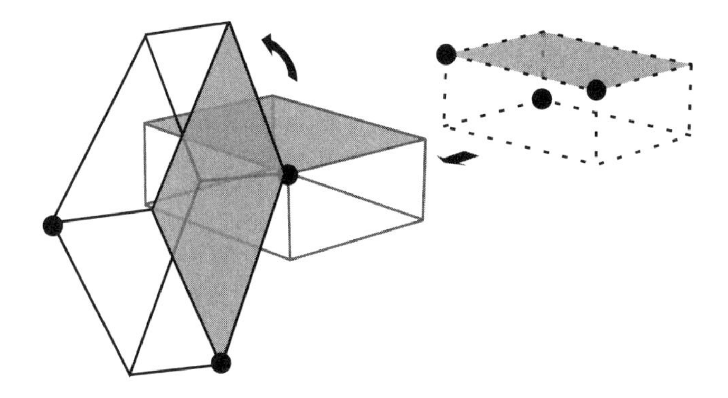

**Illustration 6.1**
The relative configuration of two observers $\mathcal{A}$ and $\mathcal{B}$ may be represented by the configuration of the virtual block corresponding to $\mathcal{A}$ relative to the reference configuration corresponding to $\mathcal{B}$.

The position and orientation of the virtual block relative to the reference configuration can be uniquely described through a combination of a pure translation and a pure rotation, given the selection of a specific point on the virtual block that is kept fixed by the pure rotation.

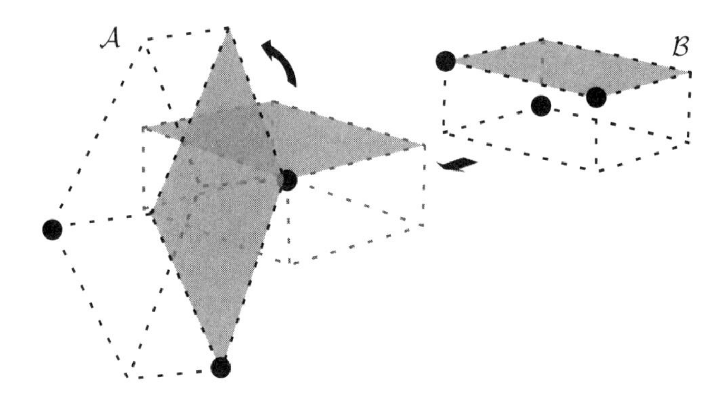


The pure translation shifts all points on the virtual block from the reference configuration of $\mathcal{B}$ to an intermediate configuration, such that the selected point coincides with the corresponding point in the reference configuration of $\mathcal{A}$.

The subsequent pure rotation is determined by the location in the final configuration of two other points on the virtual block relative to the intermediate configuration.

Let $A$ denote the selected point on the rigid body held fixed by the pure rotation and denote by $B$ and $C$ two other points on the rigid body. In Chapter 1, we found that the configuration of the rigid body relative to the reference configuration was uniquely determined by the location of the corresponding points in the final configuration relative to the reference configuration, **provided that the points $A$, $B$, and $C$ did not lie on a common straight line**. Choose $B$ and $C$, such that the separations $\overrightarrow{AB}$ and $\overrightarrow{AC}$ have unit length and are perpendicular. Then, the vectors

$$\mathbf{a}_1 = \left[\overrightarrow{AB}\right], \mathbf{a}_2 = \left[\overrightarrow{AC}\right], \text{ and } \mathbf{a}_3 = \mathbf{a}_1 \times \mathbf{a}_2$$

are the components of a triad whose orientation uniquely determines the orientation of the rigid body.

The triad introduced here is called the *reference triad* of the rigid body or of the corresponding observer. The reference triad is a triad whose orientation is fixed relative to the rigid body or fixed relative to the reference configuration of the observer.

There is no preferred choice of triad to qualify as the reference triad of a given rigid body or observer.

When describing the configuration of rigid bodies, it is common to choose triads that correspond to some geometrical feature. For example, a natural choice of reference triad of a rectangular block is three mutually perpendicular vectors of unit length that are parallel to the edges of the block. In the absence of geometrical features, such as edges or symmetries, to base the selection of reference triad on, any triad will do.

If two observers $\mathcal{A}$ and $\mathcal{B}$ share the same reference triad, then the configuration of $\mathcal{B}$ relative to $\mathcal{A}$ is described through a pure translation $\mathbf{T}_{\mathcal{A}\to\mathcal{B}}$, but no rotation. In other words,

$$\mathbf{R}_{\mathcal{A}\to\mathcal{B}} = \mathbf{I}.$$

Conversely, if

$$\mathbf{R}_{\mathcal{A}\to\mathcal{B}} = \mathbf{I},$$

then the reference triads of the two observers $\mathcal{A}$ and $\mathcal{B}$ coincide.

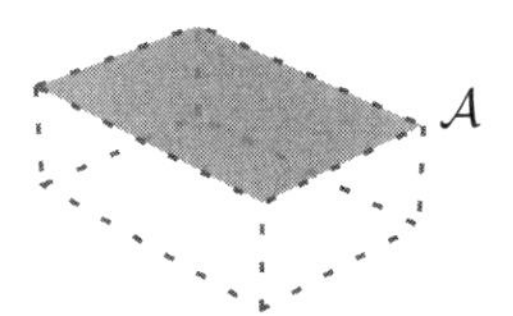


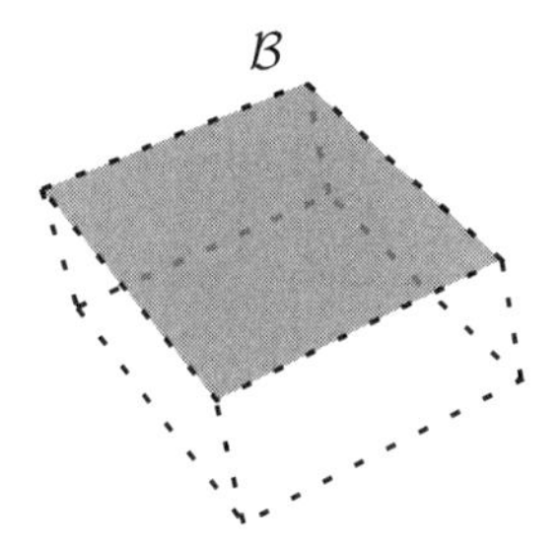


Similarly, if the reference triad of a rigid body, say a block, coincides with the reference triad of an observer $\mathcal{A}$, then the rigid body's configuration relative to $\mathcal{A}$ is described through a pure translation $\mathbf{T}_{\mathcal{A}}$, but no rotation. In other words,

$$\mathbf{R}_{\mathcal{A}} = \mathbf{I}.$$

**Illustration 6.2**

Suppose the configuration of an observer $\mathcal{B}$ relative to an observer $\mathcal{A}$ is given by a non-trivial pure rotation, but no translation, i.e.,

$$\mathbf{R}_{\mathcal{A}\to\mathcal{B}} \neq \mathbf{I}, \mathbf{T}_{\mathcal{A}\to\mathcal{B}} = \mathbf{I}.$$

Then, the reference triads of $\mathcal{A}$ and $\mathcal{B}$ do not coincide. The pure rotation $\mathbf{R}_{\mathcal{A}\to\mathcal{B}}$ contains the information necessary to rotate the reference configuration of $\mathcal{A}$ about the common reference point so that it coincides with the reference configuration of $\mathcal{B}$. The rotation $\mathbf{R}_{\mathcal{A}\to\mathcal{B}}$ is uniquely determined by the relative orientation of the reference triads of $\mathcal{A}$ and $\mathcal{B}$.

### 6.1.2 Rotations

If the reference triads of two observers $\mathcal{A}$ and $\mathcal{B}$ are denoted by $a$ and $b$, respectively, then the pure rotation $\mathbf{R}_{\mathcal{A}\to\mathcal{B}}$ is uniquely determined by the rotation matrix $R_{ab}$, where

$$R_{ab} = a^T \bullet b.$$

In particular, if $\mathbf{R}_{\mathcal{A}\to\mathcal{B}}$ corresponds to a rotation about an axis parallel to the **unit vector**

$$a\begin{pmatrix} v_1 \\ v_2 \\ v_3 \end{pmatrix}$$

by an angle $\varphi$, then

$$\begin{aligned} R_{ab} &= R(\varphi, v_1, v_2, v_3) \\ &\overset{def}{=} \begin{pmatrix} v_1^2 + (1 - v_1^2)\, c & (1-c)\, v_1 v_2 - v_3 s & (1-c)\, v_1 v_3 + v_2 s \\ (1-c)\, v_1 v_2 + v_3 s & v_2^2 + (1 - v_2^2)\, c & (1-c)\, v_2 v_3 - v_1 s \\ (1-c)\, v_1 v_3 - v_2 s & (1-c)\, v_2 v_3 + v_1 s & v_3^2 + (1 - v_3^2)\, c \end{pmatrix}, \end{aligned}$$

where $c = \cos\varphi$ and $s = \sin\varphi$.

In the previous chapter, we developed algebraic operations on rotation matrices that corresponded to the operations on pure rotations introduced in Chapter 2. For example, the correspondences

$$R_{ab} \leftrightarrow \mathbf{R}_{\mathcal{A}\to\mathcal{B}},\, R_{bc} \leftrightarrow \mathbf{R}_{\mathcal{B}\to\mathcal{C}}$$

imply that

$$R_{ac} = R_{ab} R_{bc} \leftrightarrow \mathbf{R}_{\mathcal{B}\to\mathcal{C}} \circ \mathbf{R}_{\mathcal{A}\to\mathcal{B}} = \mathbf{R}_{\mathcal{A}\to\mathcal{C}}.$$

Using the rotation matrix $R_{ab}$, we are also able to express the triad $b$ in terms of the triad $a$:

$$b = a R_{ab}$$

as well as the matrix representation of a vector $\mathbf{v}$ relative to the triad $b$, ${}^b v$, in terms of the matrix representation of $\mathbf{v}$ relative to the triad $a$, ${}^a v$:

$${}^b v = R_{ba}\, {}^a v,$$

where

$$R_{ba} = (R_{ab})^{-1} = (R_{ab})^T.$$

## 6.2 Examples

(Ex. 6.1 – Ex. 6.8)

### 6.2.1 A Still Life

Suppose we want to describe the geometry of a wireframe representation of a tetrahedron, as depicted below.

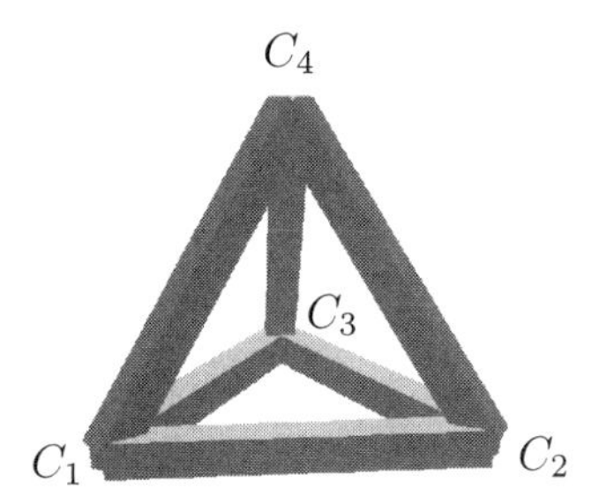


Introduce a main observer $\mathcal{W}$ with reference point $W$ at the geometric center of the tetrahedron and with reference triad $w$. Let the $i$-th edge correspond to a rigid body with reference point $E_i$ at the midpoint of the edge and reference triad $e^{(i)}$, such that the $\mathbf{e}_1^{(i)}$ basis vector is parallel to the edge.

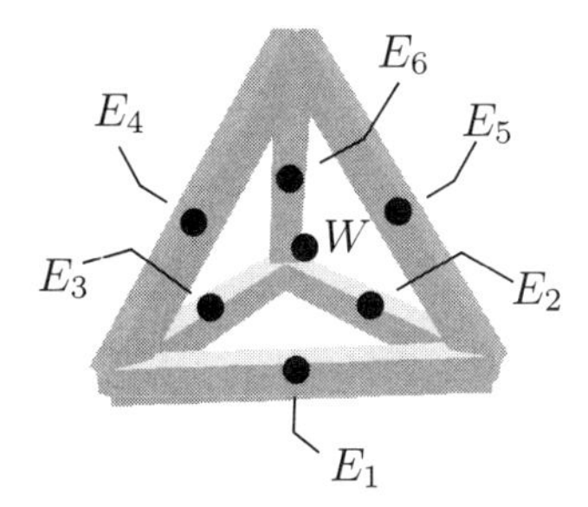


Suppose that the orientation of the tetrahedron is chosen such that the locations of each of the four corners relative to the $\mathcal{W}$ observer are given by the four position vectors

$$\mathbf{r}^{WC_1} = w\begin{pmatrix} -\frac{1}{2}l \\ -\frac{\sqrt{3}}{6}l \\ -\frac{\sqrt{6}}{12}l \end{pmatrix}, \mathbf{r}^{WC_2} = w\begin{pmatrix} \frac{1}{2}l \\ -\frac{\sqrt{3}}{6}l \\ -\frac{\sqrt{6}}{12}l \end{pmatrix},$$

$$\mathbf{r}^{WC_3} = w\begin{pmatrix} 0 \\ \frac{\sqrt{3}}{3}l \\ -\frac{\sqrt{6}}{12}l \end{pmatrix}, \text{ and } \mathbf{r}^{WC_4} = w\begin{pmatrix} 0 \\ 0 \\ \frac{\sqrt{6}}{4}l \end{pmatrix},$$

where $l$ is the length of the tetrahedron's edges.

The configuration of the $i$-th edge relative to $\mathcal{W}$ is given by a pure translation $\mathbf{T}_i$ corresponding to the position vector

$$\mathbf{r}^{WE_i}$$

and a pure rotation $\mathbf{R}_i$ corresponding to the rotation matrix

$$R_{we^{(i)}}.$$

Specifically, we find that

$$\mathbf{r}^{WE_1} = \frac{1}{2}\mathbf{r}^{WC_1} + \frac{1}{2}\mathbf{r}^{WC_2} = w\begin{pmatrix} 0 \\ -\frac{\sqrt{3}}{6}l \\ -\frac{\sqrt{6}}{12}l \end{pmatrix},$$

$$\mathbf{r}^{WE_2} = \frac{1}{2}\mathbf{r}^{WC_2} + \frac{1}{2}\mathbf{r}^{WC_3} = w\begin{pmatrix} \frac{1}{4}l \\ \frac{\sqrt{3}}{12}l \\ -\frac{\sqrt{6}}{12}l \end{pmatrix},$$

$$\mathbf{r}^{WE_3} = \frac{1}{2}\mathbf{r}^{WC_3} + \frac{1}{2}\mathbf{r}^{WC_1} = w\begin{pmatrix} -\frac{1}{4}l \\ \frac{\sqrt{3}}{12}l \\ -\frac{\sqrt{6}}{12}l \end{pmatrix},$$

$$\mathbf{r}^{WE_4} = \frac{1}{2}\mathbf{r}^{WC_1} + \frac{1}{2}\mathbf{r}^{WC_4} = w\begin{pmatrix} -\frac{1}{4}l \\ -\frac{\sqrt{3}}{12}l \\ \frac{\sqrt{6}}{12}l \end{pmatrix},$$

$$\mathbf{r}^{WE_5} = \frac{1}{2}\mathbf{r}^{WC_2} + \frac{1}{2}\mathbf{r}^{WC_4} = w\begin{pmatrix} \frac{1}{4}l \\ -\frac{\sqrt{3}}{12}l \\ \frac{\sqrt{6}}{12}l \end{pmatrix},$$

and

$$\mathbf{r}^{WE_6} = \frac{1}{2}\mathbf{r}^{WC_3} + \frac{1}{2}\mathbf{r}^{WC_4} = w\begin{pmatrix} 0 \\ \frac{\sqrt{3}}{6}l \\ \frac{\sqrt{6}}{12}l \end{pmatrix}.$$

We illustrate the computation of the rotation matrices $R_{we^{(i)}}$ by considering the $i = 3$ edge. Specifically, to compute the rotation matrix $R_{we^{(3)}}$, we construct the triad $e^{(3)}$ by requiring that the basis vector $\mathbf{e}_1^{(3)}$ be parallel to the vector

$$\mathbf{r}^{WC_1} - \mathbf{r}^{WC_3}.$$

Following the method presented in the previous chapters, we find

$$\mathbf{e}_1^{(3)} = \frac{\mathbf{r}^{WC_1} - \mathbf{r}^{WC_3}}{\|\mathbf{r}^{WC_1} - \mathbf{r}^{WC_3}\|} = w \begin{pmatrix} -\frac{1}{2} \\ -\frac{\sqrt{3}}{2} \\ 0 \end{pmatrix},$$

$$\mathbf{e}_2^{(3)} = \frac{\mathbf{e}_1^{(3)} \times \mathbf{w}_3}{\left\|\mathbf{e}_1^{(3)} \times \mathbf{w}_3\right\|} = w \begin{pmatrix} -\frac{\sqrt{3}}{2} \\ \frac{1}{2} \\ 0 \end{pmatrix},$$

and

$$\mathbf{e}_3^{(3)} = \mathbf{e}_1^{(3)} \times \mathbf{e}_2^{(3)} = w \begin{pmatrix} 0 \\ 0 \\ -1 \end{pmatrix}.$$

The rotation matrix is now given by

$$R_{we^{(3)}} = w^T \bullet e^{(3)} = \begin{pmatrix} -\frac{1}{2} & -\frac{\sqrt{3}}{2} & 0 \\ -\frac{\sqrt{3}}{2} & \frac{1}{2} & 0 \\ 0 & 0 & -1 \end{pmatrix}.$$

Similarly,

$$R_{we^{(1)}} = \begin{pmatrix} 1 & 0 & 0 \\ 0 & -1 & 0 \\ 0 & 0 & -1 \end{pmatrix}, R_{we^{(2)}} = \begin{pmatrix} -\frac{1}{2} & \frac{\sqrt{3}}{2} & 0 \\ \frac{\sqrt{3}}{2} & \frac{1}{2} & 0 \\ 0 & 0 & -1 \end{pmatrix},$$

$$R_{we^{(4)}} = \begin{pmatrix} \frac{1}{2} & \frac{1}{2} & \frac{\sqrt{2}}{2} \\ \frac{\sqrt{3}}{6} & -\frac{\sqrt{3}}{2} & \frac{\sqrt{6}}{6} \\ \frac{\sqrt{6}}{3} & 0 & -\frac{\sqrt{3}}{3} \end{pmatrix}, R_{we^{(5)}} = \begin{pmatrix} -\frac{1}{2} & \frac{1}{2} & -\frac{\sqrt{2}}{2} \\ \frac{\sqrt{3}}{6} & \frac{\sqrt{3}}{2} & \frac{\sqrt{6}}{6} \\ \frac{\sqrt{6}}{3} & 0 & -\frac{\sqrt{3}}{3} \end{pmatrix},$$

and

$$R_{we^{(6)}} = \begin{pmatrix} 0 & -1 & 0 \\ -\frac{\sqrt{3}}{3} & 0 & -\frac{\sqrt{6}}{3} \\ \frac{\sqrt{6}}{3} & 0 & -\frac{\sqrt{3}}{3} \end{pmatrix}.$$

**Illustration 6.3**

Suppose we want to model the geometry of a conical pile of rods as shown in the figure.

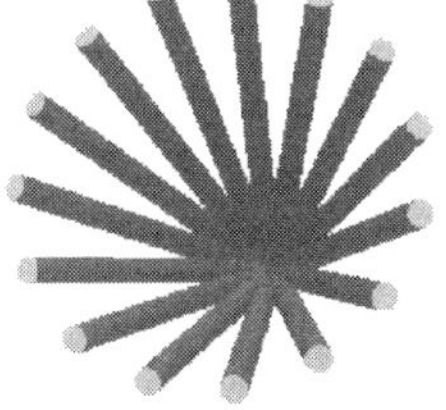

Introduce a main observer $\mathcal{W}$ with reference point $W$ at the top of the cone and reference triad $w$, such that the symmetry axis of the cone is parallel to the vector

$$\mathbf{v} = w \begin{pmatrix} \sqrt{3} \\ 1 \\ 0 \end{pmatrix}$$

and one of the rods is parallel to the vector

$$\mathbf{u} = w \begin{pmatrix} 1 \\ \sqrt{3} \\ 0 \end{pmatrix}.$$

It follows that the opening angle $\theta$ of the cone satisfies

$$\cos \frac{\theta}{2} = \frac{\mathbf{u} \bullet \mathbf{v}}{\|\mathbf{u}\| \, \|\mathbf{v}\|} = \frac{\sqrt{3}}{2},$$

i.e., $\theta = 60°$.

Let the $i$-th rod be represented by a rigid rod with reference position $R_i$ at the center of the rod and reference triad $r^{(i)}$, such that $\mathbf{r}_3^{(i)}$ is parallel to the rod.

The configuration of the $i$-th rod relative to the observer $\mathcal{W}$ is given by a pure translation $\mathbf{T}_i$ corresponding to the position vector

$$\mathbf{r}^{WR_i}$$

and a pure rotation $\mathbf{R}_i$ corresponding to the rotation matrix

$$R_{wr^{(i)}}.$$

Specifically,

$$\mathbf{r}^{WR_i} = r^{(i)} \begin{pmatrix} 0 \\ 0 \\ \frac{1}{2} l \end{pmatrix},$$

where $l$ is the length of the rod.

Let $i = 1$ correspond to the rod for which $\mathbf{r}_3^{(1)}$ is parallel to $\mathbf{u}$. In particular,

$$\mathbf{r}_3^{(1)} = \frac{\mathbf{u}}{\|\mathbf{u}\|} = w \begin{pmatrix} \frac{1}{2} \\ \frac{\sqrt{3}}{2} \\ 0 \end{pmatrix},$$

$$\mathbf{r}_1^{(1)} = \frac{\mathbf{r}_3^{(1)} \times \mathbf{w}_1}{\left\| \mathbf{r}_3^{(1)} \times \mathbf{w}_1 \right\|} = w \begin{pmatrix} 0 \\ 0 \\ -1 \end{pmatrix},$$

$$\mathbf{r}_2^{(1)} = \mathbf{r}_3^{(1)} \times \mathbf{r}_1^{(1)} = w \begin{pmatrix} -\frac{\sqrt{3}}{2} \\ \frac{1}{2} \\ 0 \end{pmatrix},$$

and thus

$$R_{wr^{(1)}} = w^T \bullet r^{(1)} = \begin{pmatrix} 0 & -\frac{\sqrt{3}}{2} & \frac{1}{2} \\ 0 & \frac{1}{2} & \frac{\sqrt{3}}{2} \\ -1 & 0 & 0 \end{pmatrix}.$$

The $i$-th rod is rotated relative to the first rod about the axis through W that is parallel to $\mathbf{v}$ by an amount

$$\frac{2\pi}{N}(i-1),$$

where $N$ is the number of rods. The matrix representation of $\mathbf{v}$ relative to the $r^{(1)}$ triad is given by

$$\begin{aligned} {}^{r^{(1)}}v &= R_{r^{(1)}w}\,{}^{w}v \\ &= \begin{pmatrix} 0 & -\frac{\sqrt{3}}{2} & \frac{1}{2} \\ 0 & \frac{1}{2} & \frac{\sqrt{3}}{2} \\ -1 & 0 & 0 \end{pmatrix}^T \begin{pmatrix} \sqrt{3} \\ 1 \\ 0 \end{pmatrix} \\ &= \begin{pmatrix} 0 \\ -1 \\ \sqrt{3} \end{pmatrix}. \end{aligned}$$

It follows that

$$R_{r^{(1)}r^{(i)}} = R\left(\frac{2\pi}{N}(i-1), 0, -\frac{1}{2}, \frac{\sqrt{3}}{2}\right),$$

where the notation $R(\varphi, v_1, v_2, v_3)$ was introduced in the previous chapter. The rotation matrix $R_{wr^{(i)}}$ is now obtained from the product

$$R_{wr^{(1)}} R_{r^{(1)}r^{(i)}}.$$

### 6.2.2 The Single Moving Body

When time-dependent changes take place in the configuration of a rigid body relative to the main observer, the recommended methodology requires the introduction of at least one auxiliary observer between the rigid body and the main observer. Specifically, the auxiliary observer is introduced in such a way that the rigid body remains stationary relative to the auxiliary observer, while the motion of the auxiliary observer relative to the main observer contains the entire time-dependence.

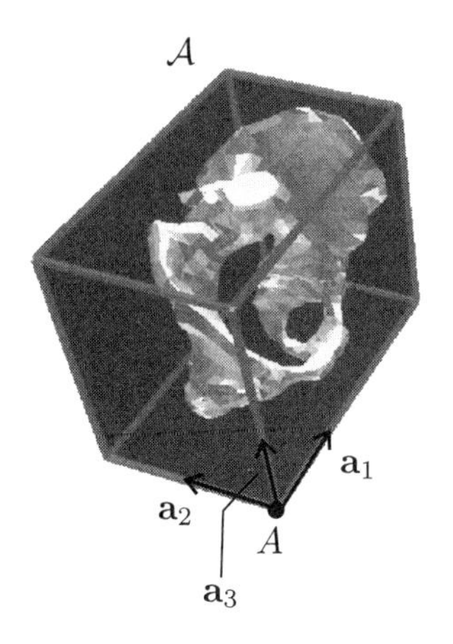


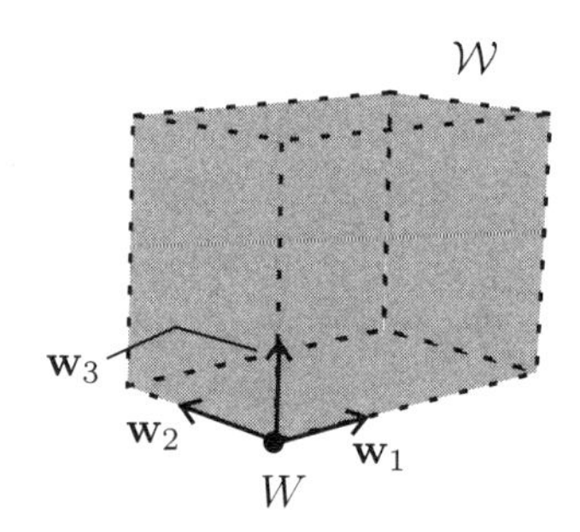


Suppose, for example, that we want to describe the configuration of a single, freely moving rigid body. Introduce a main observer $\mathcal{W}$ with reference point $W$ somewhere in space and reference triad $w$. Introduce an auxiliary observer $\mathcal{A}$, relative to which the rigid body is stationary, with reference point $A$ coinciding with some arbitrary point on the rigid body and reference triad $a$.

Assume for simplicity that the position of the rigid body relative to $\mathcal{W}$ is described by the identity translation. Then, the configuration of the observer $\mathcal{A}$ relative to $\mathcal{W}$ is given by the pure rotation $\mathbf{R}_{\mathcal{W}\to\mathcal{A}}$ corresponding to the rotation matrix

$$R_{wa}.$$

Since the rigid body's orientation is unrestricted, we can use a $1-3-1$ sequence of rotations to decompose $\mathbf{R}_{\mathcal{W}\to\mathcal{A}}$, such that

$$R_{wa} = \begin{pmatrix} 1 & 0 & 0 \\ 0 & c_1 & -s_1 \\ 0 & s_1 & c_1 \end{pmatrix} \begin{pmatrix} c_2 & -s_2 & 0 \\ s_2 & c_2 & 0 \\ 0 & 0 & 1 \end{pmatrix} \begin{pmatrix} 1 & 0 & 0 \\ 0 & c_3 & -s_3 \\ 0 & s_3 & c_3 \end{pmatrix},$$

where $c_i = \cos q_i$ and $s = \sin q_i$, and $q_1$, $q_2$, and $q_3$ are time-dependent quantities that uniquely specify the rotation matrix. Following the terminology introduced in Chapter 4, these quantities are called *configuration coordinates*, since they provide information about the configuration of the rigid body relative to the main observer as a function of time.

There are numerous other choices of decompositions of $\mathbf{R}_{\mathcal{W}\to\mathcal{A}}$ that can be specified by three configuration coordinates. The discussion in the previous chapter showed that fewer than three coordinates would not suffice to specify an arbitrary orientation. It follows that, in the absence of translations, the rigid body has three *geometric degrees of freedom*.

**Illustration 6.4**

In the previous chapter, we found that the $[i,j]$-th entry in a rotation matrix $R_{ab}$ could be expressed as

$$[R_{ab}]_{ij} = \delta_{ij} \cos q_4 + (1 - \cos q_4)\, q_i q_j - \sin q_4 \sum_{k=1}^{3} \varepsilon_{ijk} q_k,$$

where

$$\delta_{11} = \delta_{22} = \delta_{33} = \varepsilon_{123} = \varepsilon_{231} = \varepsilon_{312} = -\varepsilon_{321} = -\varepsilon_{213} = -\varepsilon_{132} = 1$$

and zero otherwise,

$$\begin{pmatrix} q_1 \\ q_2 \\ q_3 \end{pmatrix}$$

is the eigenvector of $R_{ab}$ corresponding to the eigenvalue 1, and

$$q_1^2 + q_2^2 + q_3^2 = 1.$$

This last condition is a configuration constraint on the configuration coordinates $q_1$, $q_2$, $q_3$, and $q_4$.

An alternative formulation is obtained by introducing the new configuration coordinates

$$\tilde{q}_i = q_i \sin \frac{q_4}{2},\ i = 1, 2, 3, \text{ and } \tilde{q}_4 = \cos \frac{q_4}{2}.$$

Since

$$\cos q_4 = 2 \cos^2 \frac{q_4}{2} - 1 = 1 - 2 \sin^2 \frac{q_4}{2} \text{ and } \sin q_4 = 2 \sin \frac{q_4}{2} \cos \frac{q_4}{2},$$

it follows that

$$\delta_{ij} \cos q_4 + (1 - \cos q_4)\, q_i q_j - \sin q_4 \sum_{k=1}^{3} \varepsilon_{ijk} q_k$$
$$= \delta_{ij} \left(2\tilde{q}_4^2 - 1\right) + 2\tilde{q}_i \tilde{q}_j - 2\tilde{q}_4 \sum_{k=1}^{3} \varepsilon_{ijk} \tilde{q}_k.$$

The rotation matrix thus becomes

$$\begin{pmatrix} 2\tilde{q}_4^2 + 2\tilde{q}_1^2 - 1 & 2\tilde{q}_1\tilde{q}_2 - 2\tilde{q}_4\tilde{q}_3 & 2\tilde{q}_1\tilde{q}_3 + 2\tilde{q}_4\tilde{q}_2 \\ 2\tilde{q}_2\tilde{q}_1 + 2\tilde{q}_4\tilde{q}_3 & 2\tilde{q}_4^2 + 2\tilde{q}_2^2 - 1 & 2\tilde{q}_2\tilde{q}_3 - 2\tilde{q}_4\tilde{q}_1 \\ 2\tilde{q}_3\tilde{q}_1 - 2\tilde{q}_4\tilde{q}_2 & 2\tilde{q}_3\tilde{q}_2 + 2\tilde{q}_4\tilde{q}_1 & 2\tilde{q}_4^2 + 2\tilde{q}_3^2 - 1 \end{pmatrix}.$$

The configuration coordinates $\tilde{q}_1$, $\tilde{q}_2$, $\tilde{q}_3$, and $\tilde{q}_4$ are called *Euler parameters*. Only values that satisfy the configuration constraint

$$\tilde{q}_1^2 + \tilde{q}_2^2 + \tilde{q}_3^2 + \tilde{q}_4^2 = \left(q_1^2 + q_2^2 + q_3^2\right) \sin^2 \frac{q_4}{2} + \cos^2 \frac{q_4}{2} = 1$$

correspond to actual configurations of the rigid body.

Suppose we want to describe the configuration of a rod attached at one end to a spherical joint.

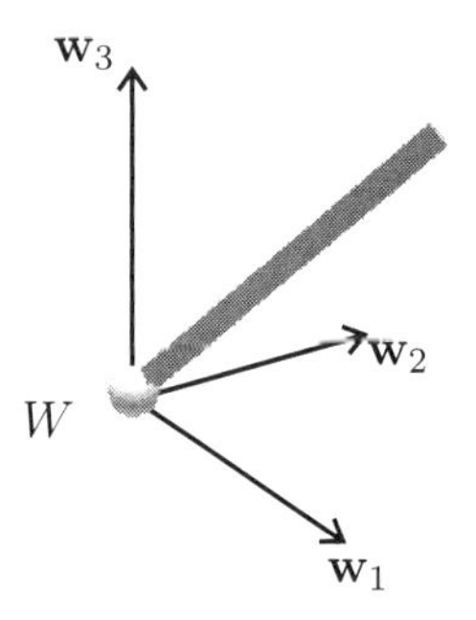


Introduce a main observer $\mathcal{W}$ with reference point $W$ at the spherical joint and reference triad $w$. Let $\mathcal{A}$ be an auxiliary observer with reference point $A$ coinciding with $W$ and reference triad $a$, such that the rod is parallel to the $\mathbf{a}_1$ vector.

The configuration of the rod relative to the observer $\mathcal{A}$ is given by a pure translation $\mathbf{T}$ corresponding to the position vector

$$\mathbf{r}^{AC},$$

where $C$ is the point at the center of the rod. By assumption,

$$\mathbf{r}^{AC} = a \begin{pmatrix} \frac{1}{2}l \\ 0 \\ 0 \end{pmatrix},$$

where $l$ is the length of the rod.

The configuration of the observer $\mathcal{A}$ relative to $\mathcal{W}$ is given by a pure rotation $\mathbf{R}_{\mathcal{W}\to\mathcal{A}}$ corresponding to the rotation matrix

$$R_{wa}.$$

The pure rotation $\mathbf{R}_{\mathcal{W}\to\mathcal{A}}$ may be decomposed into two pure rotations by the introduction of an intermediate auxiliary observer $\mathcal{B}$ with reference point $B$ coinciding with $A$ and $W$ and reference triad $b$, such that the axes held fixed by the pure rotations $\mathbf{R}_{\mathcal{W}\to\mathcal{B}}$ and $\mathbf{R}_{\mathcal{B}\to\mathcal{A}}$, respectively, are parallel to the vectors $\mathbf{w}_3$ and $\mathbf{w}_1$, respectively. In particular,

$$R_{wb} = R(q_1, 0, 0, 1) = \begin{pmatrix} \cos q_1 & -\sin q_1 & 0 \\ \sin q_1 & \cos q_1 & 0 \\ 0 & 0 & 1 \end{pmatrix}.$$

Moreover, the matrix representation of the vector $\mathbf{w}_1$ in the $b$ triad is given by

$$\begin{aligned}
{}^{b}(\mathbf{w}_1) &= R_{bw}{}^{w}(\mathbf{w}_1) \\
&= \begin{pmatrix} \cos q_1 & -\sin q_1 & 0 \\ \sin q_1 & \cos q_1 & 0 \\ 0 & 0 & 1 \end{pmatrix}^T \begin{pmatrix} 1 \\ 0 \\ 0 \end{pmatrix} \\
&= \begin{pmatrix} \cos q_1 \\ -\sin q_1 \\ 0 \end{pmatrix}
\end{aligned}$$

and thus

$$\begin{aligned}
R_{ba} &= R\left(q_2, \cos q_1, -\sin q_1, 0\right) \\
&= \begin{pmatrix} \cos q_2 \sin^2 q_1 + \cos^2 q_1 & \cos q_1 \sin q_1 (\cos q_2 - 1) & -\sin q_1 \sin q_2 \\ \cos q_1 \sin q_1 (\cos q_2 - 1) & \cos q_2 \cos^2 q_1 + \sin^2 q_1 & -\cos q_1 \sin q_2 \\ \sin q_1 \sin q_2 & \cos q_1 \sin q_2 & \cos q_2 \end{pmatrix}.
\end{aligned}$$

The configuration coordinates $q_1$ and $q_2$ suffice to describe an arbitrary orientation of the rod. In particular, the position vector

$$\begin{aligned}
\mathbf{r}^{WC} &= \mathbf{r}^{AC} \\
&= a \begin{pmatrix} \frac{1}{2}l \\ 0 \\ 0 \end{pmatrix} \\
&= w R_{wa} \begin{pmatrix} \frac{1}{2}l \\ 0 \\ 0 \end{pmatrix} \\
&= w R_{wb} R_{ba} \begin{pmatrix} \frac{1}{2}l \\ 0 \\ 0 \end{pmatrix} \\
&= w \begin{pmatrix} \frac{1}{2}l \cos q_1 \\ \frac{1}{2}l \sin q_1 \cos q_2 \\ \frac{1}{2}l \sin q_1 \sin q_2 \end{pmatrix}.
\end{aligned}$$

### 6.2.3 Degrees of Freedom

In Chapter 4, we found that a free rigid body, in the absence of rotations, has three geometric degrees of freedom. This was tantamount to the claim that:

- No fewer than three configuration coordinates would suffice to capture all possible positions of the rigid body;
- No more than three configuration coordinates would be necessary to capture all possible positions of the rigid body.

In this chapter, we found that a free rigid body, in the absence of translations, has three geometric degrees of freedom. This was tantamount to the claim that:

- No fewer than three configuration coordinates would suffice to capture all possible orientations of the rigid body;
- No more than three configuration coordinates would be necessary to capture all possible orientations of the rigid body.

From Chapter 1, we recall that the specification of the position of a rigid body is entirely independent and separate from the specification of the orientation of the rigid body. We conclude, as already demonstrated in Chapter 1, that a rigid body whose position and orientation can change arbitrarily has six geometric degrees of freedom.

When a rigid body has fewer than six geometric degrees of freedom, it is *constrained*. As discussed in Chapter 4, this implies that there are *configuration constraints*, i.e., equalities in the configuration coordinates, that restrict the choice of values for the configuration coordinates that correspond to allowable configurations of the rigid body. For example, the rigid bodies in Chapter 4 were constrained in orientation, since three configuration coordinates sufficed to describe their configuration. Similarly, the rigid bodies considered thus far in this chapter were constrained in position, since three configuration coordinates sufficed to describe their configuration.

### 6.2.4 Multiple Moving Rigid Bodies

In the absence of constraints, each rigid body in a multibody mechanism has six geometric degrees of freedom. If $N$ rigid bodies have fewer than $6N$ degrees of freedom, the mechanism is constrained.

Suppose you want to describe the geometry of a bench-based radial arm saw as shown in the figure.

There are four parts that move relative to the stationary workbench, namely, the tool arm, the tool trolley, the blade support, and the blade. A vertical cylinder attached to the workbench provides the support for the mechanism. Specifically:

- The tool arm is free to slide up and down along this cylinder as well as rotate about the cylinder;
- The tool trolley is free to slide along the arm;
- The blade support is free to rotate relative to the tool trolley about an axis parallel to the arm;
- The blade is free to rotate about an arm perpendicular to the axis of rotation of the blade support.

It follows that the radial arm saw has five geometric degrees of freedom, two that correspond to translations and three corresponding to rotations.

Introduce a main observer $\mathcal{W}$, relative to which the workbench remains stationary. Let its reference point $W$ be located at the center of the vertical cylinder and level with the table top. Let its reference triad $w$ be oriented such that the table top is parallel to the $\mathbf{w}_1$ and $\mathbf{w}_2$ basis vectors and $\mathbf{w}_3$ points away from the table top in the direction of the tool arm.

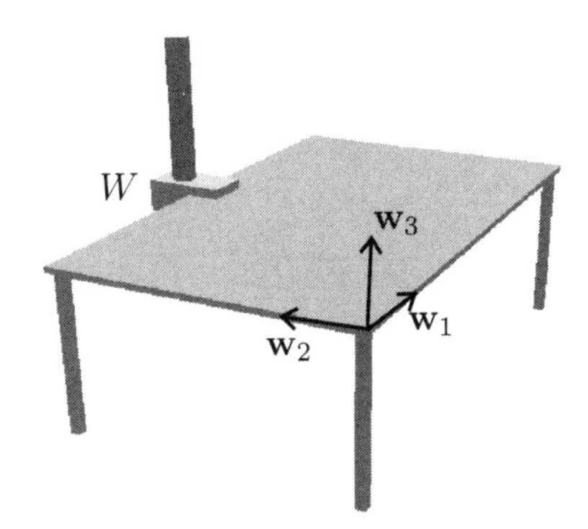


Since each of the four parts of the radial arm saw move relative to $\mathcal{W}$, the recommended methodology requires the introduction of an auxiliary observer for each part, such that the part's configuration is stationary relative to the corresponding auxiliary observer.

Introduce an auxiliary observer $\mathcal{A}$, relative to which the tool arm remains stationary. Let its reference point $A$ be located at the center of the vertical cylinder and level with the tool arm. Let its reference triad $a$ be oriented such that $\mathbf{a}_3$ equals $\mathbf{w}_3$ and $\mathbf{a}_2$ is parallel to the tool arm.

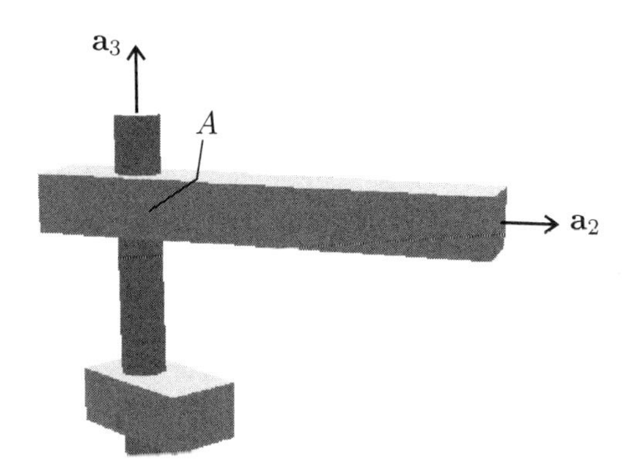


The configuration of the observer $\mathcal{A}$ relative to $\mathcal{W}$ is then given by a pure translation $\mathbf{T}_{\mathcal{W}\to\mathcal{A}}$ corresponding to the position vector

$$\mathbf{r}^{WA}$$

and a pure rotation $\mathbf{R}_{\mathcal{W}\to\mathcal{A}}$ corresponding to the rotation matrix

$$R_{wa}.$$

Specifically,

$$\mathbf{r}^{WA} = w \begin{pmatrix} 0 \\ 0 \\ q_1 \end{pmatrix},$$

since the tool arm may only slide in the $\mathbf{w}_3$ direction. Moreover,

$$R_{wa} = R(q_2, 0, 0, 1) = \begin{pmatrix} \cos q_2 & -\sin q_2 & 0 \\ \sin q_2 & \cos q_2 & 0 \\ 0 & 0 & 1 \end{pmatrix},$$

since the tool arm may only rotate about the $\mathbf{w}_3$ direction.

Introduce an auxiliary observer $\mathcal{B}$, relative to which the tool trolley remains stationary. Let its reference point $B$ be located on the center line of the tool arm and symmetric relative to the ends of the trolley. Let its reference triad $b$ equal $a$.

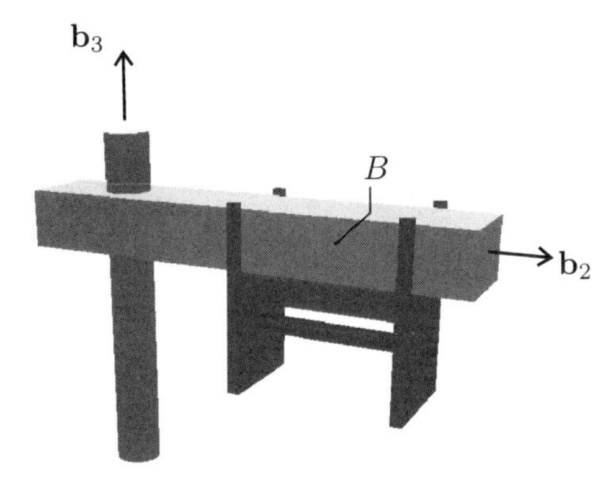


The configuration of the observer $\mathcal{B}$ relative to $\mathcal{A}$ is given by a pure translation $\mathbf{T}_{\mathcal{A}\to\mathcal{B}}$ corresponding to the position vector

$$\mathbf{r}^{AB}$$

and the identity rotation $\mathbf{R}_{\mathcal{A}\to\mathcal{B}} = \mathbf{I}$. Specifically,

$$\mathbf{r}^{AB} = a \begin{pmatrix} 0 \\ q_3 \\ 0 \end{pmatrix},$$

since the trolley may only slide along the $\mathbf{a}_2$ direction.

Introduce an auxiliary observer $\mathcal{C}$, relative to which the blade support remains stationary. Let its reference point $C$ be located at the center of the axis about which the blade support rotates. Let its reference triad $c$ be oriented such that $\mathbf{c}_2$ equals $\mathbf{a}_2$ and $\mathbf{c}_3$ points toward the center of the blade.

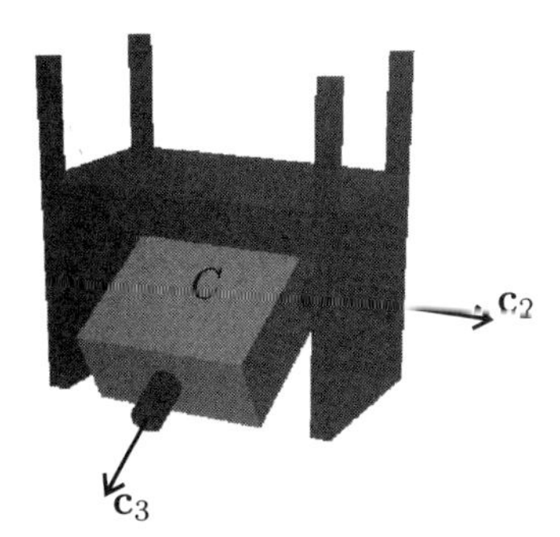


The configuration of the observer $\mathcal{C}$ relative to $\mathcal{B}$ is given by a pure translation $\mathbf{T}_{\mathcal{B}\to\mathcal{C}}$ corresponding to the position vector

$$\mathbf{r}^{BC}$$

and the pure rotation $\mathbf{R}_{\mathcal{B}\to\mathcal{C}}$ corresponding to the rotation matrix

$$R_{bc}.$$

Specifically,

$$\mathbf{r}^{BC} =_a \begin{pmatrix} 0 \\ 0 \\ -p_1 \end{pmatrix},$$

where $p_1$ is some time-independent parameter. Moreover,

$$R_{bc} = R(q_4, 0, 1, 0) = \begin{pmatrix} \cos q_4 & 0 & \sin q_4 \\ 0 & 1 & 0 \\ -\sin q_4 & 0 & \cos q_4 \end{pmatrix},$$

since the blade support rotates about the $\mathbf{a}_2$ direction.

Finally, introduce an auxiliary observer $\mathcal{D}$, relative to which the blade remains stationary. Let its reference point $D$ be at the center of the blade. Let its reference triad $d$ be such that $\mathbf{d}_3$ equals $\mathbf{c}_3$.

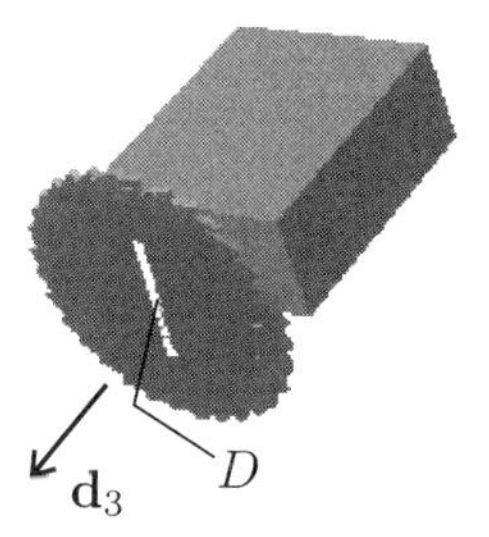


The configuration of the observer $\mathcal{D}$ relative to $\mathcal{C}$ is given by the pure translation $\mathbf{T}_{\mathcal{C}\to\mathcal{D}}$ corresponding to the position vector

$$\mathbf{r}^{CD}$$

and a pure rotation $\mathbf{R}_{\mathcal{C}\to\mathcal{D}}$ corresponding to the rotation matrix

$$R_{cd}.$$

Specifically,

$$\mathbf{r}^{CD} =_c \begin{pmatrix} 0 \\ 0 \\ p_2 \end{pmatrix},$$

where $p_2$ is some time-independent parameter. Moreover,

$$R_{cd} = R(q_5, 0, 0, 1) = \begin{pmatrix} \cos q_5 & -\sin q_5 & 0 \\ \sin q_5 & \cos q_5 & 0 \\ 0 & 0 & 1 \end{pmatrix},$$

since the blade rotates about the $\mathbf{c}_3$ direction.

## 6.3 MAMBO

The relative orientation of two observers can be uniquely described through a pure rotation, given the selection of a reference triad for each of the observers. If the observer $\mathcal{A}$ has the reference triad $a$ and the observer $\mathcal{B}$ has the reference triad $b$, then the rotation matrix $R_{ab}$ uniquely describes the pure rotation $\mathbf{R}_{\mathcal{A}\to\mathcal{B}}$ between $\mathcal{A}$ and $\mathcal{B}$.

In a MAMBO geometry description, the specification of a rotation matrix relating the reference triads of different observers is given through an **ORIENT** statement.

**Illustration 6.5**

The following extract from a MAMBO .geo file shows the use of the **ORIENT** statement to describe the relative orientation of successive observers.

```
MODULE W {
   BODY E {
      ORIENT {1/sqrt(2),-1/sqrt(2),0,0,0,-1,1/sqrt(2),1/sqrt(2),0}
      BODY F {
         ORIENT {cos(theta),0,sin(theta),0,1,0,
                 -sin(theta),0,cos(theta)}
      }
   }
}
```

Here, the orientation of the observer $\mathcal{E}$ relative to the observer $\mathcal{W}$ is given by a pure rotation $\mathbf{R}_{\mathcal{W}\to\mathcal{E}}$ corresponding to the rotation matrix

$$\mathbf{R}_{we} = \begin{pmatrix} \frac{1}{\sqrt{2}} & -\frac{1}{\sqrt{2}} & 0 \\ 0 & 0 & -1 \\ \frac{1}{\sqrt{2}} & \frac{1}{\sqrt{2}} & 0 \end{pmatrix},$$

where $e$ and $w$ are the reference triads of $\mathcal{E}$ and $\mathcal{W}$, respectively.

Similarly, the orientation of the observer $\mathcal{F}$ relative to the observer $\mathcal{E}$ is given by a pure rotation $\mathbf{R}_{\mathcal{E}\to\mathcal{F}}$ corresponding to the rotation matrix

$$\mathbf{R}_{ef} = \begin{pmatrix} \cos\theta & 0 & \sin\theta \\ 0 & 1 & 0 \\ -\sin\theta & 0 & \cos\theta \end{pmatrix},$$

where $f$ is the reference triad of $\mathcal{F}$.

By default, MAMBO interprets the absence of an **ORIENT** statement to be equivalent to the specification

```
ORIENT {1,0,0,0,1,0,0,0,1}
```

i.e., that the reference triad of the current observer coincides with that of the parent observer.

The orientation of a rigid body relative to some observer can be uniquely described through a pure rotation. If the observer $\mathcal{A}$ has the reference triad $a$ and the triad $b$ is the reference triad of the rigid body, then the rotation matrix $R_{ab}$ uniquely describes the pure rotation $\mathbf{R}_{\mathcal{A}}$ between $\mathcal{A}$ and the rigid body.

In MAMBO, the reference triad associated with a cylinder is oriented in such a way that the 3-direction is parallel to the symmetry axis of the cylinder. Similarly, the reference triad associated with a MAMBO block is oriented in such a way that the basis vectors are parallel to the edges of the block. Since a MAMBO sphere lacks any distinguishing surface features, the associated reference triad has some predefined orientation relative to the sphere.

**Illustration 6.6**

We may use the **ORIENT** statement to describe the rotation matrix relating the reference triad of an observer and the reference triad of a rigid body.

```
MODULE W {
  BODY E {
    ORIENT {1/sqrt(2),-1/sqrt(2),0,0,0,-1,1/sqrt(2),1/sqrt(2),0}
    BODY F {
      ORIENT {cos(theta),0,sin(theta),0,1,0,
              -sin(theta),0,cos(theta)}
      CYLINDER {
        ORIENT {1/2,-1/sqrt(2),1/2,
                1/sqrt(2),0,-1/sqrt(2),
                1/2, 1/sqrt(2), 1/2}
      }
    }
    BLOCK {
      ORIENT {-1,0,0,0,0,1,0,1,0}
    }
  }
}
```

Here, the orientation of the block relative to the observer $\mathcal{E}$ is given by a pure rotation $\mathbf{R}_{\mathcal{E}}$ corresponding to the rotation matrix

$$R_{eb} = \begin{pmatrix} -1 & 0 & 0 \\ 0 & 0 & 1 \\ 0 & 1 & 0 \end{pmatrix},$$

where $b$ is the reference triad of the block and $e$ is the reference triad of the $\mathcal{E}$ observer.

Similarly, the orientation of the cylinder relative to the observer $\mathcal{F}$ is given by a pure rotation $\mathbf{R}_{\mathcal{F}}$ corresponding to the rotation matrix

$$R_{fc} = \begin{pmatrix} \frac{1}{2} & -\frac{1}{\sqrt{2}} & \frac{1}{2} \\ \frac{1}{\sqrt{2}} & 0 & -\frac{1}{\sqrt{2}} \\ \frac{1}{2} & \frac{1}{\sqrt{2}} & \frac{1}{2} \end{pmatrix},$$

where $c$ is the reference triad of the cylinder and $f$ is the reference triad of the $\mathcal{F}$ observer.

The tree structure corresponding to the geometry description in the last illustration has the following form:

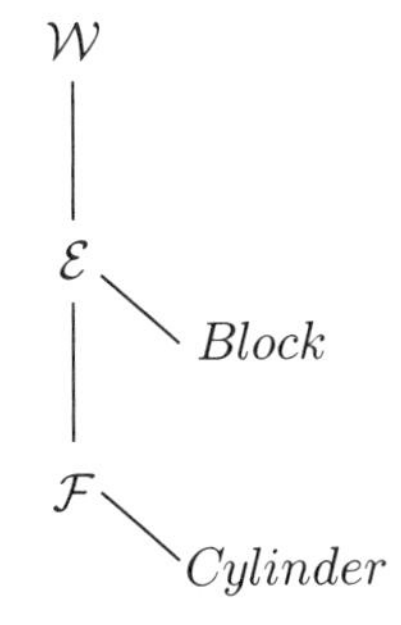


We could represent the same arrangement of rigid bodies relative to the $\mathcal{W}$ observer by relating the configuration of the block to the $\mathcal{F}$ observer.

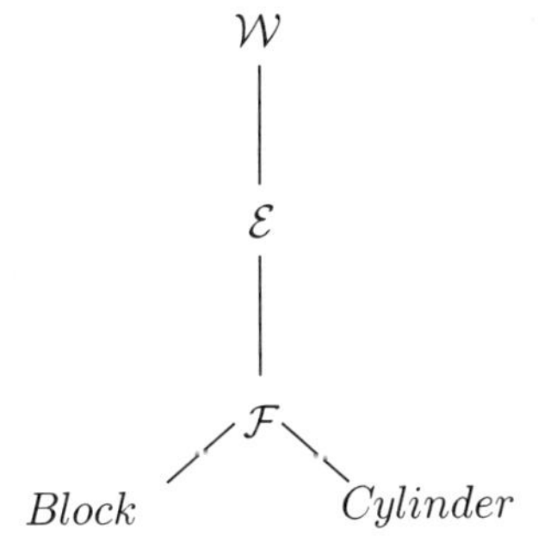


The corresponding MAMBO geometry description becomes

```
MODULE W {
   BODY E {
      ORIENT {1/sqrt(2),-1/sqrt(2),0,0,0,-1,1/sqrt(2),1/sqrt(2),0}
      BODY F {
         ORIENT {cos(theta),0,sin(theta),0,1,0,
```

```
                -sin(theta),0,cos(theta)}
            CYLINDER {
               ORIENT {1/2,-1/sqrt(2),1/2,1/sqrt(2),0,-1/sqrt(2),
                       1/2, 1/sqrt(2), 1/2}
            }
            BLOCK {
               ORIENT {-cos(theta),-sin(theta),0,0,0,1,
                       -sin(theta),cos(theta),0}
            }
         }
      }
   }
```

Here, the **ORIENT** statement relating the pure rotation between the $\mathcal{F}$ observer and the block is obtained from the following computation:

$$\begin{aligned} R_{fb} &= R_{fe}R_{eb} = (R_{ef})^T R_{eb} \\ &= \begin{pmatrix} \cos\theta & 0 & \sin\theta \\ 0 & 1 & 0 \\ -\sin\theta & 0 & \cos\theta \end{pmatrix}^T \begin{pmatrix} -1 & 0 & 0 \\ 0 & 0 & 1 \\ 0 & 1 & 0 \end{pmatrix} \\ &= \begin{pmatrix} -\cos\theta & -\sin\theta & 0 \\ 0 & 0 & 1 \\ -\sin\theta & \cos\theta & 0 \end{pmatrix}. \end{aligned}$$

The $\mathcal{F}$ observer may be entirely eliminated from the observer tree structure.

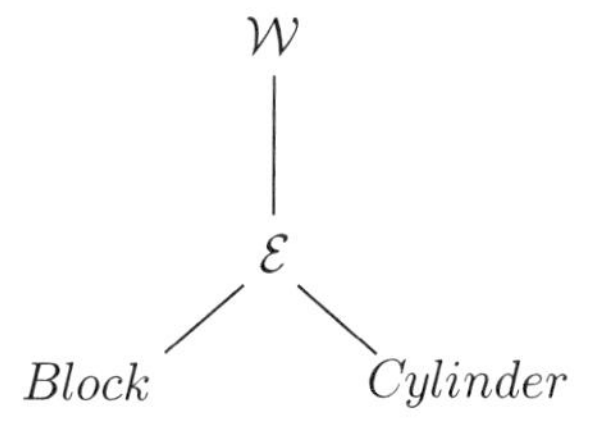


The corresponding MAMBO geometry description becomes

```
MODULE W {
   BODY E {
      ORIENT {1/sqrt(2),-1/sqrt(2),0,0,0,-1,1/sqrt(2),1/sqrt(2),0}
      CYLINDER {
         ORIENT {(cos(theta)+sin(theta))/2,
                 (-cos(theta)+sin(theta))/sqrt(2),
                 (cos(theta)+sin(theta))/2,
                 1/sqrt(2),0,-1/sqrt(2),
                 (cos(theta)-sin(theta))/2,
                 (cos(theta)+sin(theta))/sqrt(2),
                 (cos(theta)-sin(theta))/2}
```

```
        }
        BLOCK {
            ORIENT {-1,0,0,0,0,1,0,1,0}
        }
    }
}
```

Here, the **ORIENT** statement relating the pure rotation between the $\mathcal{E}$ observer and the cylinder is obtained from the following computation:

$$
\begin{aligned}
R_{ec} &= R_{ef}R_{fc} \\
&= \begin{pmatrix} \cos\theta & 0 & \sin\theta \\ 0 & 1 & 0 \\ -\sin\theta & 0 & \cos\theta \end{pmatrix} \begin{pmatrix} \frac{1}{2} & -\frac{1}{\sqrt{2}} & \frac{1}{2} \\ \frac{1}{\sqrt{2}} & 0 & -\frac{1}{\sqrt{2}} \\ \frac{1}{2} & \frac{1}{\sqrt{2}} & \frac{1}{2} \end{pmatrix} \\
&= \begin{pmatrix} \frac{1}{2}\cos\theta + \frac{1}{2}\sin\theta & -\frac{1}{\sqrt{2}}\cos\theta + \frac{1}{\sqrt{2}}\sin\theta & \frac{1}{2}\cos\theta + \frac{1}{2}\sin\theta \\ \frac{1}{\sqrt{2}} & 0 & -\frac{1}{\sqrt{2}} \\ \frac{1}{2}\cos\theta - \frac{1}{2}\sin\theta & \frac{1}{\sqrt{2}}\cos\theta + \frac{1}{\sqrt{2}}\sin\theta & \frac{1}{2}\cos\theta - \frac{1}{2}\sin\theta \end{pmatrix}.
\end{aligned}
$$

From Chapter 4, we recall the possibility of using labels and placeholders to replace actual numbers in a MAMBO geometry description. In particular,

- The MAMBO *time variable* can be changed interactively by the user during a MAMBO session and changes linearly during a simulation. The MAMBO time variable must be declared as such in the MAMBO .dyn file with a `time` statement;

- MAMBO *parameters* can be changed interactively by the user during a MAMBO session, but remain constant during a simulation. MAMBO parameters must be declared as such in the MAMBO .dyn file with a `parameter` statement;

- MAMBO *animated variables* cannot be changed interactively by the user during a MAMBO session, but change with time during a simulation. The dependence of a MAMBO animated variable on MAMBO parameters, MAMBO states, and the MAMBO time variable is declared within an `anims` block in the MAMBO .dyn file;

- MAMBO *state variables* can be changed interactively by the user during a MAMBO session and may change with time during a simulation. Any change is governed by a set of kinematic differential equations or by a MAMBO dataset. MAMBO state variables must be declared as such in the MAMBO .dyn file with a `states` statement.

All of these types of placeholders may be used in **ORIENT** statements. The only restriction is that the matrix contained within the **ORIENT**

statement should be orthogonal[1]. A non-orthogonal matrix within a MAMBO **ORIENT** matrix may lead to unexpected visual effects, desirable or undesirable.

As noted in the previous chapter, the **POINT** statement only conveys information about a matrix representation of a position vector. The corresponding reference triad is understood to be associated with the parent observer in the observer hierarchy. In the MAMBO geometry description,

```
MODULE W {
   BODY A {
      POINT {1/sqrt(2),-1,1/sqrt(2)}
      ORIENT {1/sqrt(2),-1/sqrt(2),0,0,0,-1,1/sqrt(2),1/sqrt(2),0}
      BODY B {
         POINT {-1+sin(theta),0,-1+cos(theta)}
         ORIENT {cos(theta),0,sin(theta),0,1,0,
                    -sin(theta),0,cos(theta)}
      }
   }
}
```

the configuration of the $\mathcal{A}$ observer relative to $\mathcal{W}$ is given by a pure translation $\mathbf{T}_{\mathcal{W}\to\mathcal{A}}$ corresponding to the position vector

$$\mathbf{r}^{WA} = w \begin{pmatrix} \frac{1}{\sqrt{2}} \\ -1 \\ \frac{1}{\sqrt{2}} \end{pmatrix}$$

and a pure rotation $\mathbf{R}_{\mathcal{W}\to\mathcal{A}}$ corresponding to the rotation matrix

$$R_{wa} = \begin{pmatrix} \frac{1}{\sqrt{2}} & -\frac{1}{\sqrt{2}} & 0 \\ 0 & 0 & -1 \\ \frac{1}{\sqrt{2}} & \frac{1}{\sqrt{2}} & 0 \end{pmatrix}$$

where $W$ and $A$ are the reference points of $\mathcal{W}$ and $\mathcal{A}$, respectively, and $w$ and $a$ are the reference triads of $\mathcal{W}$ and $\mathcal{A}$, respectively.

Similarly, the configuration of the $\mathcal{B}$ observer relative to $\mathcal{A}$ is given by a pure translation $\mathbf{T}_{\mathcal{A}\to\mathcal{B}}$ corresponding to the position vector

$$\mathbf{r}^{AB} = a \begin{pmatrix} -1+\sin\theta \\ 0 \\ -1+\cos\theta \end{pmatrix}$$

and a pure rotation $\mathbf{R}_{\mathcal{A}\to\mathcal{B}}$ corresponding to the rotation matrix

$$R_{ab} = \begin{pmatrix} \cos\theta & 0 & \sin\theta \\ 0 & 1 & 0 \\ -\sin\theta & 0 & \cos\theta \end{pmatrix},$$

[1]See Definition 5.1 on page 213.

where $B$ and $b$ are the reference point and reference triad, respectively, of the observer $\mathcal{B}$.

As before, we may reorganize the geometry description, such that the $\mathcal{B}$ observer is promoted to main observer.

```
MODULE B {
  BODY A {
    POINT {cos(theta)-sin(theta),0,sin(theta)+cos(theta)-1}
    ORIENT {cos(theta),0,-sin(theta),0,1,0,sin(theta),0,cos(theta)}
    BODY W {
      POINT {-1,0,-1}
      ORIENT {1/sqrt(2),0,1/sqrt(2),-1/sqrt(2),0,1/sqrt(2),0,-1,0}
    }
  }
}
```

Here, the configuration of the $\mathcal{A}$ observer relative to $\mathcal{B}$ is given by a pure translation $\mathbf{T}_{\mathcal{B}\to\mathcal{A}}$ corresponding to the position vector

$$\begin{aligned}
\mathbf{r}^{BA} &= -\mathbf{r}^{BA} = -a\begin{pmatrix} -1+\sin\theta \\ 0 \\ -1+\cos\theta \end{pmatrix} \\
&= -bR_{ba}\begin{pmatrix} -1+\sin\theta \\ 0 \\ -1+\cos\theta \end{pmatrix} \\
&= -b\left(R_{ab}\right)^T\begin{pmatrix} -1+\sin\theta \\ 0 \\ -1+\cos\theta \end{pmatrix} \\
&= b\begin{pmatrix} \cos\theta-\sin\theta \\ 0 \\ \sin\theta+\cos\theta-1 \end{pmatrix}
\end{aligned}$$

and a pure rotation $\mathbf{R}_{\mathcal{B}\to\mathcal{A}}$ corresponding to the rotation matrix

$$R_{ba} = \left(R_{ab}\right)^T = \begin{pmatrix} \cos\theta & 0 & -\sin\theta \\ 0 & 1 & 0 \\ \sin\theta & 0 & \cos\theta \end{pmatrix}.$$

Similarly, the configuration of $\mathcal{W}$ relative to $\mathcal{A}$ is given by a pure translation $\mathbf{T}_{\mathcal{A}\to\mathcal{W}}$ corresponding to the position vector

$$\begin{aligned}
\mathbf{r}^{AW} &= -\mathbf{r}^{WA} = -w\begin{pmatrix} \frac{1}{\sqrt{2}} \\ -1 \\ \frac{1}{\sqrt{2}} \end{pmatrix} \\
&= -aR_{aw}\begin{pmatrix} \frac{1}{\sqrt{2}} \\ -1 \\ \frac{1}{\sqrt{2}} \end{pmatrix}
\end{aligned}$$

$$= -a\,(R_{wa})^T \begin{pmatrix} \frac{1}{\sqrt{2}} \\ -1 \\ \frac{1}{\sqrt{2}} \end{pmatrix}$$

$$= a \begin{pmatrix} -1 \\ 0 \\ -1 \end{pmatrix}$$

and a pure rotation $\mathbf{R}_{\mathcal{A}\to\mathcal{W}}$ corresponding to the rotation matrix

$$R_{aw} = (R_{wa})^T = \begin{pmatrix} \frac{1}{\sqrt{2}} & 0 & \frac{1}{\sqrt{2}} \\ -\frac{1}{\sqrt{2}} & 0 & \frac{1}{\sqrt{2}} \\ 0 & -1 & 0 \end{pmatrix}.$$

(Ex. 6.9 – Ex. 6.10)

## 6.4 The Mambo Toolbox

### 6.4.1 Triads

Triads are represented within the Mambo toolbox by entries in the global variables `GlobalTriadDeclarations` and `GlobalTriadDefinitions`. User-initiated changes to these variables are made possible through the procedures `DeclareTriads` and `DefineTriads`.

**Illustration 6.7**
In the following Mambo toolbox session, the triads $a$, $b$, and $c$ are declared to the program.

```
> Restart():
> DeclareTriads(a,b,c):
```

The statement

```
> DefineTriads([a,b,theta,1],[b,c,Pi/2,1,0,1]):
```

establishes the rotation matrices

$$R_{ab} = R\,(\theta, 1, 0, 0) = \begin{pmatrix} 1 & 0 & 0 \\ 0 & \cos\theta & -\sin\theta \\ 0 & \sin\theta & \cos\theta \end{pmatrix}$$

and

$$R_{bc} = R\left(\frac{\pi}{2}, 1, 0, 1\right) = \begin{pmatrix} \frac{1}{2} & -\frac{1}{\sqrt{2}} & \frac{1}{2} \\ \frac{1}{\sqrt{2}} & 0 & -\frac{1}{\sqrt{2}} \\ \frac{1}{2} & \frac{1}{\sqrt{2}} & \frac{1}{2} \end{pmatrix}.$$

The effect of these statements on the global variables `GlobalTriad-Declarations` and `GlobalTriadDefinitions` is made clear by the following statements:

```
> print(GlobalTriadDeclarations);
```

$$
\text{table}([\; b = \{a,\, c\} \quad c = \{b\} \quad a = \{b\} \;])
$$

```
> print(GlobalTriadDefinitions);
```

$$
\text{table}([
$$

$$
(b,\, c) = \begin{bmatrix} \frac{1}{2} & -\frac{1}{2}\sqrt{2} & \frac{1}{2} \\ \frac{1}{2}\sqrt{2} & 0 & -\frac{1}{2}\sqrt{2} \\ \frac{1}{2} & \frac{1}{2}\sqrt{2} & \frac{1}{2} \end{bmatrix}
$$

$$
(a,\, b) = \begin{bmatrix} 1 & 0 & 0 \\ 0 & \cos(\theta) & -\sin(\theta) \\ 0 & \sin(\theta) & \cos(\theta) \end{bmatrix}
$$

$$
(c,\, b) = \begin{bmatrix} \frac{1}{2} & \frac{1}{2}\sqrt{2} & \frac{1}{2} \\ -\frac{1}{2}\sqrt{2} & 0 & \frac{1}{2}\sqrt{2} \\ \frac{1}{2} & -\frac{1}{2}\sqrt{2} & \frac{1}{2} \end{bmatrix}
$$

$$
(b,\, a) = \begin{bmatrix} 1 & 0 & 0 \\ 0 & \cos(\theta) & \sin(\theta) \\ 0 & -\sin(\theta) & \cos(\theta) \end{bmatrix}
$$

$$
])
$$

The contents of the global variables `GlobalTriadDeclarations` and `GlobalTriadDefinitions` reflect the fact that the relative orientations of the triads $a$ and $b$ and the triads $b$ and $c$, respectively, are now known. Where `GlobalTriadDeclarations` tracks all direct relations between triads, `GlobalTriadDefinitions` stores any rotation matrices between triads in `GlobalTriadDeclarations` that have been computed during a MAMBO toolbox session.

The MAMBO toolbox procedure `FindRotation` can be invoked to compute the rotation matrix between any two triads that are declared and are related. Continuing with the MAMBO session in the illustration, we find

```
> FindRotation(a,c);
```

$$\begin{bmatrix} \frac{1}{2} & -\frac{1}{2}\sqrt{2} & \frac{1}{2} \\ \frac{1}{2}\cos(\theta)\sqrt{2} - \frac{1}{2}\sin(\theta) & -\frac{1}{2}\sin(\theta)\sqrt{2} & -\frac{1}{2}\cos(\theta)\sqrt{2} - \frac{1}{2}\sin(\theta) \\ \frac{1}{2}\sin(\theta)\sqrt{2} + \frac{1}{2}\cos(\theta) & \frac{1}{2}\cos(\theta)\sqrt{2} & -\frac{1}{2}\sin(\theta)\sqrt{2} + \frac{1}{2}\cos(\theta) \end{bmatrix}$$

as follows from

$$\begin{aligned} R_{ac} &= R_{ab}R_{bc} \\ &= \begin{pmatrix} 1 & 0 & 0 \\ 0 & \cos\theta & -\sin\theta \\ 0 & \sin\theta & \cos\theta \end{pmatrix} \begin{pmatrix} \frac{1}{2} & -\frac{1}{\sqrt{2}} & \frac{1}{2} \\ \frac{1}{\sqrt{2}} & 0 & -\frac{1}{\sqrt{2}} \\ \frac{1}{2} & \frac{1}{\sqrt{2}} & \frac{1}{2} \end{pmatrix} \\ &= \begin{pmatrix} \frac{1}{2} & -\frac{1}{\sqrt{2}} & \frac{1}{2} \\ \frac{1}{\sqrt{2}}\cos\theta - \frac{1}{2}\sin\theta & -\frac{1}{\sqrt{2}}\sin\theta & -\frac{1}{\sqrt{2}}\cos\theta - \frac{1}{2}\sin\theta \\ \frac{1}{\sqrt{2}}\sin\theta + \frac{1}{2}\cos\theta & \frac{1}{\sqrt{2}}\cos\theta & -\frac{1}{\sqrt{2}}\sin\theta + \frac{1}{2}\cos\theta \end{pmatrix}. \end{aligned}$$

The global variable `GlobalTriadDefinitions` is automatically appended with the rotation matrices $R_{ac}$ and $R_{ca} = (R_{ac})^T$.

The MAMBO toolbox employs the rotation matrices between two triads to apply the `&oo`, `&xx`, `VectorLength`, and `Express` functions to MAMBO vectors with matrix representations relative to multiple triads. Their function is illustrated with the following sequence of statements:

```
> v:=MakeTranslations([a,1,0,1],[b,0,1,1]):
> w:=MakeTranslations(c,1,1,0):
> v &oo w;
```

$$1 + \frac{1}{2}\sqrt{2} + \frac{1}{2}\sin(\theta)\sqrt{2} + \frac{1}{2}\cos(\theta) + \frac{1}{2}\cos(\theta)\sqrt{2}$$

```
> Express(v,a);
```

table([
1 = table([
"Coordinates" $= [1, \cos(\theta) - \sin(\theta), 1 + \sin(\theta) + \cos(\theta)]$
"Triad" $= a$
])
"Type" = "Vector"
"Size" = 1
])

```
> v &xx w;
```

table([
1 = table([

"Coordinates" $= \left[-\frac{1}{2}\cos(\theta)\sqrt{2} + \frac{1}{2}\sin(\theta) + \frac{1}{2}\sin(\theta)\sqrt{2},\right.$
$\frac{1}{2} - \frac{1}{2}\sqrt{2} - \frac{1}{2}\sin(\theta)\sqrt{2} - \frac{1}{2}\cos(\theta) - \frac{1}{2}\cos(\theta)\sqrt{2},$
$\left.\frac{1}{2}\cos(\theta)\sqrt{2} - \frac{1}{2}\sin(\theta) - \frac{1}{2}\sin(\theta)\sqrt{2}\right]$
"Triad" $= a$
])
"Type" = "Vector"
2 = table([
"Coordinates" $= \left[\frac{1}{2}, \frac{1}{2} - \frac{1}{2}\sqrt{2}, -\frac{1}{2} + \frac{1}{2}\sqrt{2}\right]$
"Triad" $= b$
])
"Size" = 2
])

```
> VectorLength(v);
```

$$\sqrt{4 + 2\sin(\theta) + 2\cos(\theta)}$$

### 6.4.2 Observers

As discussed in Chapter 4, the `DefineObservers` procedure is used to associate a point and a triad with an observer. For example, let $A$, $B$, and $W$ and $a$, $b$, and $w$, be the reference points and reference triads, respectively, of three observers $\mathcal{A}$, $\mathcal{B}$, and $\mathcal{W}$, such that

$$\mathbf{r}^{WA} = a \begin{pmatrix} 1 \\ 2 \\ 3 \end{pmatrix}, R_{wa} = R(q_1, 1, 1, 0)\, R(q_2, 1, 0, 0),$$

$$\mathbf{r}^{AB} = a \begin{pmatrix} 0 \\ 0 \\ 1 \end{pmatrix}, \text{ and } R_{ab} = R(q_3, 0, 1, 0)\, R(q_4, 0, 0, 1).$$

The following statements provide the necessary information to the MAMBO toolbox.

```
> Restart():
> DeclareObservers(A,B,W):
> DeclareTriads(a,b,w):
> DeclarePoints(A,B,W):
> DefinePoints([W,A,a,1,2,3],[A,B,a,3]):
> DefineTriads([w,a,[q1,1,1,0],[q2,1]],
> [a,b,[q3,2],[q4,3]]):
> DefineObservers([W,W,w],[A,A,a],[B,B,b]):
```

The MAMBO toolbox procedure `FindOrientation` can be invoked to compute the rotation matrix corresponding to the pure rotation relating the orientations of two observers.

```
> FindOrientation(A,B);
```

$$\begin{bmatrix} \cos(q3)\cos(q4) & -\cos(q3)\sin(q4) & \sin(q3) \\ \sin(q4) & \cos(q4) & 0 \\ -\sin(q3)\cos(q4) & \sin(q3)\sin(q4) & \cos(q3) \end{bmatrix}$$

We may again use the `DefineNeighbors` and `GeometryOutput` commands to generate a MAMBO geometry description using the observers introduced above.

```
> DefineNeighbors([W,A],[A,B]):
> GeometryOutput(main=W);
```

```
MODULE W {
   BODY A {
      POINT
{1/2*cos(q1)+1/2+2*(1/2-1/2*cos(q1))*cos(q2)+sin(q1)*2^(1/2)
*sin(q2)-3*(1/2-1/2*cos(q1))*sin(q2)+3/2*sin(q1)*2^(1/2)*cos
(q2),1/2-1/2*cos(q1)+2*(1/2*cos(q1)+1/2)*cos(q2)-sin(q1)*2^(
1/2)*sin(q2)-3*(1/2*cos(q1)+1/2)*sin(q2)-3/2*sin(q1)*2^(1/2)
*cos(q2),-1/2*sin(q1)*2^(1/2)+sin(q1)*2^(1/2)*cos(q2)+2*cos(
```

```
q1)*sin(q2)-3/2*sin(q1)*2^(1/2)*sin(q2)+3*cos(q1)*cos(q2)}
      ORIENT
{1/2*cos(q1)+1/2,(1/2-1/2*cos(q1))*cos(q2)+1/2*sin(q1)*2^(1/
2)*sin(q2),-(1/2-1/2*cos(q1))*sin(q2)+1/2*sin(q1)*2^(1/2)*co
s(q2),1/2-1/2*cos(q1),(1/2*cos(q1)+1/2)*cos(q2)-1/2*sin(q1)*
2^(1/2)*sin(q2),-(1/2*cos(q1)+1/2)*sin(q2)-1/2*sin(q1)*2^(1/
2)*cos(q2),-1/2*sin(q1)*2^(1/2),1/2*sin(q1)*2^(1/2)*cos(q2)+
cos(q1)*sin(q2),-1/2*sin(q1)*2^(1/2)*sin(q2)+cos(q1)*cos(q2)}
      BODY B {
         POINT {0,0,1}
         ORIENT
{cos(q3)*cos(q4),-cos(q3)*sin(q4),sin(q3),sin(q4),cos(q4),0,-
sin(q3)*cos(q4),sin(q3)*sin(q4),cos(q3)}
      }
   }
}
```

**Illustration 6.8**

As in the previous chapter, we may reorganize the observers so as to promote $\mathcal{A}$ to be the main observer:

```
> GeometryOutput(main=A);

MODULE A {
   BODY W {
      POINT {-1,-2,-3}
      ORIENT
{1/2*cos(q1)+1/2,1/2-1/2*cos(q1),-1/2*sin(q1)*2^(1/2),(1/2-1
/2*cos(q1))*cos(q2)+1/2*sin(q1)*2^(1/2)*sin(q2),(1/2*cos(q1)
+1/2)*cos(q2)-1/2*sin(q1)*2^(1/2)*sin(q2),1/2*sin(q1)*2^(1/2
)*cos(q2)+cos(q1)*sin(q2),-(1/2-1/2*cos(q1))*sin(q2)+1/2*sin
(q1)*2^(1/2)*cos(q2),-(1/2*cos(q1)+1/2)*sin(q2)-1/2*sin(q1)*
2^(1/2)*cos(q2),-1/2*sin(q1)*2^(1/2)*sin(q2)+cos(q1)*cos(q2)}
   }
   BODY B {
      POINT {0,0,1}
      ORIENT
{cos(q3)*cos(q4),-cos(q3)*sin(q4),sin(q3),sin(q4),cos(q4),0,-
sin(q3)*cos(q4),sin(q3)*sin(q4),cos(q3)}
   }
}
```

or

```
> Undo():
> DefineNeighbors([W,B],[A,B]):
> GeometryOutput(main=A);

MODULE A {
   BODY B {
      POINT {0,0,1}
      ORIENT
{cos(q3)*cos(q4),-cos(q3)*sin(q4),sin(q3),sin(q4),cos(q4),0,
-sin(q3)*cos(q4),sin(q3)*sin(q4),cos(q3)}
      BODY W {
         POINT
{-cos(q3)*cos(q4)-2*sin(q4)+4*sin(q3)*cos(q4),cos(q3)*sin(q4
)-2*cos(q4)-4*sin(q3)*sin(q4),-sin(q3)-4*cos(q3)}
```

```
            ORIENT
{cos(q3)*cos(q4)*(1/2*cos(q1)+1/2)+sin(q4)*((1/2-1/2*cos(q1)
)*cos(q2)+1/2*sin(q1)*2^(1/2)*sin(q2))-sin(q3)*cos(q4)*(-(1/
2-1/2*cos(q1))*sin(q2)+1/2*sin(q1)*2^(1/2)*cos(q2)),cos(q3)*
cos(q4)*(1/2-1/2*cos(q1))+sin(q4)*((1/2*cos(q1)+1/2)*cos(q2)
-1/2*sin(q1)*2^(1/2)*sin(q2))-sin(q3)*cos(q4)*(-(1/2*cos(q1)
+1/2)*sin(q2)-1/2*sin(q1)*2^(1/2)*cos(q2)),-1/2*cos(q3)*cos(
q4)*sin(q1)*2^(1/2)+sin(q4)*(1/2*sin(q1)*2^(1/2)*cos(q2)+cos
(q1)*sin(q2))-sin(q3)*cos(q4)*(-1/2*sin(q1)*2^(1/2)*sin(q2)+
cos(q1)*cos(q2)),-cos(q3)*sin(q4)*(1/2*cos(q1)+1/2)+cos(q4)*
((1/2-1/2*cos(q1))*cos(q2)+1/2*sin(q1)*2^(1/2)*sin(q2))+sin(
q3)*sin(q4)*(-(1/2-1/2*cos(q1))*sin(q2)+1/2*sin(q1)*2^(1/2)*
cos(q2)),-cos(q3)*sin(q4)*(1/2-1/2*cos(q1))+cos(q4)*((1/2*co
s(q1)+1/2)*cos(q2)-1/2*sin(q1)*2^(1/2)*sin(q2))+sin(q3)*sin(
q4)*(-(1/2*cos(q1)+1/2)*sin(q2)-1/2*sin(q1)*2^(1/2)*cos(q2))
,1/2*cos(q3)*sin(q4)*sin(q1)*2^(1/2)+cos(q4)*(1/2*sin(q1)*2^
(1/2)*cos(q2)+cos(q1)*sin(q2))+sin(q3)*sin(q4)*(-1/2*sin(q1)
*2^(1/2)*sin(q2)+cos(q1)*cos(q2)),sin(q3)*(1/2*cos(q1)+1/2)+
cos(q3)*(-(1/2-1/2*cos(q1))*sin(q2)+1/2*sin(q1)*2^(1/2)*cos(
q2)),sin(q3)*(1/2-1/2*cos(q1))+cos(q3)*(-(1/2*cos(q1)+1/2)*s
in(q2)-1/2*sin(q1)*2^(1/2)*cos(q2)),-1/2*sin(q3)*sin(q1)*2^(
1/2)+cos(q3)*(-1/2*sin(q1)*2^(1/2)*sin(q2)+cos(q1)*cos(q2))}
        }
    }
}
```

where we used the `Undo` utility to undo the latest change in any of the global variables.

### 6.4.3 A Sample Project

Suppose you want to visualize the motion of a wireframe representation of a tetrahedron that is turning over on a stationary plane while keeping one corner fixed, as depicted below.

Introduce a main observer $\mathcal{W}$ with reference point $W$ at the stationary corner of the tetrahedron and reference triad $w$, such that the stationary plane through $W$ is spanned by the basis vectors $\mathbf{w}_1$ and $\mathbf{w}_2$. Furthermore, let $\mathcal{A}$ be an auxiliary observer, relative to which the tetrahedron is stationary, with reference point $A$ at the corner kept fixed relative to $\mathcal{W}$ and reference triad $a$, such that

$$\mathbf{r}^{AC_1} = a\begin{pmatrix} l \\ 0 \\ 0 \end{pmatrix},\ \mathbf{r}^{AC_2} = a\begin{pmatrix} \frac{1}{2}l \\ \frac{\sqrt{3}}{2}l \\ 0 \end{pmatrix},\ \text{and } \mathbf{r}^{AC_3} = a\begin{pmatrix} \frac{1}{2}l \\ \frac{\sqrt{3}}{6}l \\ \frac{\sqrt{6}}{3}l \end{pmatrix},$$

where $C_1$, $C_2$, and $C_3$ are the remaining corners of the tetrahedron and $l$ is the length of the tetrahedron's edges.

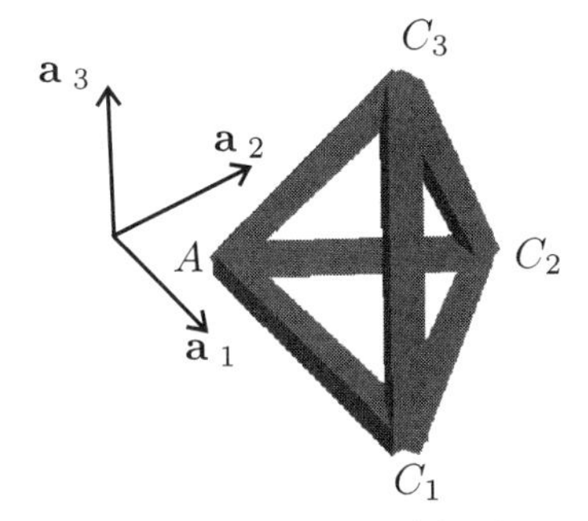


Let the $i$-th edge correspond to a rigid body with reference point $E_i$ at the midpoint of the edge and reference triad $e^{(i)}$, such that the $\mathbf{e}_1^{(i)}$ basis vector is parallel to the edge. The configuration of the $i$-th edge relative to $\mathcal{A}$ is given by a pure translation $\mathbf{T}_i$ corresponding to the position vector

$$\mathbf{r}^{AE_i}$$

and a pure rotation $\mathbf{R}_i$ corresponding to the rotation matrix

$$R_{ae^{(i)}}.$$

Specifically, we find that

$$\begin{aligned}
\mathbf{r}^{AE_1} &= \frac{1}{2}\mathbf{r}^{AC_1}, \\
\mathbf{r}^{AE_2} &= \frac{1}{2}\mathbf{r}^{AC_2}, \\
\mathbf{r}^{AE_3} &= \frac{1}{2}\mathbf{r}^{AC_3}, \\
\mathbf{r}^{AE_4} &= \frac{1}{2}\mathbf{r}^{AC_1} + \frac{1}{2}\mathbf{r}^{AC_2}, \\
\mathbf{r}^{AE_5} &= \frac{1}{2}\mathbf{r}^{AC_1} + \frac{1}{2}\mathbf{r}^{AC_3},
\end{aligned}$$

and

$$\mathbf{r}^{AE_6} = \frac{1}{2}\mathbf{r}^{AC_2} + \frac{1}{2}\mathbf{r}^{AC_3}.$$

To compute the rotation matrices

$$R_{ae^{(i)}} = a^T \bullet e^{(i)},$$

we proceed by constructing the triads $e^{(i)}$, such that

$$\mathbf{e}_1^{(1)} = \frac{\mathbf{r}^{AC_1}}{\|\mathbf{r}^{AC_1}\|},$$
$$\mathbf{e}_1^{(2)} = \frac{\mathbf{r}^{AC_2}}{\|\mathbf{r}^{AC_2}\|},$$
$$\mathbf{e}_1^{(3)} = \frac{\mathbf{r}^{AC_3}}{\|\mathbf{r}^{AC_3}\|},$$
$$\mathbf{e}_1^{(4)} = \frac{\mathbf{r}^{AC_2} - \mathbf{r}^{AC_1}}{\|\mathbf{r}^{AC_2} - \mathbf{r}^{AC_1}\|},$$
$$\mathbf{e}_1^{(5)} = \frac{\mathbf{r}^{AC_3} - \mathbf{r}^{AC_1}}{\|\mathbf{r}^{AC_3} - \mathbf{r}^{AC_1}\|},$$

and

$$\mathbf{e}_1^{(6)} = \frac{\mathbf{r}^{AC_3} - \mathbf{r}^{AC_2}}{\|\mathbf{r}^{AC_2} - \mathbf{r}^{AC_2}\|}.$$

The remaining basis vectors $\mathbf{e}_2^{(i)}$ and $\mathbf{e}_3^{(i)}$ are then obtained from the formula

$$\mathbf{e}_2^{(i)} = \frac{\mathbf{e}_1^{(i)} \times \mathbf{a}_3}{\left\|\mathbf{e}_1^{(i)} \times \mathbf{a}_3\right\|}$$

and

$$\mathbf{e}_3^{(i)} = \mathbf{e}_1^{(i)} \times \mathbf{e}_2^{(i)}.$$

The MAMBO toolbox statements

```
> Restart():
> DeclareObservers(W,A):
> DeclarePoints(W,A,seq(cat(E,i),i=1..6),
> seq(cat(C,i),i=1..3)):
> DeclareTriads(w,a,seq(cat(e,i),i=1..6)):
> DefineObservers([W,W,w],[A,A,a]):
> DefineNeighbors([W,A]):
> DefinePoints([A,C1,a,l,0,0],
> [A,C2,a,1/2*l,sqrt(3)/2*l,0],
> [A,C3,a,1/2*l,sqrt(3)/6*l,sqrt(6)/3*l]):
```

```
> DefinePoints([A,E1,(1/2) &** FindTranslation(A,C1)],
> [A,E2,(1/2) &** FindTranslation(A,C2)],
> [A,E3,(1/2) &** FindTranslation(A,C3)],
> [A,E4,(1/2) &** (FindTranslation(A,C1)
> &++ FindTranslation(A,C2))],
> [A,E5,(1/2) &** (FindTranslation(A,C1)
> &++ FindTranslation(A,C3))],
> [A,E6,(1/2) &** (FindTranslation(A,C2)
> &++ FindTranslation(A,C3))]):
> b1:=(1/l) &** FindTranslation(A,C1):
> b2:=(1/VectorLength(b1 &xx MakeTranslations(a,3)))
> &** (b1 &xx MakeTranslations(a,3)):
> b3:=b1 &xx b2:
> DefineTriads(a,e1,matrix(3,3,(i,j)->
> MakeTranslations(a,i) &oo cat(b,j))):
> b1:=(1/l) &** FindTranslation(A,C2):
> b2:=(1/VectorLength(b1 &xx MakeTranslations(a,3)))
> &** (b1 &xx MakeTranslations(a,3)):
> b3:=b1 &xx b2:
> DefineTriads(a,e2,matrix(3,3,(i,j)->
> MakeTranslations(a,i) &oo cat(b,j))):
> b1:=(1/l) &** FindTranslation(A,C3):
> b2:=(1/VectorLength(b1 &xx MakeTranslations(a,3)))
> &** (b1 &xx MakeTranslations(a,3)):
> b3:=b1 &xx b2:
> DefineTriads(a,e3,matrix(3,3,(i,j)->
> MakeTranslations(a,i) &oo cat(b,j))):
> b1:=(1/l) &** FindTranslation(C1,C2):
> b2:=(1/VectorLength(b1 &xx MakeTranslations(a,3)))
> &** (b1 &xx MakeTranslations(a,3)):
> b3:=b1 &xx b2:
> DefineTriads(a,e4,matrix(3,3,(i,j)->
> MakeTranslations(a,i) &oo cat(b,j))):
> b1:=(1/l) &** FindTranslation(C1,C3):
> b2:=(1/VectorLength(b1 &xx MakeTranslations(a,3)))
> &** (b1 &xx MakeTranslations(a,3)):
> b3:=b1 &xx b2:
> DefineTriads(a,e5,matrix(3,3,(i,j)->
> MakeTranslations(a,i) &oo cat(b,j))):
> b1:=(1/l) &** FindTranslation(C2,C3):
> b2:=(1/VectorLength(b1 &xx MakeTranslations(a,3)))
> &** (b1 &xx MakeTranslations(a,3)):
> b3:=b1 &xx b2:
> DefineTriads(a,e6,matrix(3,3,(i,j)->
> MakeTranslations(a,i) &oo cat(b,j))):
```

establish the corresponding geometry. To visualize the wireframe representation of the tetrahedron, we need to add MAMBO objects to the geometry description. The MAMBO toolbox procedure `DefineObjects` associates the desired objects with the appropriate observer. In the following MAMBO toolbox statement, six MAMBO blocks are directly related to the $\mathcal{A}$ observer with reference points given by the $E_i$'s and reference triads given by the $e^{(i)}$'s, and one MAMBO block representing the stationary plane is directly related to the $\mathcal{W}$ observer with reference point and reference triad coinciding with those of $\mathcal{W}$.

```
> DefineObjects(seq([A,'Block',point=cat(E,i),
> orient=cat(e,i),xlength=l,ylength=l/10,zlength=l/10,
> color=green],i=1..6),[W,'Block',xlength=5*l,
> ylength=5*l,zlength=l/100,color=white]):
```

The coincidence of the reference points $A$ and $W$ of the auxiliary observer $\mathcal{A}$ and the main observer $\mathcal{W}$ implies that the position of $\mathcal{A}$ relative to $\mathcal{W}$ is given by the identity translation or, alternatively, that

$$\mathbf{r}^{WA} = \mathbf{0}.$$

Finally, we shall assume that the orientation of $\mathcal{A}$ relative to $\mathcal{W}$ is given by the pure rotation $\mathbf{R}_{\mathcal{W}\to\mathcal{A}}$ corresponding to a rotation about a direction parallel to $\mathbf{w}_3$ by an angle $q_1$ followed by a rotation about a direction parallel to $\mathbf{r}^{AC_2}$ by an angle $q_2$. It follows that

$$R_{wa} = R(q_1, 0, 0, 1)\, R\left(q_2, \frac{1}{2}, \frac{\sqrt{3}}{2}, 0\right).$$

Continuing with the same MAMBO toolbox session, these definitions are achieved by the statements

```
> DefinePoints(W,A,NullVector()):
> DefineTriads(w,a,[q1,3],[q2,1/2,sqrt(3)/2,0]):
```

The statement

```
> GeometryOutput(main=W,
> filename="flippingtetrahedron.geo");
```

exports the resulting geometry hierarchy to the file `flippingtetrahedron.geo`.

Note that the angle between any two edges meeting at a corner of the tetrahedron is given by

$$\arccos \frac{\mathbf{r}^{AC_1} \bullet \mathbf{r}^{AC_2}}{\|\mathbf{r}^{AC_1}\|\,\|\mathbf{r}^{AC_2}\|} = \frac{\pi}{3}.$$

Similarly, the angle between any two faces of the tetrahedron is given by

$$\arccos \frac{\mathbf{a}_2 \bullet \mathbf{r}^{E_1C_3}}{\|\mathbf{a}_2\|\,\|\mathbf{r}^{E_1C_3}\|} = \arccos\left(\frac{1}{3}\right).$$

It follows that a visually satisfactory animation is obtained by increasing $q_1$ discretely by $\frac{\pi}{3}$ every second while $q_2$ decreases continuously from 0 to $\arccos\left(\frac{1}{3}\right) - \pi$ during every whole second as shown in the figure.

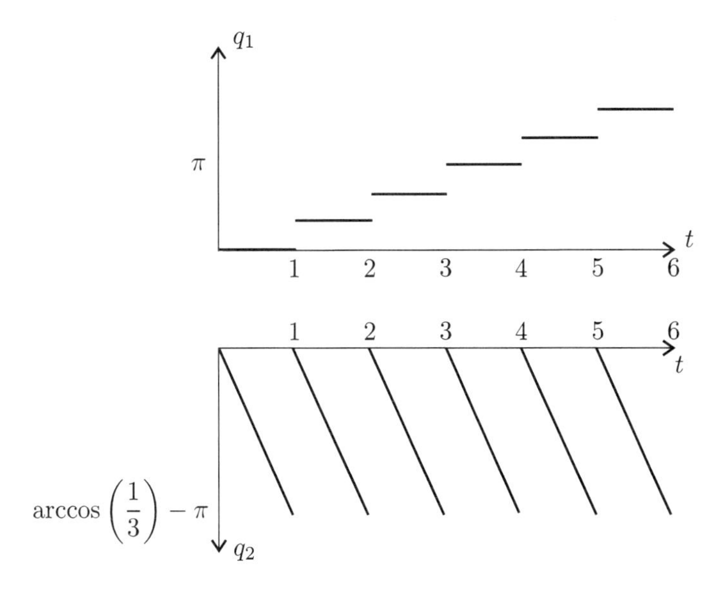


In other words,

$$q_1(t) = \frac{\pi}{3} \lfloor t \rfloor$$

and

$$q_2(t) = \left(\arccos\left(\frac{1}{3}\right) - \pi\right)(t - \lfloor t \rfloor),$$

where $\lfloor t \rfloor$ equals the integer part of $t$. Since both $q_1(t)$ and $q_2(t)$ are discontinuous, we cannot formulate a set of kinematic differential equations to govern their evolution. Instead, we treat $q_1$ and $q_2$ as MAMBO animated variables as suggested in the following MAMBO toolbox statement:

```
> MotionOutput(anims=[q1=Pi/3*floor(t),
> q2=(acos(1/3)-Pi)*(t-floor(t))],parameters=[l=1],
> filename="floppingtetrahedron.dyn");
```

where we note the use of C syntax `acos` for the arccos function and `floor` for the $\lfloor \cdot \rfloor$ function.

Although the motion that results from the definitions of $q_1(t)$ and $q_2(t)$ has the visual appearance that we desire, this is only an illusion. In fact, the tetrahedron repeatedly rotates about the same edge instead of switching to a new edge whenever a new face becomes parallel with the stationary plane. It is by the symmetry of the tetrahedron that the discrete changes in $q_1$ are able to generate the appearance of a switch. A more satisfying approach would be one that constrained alternating edges to be stationary relative to $\mathcal{W}$. We will develop the methodology for achieving this in Chapter 9.

## 6.5 Exercises

**Exercise 6.1** The configuration of a thin disk of radius $\rho$ is constrained in such a way that the disk makes tangential contact with a planar surface through the reference point $W$ of an observer $\mathcal{W}$ and spanned by the $\mathbf{w}_1$ and $\mathbf{w}_2$ basis vectors in the corresponding reference triad $w$. Formulate the corresponding configuration constraints.

**Solution.** Let $\mathcal{A}$ be an auxiliary observer, relative to which the disk remains stationary, with reference point $A$ at the center of the disk and reference triad $a$, such that $\mathbf{a}_3$ is perpendicular to the plane of the disk. Then

$$\mathbf{r}^{WA} = w \begin{pmatrix} q_1 \\ q_2 \\ q_3 \end{pmatrix}$$

and $R_{wa}$

$$= R(q_4, 0, 0, 1)\, R(q_5, 1, 0, 0)\, R(q_6, 0, 0, 1)$$

describe an arbitrary configuration of the disk relative to the observer $\mathcal{W}$.

That the disk makes tangential contact with the planar surface spanned by $\mathbf{w}_1$ and $\mathbf{w}_2$ implies that there is a point $P$ on the rim of the disk that is in contact with the plane and that the tangent direction to the rim of the disk at $P$ is parallel to the planar surface. Thus, if

$$\mathbf{r}^{AP} = a \begin{pmatrix} \rho\cos q_7 \\ \rho\sin q_7 \\ 0 \end{pmatrix}$$

for some $q_7$, then $P$ is in contact with the plane if

$$\begin{aligned} 0 &= \mathbf{r}^{WP} \bullet \mathbf{w}_3 = \left(\mathbf{r}^{WA} + \mathbf{r}^{AP}\right) \bullet \mathbf{w}_3 \\ &= q_3 + \rho \sin q_5 \sin(q_6 + q_7). \end{aligned}$$

Moreover, since the vector $\mathbf{w}_3 \times \mathbf{a}_3$ is parallel to both the planar surface and the plane of the disk, it follows that

$$\begin{aligned} 0 &= \mathbf{r}^{AP} \bullet (\mathbf{w}_3 \times \mathbf{a}_3) \\ &= \rho \sin q_5 \cos(q_6 + q_7) \end{aligned}$$

guarantees that the contact at $P$ is tangential.

**Exercise 6.2** Show that any values of $q_1$ through $q_7$, such that $\sin q_5 \neq 0$, that satisfy the configuration constraints in the previous exercise are regular relative to the pair $q_3$ and $q_6$.

**Solution.** Here,

$$\begin{aligned} f_1(q_1, \ldots, q_7) &= q_3 + \rho \sin q_5 \sin(q_6 + q_7), \\ f_2(q_1, \ldots, q_7) &= \rho \sin q_5 \cos(q_6 + q_7), \end{aligned}$$

and thus

$$\begin{vmatrix} \frac{\partial f_1}{\partial q_3} & \frac{\partial f_1}{\partial q_6} \\ \frac{\partial f_2}{\partial q_3} & \frac{\partial f_2}{\partial q_6} \end{vmatrix} = -\rho \sin q_5 \sin(q_6 + q_7),$$

which equals zero if $\sin q_5 = 0$ or $\sin(q_6 + q_7) = 0$. But, if $\sin q_5 \neq 0$, then $f_2 = 0$ only if $\cos(q_6 + q_7) = 0$, i.e., $\sin(q_6 + q_7) = \pm 1$ and the claim follows.

**Exercise 6.3** Use the MAMBO toolbox to repeat the discussion in the previous two exercises.

**Solution.** The following MAMBO toolbox statements

```
> Restart():
> DeclareObservers(W,A):
> DeclarePoints(W,A,P):
> DeclareTriads(w,a):
> DefineObservers([W,W,w],
> [A,A,a]):
> DefinePoints([W,A,w,q1,q2,q3],
> [A,P,a,rho*cos(q7),
> rho*sin(q7),0]):
> DefineTriads([w,a,[q4,3],
> [q5,1],[q6,3]]):
```

define the basic geometry. The constraints are obtained through the statements

```
> f1:=FindTranslation(W,P) &oo
> MakeTranslations(w,3):
> f2:=FindTranslation(A,P) &oo
> (MakeTranslations(w,3) &xx
> MakeTranslations(a,3)):
```

Finally, the determinant is computed through the statement

```
> factor(det(matrix(2,2,
> [[diff(f1,q3),diff(f1,q6)],
> [diff(f2,q3),diff(f2,q6)]])));
```

$-\rho\sin(q5)\,(\cos(q7)\sin(q6)+\sin(q7)\cos(q6))$

**Exercise 6.4** Show that a thin disk that makes tangential contact with a planar surface has five geometric degrees of freedom, as long as it is not parallel to the surface.

**Solution.** In the previous exercises, seven configuration coordinates were introduced to describe the configuration of the thin disk. We found that, as long as $\sin q_5 \neq 0$, the corresponding configuration constraints could be solved for two of the configuration coordinates in terms of the remaining configuration coordinates. It follows that, as long as $\sin q_5 \neq 0$, the configuration of the thin disk can be described using only five configuration coordinates. When $\sin q_5 = 0$, $\mathbf{a}_3 \times \mathbf{w}_3 = \mathbf{0}$, i.e., the disk is parallel to the planar surface. It follows that the thin disk has five geometric degrees of freedom, as long as it is not parallel to the planar surface.

**Exercise 6.5** The configuration of a thin disk of radius $\rho$ is constrained in such a way that the disk makes tangential contact with a spherical surface of radius $R$ centered on the reference point $W$ of an observer $\mathcal{W}$. Find the corresponding configuration constraints.

**Solution.** Let $\mathcal{A}$ be an auxiliary observer, relative to which the disk remains stationary, with reference point $A$ at the center of the disk and reference triad $a$, such that $\mathbf{a}_3$ is perpendicular to the plane of the disk. Then

$$\mathbf{r}^{WA} = w \begin{pmatrix} q_1 \\ q_2 \\ q_3 \end{pmatrix}$$

and $R_{wa}$

$$= R(q_4, 0, 0, 1)\, R(q_5, 1, 0, 0)\, R(q_6, 0, 0, 1)$$

describe an arbitrary configuration of the disk relative to the observer $\mathcal{W}$.

That the disk makes tangential contact with the spherical surface implies that there is a point $P$ on the rim of the disk that is in contact with a point $P'$ on the surface of the sphere and that the tangent direction to the rim of the disk at $P$ is tangential to the spherical surface at $P'$. Thus, if

$$\mathbf{r}^{AP} = a \begin{pmatrix} \rho\cos q_7 \\ \rho\sin q_7 \\ 0 \end{pmatrix} \text{ and}$$

$$\mathbf{r}^{WP'} = w \begin{pmatrix} R\sin q_8 \cos q_9 \\ R\sin q_8 \sin q_9 \\ R\cos q_8 \end{pmatrix}$$

for some $q_7$, $q_8$, and $q_9$, then $P$ is in contact with $P'$ if

$$0 = \left(\mathbf{r}^{WP} - \mathbf{r}^{WP'}\right) \bullet \mathbf{w}_1,$$

$$0 = \left(\mathbf{r}^{WP} - \mathbf{r}^{WP'}\right) \bullet \mathbf{w}_2,$$

$$0 = \left(\mathbf{r}^{WP} - \mathbf{r}^{WP'}\right) \bullet \mathbf{w}_3.$$

Moreover, since the vector $\mathbf{r}^{WP'} \times \mathbf{a}_3$ is tangential to the spherical surface and parallel to the plane of the disk, it follows that

$$0 = \mathbf{r}^{AP} \bullet \left(\mathbf{r}^{WP'} \times \mathbf{a}_3\right)$$

guarantees that the contact at $P$ is tangential.

The following MAMBO toolbox statements compute the corresponding configuration constraints in terms of the nine configuration coordinates:

```
> Restart():
> DeclareObservers(W,A):
> DeclarePoints(W,A,P,Pp):
> DeclareTriads(w,a):
> DefineObservers([W,W,w],
> [A,A,a]):
> DefinePoints([W,A,w,q1,q2,q3],
> [A,P,a,rho*cos(q7),
> rho*sin(q7),0],
> [W,Pp,w,R*sin(q8)*cos(q9),
> R*sin(q8)*sin(q9),R*cos(q8)]):
> DefineTriads([w,a,[q4,3],
> [q5,1],[q6,3]]):
> f1:=(FindTranslation(W,P) &--
> FindTranslation(W,Pp)) &oo
> MakeTranslations(w,1)=0:
> f2:=(FindTranslation(W,P) &--
> FindTranslation(W,Pp)) &oo
> MakeTranslations(w,2)=0:
> f3:=(FindTranslation(W,P) &--
> FindTranslation(W,Pp)) &oo
> MakeTranslations(w,3)=0:
> f4:=(FindTranslation(W,Pp) &xx
> MakeTranslations(a,3)) &oo
> FindTranslation(A,P)=0:
```

**Exercise 6.6** Find the number of geometric degrees of freedom of the disk in the previous exercise.

**Exercise 6.7** The configuration of a thin disk of radius $\rho$ is constrained in such a way that the disk makes tangential contact with a cylindrical surface of radius $R$ centered on the reference point $W$ of an observer $\mathcal{W}$ and parallel to the basis vector $\mathbf{w}_3$ of the corresponding reference triad $w$. Find the corresponding configuration constraints and determine the number of geometric degrees of freedom of the disk.

**Exercise 6.8** Analyze the following mechanisms to determine the number of geometric degrees of freedom:

a) A unicycle
b) A bicycle
c) An office chair
d) A can opener
e) A pair of scissors
f) An umbrella

**Exercise 6.9** Use the MAMBO toolbox to formulate a MAMBO geometry description of the following objects and visualize using MAMBO. You may find the information in the MAMBO reference manual regarding the geometric properties of MAMBO spheres, blocks, and cylinders helpful.

a) Icosahedron
b) Spider web
c) Hexagonal honeycomb
d) Bridge truss
e) Circle of rectangular chairs
f) Brick tower
g) Bird feather

**Exercise 6.10** For each of the scenes below, use the MAMBO toolbox to formulate a MAMBO geometry description and implement different animation sequences in MAMBO.

a) Operating window shades
b) A grandfather's clock
c) A handheld fan
d) A turntable with play arm
e) Dialing on an analog dialing pad
f) Sanding with an orbital sander

## Summary of notation

Lower-case $q$'s with various subscripts and decorations, such as $q_1$ and $\tilde{q}_3$, were used in this chapter to denote configuration coordinates.

## Summary of terminology

A selected triad that is stationary relative to a rigid body or an observer is called a *reference triad.* (Page 250)

A variable that is used to describe the configuration of a mechanism is called a *configuration coordinate.* (Page 258)

The number of *geometric degrees of freedom* of a mechanism is the smallest number of configuration coordinates necessary to describe the configuration of the mechanism. (Page 258)

A mechanism is *constrained* if its configuration is limited. (Page 262)

An equation in the configuration coordinates that corresponds to a constraint on the configuration of a mechanism is called a *configuration constraint.* (Page 262)

In MAMBO, the orientation of an observer or a rigid body relative to an observer is given through an **ORIENT** statement in the .geo file. (Page 267)

In MAMBO, parameters are declared through a `parameters` statement in the .dyn file. (Page 271)

In MAMBO, the time variable is labeled through a `time` statement in the .dyn file. (Page 271)

In MAMBO, animated variables are declared through an `anims` block in the .dyn file. (Page 271)

In MAMBO, states are declared through a `states` statement in the .dyn file. (Page 271)

In the MAMBO toolbox, the global variable `GlobalTriadDeclarations` contains the names of all declared triads. (Page 274)

In the MAMBO toolbox, the global variable `GlobalTriadDefinitions` contains rotation matrices relating declared triads. (Page 274)

In the MAMBO toolbox, the procedure `DeclareTriads` appends `GlobalTriadDeclarations` with any number of triad labels. (Page 274)

In the MAMBO toolbox, the procedure `DefineTriads` appends `GlobalTriadDefinitions` with any number of rotation matrices relating declared triads. (Page 274)

(Page 276) In the MAMBO toolbox, the procedure `FindRotation` is used to find the rotation matrix between two declared triads.

(Page 276) In the MAMBO toolbox, the procedures `&oo`, `&xx`, `VectorLength`, and `Express` return the vector dot product, the vector cross product, the length of a vector, and the vector expressed relative to a given triad, respectively.

(Page 278) In the MAMBO toolbox, the procedure `DefineObservers` appends `GlobalObserverDefinitions` with any number of associations between observers and pairs of declared reference points and declared reference triads.

(Page 278) In the MAMBO toolbox, the procedure `FindOrientation` is used to find the rotation matrix between the reference triads of two observers.

(Page 279) In the MAMBO toolbox, the procedure `Undo` is used to undo the latest change to any of the global variables.

# Chapter 7

# Review

*wherein the reader learns of:*

- *Combining the elements developed in previous chapters into a general methodology for describing the geometry of a multibody mechanism.*

Practicum

This chapter is intended to give you a breather; to let you collect your thoughts and assess the global strategy for analyzing multibody mechanisms that has been developed in previous chapters. The examples in this chapter represent the highest level of complexity you are likely to ever encounter.

After completing this chapter, you are encouraged to refer to the list of sample projects in the Appendix and to attempt to implement these in MAMBO. There is no expectation that you will be able to successfully formulate the correct geometry description on your first try. MAMBO allows you to experiment and to try different ideas. A final description along the lines of this chapter grows out of such experimentation.

# 7.1 Terminology and Notation

## 7.1.1 Configurations

At the outset of Chapter 1, I introduced the *configuration* of a rigid body as its *position* and *orientation* relative to some *reference configuration.* Neither the position, the orientation, nor the reference configuration were well-defined notions. Instead, I appealed to your intuitive understanding of these concepts based on everyday experiences and abstractions.

The idea of position was meant to suggest a general location in space, whereas the idea of orientation was meant to suggest a general attitude in space. Even without very specific definitions of these ideas, some thought established the need for referring to some basis of comparison, without which no well-defined meaning could be attached to the terms position and orientation. This basis of comparison was the reference configuration.

Subsequent material in Chapter 1 was intended to narrow down the ideas of position and orientation. This allowed for quantitative statements that would be necessary for communicating experimental observations or for implementing a multibody mechanism in a computer-visualization software. In particular, the ideas of positions and orientations were clarified by considering *changes* in configuration. These changes were categorized as *pure translations*, *pure rotations*, and combinations thereof.

## 7.1.2 Rigid-body Operations

On page 4, a pure translation was defined as an operation on a rigid body that shifts all the points on the body by an equal amount along parallel paths. I stated that there is **no change in the orientation** of a rigid body undergoing a pure translation. This was the first hint on as to how to interpret the notion of orientation.

On page 9, a pure rotation was defined as an operation on a rigid body that holds at least one point on the rigid body fixed. I stated that there is **no change in the position** of a rigid body undergoing a pure rotation. This was the first hint as to how to interpret the notion of position.

The implications of these definitions are that two identical rigid bodies have the same position if they have at least one point in common. Similarly, two identical rigid bodies have the same orientation if all points on one rigid body are shifted by an equal amount along a common direction from the corresponding points on the other rigid body.

I believe these definitions make intuitive sense. For example, if $\mathcal{B}_1$, $\mathcal{B}_2$, and $\mathcal{B}_3$ are three identical rigid bodies, such that $\mathcal{B}_1$ and $\mathcal{B}_2$ have the same orientation and $\mathcal{B}_2$ and $\mathcal{B}_3$ have the same orientation, then the rigid bodies $\mathcal{B}_1$ and $\mathcal{B}_3$ also have the same orientation. The discussion on pages 7 and 8 expressed this observation in terms of pure translations, namely that **the composition of two consecutive pure translations is equivalent to a single pure translation**.

But these definitions also challenge some obvious assumptions. For example, if $\mathcal{B}_1$, $\mathcal{B}_2$, and $\mathcal{B}_3$ are three identical rigid bodies, such that $\mathcal{B}_1$ and $\mathcal{B}_2$ have the same position and $\mathcal{B}_2$ and $\mathcal{B}_3$ have the same position, then the rigid bodies $\mathcal{B}_1$ and $\mathcal{B}_3$ **do not necessarily** have the same position. The discussion on pages 10 and 11 expressed this observation in terms of pure rotations, namely that **the composition of two consecutive pure rotations is not necessarily equivalent to a single pure rotation**.

A resolution to this paradox was offered on page 11, where we chose to require that all pure rotations keep the same point fixed. With this added condition, the statement that $\mathcal{B}_1$ and $\mathcal{B}_2$ have the same position and that $\mathcal{B}_2$ and $\mathcal{B}_3$ have the same position would imply that $\mathcal{B}_1$ and $\mathcal{B}_3$ also have the same position.

Several important properties of pure translations and pure rotations were described in detail on pages 4 through 21.

Specifically, any sequence of pure translations and pure rotations is *equivalent* to a single pure translation and a single pure rotation. Conversely, any arbitrary configuration can be *decomposed* into a sequence of pure translations and pure rotations. This result established the significance of pure translations and pure rotations to describe arbitrary configurations. Put differently, **the collection of pure translations and the collection of pure rotations are the elementary particles of multibody mechanics**. There are no changes in configuration that cannot be described in terms of a combination of a single pure translation and a single pure rotation. Having chosen a *particular point* on the rigid body to be kept fixed by any pure rotation, **the decomposition into a single pure translation and a single pure rotation was found to be unique**.

Of equal significance was the observation that the order in which a pure rotation and a pure translation are applied to a rigid body makes no difference to the final configuration. The meaning of the pure rotation does not depend on the pure translation, and vice versa. Put differently, **given the selection of a particular point on the rigid body to be kept fixed by any pure rotation, the position and the orientation of a rigid body can be specified independently of one another**. If I say that a rigid body has a certain orientation, this has no implications whatsoever on its position. If you say that a rigid body has a certain position, this has no implications whatsoever on its orientation.

### 7.1.3 Observers

You could use the notions of position, orientation, pure translation, and pure rotation to describe the configurations of all rigid bodies in your environment relative to some reference configuration. Certainly, if all rigid bodies in your environment were disjoint, independent objects, this would be a reasonable and prudent course of action. But, if there were connections between different bodies that rendered their configurations dependent, this would not necessarily be the most attractive approach.

Suppose you wanted to describe the configurations of the various parts of your hand relative to some reference configuration. This would require specifying the position and orientation of each of the parts – the palm, the digital segments, and so on – relative to the reference position and reference orientation. But, the configurations of the digits are not independent from the configuration of the palm. As the palm moves through space, the digits are constrained to follow along, reducing their mobility relative to the palm to a pure rotation. Similarly, the digits are naturally split into segments, each of which can move relative to the preceding one through a pure rotation.

It appears more natural to describe the configurations of the digits relative to the palm and the configuration of the palm relative to you. The configurations of the digits relative to you would then follow by combining the pure translations and pure rotations corresponding to each of the two separate steps.

The idea of breaking down the description of the configuration into more manageable steps requires the introduction of intermediate reference configurations. Toward the end of Chapter 1 and throughout Chapter 2, I employed the notion of *observers* to represent such reference configurations. Every mechanism was described using a *main observer* and any number of *auxiliary observers.*

On page 50, I introduced tree structures to represent a conceptual arrangement of the observers and the physical objects they were used to describe. I stressed the possibility of using any number of different observer arrangements to represent the same geometry. Some general rules were also proposed for you to follow in describing your own mechanisms of choice.

### 7.1.4 Vectors

A computationally oriented formalism for pure translations was introduced in Chapter 3. I showed how every pure translation corresponds to a collection of straight-line segments between points in the initial configuration and the corresponding points in the final configuration. These straight-line segments:

- Have the same length;
- Are parallel;
- Have the same heading.

Every one of these segments could be taken to represent the pure translation. The length, direction, and heading of the segment would correspond to the same properties of the pure translation.

On page 96, these collections of straight-line segments were named *vectors*. I presented mathematical operations on these vectors that corresponded to the algebraic operations on pure translations that were introduced in Chapter 2. Specifically:

- *Vectors could be multiplied by real numbers* to generate scaled versions of the corresponding pure translations (page 98);
- *Vectors could be summed* to generate compositions of the corresponding pure translations (page 99).

I also introduced two vector products, namely the *dot product* (page 103) and the *cross product* (page 105), with which vectors and the corresponding pure translations could be compared with respect to heading and direction. The *length* of a vector was defined as the magnitude of the shift of the corresponding pure translation (page 97).

Particularly useful are triplets of mutually perpendicular vectors of unit length. These are called *orthonormal bases* (page 111). Specifically, if $\mathbf{v}$ is an arbitrary vector and the triplet $\{\mathbf{a}_1, \mathbf{a}_2, \mathbf{a}_3\}$ is an orthonormal basis, then there exists a **unique** triplet of real numbers $v_1$, $v_2$, and $v_3$, such that

$$\mathbf{v} = v_1\mathbf{a}_1 + v_2\mathbf{a}_2 + v_3\mathbf{a}_3$$

and

$$v_i = \mathbf{a}_i \bullet \mathbf{v},\ i = 1, 2, 3.$$

The real numbers $v_1$, $v_2$, and $v_3$ are called the *coordinates of the vector* $\mathbf{v}$ *relative to the basis* $\{\mathbf{a}_1, \mathbf{a}_2, \mathbf{a}_3\}$. On page 114, I introduced the notation

$$a = \begin{pmatrix} \mathbf{a}_1 & \mathbf{a}_2 & \mathbf{a}_3 \end{pmatrix}$$

and

$$^a v = \begin{pmatrix} \mathbf{a}_1 \bullet \mathbf{v} \\ \mathbf{a}_2 \bullet \mathbf{v} \\ \mathbf{a}_3 \bullet \mathbf{v} \end{pmatrix}$$

to represent an orthonormal basis $a$ and the *matrix representation of the vector* $\mathbf{v}$ *relative to the basis* $a$, and extended the rules of normal matrix multiplication to allow for statements like

$$\mathbf{v} = a\ {}^a v.$$

All operations on vectors could be reduced to matrix operations on the matrix representations of the vectors relative to some orthonormal basis. In particular, by restricting attention to right-handed, orthonormal bases, so-called *triads*, I showed that

$$\alpha\mathbf{v}+\beta\mathbf{w} = a\left(\alpha\,{}^{a}v + \beta\,{}^{a}w\right),$$
$$\mathbf{v}\bullet\mathbf{w} = \left({}^{a}v\right)^{T}\,{}^{a}w,$$
$$\|\mathbf{v}\| = \sqrt{{}^{a}v_1^2 + {}^{a}v_2^2 + {}^{a}v_3^2},$$

and

$$\mathbf{v}\times\mathbf{w} = \begin{vmatrix} \mathbf{a}_1 & \mathbf{a}_2 & \mathbf{a}_3 \\ {}^{a}v_1 & {}^{a}v_2 & {}^{a}v_3 \\ {}^{a}w_1 & {}^{a}w_2 & {}^{a}w_3 \end{vmatrix},$$

where $a$ is a triad.

The expressions above assumed that both vectors were expressed relative to the same triad. When this was not the case, I showed you how to translate between matrix representations relative to different triads. In particular, with the definition on page 207 of the *rotation matrix* $R_{ab}$ between the triads $a$ and $b$

$$R_{ab} = a^{T}\bullet b = \begin{pmatrix} \mathbf{a}_1\bullet\mathbf{b}_1 & \mathbf{a}_1\bullet\mathbf{b}_2 & \mathbf{a}_1\bullet\mathbf{b}_3 \\ \mathbf{a}_2\bullet\mathbf{b}_1 & \mathbf{a}_2\bullet\mathbf{b}_2 & \mathbf{a}_2\bullet\mathbf{b}_3 \\ \mathbf{a}_3\bullet\mathbf{b}_1 & \mathbf{a}_3\bullet\mathbf{b}_2 & \mathbf{a}_3\bullet\mathbf{b}_3 \end{pmatrix},$$

I derived the following relations:

$$R_{ba} = \left(R_{ab}\right)^{T} = \left(R_{ab}\right)^{-1},$$
$$b = aR_{ab},\; a = bR_{ba},$$
$${}^{a}v = R_{ab}\,{}^{b}v,$$

and

$${}^{b}v = R_{ba}\,{}^{a}v.$$

Moreover, if $c$ was a third triad, then

$$R_{ac} = R_{ab}R_{bc}$$

represented a natural decomposition.

While vectors correspond in a direct way to pure translations, rotation matrices correspond in a direct way to pure rotations. Since the orientation of a rigid body is uniquely determined by the orientation of a triad rigidly attached to the body, changes in orientation due to pure rotations are quantifiable in terms of the rotation matrix between the initial and final orientations of the triad.

In an exercise in Chapter 1, it was observed that every pure rotation was equivalent to a rotation about some fixed axis by a given angle. The analogous statement for rotation matrices was made in Chapter 5, where we found that every rotation matrix can be generated by the expression

$$R(\varphi, v_1, v_2, v_3),$$

where $v_1$, $v_2$, $v_3$ were the coordinates relative to the initial triad of a vector of unit length parallel to the axis of the pure rotation and $\varphi$ was the angle of rotation.

On pages 221 through 229, I described some special (and very useful) examples of rotation matrices arising from *rotations about basis vectors* and leading to the concept of *Euler angles*. I also showed how to construct rotation matrices to align a triad vector with a given direction.

### 7.1.5 Configuration Coordinates and Constraints

The configuration of a rigid body relative to an observer is uniquely determined by a pure translation and a pure rotation, given the selection of a point on the rigid body that is held fixed by any pure rotation. This point is called the *reference point* of the rigid body. Similarly, the relative configuration of two observers is uniquely determined by a pure translation and a pure rotation, given the selection of reference points for each of the observers.

If $A$ and $B$ are the reference points of two observers $\mathcal{A}$ and $\mathcal{B}$, then the pure translation $\mathbf{T}_{\mathcal{A}\to\mathcal{B}}$ corresponds to the position vector

$$\mathbf{r}^{AB}.$$

The pure translation is uniquely determined by the relative location of the points $A$ and $B$.

The pure rotation $\mathbf{R}_{\mathcal{A}\to\mathcal{B}}$ is uniquely determined by the relative orientation of two *reference triads* $a$ and $b$, whose orientations are fixed relative to $\mathcal{A}$ and $\mathcal{B}$, respectively. In particular, the pure rotation corresponds to the rotation matrix

$$R_{ab}.$$

Any quantities that appear in the position vector $\mathbf{r}^{AB}$ or the rotation matrix $R_{ab}$, and that change when the relative configuration of the observers changes, are called *configuration coordinates*. The smallest number of configuration coordinates that are necessary to describe the configuration of a mechanism equals its *number of geometric degrees of freedom*.

In Chapters 4 and 6, I showed you how three configuration coordinates suffice to describe any arbitrary position of a rigid body, while three separate configuration coordinates suffice to describe any arbitrary orientation of a rigid body. I also showed you that three configuration coordinates are necessary to describe the position of a freely moving rigid body and that three configuration coordinates are necessary to describe the orientation of a freely moving rigid body. A freely moving rigid body thus has six geometric degrees of freedom.

When a mechanism consisting of $N$ rigid bodies has fewer than $6N$ geometric degrees of freedom, it is *constrained*. To constrain a mechanism is equivalent to imposing equalities – *configuration constraints* – that the configuration coordinates used to describe the unconstrained mechanism must satisfy. In Chapter 4, I showed how configuration constraints usually imply that it is theoretically possible to express one or several of the configuration coordinates in terms of the others. As we shall see later in this chapter and in more detail in Chapter 9, it may nevertheless be more **practical** to retain more configuration coordinates than the number of geometric degrees of freedom.

## 7.2 Modeling Algorithm

Throughout the previous chapters, I have advocated the following algorithm for arriving at a complete description of the geometry of a multibody mechanism:

**Step 1.** Identify all constituent rigid bodies. In doing this, I recognize that a rigid body may consist of multiple parts, each of which is a separate rigid body. However, the multiple parts of a rigid body are assumed to be stationary relative to each other. They move as a union relative to all other constituent rigid bodies.

**Step 2.** Introduce a reference point and a reference triad for each constituent rigid body. I usually pick some point that has particular significance for the geometry, say a symmetry point of the rigid body. Similarly, I will pick a triad for which at least one basis vector is parallel to some symmetry line of the rigid body.

**Step 3.** Introduce a main observer, relative to which all configurations are ultimately described. As suggested in Chapter 2, the choice of main observer is motivated by the purpose of the modeling, whether primarily graphics- or physics-oriented. I often pick a reference point and a reference triad of the main observer, such that it is related to the geometry of some object that is stationary relative to the main observer.

**Step 4.** Introduce a separate auxiliary observer for each rigid body whose configuration may change relative to the main observer. I pick the reference point and the reference triad of the auxiliary observer, such that the rigid body remains stationary relative to the auxiliary observer. It is not necessary that the reference point of the auxiliary observer coincides with any point on the corresponding rigid body.

**Step 5.** Arrange the observers and rigid bodies in a tree structure with the main observer as the top node, the auxiliary observers as internal nodes, and the rigid bodies as leaf nodes. I often organize the auxiliary observers to reflect the presence of mechanical joints that restrict the relative motions between different auxiliary observers. This is analogous to the discussion of describing the configurations of the digits on the hand relative to you by describing the digits' configurations relative to the palm and the configuration of the palm relative to you.

**Step 6.** Introduce configuration coordinates to quantify the position vectors and rotation matrices that relate the positions and orientations of successive nodes in the tree structure. I recommend simplicity over cleverness. Often the simplest solution is quite sufficient and will enhance the understanding over a particularly clever solution that may be detrimental to the understanding. I expect that you will have experienced both possibilities when looking at the various examples throughout the text.

**Step 7.** Identify any configuration constraints that restrict the allowable values for the configuration coordinates to actually correspond to geometrically correct configurations of the mechanism.

In the next several chapters, we will add to this algorithm to enable the simulation and animation of geometrically correct and physically realistic motions of the multibody mechanism.

## 7.3 A Bicycle

The algorithm in the previous section establishes the complete description of the instantaneous geometry of a multibody mechanism and can be used in a very crude way to generate motions. In this section, we return to the bicycle introduced in Chapter 2 and implement the modeling algorithm as suggested.

### 7.3.1 Constituent Rigid Bodies

The bicycle shown below consists of four distinct rigid bodies, namely the rear wheel, the front wheel, the steering column, and the frame.

Clearly, each of these bodies consists of multiple parts, but these parts are rigidly attached to each other. Missing from this geometry are the pedals. Their inclusion would result in at least one additional rigid body. If each of the pedals were allowed to spin relative to the pedal assembly, at least two more rigid bodies would need to be included. For the present discussion, we restrict attention to the four rigid bodies identified at the top of the section.

### 7.3.2 Reference Points and Reference Triads

Let the reference point $A_{\text{rear wheel}}$ of the rear wheel be located at the center of the rear wheel. Choose the reference triad $a^{(\text{rear wheel})}$ of the rear wheel, such that the wheel axis is parallel to the vector $\mathbf{a}_3^{(\text{rear wheel})}$.

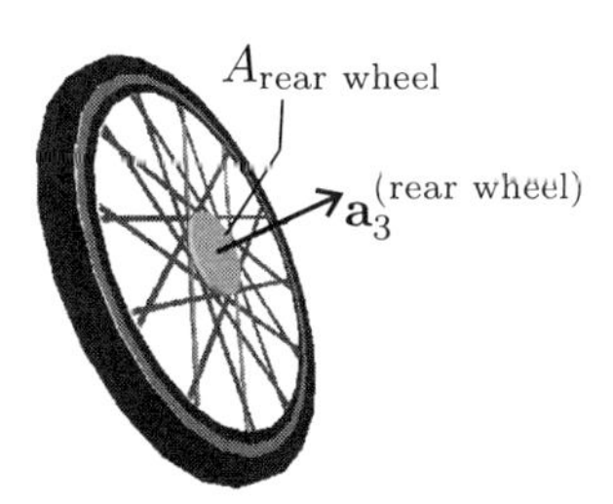


Let the reference point $A_{\text{front wheel}}$ of the front wheel be located at the center of the front wheel. Choose the reference triad $a^{(\text{front wheel})}$ of the front wheel, such that the wheel axis is parallel to the vector $\mathbf{a}_3^{(\text{front wheel})}$.

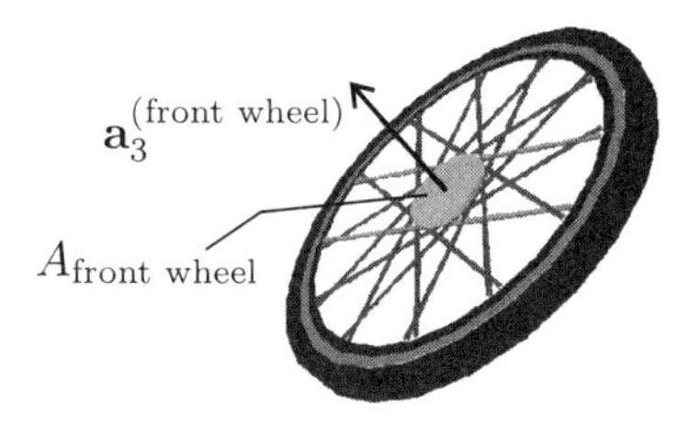


Let the reference point $A_{\text{frame}}$ of the frame coincide with $A_{\text{rear wheel}}$. Choose the reference triad $a^{(\text{frame})}$ of the frame, such that $\mathbf{a}_3^{(\text{frame})}$ equals $\mathbf{a}_3^{(\text{rear wheel})}$ and $\mathbf{a}_1^{(\text{frame})}$ is parallel to the forward direction of the bicycle saddle.

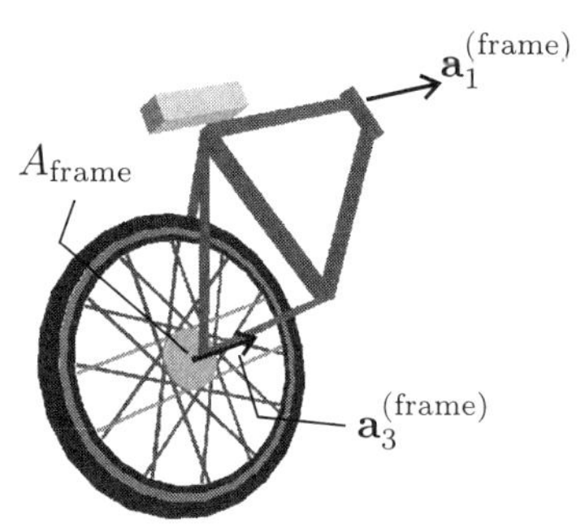


Let the reference point $A_{\text{steering}}$ of the steering column coincide with $A_{\text{front wheel}}$. Choose the reference triad $a^{(\text{steering})}$ of the steering column, such that $\mathbf{a}_3^{(\text{steering})}$ equals $\mathbf{a}_3^{(\text{front wheel})}$ and $\mathbf{a}_1^{(\text{steering})}$ is parallel to the axis of rotation of the steering column.

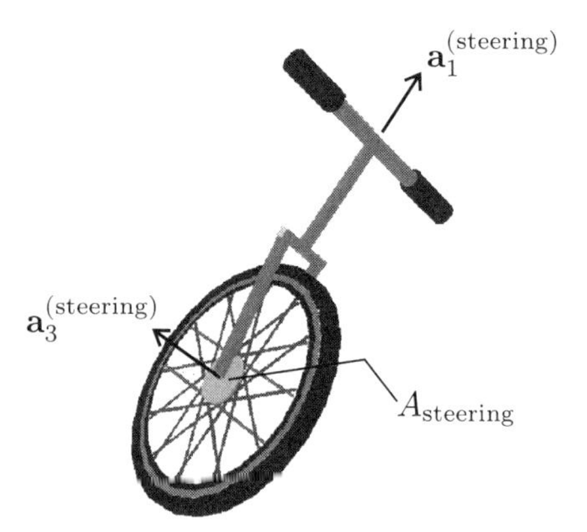


### 7.3.3 Main Observer

As we are primarily interested in arriving at a geometry description that we may implement in the MAMBO application, we introduce a main observer $\mathcal{W}$ with reference point $W$ and reference triad $w$ corresponding to the graphics application's internal reference configuration. Below, we will introduce a plane that will be stationary relative to $\mathcal{W}$ to serve as a basic background for the motion of the bicycle.

### 7.3.4 Auxiliary Observers

Since each of the four rigid bodies introduced above may change configuration relative to the main observer, introduce four auxiliary observers $\mathcal{A}_{\text{rear wheel}}$, $\mathcal{A}_{\text{front wheel}}$, $\mathcal{A}_{\text{frame}}$, and $\mathcal{A}_{\text{steering}}$, relative to which the rear wheel, front wheel, frame, and steering column, respectively, are stationary. For simplicity, let the corresponding reference points and reference triads agree with those selected for the corresponding rigid bodies.

### 7.3.5 Tree Structures

Consider the following tree structure representing the conceptual arrangement of observers and rigid bodies.

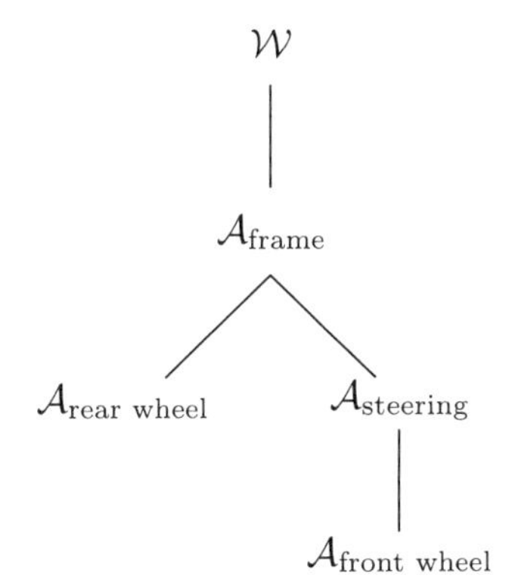


This description will be complete by specifying the pure translations and pure rotations corresponding to each of the direct connections between successive observers or between observers and rigid bodies. In particular, since the reference points and reference triads of the auxiliary observers coincide with those of the corresponding rigid bodies, the pure translations and pure rotations between the observers and the rigid bodies are the identity translation and identity rotation. Naturally, each rigid body consists of multiple parts whose configurations relative to the auxiliary observers are described by non-trivial pure translations and pure rotations. These are time-independent, however, and their description is implemented as in the discussion on still lives in Chapters 2, 4, and 6.

### 7.3.6 Configuration Coordinates

The configuration of the observer $\mathcal{A}_{\text{frame}}$ relative to the main observer $\mathcal{W}$ is described by a pure translation $\mathbf{T}_{\mathcal{W}\to\mathcal{A}_{\text{frame}}}$ corresponding to the position vector

$$\mathbf{r}^{WA_{\text{frame}}} = w \begin{pmatrix} q_1 \\ q_2 \\ q_3 \end{pmatrix}$$

and a pure rotation $\mathbf{R}_{\mathcal{W}\to\mathcal{A}_{\text{frame}}}$ corresponding to the rotation matrix

$$\begin{aligned} R_{wa^{(\text{frame})}} &= R(q_4,0,0,1)\, R(q_5,1,0,0)\, R(q_6,0,0,1) \\ &= \begin{pmatrix} c_4c_6 - s_4c_5s_6 & -c_4s_6 - s_4c_5c_6 & s_4s_5 \\ s_4c_6 + c_4c_5s_6 & -s_4s_6 + c_4c_5c_6 & -c_4s_5 \\ s_5s_6 & s_5c_6 & c_5 \end{pmatrix}, \end{aligned}$$

where $c_i = \cos q_i$ and $s = \sin q_i$, $i = 1,2,3$. The configuration coordinates $q_4$, $q_5$, and $q_6$ are Euler angles corresponding to a $3-1-3$ decomposition of the pure rotation $\mathbf{R}_{\mathcal{W}\to\mathcal{A}_{\text{frame}}}$.

The configuration of the observer $\mathcal{A}_{\text{rear wheel}}$ relative to the observer $\mathcal{A}_{\text{frame}}$ is described by a pure translation $\mathbf{T}_{\mathcal{A}_{\text{frame}}\to\mathcal{A}_{\text{rear wheel}}}$ corresponding to the position vector

$$\mathbf{r}^{A_{\text{frame}}A_{\text{rear wheel}}} = a^{(\text{frame})} \begin{pmatrix} 0 \\ 0 \\ 0 \end{pmatrix} = \mathbf{0}$$

and a pure rotation $\mathbf{R}_{\mathcal{A}_{\text{frame}}\to\mathcal{A}_{\text{rear wheel}}}$ corresponding to the rotation matrix

$$\begin{aligned} R_{a^{(\text{frame})}a^{(\text{rear wheel})}} &= R(q_7,0,0,1) \\ &= \begin{pmatrix} \cos q_7 & -\sin q_7 & 0 \\ \sin q_7 & \cos q_7 & 0 \\ 0 & 0 & 1 \end{pmatrix}. \end{aligned}$$

The configuration of the observer $\mathcal{A}_{\text{steering}}$ relative to the observer $\mathcal{A}_{\text{frame}}$ is described by a pure translation $\mathbf{T}_{\mathcal{A}_{\text{frame}}\to\mathcal{A}_{\text{steering}}}$ corresponding to the position vector

$$\mathbf{r}^{A_{\text{frame}}A_{\text{steering}}}$$

and a pure rotation $\mathbf{R}_{\mathcal{A}_{\text{frame}}\to\mathcal{A}_{\text{steering}}}$ corresponding to the rotation matrix

$$R_{a^{(\text{frame})}a^{(\text{steering})}}.$$

To compute $R_{a^{(\text{frame})}a^{(\text{steering})}}$, let the vector

$$\mathbf{v} = a^{(\text{frame})} \begin{pmatrix} p_1 \\ p_2 \\ 0 \end{pmatrix}$$

be parallel to the axis of rotation of the steering column, where $p_1, p_2 \neq 0$ are some constants. We may construct a triad $b$, such that $\mathbf{b}_1$ is parallel to $\mathbf{v}$ by implementing the methodology described in Chapter 3. In particular,

$$\mathbf{b}_1 = \frac{\mathbf{v}}{\|\mathbf{v}\|} = a^{(\text{frame})} \begin{pmatrix} \frac{p_1}{\sqrt{p_1^2+p_2^2}} \\ \frac{p_2}{\sqrt{p_1^2+p_2^2}} \\ 0 \end{pmatrix},$$

$$\mathbf{b}_2 = \frac{\mathbf{a}_3^{(\text{frame})} \times \mathbf{b}_1}{\left\|\mathbf{a}_3^{(\text{frame})} \times \mathbf{b}_1\right\|} = a^{(\text{frame})} \begin{pmatrix} -\frac{p_2}{\sqrt{p_1^2+p_2^2}} \\ \frac{p_1}{\sqrt{p_1^2+p_2^2}} \\ 0 \end{pmatrix},$$

and

$$\mathbf{b}_3 = \mathbf{b}_1 \times \mathbf{b}_2 = a^{(\text{frame})} \begin{pmatrix} 0 \\ 0 \\ 1 \end{pmatrix}.$$

The corresponding rotation matrix is given by

$$R_{a^{(\text{frame})}b} = \begin{pmatrix} \frac{p_1}{\sqrt{p_1^2+p_2^2}} & -\frac{p_2}{\sqrt{p_1^2+p_2^2}} & 0 \\ \frac{p_2}{\sqrt{p_1^2+p_2^2}} & \frac{p_1}{\sqrt{p_1^2+p_2^2}} & 0 \\ 0 & 0 & 1 \end{pmatrix}.$$

The rotation matrix $R_{a^{(\text{frame})}a^{(\text{steering})}}$ is now obtained from the product

$$\begin{aligned} R_{a^{(\text{frame})}a^{(\text{steering})}} &= R_{a^{(\text{frame})}b} R\left(q_8, 1, 0, 0\right) \\ &= \begin{pmatrix} \frac{p_1}{\sqrt{p_1^2+p_2^2}} & -\frac{p_2}{\sqrt{p_1^2+p_2^2}} \cos q_8 & \frac{p_2}{\sqrt{p_1^2+p_2^2}} \sin q_8 \\ \frac{p_2}{\sqrt{p_1^2+p_2^2}} & \frac{p_1}{\sqrt{p_1^2+p_2^2}} \cos q_8 & -\frac{p_1}{\sqrt{p_1^2+p_2^2}} \sin q_8 \\ 0 & \sin q_8 & \cos q_8 \end{pmatrix}. \end{aligned}$$

The position vector $\mathbf{r}^{A_{\text{frame}}A_{\text{steering}}}$ is given by the expression

$$\begin{aligned} \mathbf{r}^{A_{\text{frame}}A_{\text{steering}}} &= a^{(\text{frame})} \begin{pmatrix} p_3 \\ p_4 \\ 0 \end{pmatrix} + a^{(\text{steering})} \begin{pmatrix} p_5 \\ p_6 \\ 0 \end{pmatrix} \\ &= a^{(\text{frame})} \left[ \begin{pmatrix} p_3 \\ p_4 \\ 0 \end{pmatrix} + R_{a^{(\text{frame})}a^{(\text{steering})}} \begin{pmatrix} p_5 \\ p_6 \\ 0 \end{pmatrix} \right] \\ &= a^{(\text{frame})} \begin{pmatrix} p_3 + \frac{p_1 p_5}{\sqrt{p_1^2+p_2^2}} - \frac{p_2 p_6}{\sqrt{p_1^2+p_2^2}} \cos q_8 \\ p_4 + \frac{p_2 p_5}{\sqrt{p_1^2+p_2^2}} + \frac{p_1 p_6}{\sqrt{p_1^2+p_2^2}} \cos q_8 \\ p_6 \sin q_8 \end{pmatrix} \end{aligned}$$

corresponding to a combination of the vectors from the centers of the wheels to a point on the steering column.

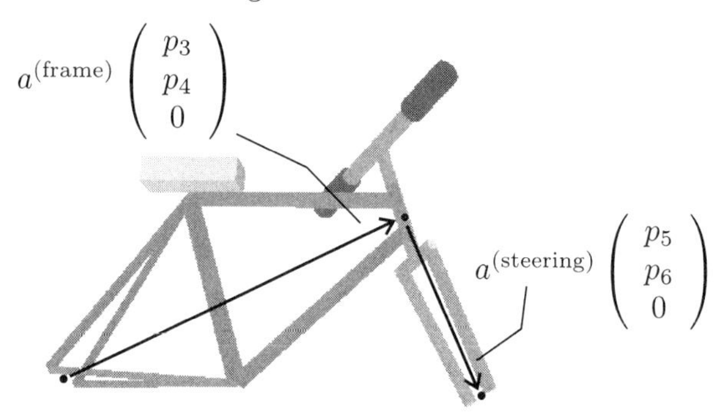


Finally, the configuration of the observer $\mathcal{A}_{\text{front wheel}}$ relative to the observer $\mathcal{A}_{\text{steering}}$ is described by a pure translation $\mathbf{T}_{\mathcal{A}_{\text{steering}}\to\mathcal{A}_{\text{front wheel}}}$ corresponding to the position vector

$$\mathbf{r}^{\mathcal{A}_{\text{steering}}\mathcal{A}_{\text{front wheel}}} = a^{(\text{steering})}\begin{pmatrix} 0 \\ 0 \\ 0 \end{pmatrix} = \mathbf{0}$$

and a pure rotation $\mathbf{R}_{\mathcal{A}_{\text{steering}}\to\mathcal{A}_{\text{front wheel}}}$ corresponding to the rotation matrix

$$\begin{aligned} R_{a^{(\text{steering})}a^{(\text{front wheel})}} &= R\left(q_9, 0, 0, 1\right) \\ &= \begin{pmatrix} \cos q_9 & -\sin q_9 & 0 \\ \sin q_9 & \cos q_9 & 0 \\ 0 & 0 & 1 \end{pmatrix}. \end{aligned}$$

### 7.3.7 Configuration Constraints

It remains to consider any constraints on the configuration of the bicycle that would reduce the number of geometric degrees of freedom from the nine found in the previous section. In particular, we shall consider constraining the rear and front wheels to remain in contact with a plane representing the ground through the reference point $W$ of the main observer $\mathcal{W}$ and parallel to the basis vectors $\mathbf{w}_1$ and $\mathbf{w}_2$.

Denote by $P_{\text{rear contact}}$ the point on the rear wheel that makes contact with the plane and at which point the tangent direction to the rear wheel is parallel to the plane. Denote by $P_{\text{front contact}}$ the point on the front wheel that makes contact with the plane and at which point the tangent direction to the front wheel is parallel to the plane. It is convenient to introduce two additional configuration coordinates, such that

$$^{\mathcal{A}_{\text{rear wheel}}}P_{\text{rear contact}} = \begin{pmatrix} R\cos q_{10} \\ R\sin q_{10} \\ 0 \end{pmatrix}$$

and

$$\mathcal{A}_{\text{front wheel}} P_{\text{front contact}} = \begin{pmatrix} R\cos q_{11} \\ R\sin q_{11} \\ 0 \end{pmatrix}$$

where $R$ is the wheel radius. Then, contact with the plane is ensured if

$$\mathbf{r}^{W P_{\text{rear contact}}} \bullet \mathbf{w}_3 = 0$$

and

$$\mathbf{r}^{W P_{\text{front contact}}} \bullet \mathbf{w}_3 = 0.$$

Since the cross products

$$\mathbf{w}_3 \times \mathbf{a}_3^{(\text{rear wheel})}$$

and

$$\mathbf{w}_3 \times \mathbf{a}_3^{(\text{front wheel})}$$

are parallel to the line of intersection between the stationary plane and the planes of the rear and front wheels, respectively, the tangent directions to the rear and front wheels at the point of contact are parallel to the plane if

$$\mathbf{r}^{A_{\text{rear wheel}} P_{\text{rear contact}}} \bullet \left(\mathbf{w}_3 \times \mathbf{a}_3^{(\text{rear wheel})}\right) = 0$$

and

$$\mathbf{r}^{A_{\text{front wheel}} P_{\text{front contact}}} \bullet \left(\mathbf{w}_3 \times \mathbf{a}_3^{(\text{front wheel})}\right) = 0.$$

The corresponding configuration constraints in terms of the configuration coordinates are quite lengthy and algebraically complicated. Below, we show the sequence of MAMBO toolbox commands that can be used to compute them explicitly.

### 7.3.8 MAMBO

The following MAMBO toolbox statements invoke the `DeclareObservers`, `DeclarePoints`, `DeclareTriads`, `DefineObservers`, `DefinePoints`, `DefineTriads`, and `DefineNeighbors` procedures to introduce all the relevant information to generate a MAMBO geometry description for the auxiliary observers representing the bicycle.

```
> Restart():
> DeclareObservers(W,Arear,Afront,Aframe,Asteer):
> DeclarePoints(W,Arear,Afront,Aframe,Asteer):
> DeclareTriads(w,arear,afront,aframe,asteer):
> DefineObservers([W,W,w],[Arear,Arear,arear],
> [Afront,Afront,afront],[Aframe,Aframe,aframe],
> [Asteer,Asteer,asteer]):
> DefinePoints([W,Aframe,w,q1,q2,q3]):
> DefineTriads([w,aframe,[q4,3],[q5,1],[q6,3]]):
> DefinePoints([Aframe,Arear,NullVector()]):
> DefineTriads([aframe,arear,[q7,3]]):
> v:=MakeTranslations(aframe,p1,p2,0):
> aframe1:=MakeTranslations(aframe,1):
> aframe2:=MakeTranslations(aframe,2):
> aframe3:=MakeTranslations(aframe,3):
> b1:=(1/VectorLength(v)) &** v:
> b2:=(1/VectorLength(aframe3 &xx b1))
> &** (aframe3 &xx b1):
> b3:=b1 &xx b2:
> DefineTriads([aframe,asteer,
> [matrix(3,3,(i,j)->cat(aframe,i) &oo cat(b,j))],
> [q8,1]]):
> DefinePoints([Aframe,Asteer,[aframe,p3,p4,0],
> [asteer,p5,p6,0]]):
> DefinePoints([Asteer,Afront,NullVector()]):
> DefineTriads([asteer,afront,q9,3]):
> DefineNeighbors([W,Aframe],[Aframe,Arear],
> [Aframe,Asteer],[Asteer,Afront]):
> GeometryOutput(main=W);
```

```
MODULE W {
   BODY Aframe {
      POINT {q1,q2,q3}
      ORIENT
{cos(q4)*cos(q6)-sin(q4)*cos(q5)*sin(q6),-cos(q4)*sin(q6)-si
n(q4)*cos(q5)*cos(q6),sin(q4)*sin(q5),sin(q4)*cos(q6)+cos(q4
)*cos(q5)*sin(q6),-sin(q4)*sin(q6)+cos(q4)*cos(q5)*cos(q6),-
cos(q4)*sin(q5),sin(q5)*sin(q6),sin(q5)*cos(q6),cos(q5)}
      BODY Asteer {
         POINT
{p3+p5/(p1^2+p2^2)^(1/2)*p1-p6/(1/(p1^2+p2^2)*p2^2+1/(p1^2+p
2^2)*p1^2)^(1/2)/(p1^2+p2^2)^(1/2)*p2*cos(q8),p4+p5/(p1^2+p2
^2)^(1/2)*p2+p6/(1/(p1^2+p2^2)*p2^2+1/(p1^2+p2^2)*p1^2)^(1/2
)/(p1^2+p2^2)^(1/2)*p1*cos(q8),p6*(1/(p1^2+p2^2)*p1^2/(1/(p1
^2+p2^2)*p2^2+1/(p1^2+p2^2)*p1^2)^(1/2)+1/(p1^2+p2^2)*p2^2/(
1/(p1^2+p2^2)*p2^2+1/(p1^2+p2^2)*p1^2)^(1/2))*sin(q8)}
         ORIENT
{1/(p1^2+p2^2)^(1/2)*p1,-1/(1/(p1^2+p2^2)*p2^2+1/(p1^2+p2^2)
*p1^2)^(1/2)/(p1^2+p2^2)^(1/2)*p2*cos(q8),1/(1/(p1^2+p2^2)*p
2^2+1/(p1^2+p2^2)*p1^2)^(1/2)/(p1^2+p2^2)^(1/2)*p2*sin(q8),1
/(p1^2+p2^2)^(1/2)*p2,1/(1/(p1^2+p2^2)*p2^2+1/(p1^2+p2^2)*p1
^2)^(1/2)/(p1^2+p2^2)^(1/2)*p1*cos(q8),-1/(1/(p1^2+p2^2)*p2^
2+1/(p1^2+p2^2)*p1^2)^(1/2)/(p1^2+p2^2)^(1/2)*p1*sin(q8),0,(
1/(p1^2+p2^2)*p1^2/(1/(p1^2+p2^2)*p2^2+1/(p1^2+p2^2)*p1^2)^(
1/2)+1/(p1^2+p2^2)*p2^2/(1/(p1^2+p2^2)*p2^2+1/(p1^2+p2^2)*p1
^2)^(1/2))*sin(q8),(1/(p1^2+p2^2)*p1^2/(1/(p1^2+p2^2)*p2^2+1
```

```
/(p1^2+p2^2)*p1^2)^(1/2)+1/(p1^2+p2^2)*p2^2/(1/(p1^2+p2^2)*p
2^2+1/(p1^2+p2^2)*p1^2)^(1/2))*cos(q8)}
            BODY Afront {
               POINT {0,0,0}
               ORIENT {cos(q9),-sin(q9),0,
                              sin(q9),cos(q9),0,0,0,1}
            }
        }
        BODY Arear {
            POINT {0,0,0}
            ORIENT {cos(q7),-sin(q7),0,
                           sin(q7),cos(q7),0,0,0,1}
        }
    }
}
```

This geometry description may now be exported into a MAMBO .geo file. MAMBO objects can be added at any level through the use of **BLOCK**, **CYLINDER**, and **SPHERE** statements. Alternatively, the `DefineObjects` procedure can be invoked to associate the objects with their parent observers within the MAMBO toolbox session prior to exporting the geometry hierarchy with `GeometryOutput`.

The configuration constraints formulated above are implemented in the MAMBO toolbox using the following statements:

```
> DeclarePoints(Prear,Pfront):
> DefinePoints(
> [Arear,Prear,arear,R*cos(q10),R*sin(q10),0],
> [Afront,Pfront,afront,R*cos(q11),R*sin(q11),0]):
> f1:=simplify(FindTranslation(Arear,Prear) &oo
> (MakeTranslations(w,3) &xx
> MakeTranslations(arear,3)))=0:
> f2:=FindTranslation(W,Prear) &oo
> MakeTranslations(w,3)=0:
> f3:=simplify(FindTranslation(Afront,Pfront) &oo
> (MakeTranslations(w,3) &xx
> MakeTranslations(afront,3)))=0:
> f4:=FindTranslation(W,Pfront) &oo
> MakeTranslations(w,3)=0:
```

By substituting values for the geometric parameters $p_1$, $p_2$, $p_3$, $p_4$, $p_5$, $p_6$, and $R$, it is possible to use these constraints to find allowable values for the configuration coordinates $q_i$, $i = 1, \ldots, 11$ that correspond to a configuration of the bicycle in which the wheels are in tangential contact with the ground. A simple method is to use trial and error within MAMBO to find approximate allowable values for the configuration coordinates and to subsequently invoke a numerical equation-solving algorithm with the approximate values as initial guesses.

Suppose, for example, that we have chosen

$$p_1 = -0.348968837, p_2 = 0.937134317,$$
$$p_3 = 2.593718433R, p_4 = 1.354840027R,$$
$$p_5 = -1.520201636R, p_6 = -0.2R,$$

where the wheel radius $R = 0.5$. As $q_1$ and $q_2$ correspond to the absolute position of the bicycle along the stationary plane, these configuration coordinates are entirely unconstrained. Similarly, as $q_4$ corresponds to the rotation of the frame about an axis perpendicular to the stationary plane, this configuration coordinate is entirely unconstrained. Finally, as $q_7$ and $q_9$ correspond to the amounts of rotation of the rear and front wheels relative to the frame and steering column, respectively, these configuration coordinates are entirely unconstrained. For simplicity, we let

$$q_1 = q_2 = q_4 = q_7 = q_9 = 0.$$

Since there are four constraint equations, the actual number of geometric degrees of freedom of the bicycle is only seven. It follows that we may assign values to two of the remaining configuration coordinates and then use the configuration constraints to solve for the remaining configuration coordinates. Here, we choose to assign values to $q_5$ corresponding to the lateral tilt of the frame and $q_8$ corresponding to the amount of rotation of the steering column:

$$q_5 = 1, q_8 = 0.7.$$

Using MAMBO, we may now manually adjust the remaining configuration coordinates $q_3$, $q_6$, $q_{10}$, and $q_{11}$ until the configuration constraints appear to be approximately satisfied. Indeed, the following choice of values yields a visually satisfactory approximation:

$$q_3 \approx 0.4, q_6 \approx 0, q_{10} \approx -1.5, q_{11} \approx 3.14$$

as is confirmed by back substitution into the constraint equations:

```
> evalf(subs(p1=-.348968837,p2=.937134317,
> p3=2.593718433*R,p4=1.354840027*R,p5=-1.520201636*R,
> p6=-.2*R,R=.5,q1=0,q2=0,q4=0,q5=1,q7=0,q8=.7,q9=0,
> q6=0,q10=-1.5,q11=3.14,q3=.4,{f1,f2,f3,f4}));
```

$$\{.02976165138 = 0, -.0196815444 = 0,$$
$$-.0623671670 = 0, -.03589913905 = 0\}$$

Numerical values for $q_3$, $q_6$, $q_{10}$, and $q_{11}$ that more accurately satisfy the constraint equations may be obtained using the MAPLE function `fsolve`. In particular, in the statements below, the `fsolve` function is invoked with the previously found approximate values as initial guesses.

```
> eq:=subs(p1=-.348968837,p2=.937134317,
> p3=2.593718433*R,p4=1.354840027*R,
> p5=-1.520201636*R,p6=-.2*R,R=.5,q1=0,
> q2=0,q4=0,q5=1,q7=0,q8=.7,q9=0,{f1,f2,f3,f4}):
> fsolve(eq,{q3=.4,q6=0,q10=-1.5,q11=3.14});
```

$$\{q3 = .4207354924,\ q6 = .01280053711,\\ q10 = -1.583596864,\ q11 = 3.288069682\}$$

The procedure outlined above may now be repeated for other choices of values for the configuration coordinates $q_5$ and $q_8$ (with arbitrary choices of values for $q_1$, $q_2$, $q_4$, $q_7$, and $q_9$) to obtain multiple configurations that satisfy the configuration constraints. A sequence of values for the full set of configuration coordinates that would give the appearance of a smooth motion could be generated by making only small changes in these seven configuration coordinates between each frame. For each new frame, the values found in the previous frame for $q_3$, $q_6$, $q_{10}$, and $q_{11}$ could then be used as initial guesses in the call to `fsolve`. Maple `for`-loops could be used to advantage in iterating this procedure.

The procedure described in the previous paragraph is clearly a cumbersome approach to generating an extensive MAMBO dataset (a MAMBO .sds file) for purposes of animation. Moreover, it actually fails to account for the constraints on the *change* of the configuration coordinates that follow from the requirement that the wheels roll on the stationary surface without slipping. Beginning in the next chapter and culminating in Chapter 9, we will develop a more efficient and comprehensive approach that effectively addresses both of these concerns.

## 7.4 A Desk Lamp

As a final example, we return to the desk lamp introduced in Chapter 2 and implement the modeling algorithm as suggested above.

### 7.4.1 Constituent Rigid Bodies

The desk lamp shown on the previous page consists of six distinct rigid bodies, namely, the base, the lower, middle, and upper beams, the bracket, and the lamp shade. Clearly, each of these bodies consists of multiple parts, but these parts are rigidly attached to each other and can thus be considered as single bodies for the purposes of developing a specification of the mechanism's configuration and any inherent constraints.

### 7.4.2 Reference Points and Reference Triads

Let the reference point $A_{\text{base}}$ of the base be located at the top of the base centered between the vertical posts. Choose the reference triad $a^{(\text{base})}$ of the base, such that the vertical posts are parallel to the vector $\mathbf{a}_3^{(\text{base})}$ and are separated in a direction parallel to the vector $\mathbf{a}_2^{(\text{base})}$.

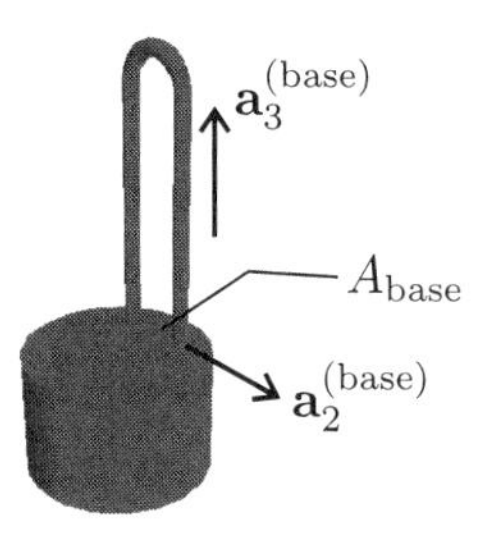


Let the reference point $A_{\text{lower beam}}$ of the lower beam be located at the center of the horizontal bar connecting the beam to the base. Choose the reference triad $a^{(\text{lower beam})}$ of the lower beam, such that the beam is parallel to the vector $\mathbf{a}_3^{(\text{lower beam})}$ and the horizontal bar is parallel to $\mathbf{a}_2^{(\text{lower beam})}$.

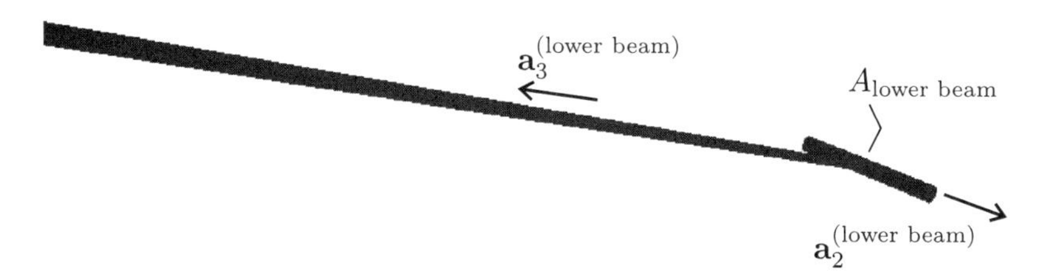


Let the reference point $A_{\text{middle beam}}$ of the middle beam be located at the center of the horizontal bar connecting the beam to the base. Choose the reference triad $a^{(\text{middle beam})}$ of the middle beam, such that the beam is parallel to the vector $\mathbf{a}_3^{(\text{middle beam})}$ and the horizontal bar is parallel to $\mathbf{a}_2^{(\text{middle beam})}$.

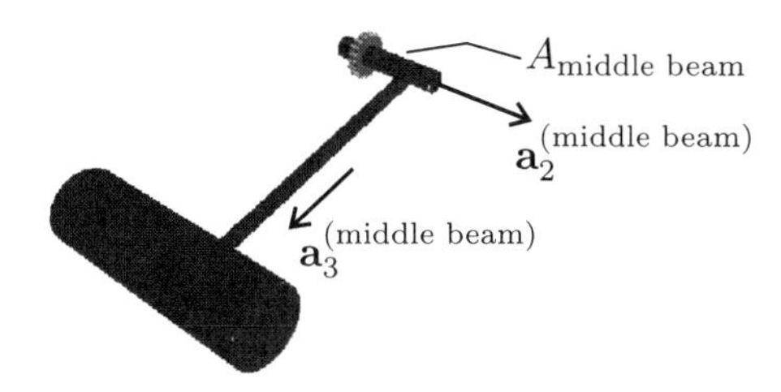


Let the reference point $A_{\text{upper beam}}$ of the upper beam be located at the center of the horizontal bar connecting the beam to the base. Choose the reference triad $a^{(\text{upper beam})}$ of the upper beam, such that the beam is parallel to the vector $\mathbf{a}_3^{(\text{upper beam})}$ and the horizontal bar is parallel to $\mathbf{a}_2^{(\text{upper beam})}$.

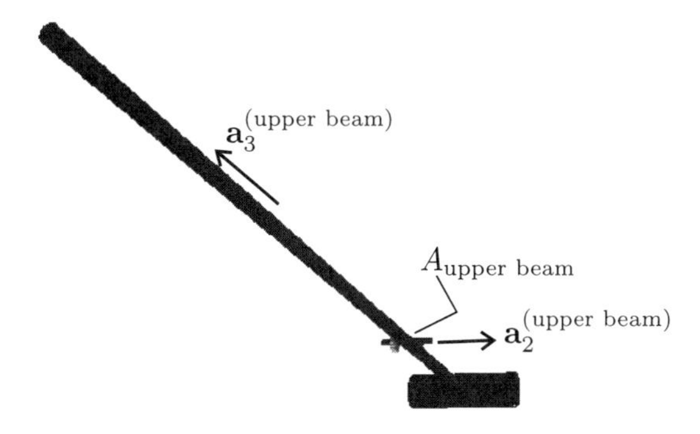


Let the reference point $A_{\text{bracket}}$ of the bracket coincide with the hinge joint connecting the bracket to the lower beam. Choose the reference triad $a^{(\text{bracket})}$ of the bracket, such that the line between the hinge joints connecting the bracket to the lower beam and the lamp shade, respectively, is parallel to $\mathbf{a}_3^{(\text{bracket})}$ and the hinge axes are parallel to $\mathbf{a}_2^{(\text{bracket})}$.

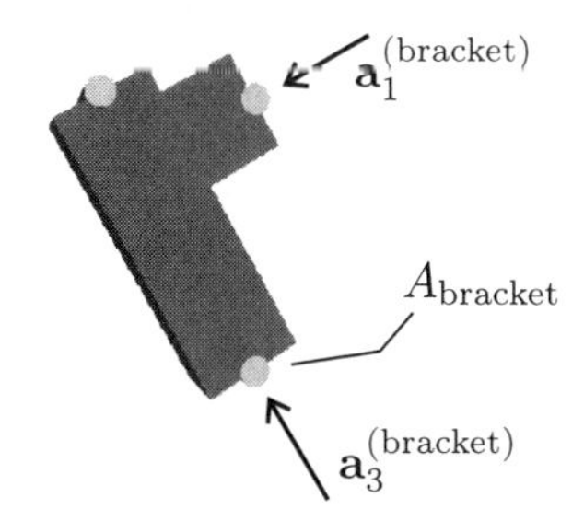

Let the reference point $A_{\text{lamp shade}}$ of the lamp shade coincide with the hinge joint connecting the lamp shade to the bracket. Choose the reference triad $a^{(\text{lamp shade})}$ of the lamp shade, such that the symmetry axis of the lamp shade is parallel to $\mathbf{a}_3^{(\text{lamp shade})}$ and the hinge axis is parallel to $\mathbf{a}_2^{(\text{lamp shade})}$.

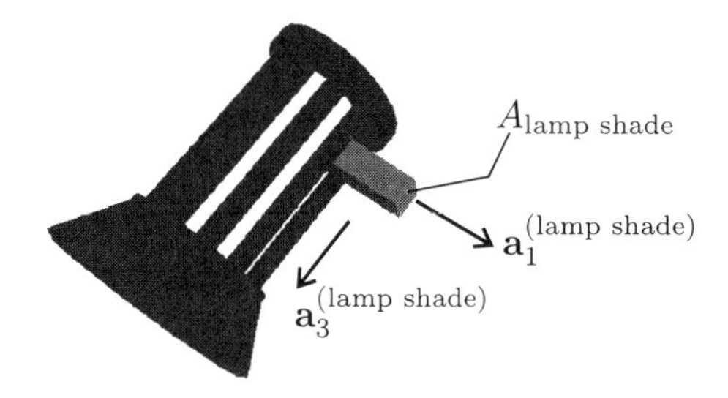


### 7.4.3 Main Observer

As we are primarily interested in arriving at a geometry description that we may implement in the MAMBO application, we introduce a main observer $\mathcal{W}$ with reference point $W$ and reference triad $w$ corresponding to the graphics application's internal reference configuration.

### 7.4.4 Auxiliary Observers

Since each of the six rigid bodies introduced above may change configuration relative to the main observer, introduce six auxiliary observers $\mathcal{A}_{\text{base}}$, $\mathcal{A}_{\text{lower beam}}$, $\mathcal{A}_{\text{middle beam}}$, $\mathcal{A}_{\text{upper beam}}$, $\mathcal{A}_{\text{bracket}}$, and $\mathcal{A}_{\text{lamp shade}}$, relative to which the base, the lower, middle, and upper beams, the bracket, and the lamp shade, respectively, are stationary. For simplicity, let the corresponding reference points and reference triads agree with those selected for the corresponding rigid bodies.

### 7.4.5 Tree Structures

Consider the tree structure on the next page, representing the conceptual arrangement of observers and rigid bodies.

This description will be complete by specifying the pure translations and pure rotations corresponding to each of the direct connections between successive observers or between observers and rigid bodies. In particular, since the reference points and reference triads of the auxiliary observers coincide with those of the corresponding rigid bodies, the pure translations and pure rotations between the observers and the rigid bodies are the identity translation and identity rotation. Naturally, each rigid body consists of multiple parts whose configurations relative to the auxiliary observers are described by non-trivial pure translations and pure rotations. These are time-independent, however, and their description is implemented as in the discussion on still lives in Chapters 2, 4, and 6.

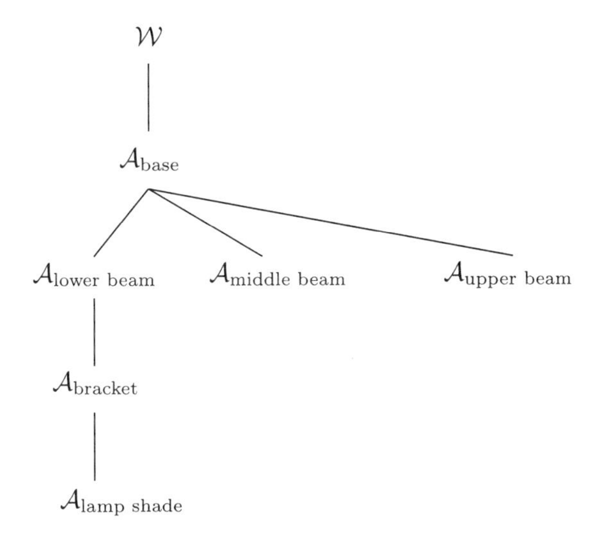


### 7.4.6 Configuration Coordinates

The configuration of the observer $\mathcal{A}_{\text{base}}$ relative to the main observer $\mathcal{W}$ is described by a pure translation $\mathbf{T}_{\mathcal{W}\to\mathcal{A}_{\text{base}}}$ corresponding to the position vector

$$\mathbf{r}^{WA_{\text{base}}} = w \begin{pmatrix} q_1 \\ q_2 \\ q_3 \end{pmatrix}$$

and a pure rotation $\mathbf{R}_{\mathcal{W}\to\mathcal{A}_{\text{base}}}$ corresponding to the rotation matrix

$$\begin{aligned} R_{wa^{(\text{base})}} &= R(q_4, 0, 0, 1)\, R(q_5, 1, 0, 0)\, R(q_6, 0, 0, 1) \\ &= \begin{pmatrix} c_4c_6 - s_4c_5s_6 & -c_4s_6 - s_4c_5c_6 & s_4s_5 \\ s_4c_6 + c_4c_5s_6 & -s_4s_6 + c_4c_5c_6 & -c_4s_5 \\ s_5s_6 & s_5c_6 & c_5 \end{pmatrix}, \end{aligned}$$

where $c_i = \cos q_i$ and $s = \sin q_i$, $i = 1, 2, 3$. The configuration coordinates $q_4$, $q_5$, and $q_6$ are Euler angles corresponding to a $3-1-3$ decomposition of the pure rotation $\mathbf{R}_{\mathcal{W}\to\mathcal{A}_{\text{base}}}$.

The configuration of the observer $\mathcal{A}_{\text{lower beam}}$ relative to the observer $\mathcal{A}_{\text{base}}$ is described by a pure translation $\mathbf{T}_{\mathcal{A}_{\text{base}}\to\mathcal{A}_{\text{lower beam}}}$ corresponding to the position vector

$$\mathbf{r}^{A_{\text{base}}A_{\text{lower beam}}} = a^{(\text{base})} \begin{pmatrix} 0 \\ 0 \\ p_1 \end{pmatrix}$$

and a pure rotation $\mathbf{R}_{\mathcal{A}_{\text{base}}\to\mathcal{A}_{\text{lower beam}}}$ corresponding to the rotation matrix

$$\begin{aligned} R_{a^{(\text{base})}a^{(\text{lower beam})}} &= R(q_7, 0, 1, 0) \\ &= \begin{pmatrix} \cos q_7 & 0 & \sin q_7 \\ 0 & 1 & 0 \\ -\sin q_7 & 0 & \cos q_7 \end{pmatrix}. \end{aligned}$$

The configuration of the observer $\mathcal{A}_{\text{middle beam}}$ relative to the observer $\mathcal{A}_{\text{base}}$ is described by a pure translation $\mathbf{T}_{\mathcal{A}_{\text{base}}\to\mathcal{A}_{\text{middle beam}}}$ corresponding to the position vector

$$\mathbf{r}^{A_{\text{base}}A_{\text{middle beam}}} = a^{(\text{base})}\begin{pmatrix} 0 \\ 0 \\ p_2 \end{pmatrix}$$

and a pure rotation $\mathbf{R}_{\mathcal{A}_{\text{base}}\to\mathcal{A}_{\text{middle beam}}}$ corresponding to the rotation matrix

$$\begin{aligned} R_{a^{(\text{base})}a^{(\text{middle beam})}} &= R(q_8, 0, 1, 0) \\ &= \begin{pmatrix} \cos q_8 & 0 & \sin q_8 \\ 0 & 1 & 0 \\ -\sin q_8 & 0 & \cos q_8 \end{pmatrix}. \end{aligned}$$

The configuration of the observer $\mathcal{A}_{\text{upper beam}}$ relative to the observer $\mathcal{A}_{\text{base}}$ is described by a pure translation $\mathbf{T}_{\mathcal{A}_{\text{base}}\to\mathcal{A}_{\text{upper beam}}}$ corresponding to the position vector

$$\mathbf{r}^{A_{\text{base}}A_{\text{upper beam}}} = a^{(\text{base})}\begin{pmatrix} 0 \\ 0 \\ p_3 \end{pmatrix}$$

and a pure rotation $\mathbf{R}_{\mathcal{A}_{\text{base}}\to\mathcal{A}_{\text{upper beam}}}$ corresponding to the rotation matrix

$$\begin{aligned} R_{a^{(\text{base})}a^{(\text{upper beam})}} &= R(q_9, 0, 1, 0) \\ &= \begin{pmatrix} \cos q_9 & 0 & \sin q_9 \\ 0 & 1 & 0 \\ -\sin q_9 & 0 & \cos q_9 \end{pmatrix}. \end{aligned}$$

The configuration of the observer $\mathcal{A}_{\text{bracket}}$ relative to the observer $\mathcal{A}_{\text{lower beam}}$ is described by a pure translation $\mathbf{T}_{\mathcal{A}_{\text{lower beam}}\to\mathcal{A}_{\text{bracket}}}$ corresponding to the position vector

$$\mathbf{r}^{A_{\text{lower beam}}A_{\text{bracket}}} = a^{(\text{lower beam})}\begin{pmatrix} 0 \\ 0 \\ p_4 \end{pmatrix}$$

and a pure rotation $\mathbf{R}_{\mathcal{A}_{\text{lower beam}} \to \mathcal{A}_{\text{bracket}}}$ corresponding to the rotation matrix

$$\begin{aligned} R_{a^{(\text{lower beam})} a^{(\text{bracket})}} &= R(q_{10}, 0, 1, 0) \\ &= \begin{pmatrix} \cos q_{10} & 0 & \sin q_{10} \\ 0 & 1 & 0 \\ -\sin q_{10} & 0 & \cos q_{10} \end{pmatrix}. \end{aligned}$$

Finally, the configuration of the observer $\mathcal{A}_{\text{lamp shade}}$ relative to the observer $\mathcal{A}_{\text{bracket}}$ is described by a pure translation $\mathbf{T}_{\mathcal{A}_{\text{bracket}} \to \mathcal{A}_{\text{lamp shade}}}$ corresponding to the position vector

$$\mathbf{r}^{A_{\text{bracket}} A_{\text{lamp shade}}} = a^{(\text{bracket})} \begin{pmatrix} 0 \\ 0 \\ p_5 \end{pmatrix}$$

and a pure rotation $\mathbf{R}_{\mathcal{A}_{\text{bracket}} \to \mathcal{A}_{\text{lamp shade}}}$ corresponding to the rotation matrix

$$\begin{aligned} R_{a^{(\text{bracket})} a^{(\text{lamp shade})}} &= R(q_{11}, 0, 1, 0) \\ &= \begin{pmatrix} \cos q_{11} & 0 & \sin q_{11} \\ 0 & 1 & 0 \\ -\sin q_{11} & 0 & \cos q_{11} \end{pmatrix}. \end{aligned}$$

### 7.4.7 Configuration Constraints

It remains for us to consider any constraints on the configuration of the desk lamp that would reduce the number of geometric degrees of freedom from the eleven found in the previous section. In particular, we shall consider constraining the rotations of the bracket and upper beam to respect the constraint imposed by the corresponding hinge joint.

Denote by $H_{\text{bracket}}$ and $H_{\text{upper beam}}$ the points on the bracket and upper beam, respectively, that coincide with the corresponding hinge joint. It follows that

$$\mathbf{r}^{A_{\text{upper beam}} H_{\text{upper beam}}} = a^{(\text{upper beam})} \begin{pmatrix} 0 \\ 0 \\ p_6 \end{pmatrix}$$

and

$$\mathbf{r}^{A_{\text{bracket}} H_{\text{bracket}}} = a^{(\text{bracket})} \begin{pmatrix} -p_7 \\ 0 \\ p_8 \end{pmatrix}.$$

That the points $H_{\text{bracket}}$ and $H_{\text{upper beam}}$ coincide with the corresponding hinge joint implies that the position vector representing their separation must be zero, i.e.,

$$\mathbf{r}^{H_{\text{bracket}} H_{\text{upper beam}}} = \mathbf{0}.$$

Since, by construction, the points $H_{\text{bracket}}$ and $H_{\text{upper beam}}$ automatically lie in a plane spanned by $\mathbf{a}_1^{(\text{base})}$ and $\mathbf{a}_3^{(\text{base})}$, this condition is equivalent to the following configuration constraints:

$$\mathbf{r}^{H_{\text{bracket}}H_{\text{upper beam}}} \bullet \mathbf{a}_1^{(\text{base})} = 0,$$
$$\mathbf{r}^{H_{\text{bracket}}H_{\text{upper beam}}} \bullet \mathbf{a}_3^{(\text{base})} = 0.$$

The corresponding configuration constraints in terms of the configuration coordinates are quite lengthy and algebraically complicated. Below, we show the sequence of MAMBO toolbox commands that can be used to compute them explicitly.

### 7.4.8 MAMBO

The following MAMBO toolbox statements invoke the `DeclareObservers`, `DeclarePoints`, `DeclareTriads`, `DefineObservers`, `DefinePoints`, `DefineTriads`, and `DefineNeighbors` procedures to introduce all the relevant information to generate a MAMBO geometry description for the auxiliary observers representing the desk lamp.

```
> Restart():

> DeclareObservers(W,Base,Lamp,Bracket,UpperBeam,
> LowerBeam,MiddleBeam):
> DeclarePoints(W,Base,Lamp,Bracket,UpperBeam,
> LowerBeam,MiddleBeam):
> DeclareTriads(w,base,lamp,bracket,upperbeam,
> lowerbeam,middlebeam):
> DefineNeighbors([W,Base],[Base,LowerBeam],
> [LowerBeam,Bracket],[Bracket,Lamp],
> [Base,MiddleBeam],[Base,UpperBeam]):

> DefineObservers([W,W,w],[Base,Base,base],
> [LowerBeam,LowerBeam,lowerbeam],
> [MiddleBeam,MiddleBeam,middlebeam],
> [UpperBeam,UpperBeam,upperbeam],
> [Bracket,Bracket,bracket],[Lamp,Lamp,lamp]):

> DefinePoints([W,Base,w,q1,q2,q3],
> [Base,LowerBeam,base,0,0,p1],
> [Base,MiddleBeam,base,0,0,p2],
> [Base,UpperBeam,base,0,0,p3],
> [LowerBeam,Bracket,lowerbeam,0,0,p4],
> [Bracket,Lamp,bracket,0,0,p5]):

> DefineTriads([w,base,[q4,3],[q5,1],[q6,3]],
> [base,lowerbeam,q7,2],[lowerbeam,bracket,q10,2],
> [bracket,lamp,q11,2],[base,middlebeam,q8,2],
> [base,upperbeam,q9,2]):

> GeometryOutput(main=W);
```

```
MODULE W {
   BODY Base {
      Point {q1,q2,q3}
      Orient
{cos(q4)*cos(q6)-sin(q4)*cos(q5)*sin(q6),-cos(q4)*sin(q6)-si
n(q4)*cos(q5)*cos(q6),sin(q4)*sin(q5),sin(q4)*cos(q6)+cos(q4
)*cos(q5)*sin(q6),-sin(q4)*sin(q6)+cos(q4)*cos(q5)*cos(q6),-
cos(q4)*sin(q5),sin(q5)*sin(q6),sin(q5)*cos(q6),cos(q5)}
      BODY MiddleBeam {
         Point {0,0,p2}
         Orient {cos(q8),0,sin(q8),0,1,0,-sin(q8),0,cos(q8)}
      }
      BODY UpperBeam {
         Point {0,0,p3}
         Orient {cos(q9),0,sin(q9),0,1,0,-sin(q9),0,cos(q9)}
      }
      BODY LowerBeam {
         Point {0,0,p1}
         Orient {cos(q7),0,sin(q7),0,1,0,-sin(q7),0,cos(q7)}
         BODY Bracket {
            Point {0,0,p4}
            Orient {cos(q10),0,sin(q10),0,1,0,
                            -sin(q10),0,cos(q10)}
            BODY Lamp {
               Point {0,0,p5}
               Orient {cos(q11),0,sin(q11),0,1,0,
                               -sin(q11),0,cos(q11)}
            }
         }
      }
   }
}
```

This geometry description may now be exported into a MAMBO .geo file. MAMBO objects can be added at any level through the use of **BLOCK**, **CYLINDER**, and **SPHERE** statements. Alternatively, the `DefineObjects` procedure can be invoked to associate the objects with their parent observers within the MAMBO toolbox session prior to exporting the geometry hierarchy with `GeometryOutput`.

The configuration constraints formulated above are implemented in the MAMBO toolbox using the following statements:

```
> DeclarePoints(HUpperBeam,HBracket):
> DefinePoints(
> [UpperBeam,HUpperBeam,upperbeam,0,0,p6],
> [Bracket,HBracket,bracket,-p7,0,p8]):
> f1:=FindTranslation(HBracket,HUpperBeam) &oo
> MakeTranslations(base,1)=0:
> f2:=FindTranslation(HBracket,HUpperBeam) &oo
> MakeTranslations(base,3)=0:
```

By substituting values for the geometric parameters $p_1$, $p_2$, $p_3$, $p_4$, $p_5$, $p_6$, $p_7$, and $p_8$, it is possible to use these constraints to find allowable values for the configuration coordinates $q_i$, $i = 1, \ldots, 11$ that correspond to a configuration of the desk lamp in which the upper beam connects to the bracket at the appropriate hinge joint. As in the case of the bicycle, we may use trial and error within MAMBO to find approximate allowable values for the configuration coordinates and subsequently invoke a numerical equation-solving algorithm with the approximate values as initial guesses.

Suppose, for example, that we have chosen

$$p_1 = 11, p_2 = 12.5, p_3 = 14, p_4 = 45,$$
$$p_5 = 3.5, p_6 = 45, p_7 = 1.5, p_8 = 2.6.$$

As $q_1$, $q_2$, $q_3$, $q_4$, $q_5$, and $q_6$ correspond to the position and orientation of the base relative to the main observer $\mathcal{W}$, these configuration coordinates are entirely unconstrained. For simplicity, we let

$$q_1 = q_2 = q_3 = q_4 = q_5 = q_6 = 0.$$

Similarly, as $q_8$ corresponds to the rotation of the middle beam relative to the base, this configuration coordinate is entirely unconstrained. However, as the middle beam provides a counterbalance to the upper beam, we give it an initial orientation given by

$$q_8 = 1.8.$$

Finally, as $q_{11}$ corresponds to the rotation of the lamp shade relative to the bracket, this configuration coordinate is entirely unconstrained. Here, we let

$$q_{11} = 2.$$

Since there are two constraint equations, the actual number of geometric degrees of freedom of the bicycle is only nine. It follows that we may assign values to one of the remaining configuration coordinates and then use the configuration constraints to solve for the remaining configuration coordinates. Here, we choose to assign values to $q_9$ corresponding to the orientation of the upper beam relative to the base:

$$q_9 = 1.2.$$

Using MAMBO, we may now manually adjust the remaining configuration coordinates $q_7$ and $q_{10}$ until the configuration constraints appear to be approximately satisfied. Indeed, the following choice of values yields a visually satisfactory approximation:

$$q_7 \approx 1.2, q_{10} \approx -0.67$$

as is confirmed by back substitution into the constraint equations:

```
> evalf(subs(q1=0,q2=0,q3=0,q4=0,q5=0,q6=0,q7=1.2,
> q8=1.8,q9=1.2,q10=-.67,q11=2,p1=11,p2=12.5,p3=14,
> p4=45,p5=3.5,p6=45,p7=1.5,p8=2.6,{f1,f2}));
```

$$\{-.020176081 = 0,\ -.001598395 = 0\}$$

Numerical values for $q_7$ and $q_{10}$ that more accurately satisfy the constraint equations may be obtained using the MAPLE function `fsolve`. In particular, in the statement below, the `fsolve` function is invoked with the previously found approximate values as initial guesses.

```
> fsolve(subs(q1=0,q2=0,q3=0,q4=0,q5=0,q6=0,q8=1.8,
> q9=1.2,q11=2,p1=11,p2=12.5,p3=14,p4=45,p5=3.5,p6=45,
> p7=1.5,p8=2.6,{f1,f2}),{q7=1.2,q10=-.67});
```

$$\{q7 = 1.200039725,\ q10 = -.6769771893\}$$

The procedure outlined above may now be repeated for other choices of values for the configuration coordinate $q_9$ (with arbitrary choices of values for $q_1$, $q_2$, $q_3$, $q_4$, $q_5$, $q_6$, $q_8$, and $q_{11}$) to obtain multiple configurations that satisfy the configuration constraints. As in the case of the bicycle, a sequence of values for the full set of configuration coordinates that would give the appearance of a smooth motion could be generated by making only small changes in these nine configuration coordinates between each frame. For each new frame, the values found in the previous frame for $q_7$ and $q_{10}$ could then be used as initial guesses in the call to `fsolve`. Maple `for`-loops could be used to advantage in iterating this procedure.

The procedure described in the previous paragraph is clearly a cumbersome approach to generating an extensive MAMBO dataset (a MAMBO .sds file) for purposes of animation. As was the case with the bicycle, it actually fails to account for the constraints on the *change* of the configuration coordinates that follow from the requirement that the teeth of the spur gears on the middle and upper beams not be allowed to pass through one another. Beginning in the next chapter and culminating in Chapter 9, we will develop a more efficient and comprehensive approach that effectively addresses both of these concerns.

# Chapter 8

# Velocities

*wherein the reader learns of:*

- *Describing the instantaneous state of motion of a rigid body or observer relative to an observer;*
- *Differentiating vectors with respect to time relative to different observers;*
- *Reconstructing the configuration as a function of time from knowledge of the linear and angular velocities.*

## Practicum

Visualize the notions of linear and angular velocity as introduced in this chapter by including cylinders representing the corresponding vectors in a MAMBO geometry description. Note, in particular, the association between the linear and angular velocity vectors and the instantaneous directions of translation and rotation of any rigid object relative to some observer.

Consider using several cylinders to represent the linear and angular velocities of a rigid body relative to different observers. Use these to build your intuition and to confirm your ability to accurately compute time derivatives.

# 8.1 Motion

(Ex. 8.1 – Ex. 8.4)

The configuration of a rigid body relative to an observer is given by a pure translation **T** and a pure rotation **R**. If the pure translation and/or the pure rotation change with time relative to the observer, i.e., if the configuration of the rigid body relative to the observer is time-dependent, then the rigid body is *moving* relative to the observer. If the rigid body is not moving relative to the observer, we say that it is *stationary* relative to the observer.

**Illustration 8.1**

In previous chapters, we found that the pure translation and pure rotation that describe the configuration of a rigid body may be different for different observers. Since the only perception an observer has of a rigid body is in terms of its configuration, we were hard put to suggest an *absolute* notion of configuration. All statements about the configuration of a rigid body were made relative to some observer.

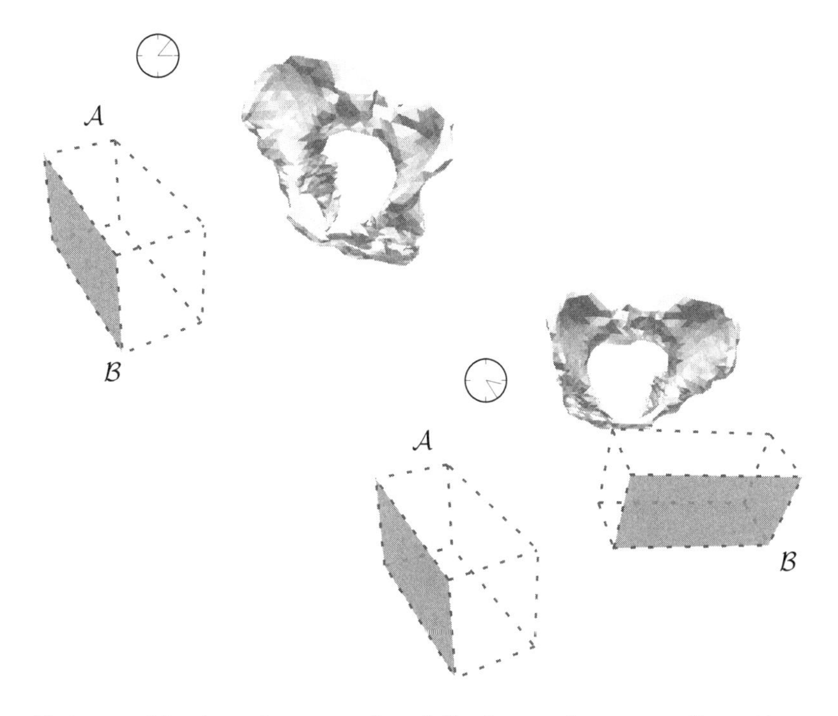


Now, consider two observers $\mathcal{A}$ and $\mathcal{B}$ whose reference configurations happen to coincide at some moment in time. At that very moment, the two observers agree with each other about the position and orientation of any arbitrary rigid body in their environment. However, unless the reference configurations continue to coincide, the two observers may disagree with each other about the time-dependence of the position and orientation of any arbitrary rigid body. For a rigid body that is stationary relative to $\mathcal{B}$, any motion of $\mathcal{B}$ relative to $\mathcal{A}$ implies that the rigid body is moving relative $\mathcal{A}$ and so on.

You perceive motion by looking for changes with time in the position and orientation of a rigid body or of an observer relative to some other observer. By considering changes over smaller and smaller intervals of time $[t_0, t_0 + h]$ based at some specific moment $t = t_0$, you can quantify the instantaneous "tendency" of motion. This idea was mentioned already in Chapter 1, with the introduction of the notions of instantaneous directions of translation and rotation and the associated linear and angular speeds.

Specifically, let $\mathbf{T}_t$ and $\mathbf{R}_t$ describe the position and orientation of a rigid body relative to some observer $\mathcal{O}$ at time $t$. Then, the change in position relative to $\mathcal{O}$ over the time interval $[t_0, t_0 + h]$ is given by the pure translation

$$\mathbf{T}_{t_0 \to t_0+h} = \mathbf{T}_{t_0+h} \circ (\mathbf{T}_{t_0})^{-1}.$$

Similarly, the change in orientation relative to $\mathcal{O}$ over the time interval $[t_0, t_0 + h]$ is given by the pure rotation

$$\mathbf{R}_{t_0 \to t_0+h} = \mathbf{R}_{t_0+h} \circ (\mathbf{R}_{t_0})^{-1}.$$

As $h \to 0$, we expect that $\mathbf{T}_{t_0 \to t_0+h}$ and $\mathbf{R}_{t_0 \to t_0+h}$ will both approach the identity operation $\mathbf{I}$. After all, there is no change in position or orientation for $h = 0$.

For each $h > 0$, the pure translation $\mathbf{T}_{t_0 \to t_0+h}$ corresponds to a shift of all points on the rigid body an equal distance in a common direction relative to $\mathcal{O}$. How do the shifting distance and shifting direction relative to $\mathcal{O}$ vary as $h \to 0$?

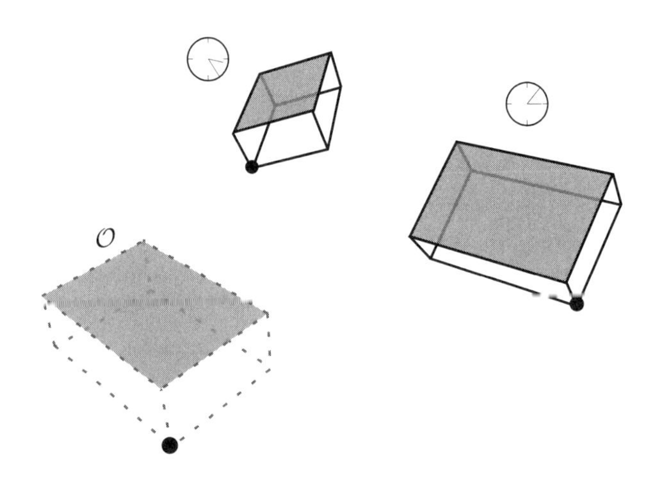

For each $h > 0$, the pure rotation $\mathbf{R}_{t_0 \to t_0+h}$ corresponds to a rotation of the rigid body by some angle about a fixed axis relative to $\mathcal{O}$. How do the turning angle and turning direction relative to $\mathcal{O}$ vary as $h \to 0$?

Both the shifting distance $d$ and the turning angle $\varphi$ relative to $\mathcal{O}$ can be thought of as functions of $h$. Since these functions approach the unique limit 0 as $h \to 0$, we expect that

$$d \approx hv\,(t_0) \text{ and } \varphi \approx h\omega\,(t_0)$$

for sufficiently small $h$, where the quantities $v\,(t_0)$ and $\omega\,(t_0)$ are called the *linear and angular speeds*, respectively, at $t = t_0$ of the rigid body relative to the observer $\mathcal{O}$.

The directions of translation and rotation relative to $\mathcal{O}$ can also be thought of as functions of $h$. These directions are not defined for $h = 0$, since a shift by a zero distance or a rotation by a zero angle yields the same result for any direction. Nevertheless, the directions typically have well-defined limits as $h \to 0$. Specifically, the limiting direction relative to $\mathcal{O}$ of the pure translation $\mathbf{T}_{t_0 \to t_0+h}$ as $h \to 0$ is called the *instantaneous direction of translation* at $t = t_0$ of the rigid body relative to the observer $\mathcal{O}$. Similarly, the limiting direction of the pure rotation $\mathbf{R}_{t_0 \to t_0+h}$ as $h \to 0$ is called the *instantaneous direction of rotation* at $t = t_0$ of the rigid body relative to the observer $\mathcal{O}$.

In the next two sections, we will derive quantitative expressions for computing the linear and angular speeds and the corresponding instantaneous directions of motion.

### 8.1.1 Linear Velocity

Let $\mathcal{A}$ and $\mathcal{B}$ be two observers with reference points $A$ and $B$ and reference triads $a$ and $b$, respectively. The position of $\mathcal{B}$ relative to $\mathcal{A}$ is given by the pure translation $\mathbf{T}_{\mathcal{A} \to \mathcal{B}}$ corresponding to the position vector $\mathbf{r}^{AB}$. If the position of $\mathcal{B}$ relative to $\mathcal{A}$ is changing with time, it follows that the matrix representation

$${}^{a}\left(\mathbf{r}^{AB}\right)$$

is time-dependent. If the vector

$$a\;{}^{a}\left(\mathbf{r}^{AB}\right)(t_0)$$

describes the position of the observer $\mathcal{B}$ relative to $\mathcal{A}$ at time $t = t_0$ and the vector

$$a\;{}^{a}\left(\mathbf{r}^{AB}\right)(t_0 + h)$$

describes the position of the observer $\mathcal{B}$ relative to $\mathcal{A}$ at time $t = t_0 + h$, then the vector

$$a\;{}^{a}\left(\mathbf{r}^{AB}\right)(t_0 + h) - a\;{}^{a}\left(\mathbf{r}^{AB}\right)(t_0)$$

describes the *change* in position of the observer $\mathcal{B}$ relative to $\mathcal{A}$ over the interval $[t_0, t_0 + h]$.

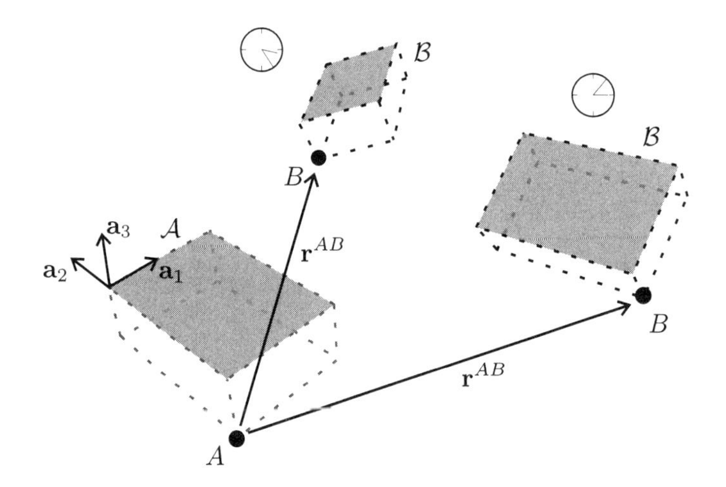


For sufficiently small $h$, we can use Taylor's theorem and expand the matrix representation ${}^a\left(\mathbf{r}^{AB}\right)(t_0 + h)$ in powers of $h$:

$$ {}^a\left(\mathbf{r}^{AB}\right)(t_0 + h) \approx {}^a\left(\mathbf{r}^{AB}\right)(t_0) + h \left. \frac{d\,{}^a\left(\mathbf{r}^{AB}\right)}{dt}(t) \right|_{t=t_0} . $$

It follows that

$$ a\,{}^a\left(\mathbf{r}^{AB}\right)(t_0 + h) - a\,{}^a\left(\mathbf{r}^{AB}\right)(t_0) \approx ha \left[ \left. \frac{d\,{}^a\left(\mathbf{r}^{AB}\right)}{dt}(t) \right|_{t=t_0} \right] . $$

For sufficiently small $h$, the change in position is given by the product of $h$ and the $h$-independent vector

$$ a \left[ \left. \frac{d\,{}^a\left(\mathbf{r}^{AB}\right)}{dt}(t) \right|_{t=t_0} \right] $$

called the *linear velocity at time* $t = t_0$ *of the observer* $\mathcal{B}$ *relative to* $\mathcal{A}$ and denoted by ${}^{\mathcal{A}}\mathbf{v}^{\mathcal{B}}(t_0)$.

**Illustration 8.2**

By definition, the linear velocity of the observer $\mathcal{B}$ relative to $\mathcal{A}$ is a vector. It does not correspond to a pure translation, however. In fact, the quantity

$$ \left\| {}^{\mathcal{A}}\mathbf{v}^{\mathcal{B}} \right\| \stackrel{def}{=} \sqrt{ {}^{\mathcal{A}}\mathbf{v}^{\mathcal{B}} \bullet {}^{\mathcal{A}}\mathbf{v}^{\mathcal{B}} } $$

equals the *linear speed of the observer* $\mathcal{B}$ *relative to* $\mathcal{A}$.

The direction of the linear velocity vector is the limiting direction relative to $\mathcal{A}$ of the pure translation $\mathbf{T}_{t \to t+h}$ from the configuration at time $t$ to the configuration at time $t+h$ as $h \to 0$.

The linear velocity vector thus contains information about the instantaneous direction and instantaneous speed of the translational motion of observer $\mathcal{B}$ relative to $\mathcal{A}$.

### 8.1.2 Angular Velocity

Let $\mathcal{A}$ and $\mathcal{B}$ be two observers with reference points $A$ and $B$ and reference triads $a$ and $b$, respectively. The orientation of $\mathcal{B}$ relative to $\mathcal{A}$ is given by the pure rotation $\mathbf{R}_{\mathcal{A} \to \mathcal{B}}$ corresponding to the rotation matrix $R_{ab}$. If the orientation of $\mathcal{B}$ relative to $\mathcal{A}$ is changing with time, it follows that the rotation matrix $R_{ab}$ is time-dependent. If

$$R_{ab}(t_0)$$

describes the orientation of the observer $\mathcal{B}$ relative to $\mathcal{A}$ at time $t = t_0$ and

$$R_{ab}(t_0 + h)$$

describes the orientation of the observer $\mathcal{B}$ relative to $\mathcal{A}$ at time $t = t_0 + h$, then the matrix product

$$(R_{ab}(t_0))^{-1} R_{ab}(t_0 + h) = R_{ba}(t_0) R_{ab}(t_0 + h)$$

describes the *change* in orientation of the observer $\mathcal{B}$ relative to $\mathcal{A}$ over the interval $[t_0, t_0 + h]$.

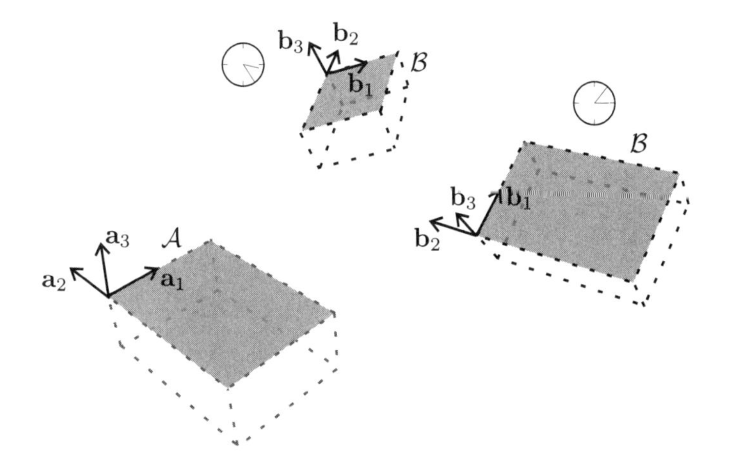

For sufficiently small $h$, we can use Taylor's theorem and expand the rotation matrix $R_{ab}(t_0+h)$ in powers of $h$:

$$R_{ab}(t_0+h) \approx R_{ab}(t_0) + h\left.\frac{dR_{ab}}{dt}(t)\right|_{t=t_0}.$$

It follows that

$$R_{ba}(t_0)\, R_{ab}(t_0+h) \approx Id + hR_{ba}(t_0)\left.\frac{dR_{ab}}{dt}(t)\right|_{t=t_0},$$

where $Id$ is the identity matrix. The result of Exercise 8.1 shows that

$$R_{ba}(t_0)\left.\frac{dR_{ab}}{dt}(t)\right|_{t=t_0} = \begin{pmatrix} 0 & -\omega_3(t_0) & \omega_2(t_0) \\ \omega_3(t_0) & 0 & -\omega_1(t_0) \\ -\omega_2(t_0) & \omega_1(t_0) & 0 \end{pmatrix}$$

for some functions $\omega_1(t)$, $\omega_2(t)$, and $\omega_3(t)$. The vector

$$b\begin{pmatrix} \omega_1(t_0) \\ \omega_2(t_0) \\ \omega_3(t_0) \end{pmatrix}$$

is called the *angular velocity at time* $t = t_0$ *of the observer* $\mathcal{B}$ *relative to* $\mathcal{A}$ and is denoted by ${}^{\mathcal{A}}\boldsymbol{\omega}^{\mathcal{B}}(t_0)$.

1

**Illustration 8.3**

By definition, the angular velocity of the observer $\mathcal{B}$ relative to $\mathcal{A}$ is a vector. It does not correspond to a pure translation, however. In particular, define the quantity

$$\left\|{}^{\mathcal{A}}\boldsymbol{\omega}^{\mathcal{B}}\right\| \overset{def}{=} \sqrt{{}^{\mathcal{A}}\boldsymbol{\omega}^{\mathcal{B}} \bullet {}^{\mathcal{A}}\boldsymbol{\omega}^{\mathcal{B}}}$$

and consider the rotation matrix

$$R\left(h\left\|{}^{\mathcal{A}}\boldsymbol{\omega}^{\mathcal{B}}\right\|, \frac{\omega_1}{\left\|{}^{\mathcal{A}}\boldsymbol{\omega}^{\mathcal{B}}\right\|}, \frac{\omega_2}{\left\|{}^{\mathcal{A}}\boldsymbol{\omega}^{\mathcal{B}}\right\|}, \frac{\omega_3}{\left\|{}^{\mathcal{A}}\boldsymbol{\omega}^{\mathcal{B}}\right\|}\right)$$

in terms of the notation introduced in Chapter 5. Then, since

$$\cos\left(h\left\|{}^{\mathcal{A}}\boldsymbol{\omega}^{\mathcal{B}}\right\|\right) \approx 1 \text{ and } \sin\left(h\left\|{}^{\mathcal{A}}\boldsymbol{\omega}^{\mathcal{B}}\right\|\right) \approx h\left\|{}^{\mathcal{A}}\boldsymbol{\omega}^{\mathcal{B}}\right\|$$

for small $h$,

$$\begin{aligned}
&\left[R\left(h\left\|{}^{\mathcal{A}}\boldsymbol{\omega}^{\mathcal{B}}\right\|, \frac{\omega_1}{\left\|{}^{\mathcal{A}}\boldsymbol{\omega}^{\mathcal{B}}\right\|}, \frac{\omega_2}{\left\|{}^{\mathcal{A}}\boldsymbol{\omega}^{\mathcal{B}}\right\|}, \frac{\omega_3}{\left\|{}^{\mathcal{A}}\boldsymbol{\omega}^{\mathcal{B}}\right\|}\right)\right]_{ij} \\
&\quad = \delta_{ij}\cos\left(h\left\|{}^{\mathcal{A}}\boldsymbol{\omega}^{\mathcal{B}}\right\|\right) + \left(1-\cos\left(h\left\|{}^{\mathcal{A}}\boldsymbol{\omega}^{\mathcal{B}}\right\|\right)\right)\frac{\omega_i\omega_j}{\left\|{}^{\mathcal{A}}\boldsymbol{\omega}^{\mathcal{B}}\right\|^2} \\
&\qquad -\sin\left(h\left\|{}^{\mathcal{A}}\boldsymbol{\omega}^{\mathcal{B}}\right\|\right)\sum_{k=1}^{3}\varepsilon_{ijk}\frac{\omega_k}{\left\|{}^{\mathcal{A}}\boldsymbol{\omega}^{\mathcal{B}}\right\|} \\
&\quad \approx \delta_{ij} - h\sum_{k=1}^{3}\varepsilon_{ijk}\omega_k.
\end{aligned}$$

But this equals

$$\left[Id + hR_{ba}\frac{dR_{ab}}{dt}\right]_{ij}$$

and we conclude that

$$[R_{ba}(t)\, R_{ab}(t+h)] \approx R\left(h\left\|{}^{\mathcal{A}}\boldsymbol{\omega}^{\mathcal{B}}(t)\right\|, \frac{\omega_1(t)}{\left\|{}^{\mathcal{A}}\boldsymbol{\omega}^{\mathcal{B}}(t)\right\|}, \frac{\omega_2(t)}{\left\|{}^{\mathcal{A}}\boldsymbol{\omega}^{\mathcal{B}}(t)\right\|}, \frac{\omega_3(t)}{\left\|{}^{\mathcal{A}}\boldsymbol{\omega}^{\mathcal{B}}(t)\right\|}\right).$$

It follows that the quantity

$$\left\|{}^{\mathcal{A}}\boldsymbol{\omega}^{\mathcal{B}}\right\|$$

equals the *angular speed of the observer $\mathcal{B}$ relative to $\mathcal{A}$* and that ${}^{\mathcal{A}}\boldsymbol{\omega}^{\mathcal{B}}$ is parallel to the instantaneous direction of rotation of $\mathcal{B}$ relative to $\mathcal{A}$.

### 8.1.3 Notation and Review

The derivative of an arbitrary vector $\mathbf{r}$ relative to a triad $a$ is defined as the vector

$$\frac{{}^{a}d\mathbf{r}}{dt} \stackrel{def}{=} a\frac{d\,{}^{a}r(t)}{dt}.$$

The superscript to the left of the differentiation symbol reflects the choice of triad on the right-hand side. The derivative of the same vector $\mathbf{r}$ relative to the triad $b$ is then given by the vector

$$\frac{{}^{b}d\mathbf{r}}{dt} = b\frac{d\,{}^{b}r(t)}{dt}.$$

The differentiation symbol on the right-hand side **does not** use a left superscript, since it corresponds to a standard derivative of a numerical function of $t$. The above expressions define the differentiation operator

$$\frac{{}^{a}d}{dt}.$$

As we shall see below, in general,

$$\frac{{}^{b}d\mathbf{r}}{dt} \neq \frac{{}^{a}d\mathbf{r}}{dt}.$$

If $a$ is the reference triad of an observer $\mathcal{A}$, then we can define

$$\frac{{}^{\mathcal{A}}d\mathbf{r}}{dt} \stackrel{def}{=} \frac{{}^{a}d\mathbf{r}}{dt}$$

for an arbitrary vector $\mathbf{r}$. Since the reference point of $\mathcal{A}$ is irrelevant to the computation of this derivative, no new information is suggested by this notation. I will not use this further.

If $\mathcal{A}$ and $\mathcal{B}$ are two observers with reference points $A$ and $B$ and reference triads $a$ and $b$, then the derivative

$$\frac{{}^a d\mathbf{r}^{AB}}{dt}$$

of the position vector $\mathbf{r}^{AB}$ relative to the $a$ triad is defined as the *linear velocity of* $\mathcal{B}$ *relative to* $\mathcal{A}$. Similarly,

$$\frac{{}^b d\mathbf{r}^{BA}}{dt}$$

is the linear velocity of $\mathcal{A}$ relative to $\mathcal{B}$. The derivatives

$$\frac{{}^a d\mathbf{r}^{BA}}{dt} \text{ and } \frac{{}^b d\mathbf{r}^{AB}}{dt}$$

have **no immediate** physical meaning.

The linear velocity of $\mathcal{B}$ relative to $\mathcal{A}$ is denoted by ${}^{\mathcal{A}}\mathbf{v}^{\mathcal{B}}$. This text consistently uses a lower-case $\mathbf{v}$ to denote a linear velocity. The left and right superscripts correspond to the two observers and appear in an order that suggests a direction from left to right.

By definition, the linear velocity of $\mathcal{B}$ relative to $\mathcal{A}$ is given by the derivative

$$\frac{{}^a d\mathbf{r}^{AB}}{dt} = a\frac{d\,{}^a\left(\mathbf{r}^{AB}\right)}{dt} = a\frac{d\,{}^{\mathcal{A}}B}{dt},$$

where

$${}^{\mathcal{A}}B$$

is the coordinate representation of the point $B$ relative to the $\mathcal{A}$ observer. More generally, we introduce the notation

$$\frac{{}^{\mathcal{A}}dP}{dt} \stackrel{def}{=} a\frac{d\,{}^{\mathcal{A}}P}{dt} = a\frac{d\,{}^a\left(\mathbf{r}^{AP}\right)}{dt} = \frac{{}^a d\mathbf{r}^{AP}}{dt}$$

to denote the *velocity of the point* $P$ *relative to the observer* $\mathcal{A}$. It follows that

$$\frac{{}^{\mathcal{A}}dB}{dt} = {}^{\mathcal{A}}\mathbf{v}^{\mathcal{B}},$$

where $B$ is the reference point of the $\mathcal{B}$ observer.

The angular velocity of $\mathcal{B}$ relative to $\mathcal{A}$ is the vector ${}^{\mathcal{A}}\boldsymbol{\omega}^{\mathcal{B}}$, such that

$$^{b}\left({}^{\mathcal{A}}\boldsymbol{\omega}^{\mathcal{B}}\right) = \begin{pmatrix} \omega_1 \\ \omega_2 \\ \omega_3 \end{pmatrix},$$

where

$$R_{ba}\frac{dR_{ab}}{dt} = \begin{pmatrix} 0 & -\omega_3 & \omega_2 \\ \omega_3 & 0 & -\omega_1 \\ -\omega_2 & \omega_1 & 0 \end{pmatrix}.$$

This text consistently uses the lower-case Greek letter $\boldsymbol{\omega}$ to denote an angular velocity. The left and right superscripts play the same role as in the case of the linear velocity. Contrary to the case of the linear velocity, only the reference triads of $\mathcal{A}$ and $\mathcal{B}$ are involved in the computation of ${}^{\mathcal{A}}\boldsymbol{\omega}^{\mathcal{B}}$. To reflect this fact, I will use the notation

$$^{a}\boldsymbol{\omega}^{b}$$

for the angular velocity vector between the triad $a$ and the triad $b$.

## 8.2 Different Viewpoints

(Ex. 8.5 – Ex. 8.12)

Let $\mathcal{A}$ and $\mathcal{B}$ be two observers with reference points $A$ and $B$ and reference triads $a$ and $b$ and consider the difference

$$\begin{aligned} \frac{^{a}d\mathbf{r}}{dt} - \frac{^{b}d\mathbf{r}}{dt} &= a\frac{d\,{}^{a}r}{dt} - b\frac{d\,{}^{b}r}{dt} \\ &= a\frac{d}{dt}\left[R_{ab}{}^{b}r\right] - b\frac{d\,{}^{b}r}{dt} \\ &= \left(aR_{ab} - b\right)\frac{d\,{}^{b}r}{dt} + a\frac{dR_{ab}}{dt}\,{}^{b}r \\ &= bR_{ba}\frac{dR_{ab}}{dt}\,{}^{b}r \end{aligned}$$

for some arbitrary vector $\mathbf{r}$. The result of Exercise 8.5 shows that this equals the cross product between ${}^{a}\boldsymbol{\omega}^{b}$ and $\mathbf{r}$, i.e.,

$$\frac{^{a}d\mathbf{r}}{dt} = \frac{^{b}d\mathbf{r}}{dt} + {}^{a}\boldsymbol{\omega}^{b} \times \mathbf{r}.$$

**Illustration 8.4**

Suppose that $\mathbf{r}^{AP}$ and $\mathbf{r}^{BP}$ are the position vectors from the reference points of the observers $\mathcal{A}$ and $\mathcal{B}$ to some point $P$ on a rigid body.

1

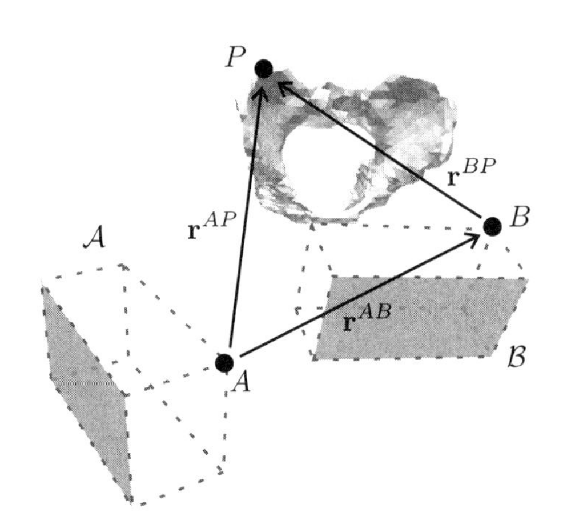


Then, the velocity of the point $P$ relative to the observer $\mathcal{A}$ is given by

$$\frac{{}^{\mathcal{A}}dP}{dt} = \frac{{}^{a}d\mathbf{r}^{AP}}{dt}.$$

Similarly, the velocity of the point $P$ relative to the observer $\mathcal{B}$ is given by

$$\frac{{}^{\mathcal{B}}dP}{dt} = \frac{{}^{b}d\mathbf{r}^{BP}}{dt}.$$

But,

$$\mathbf{r}^{AP} = \mathbf{r}^{AB} + \mathbf{r}^{BP}$$

and thus

$$\begin{aligned}
\frac{{}^{\mathcal{A}}dP}{dt} - \frac{{}^{\mathcal{B}}dP}{dt} &= \frac{{}^{a}d\mathbf{r}^{AP}}{dt} - \frac{{}^{b}d\mathbf{r}^{BP}}{dt} \\
&= \frac{{}^{a}d}{dt}\left(\mathbf{r}^{AB} + \mathbf{r}^{BP}\right) - \frac{{}^{b}d\mathbf{r}^{BP}}{dt} \\
&= \frac{{}^{a}d\mathbf{r}^{AB}}{dt} + {}^{a}\boldsymbol{\omega}^{b} \times \mathbf{r}^{BP} \\
&= {}^{\mathcal{A}}\mathbf{v}^{\mathcal{B}} + {}^{a}\boldsymbol{\omega}^{b} \times \mathbf{r}^{BP}.
\end{aligned}$$

The velocity of the point $P$ relative to the observer $\mathcal{B}$ equals the velocity of the point $P$ relative to the observer $\mathcal{A}$ if and only if

$${}^{\mathcal{A}}\mathbf{v}^{\mathcal{B}} + {}^{a}\boldsymbol{\omega}^{b} \times \mathbf{r}^{BP} = \mathbf{0}.$$

Recall the relation between the derivatives of a vector relative to two different triads:

$$\frac{{}^{a}d\mathbf{r}}{dt} - \frac{{}^{b}d\mathbf{r}}{dt} = {}^{a}\boldsymbol{\omega}^{b} \times \mathbf{r}.$$

Switch the role of the two triads $a$ and $b$ to yield

$$\frac{{}^{b}d\mathbf{r}}{dt} - \frac{{}^{a}d\mathbf{r}}{dt} = {}^{b}\boldsymbol{\omega}^{a} \times \mathbf{r}.$$

Adding these two equations shows that

$$\mathbf{0} = \left({}^{a}\boldsymbol{\omega}^{b} + {}^{b}\boldsymbol{\omega}^{a}\right) \times \mathbf{r}$$

for an arbitrary vector $\mathbf{r}$. But this is possible only if

$${}^{a}\boldsymbol{\omega}^{b} + {}^{b}\boldsymbol{\omega}^{a} = \mathbf{0},$$

i.e.,

$${}^{b}\boldsymbol{\omega}^{a} = -{}^{a}\boldsymbol{\omega}^{b}.$$

**Illustration 8.5**

In the previous illustration, we found that

$$\frac{{}^{\mathcal{A}}dP}{dt} - \frac{{}^{\mathcal{B}}dP}{dt} = {}^{\mathcal{A}}\mathbf{v}^{\mathcal{B}} + {}^{a}\boldsymbol{\omega}^{b} \times \mathbf{r}^{BP},$$

where ${}^{\mathcal{A}}\frac{dP}{dt}$ and ${}^{\mathcal{B}}\frac{dP}{dt}$ were the velocities of a point $P$ on a rigid body relative to two observers $\mathcal{A}$ and $\mathcal{B}$, with reference points $A$ and $B$ and reference triads $a$ and $b$, respectively. By switching the roles of the observers, we find

$$\begin{aligned}\frac{{}^{\mathcal{B}}dP}{dt} - \frac{{}^{\mathcal{A}}dP}{dt} &= {}^{\mathcal{B}}\mathbf{v}^{\mathcal{A}} + {}^{b}\boldsymbol{\omega}^{a} \times \mathbf{r}^{AP} \\ &= {}^{\mathcal{B}}\mathbf{v}^{\mathcal{A}} - {}^{a}\boldsymbol{\omega}^{b} \times \left(\mathbf{r}^{AB} + \mathbf{r}^{BP}\right).\end{aligned}$$

Adding these two equations shows that

$$\mathbf{0} = {}^{\mathcal{A}}\mathbf{v}^{\mathcal{B}} + {}^{\mathcal{B}}\mathbf{v}^{\mathcal{A}} - {}^{a}\boldsymbol{\omega}^{b} \times \mathbf{r}^{AB},$$

i.e.,

$${}^{\mathcal{B}}\mathbf{v}^{\mathcal{A}} = -{}^{\mathcal{A}}\mathbf{v}^{\mathcal{B}} + {}^{a}\boldsymbol{\omega}^{b} \times \mathbf{r}^{AB}.$$

1

Now, let $\mathcal{A}$, $\mathcal{B}$, and $\mathcal{C}$ be three different observers with reference points $A$, $B$, and $C$, respectively, and reference triads $a$, $b$, and $c$, respectively. From the previous discussion, we recall that

$$\frac{{}^{a}d\mathbf{r}}{dt} - \frac{{}^{b}d\mathbf{r}}{dt} = {}^{a}\boldsymbol{\omega}^{b} \times \mathbf{r},$$

$$\frac{{}^{b}d\mathbf{r}}{dt} - \frac{{}^{c}d\mathbf{r}}{dt} = {}^{b}\boldsymbol{\omega}^{c} \times \mathbf{r},$$

and

$$\frac{{}^{a}d\mathbf{r}}{dt} - \frac{{}^{c}d\mathbf{r}}{dt} = {}^{a}\boldsymbol{\omega}^{c} \times \mathbf{r}.$$

Adding the first two relations and subtracting the last relation shows that

$$\mathbf{0} = \left({}^{a}\boldsymbol{\omega}^{b} + {}^{b}\boldsymbol{\omega}^{c} - {}^{a}\boldsymbol{\omega}^{c}\right) \times \mathbf{r}$$

for an arbitrary vector $\mathbf{r}$. But this is only possible if

$${}^{a}\boldsymbol{\omega}^{b} + {}^{b}\boldsymbol{\omega}^{c} - {}^{a}\boldsymbol{\omega}^{c} = \mathbf{0},$$

i.e.,

$${}^{a}\boldsymbol{\omega}^{c} = {}^{a}\boldsymbol{\omega}^{b} + {}^{b}\boldsymbol{\omega}^{c}.$$

**Illustration 8.6**

The result of a previous illustration shows that

$$\frac{{}^{\mathcal{A}}dP}{dt} - \frac{{}^{\mathcal{B}}dP}{dt} = {}^{\mathcal{A}}\mathbf{v}^{\mathcal{B}} + {}^{a}\boldsymbol{\omega}^{b} \times \mathbf{r}^{BP},$$
$$\frac{{}^{\mathcal{B}}dP}{dt} - \frac{{}^{\mathcal{C}}dP}{dt} = {}^{\mathcal{B}}\mathbf{v}^{\mathcal{C}} + {}^{b}\boldsymbol{\omega}^{c} \times \mathbf{r}^{CP},$$

and

$$\frac{{}^{\mathcal{A}}dP}{dt} - \frac{{}^{\mathcal{C}}dP}{dt} = {}^{\mathcal{A}}\mathbf{v}^{\mathcal{C}} + {}^{a}\boldsymbol{\omega}^{c} \times \mathbf{r}^{CP}.$$

Adding the first two relations and subtracting the last relation shows that

$$\begin{aligned}
\mathbf{0} &= {}^{\mathcal{A}}\mathbf{v}^{\mathcal{B}} + {}^{\mathcal{B}}\mathbf{v}^{\mathcal{C}} - {}^{\mathcal{A}}\mathbf{v}^{\mathcal{C}} \\
&\quad + {}^{a}\boldsymbol{\omega}^{b} \times \mathbf{r}^{BP} + {}^{b}\boldsymbol{\omega}^{c} \times \mathbf{r}^{CP} - {}^{a}\boldsymbol{\omega}^{c} \times \mathbf{r}^{CP} \\
&= {}^{\mathcal{A}}\mathbf{v}^{\mathcal{B}} + {}^{\mathcal{B}}\mathbf{v}^{\mathcal{C}} - {}^{\mathcal{A}}\mathbf{v}^{\mathcal{C}} \\
&\quad + {}^{a}\boldsymbol{\omega}^{b} \times \left(\mathbf{r}^{BC} + \mathbf{r}^{CP}\right) + {}^{b}\boldsymbol{\omega}^{c} \times \mathbf{r}^{CP} \\
&\quad - {}^{a}\boldsymbol{\omega}^{b} \times \mathbf{r}^{CP} - {}^{b}\boldsymbol{\omega}^{c} \times \mathbf{r}^{CP} \\
&= {}^{\mathcal{A}}\mathbf{v}^{\mathcal{B}} + {}^{\mathcal{B}}\mathbf{v}^{\mathcal{C}} - {}^{\mathcal{A}}\mathbf{v}^{\mathcal{C}} + {}^{a}\boldsymbol{\omega}^{b} \times \mathbf{r}^{BC},
\end{aligned}$$

i.e.,

$$^{\mathcal{A}}\mathbf{v}^{\mathcal{C}} = {}^{\mathcal{A}}\mathbf{v}^{\mathcal{B}} + {}^{\mathcal{B}}\mathbf{v}^{\mathcal{C}} + {}^{a}\boldsymbol{\omega}^{b} \times \mathbf{r}^{BC}.$$

## 8.3 Integration

(Ex. 8.13)

The position of a rigid body relative to some observer $\mathcal{A}$ is given by a pure translation $\mathbf{T}_{\mathcal{A}}$ corresponding to the position vector $\mathbf{r}^{AB}$, where $A$ and $B$ are the reference points of the observer $\mathcal{A}$ and the rigid body, respectively. If the rigid body is unconstrained, three configuration coordinates are necessary and sufficient to describe its position. For example,

$$\mathbf{r}^{AB} = a \begin{pmatrix} q_1 \\ q_2 \\ q_3 \end{pmatrix},$$

where $q_1$, $q_2$, and $q_3$ are the Cartesian coordinates of the point $B$ in a coordinate system with origin at $A$ and axes parallel to the basis vectors of the reference triad $a$ of the observer $\mathcal{A}$, i.e.,

$$^{\mathcal{A}}B = \begin{pmatrix} q_1 \\ q_2 \\ q_3 \end{pmatrix}.$$

From the definition of the linear velocity of the rigid body relative to the observer $\mathcal{A}$, we have

$$\begin{aligned} \frac{^{\mathcal{A}}dB}{dt} &= \frac{^{a}d\mathbf{r}^{AB}}{dt} \\ &= a\frac{d\ {}^{a}\left(\mathbf{r}^{AB}\right)}{dt} \\ &= a\begin{pmatrix} \dot{q}_1 \\ \dot{q}_2 \\ \dot{q}_3 \end{pmatrix}, \end{aligned}$$

where the dot superscript denotes differentiation with respect to $t$. If we know the configuration coordinates $q_1(t)$, $q_2(t)$, and $q_3(t)$ as functions of time, then we can compute their derivatives as functions of time, and therefore the linear velocity of the rigid body relative to $\mathcal{A}$ as a function of time.

Conversely, suppose that

$$\frac{^{\mathcal{A}}dB}{dt} = a\begin{pmatrix} u_1(t) \\ u_2(t) \\ u_3(t) \end{pmatrix},$$

1

where $u_1(t)$, $u_2(t)$, and $u_3(t)$ are known functions of time. It follows that

$$\dot{q}_1 = u_1(t),$$
$$\dot{q}_2 = u_2(t),$$

and

$$\dot{q}_3 = u_3(t).$$

or, equivalently,

$$\begin{pmatrix} 1 & 0 & 0 \\ 0 & 1 & 0 \\ 0 & 0 & 1 \end{pmatrix} \begin{pmatrix} \dot{q}_1 \\ \dot{q}_2 \\ \dot{q}_3 \end{pmatrix} = \begin{pmatrix} u_1(t) \\ u_2(t) \\ u_3(t) \end{pmatrix}.$$

The coefficient matrix on the left-hand side is always invertible, since

$$\begin{vmatrix} 1 & 0 & 0 \\ 0 & 1 & 0 \\ 0 & 0 & 1 \end{vmatrix} = 1 \neq 0.$$

The differential equations in the unknown functions $q_1(t)$, $q_2(t)$, and $q_3(t)$ are therefore non-singular for all $q_1$, $q_2$, and $q_3$. Given initial conditions $q_1(t_0)$, $q_2(t_0)$, and $q_3(t_0)$, the differential equations can be solved for $q_1(t)$, $q_2(t)$, and $q_3(t)$. The result may be substituted into the expression for the position vector

$$\mathbf{r}^{AB} = a \begin{pmatrix} q_1(t) \\ q_2(t) \\ q_3(t) \end{pmatrix}$$

or the coordinate representation of $B$ relative to $\mathcal{A}$:

$${}^{\mathcal{A}}B = \begin{pmatrix} q_1(t) \\ q_2(t) \\ q_3(t) \end{pmatrix}.$$

**Illustration 8.7**

In terms of spherical coordinates,

$$\mathbf{r}^{AB} = a \begin{pmatrix} \tilde{q}_1 \sin \tilde{q}_2 \cos \tilde{q}_3 \\ \tilde{q}_1 \sin \tilde{q}_2 \sin \tilde{q}_3 \\ \tilde{q}_1 \cos \tilde{q}_2 \end{pmatrix},$$

i.e.,

$${}^{\mathcal{A}}B = \begin{pmatrix} \tilde{q}_1 \sin \tilde{q}_2 \cos \tilde{q}_3 \\ \tilde{q}_1 \sin \tilde{q}_2 \sin \tilde{q}_3 \\ \tilde{q}_1 \cos \tilde{q}_2 \end{pmatrix}.$$

The velocity of the point $B$ relative to $\mathcal{A}$ now becomes

$$\begin{aligned}
\frac{{}^{\mathcal{A}}dB}{dt} &= \frac{{}^{a}d\mathbf{r}^{AB}}{dt} \\
&= a\frac{d\,{}^{a}\left(\mathbf{r}^{AB}\right)}{dt} \\
&= a\begin{pmatrix} \dot{\tilde{q}}_1 \sin\tilde{q}_2 \cos\tilde{q}_3 + \tilde{q}_1\dot{\tilde{q}}_2 \cos\tilde{q}_2 \cos\tilde{q}_3 - \tilde{q}_1\dot{\tilde{q}}_3 \sin\tilde{q}_2 \sin\tilde{q}_3 \\ \dot{\tilde{q}}_1 \sin\tilde{q}_2 \sin\tilde{q}_3 + \tilde{q}_1\dot{\tilde{q}}_2 \cos\tilde{q}_2 \sin\tilde{q}_3 + \tilde{q}_1\dot{\tilde{q}}_3 \sin\tilde{q}_2 \cos\tilde{q}_3 \\ \dot{\tilde{q}}_1 \cos\tilde{q}_2 - \tilde{q}_1\dot{\tilde{q}}_2 \sin\tilde{q}_2 \end{pmatrix}.
\end{aligned}$$

If we are given $\tilde{q}_1(t)$, $\tilde{q}_2(t)$, and $\tilde{q}_3(t)$, we can compute the linear velocity of the rigid body relative to the observer $\mathcal{A}$ as a function of time.

Conversely, suppose that

$$\frac{{}^{\mathcal{A}}dB}{dt} = a\begin{pmatrix} u_1(t) \\ u_2(t) \\ u_3(t) \end{pmatrix},$$

where $u_1(t)$, $u_2(t)$, and $u_3(t)$ are known functions of time. It follows that

$$\begin{aligned}
\dot{\tilde{q}}_1 \sin\tilde{q}_2 \cos\tilde{q}_3 + \tilde{q}_1\dot{\tilde{q}}_2 \cos\tilde{q}_2 \cos\tilde{q}_3 - \tilde{q}_1\dot{\tilde{q}}_3 \sin\tilde{q}_2 \sin\tilde{q}_3 &= u_1(t), \\
\dot{\tilde{q}}_1 \sin\tilde{q}_2 \sin\tilde{q}_3 + \tilde{q}_1\dot{\tilde{q}}_2 \cos\tilde{q}_2 \sin\tilde{q}_3 + \tilde{q}_1\dot{\tilde{q}}_3 \sin\tilde{q}_2 \cos\tilde{q}_3 &= u_2(t),
\end{aligned}$$

and

$$\dot{\tilde{q}}_1 \cos\tilde{q}_2 - \tilde{q}_1\dot{\tilde{q}}_2 \sin\tilde{q}_2 = u_3(t),$$

or, equivalently,

$$\begin{pmatrix} \sin\tilde{q}_2\cos\tilde{q}_3 & \tilde{q}_1\cos\tilde{q}_2\cos\tilde{q}_3 & -\tilde{q}_1\sin\tilde{q}_2\sin\tilde{q}_3 \\ \sin\tilde{q}_2\sin\tilde{q}_3 & \tilde{q}_1\cos\tilde{q}_2\sin\tilde{q}_3 & \tilde{q}_1\sin\tilde{q}_2\cos\tilde{q}_3 \\ \cos\tilde{q}_2 & -\tilde{q}_1\sin\tilde{q}_2 & 0 \end{pmatrix}\begin{pmatrix} \dot{\tilde{q}}_1 \\ \dot{\tilde{q}}_2 \\ \dot{\tilde{q}}_3 \end{pmatrix} = \begin{pmatrix} u_1(t) \\ u_2(t) \\ u_3(t) \end{pmatrix}.$$

The coefficient matrix on the left-hand side is invertible as long as

$$\begin{vmatrix} \sin\tilde{q}_2\cos\tilde{q}_3 & \tilde{q}_1\cos\tilde{q}_2\cos\tilde{q}_3 & -\tilde{q}_1\sin\tilde{q}_2\sin\tilde{q}_3 \\ \sin\tilde{q}_2\sin\tilde{q}_3 & \tilde{q}_1\cos\tilde{q}_2\sin\tilde{q}_3 & \tilde{q}_1\sin\tilde{q}_2\cos\tilde{q}_3 \\ \cos\tilde{q}_2 & -\tilde{q}_1\sin\tilde{q}_2 & 0 \end{vmatrix} = \tilde{q}_1^2\sin\tilde{q}_2 \neq 0.$$

At the singular points $\tilde{q}_1 = 0$ or $\tilde{q}_2 = 0, \pi$, the matrix inverse

$$\begin{pmatrix} \sin\tilde{q}_2\cos\tilde{q}_3 & \tilde{q}_1\cos\tilde{q}_2\cos\tilde{q}_3 & -\tilde{q}_1\sin\tilde{q}_2\sin\tilde{q}_3 \\ \sin\tilde{q}_2\sin\tilde{q}_3 & \tilde{q}_1\cos\tilde{q}_2\sin\tilde{q}_3 & \tilde{q}_1\sin\tilde{q}_2\cos\tilde{q}_3 \\ \cos\tilde{q}_2 & -\tilde{q}_1\sin\tilde{q}_2 & 0 \end{pmatrix}^{-1}$$

does not exist and we cannot compute $\dot{\tilde{q}}_1$, $\dot{\tilde{q}}_2$, or $\dot{\tilde{q}}_3$.

The differential equations in the unknown functions $\tilde{q}_1(t)$, $\tilde{q}_2(t)$, and $\tilde{q}_3(t)$ are non-singular for $\tilde{q}_1 \neq 0$ and $\tilde{q}_2 \neq n\pi$ for any integer $n$. Given initial conditions $\tilde{q}_1(t_0) \neq 0$, $\tilde{q}_2(t_0) \neq n\pi$, and $\tilde{q}_3(t_0)$, the differential equations can be solved for $\tilde{q}_1(t)$, $\tilde{q}_2(t)$, and $\tilde{q}_3(t)$ until a time $t^*$ when $\tilde{q}_1(t^*) = 0$ or $\tilde{q}_2(t^*) = n\pi$ for some integer $n$.

The result may be substituted into the expression for the position vector

$$\mathbf{r}^{AB} = a \begin{pmatrix} \tilde{q}_1(t)\sin\tilde{q}_2(t)\cos\tilde{q}_3(t) \\ \tilde{q}_1(t)\sin\tilde{q}_2(t)\sin\tilde{q}_3(t) \\ \tilde{q}_1(t)\cos\tilde{q}_2(t) \end{pmatrix}$$

or the coordinate representation of $B$ relative to $\mathcal{A}$:

$$^{\mathcal{A}}B = \begin{pmatrix} \tilde{q}_1(t)\sin\tilde{q}_2(t)\cos\tilde{q}_3(t) \\ \tilde{q}_1(t)\sin\tilde{q}_2(t)\sin\tilde{q}_3(t) \\ \tilde{q}_1(t)\cos\tilde{q}_2(t) \end{pmatrix}.$$

The orientation of a rigid body relative to some observer $\mathcal{A}$ is given by a pure rotation $\mathbf{R}_{\mathcal{A}}$ corresponding to the rotation matrix $R_{ab}$, where $a$ and $b$ are the reference triads of the observer $\mathcal{A}$ and the rigid body, respectively. If the rigid body is unconstrained, three configuration coordinates are necessary and sufficient to describe its orientation. For example, let $q_1$, $q_2$, and $q_3$ be the Euler angles corresponding to a $3-1-3$ sequence of rotations:

$$\begin{aligned} R_{ab} &= R(q_1,0,0,1)\,R(q_2,1,0,0)\,R(q_3,0,0,1) \\ &= \begin{pmatrix} c_1c_3 - s_1c_2s_3 & -c_1s_3 - s_1c_2c_3 & s_1s_2 \\ s_1c_3 + c_1c_2s_3 & -s_1s_3 + c_1c_2c_3 & -c_1s_2 \\ s_2s_3 & s_2c_3 & c_2 \end{pmatrix}, \end{aligned}$$

where $c_i = \cos q_i$ and $s_i = \sin q_i$ for $i = 1, 2, 3$.

From the definition of the angular velocity between the triad $a$ and the triad $b$, we have

$$^{a}\boldsymbol{\omega}^{b} = b \begin{pmatrix} \omega_1 \\ \omega_2 \\ \omega_3 \end{pmatrix},$$

where

$$R_{ba}\frac{dR_{ab}}{dt} = \begin{pmatrix} 0 & -\omega_3 & \omega_2 \\ \omega_3 & 0 & -\omega_1 \\ -\omega_2 & \omega_1 & 0 \end{pmatrix}.$$

But (after much simplification)

$$R_{ba}\frac{dR_{ab}}{dt} = \begin{pmatrix} 0 & -\dot{q}_1c_2 - \dot{q}_3 & \dot{q}_1s_2c_3 - \dot{q}_2s_3 \\ \dot{q}_1c_2 + \dot{q}_3 & 0 & -\dot{q}_1s_2s_3 - \dot{q}_2c_3 \\ -\dot{q}_1s_2c_3 + \dot{q}_2s_3 & \dot{q}_1s_2s_3 + \dot{q}_2c_3 & 0 \end{pmatrix},$$

i.e.,

$$^{a}\boldsymbol{\omega}^{b} = b \begin{pmatrix} \dot{q}_1 s_2 s_3 + \dot{q}_2 c_3 \\ \dot{q}_1 s_2 c_3 - \dot{q}_2 s_3 \\ \dot{q}_1 c_2 + \dot{q}_3 \end{pmatrix}.$$

If we know the configuration coordinates $q_1(t)$, $q_2(t)$, and $q_3(t)$ as functions of time, then we can compute their derivatives as functions of time and, consequently, the angular velocity between the triad $a$ and the triad $b$ as a function of time.

Conversely, suppose that

$$^{a}\boldsymbol{\omega}^{b} = b \begin{pmatrix} u_1(t) \\ u_2(t) \\ u_3(t) \end{pmatrix},$$

where $u_1(t)$, $u_2(t)$, and $u_3(t)$ are known functions of time. It follows that

$$\dot{q}_1 s_2 s_3 + \dot{q}_2 c_3 = u_1(t),$$
$$\dot{q}_1 s_2 c_3 - \dot{q}_2 s_3 = u_2(t),$$

and

$$\dot{q}_1 c_2 + \dot{q}_3 = u_3(t),$$

or, equivalently,

$$\begin{pmatrix} s_2 s_3 & c_3 & 0 \\ s_2 c_3 & -s_3 & 0 \\ c_2 & 0 & 1 \end{pmatrix} \begin{pmatrix} \dot{q}_1 \\ \dot{q}_2 \\ \dot{q}_3 \end{pmatrix} = \begin{pmatrix} u_1(t) \\ u_2(t) \\ u_3(t) \end{pmatrix}.$$

The coefficient matrix on the left-hand side is non-singular as long as

$$\begin{vmatrix} s_2 s_3 & c_3 & 0 \\ s_2 c_3 & -s_3 & 0 \\ c_2 & 0 & 1 \end{vmatrix} = -\sin q_2 \neq 0.$$

It follows that the differential equations in the unknown functions $q_1(t)$, $q_2(t)$, and $q_3(t)$ are non-singular for $q_2 \neq n\pi$ for any integer $n$. Given initial conditions $q_1(t_0)$, $q_2(t_0) \neq n\pi$, and $q_3(t_0)$, the differential equations can be solved for $q_1(t)$, $q_2(t)$, and $q_3(t)$, until such a time $t^*$ when $q_2(t^*) = n\pi$ for some integer $n$.

The result can be substituted back into the rotation matrix

$$R_{ab} = \begin{pmatrix} c_1(t)c_3(t) - s_1(t)c_2(t)s_3(t) & -c_1(t)s_3(t) - s_1(t)c_2(t)c_3(t) & s_1(t)s_2(t) \\ s_1(t)c_3(t) + c_1(t)c_2(t)s_3(t) & -s_1(t)s_3(t) + c_1(t)c_2(t)c_3(t) & -c_1(t)s_2(t) \\ s_2(t)s_3(t) & s_2(t)c_3(t) & c_2(t) \end{pmatrix}$$

to yield the orientation of the rigid body relative to the observer $\mathcal{A}$.

(Ex. 8.14 – Ex. 8.15)

## 8.4 Examples

### 8.4.1 Rotations about $\mathbf{a}_i$

Let $\mathcal{A}$ and $\mathcal{B}$ be two observers with reference points $A$ and $B$ and reference triads $a$ and $b$, respectively, such that $\mathbf{r}^{AB}$ is constant and

$$R_{ab} = R(\varphi, 1, 0, 0) = \begin{pmatrix} 1 & 0 & 0 \\ 0 & \cos\varphi & -\sin\varphi \\ 0 & \sin\varphi & \cos\varphi \end{pmatrix}$$

corresponding to a pure rotation by an angle $\varphi$ about an axis parallel to the $\mathbf{a}_1$ basis vector.

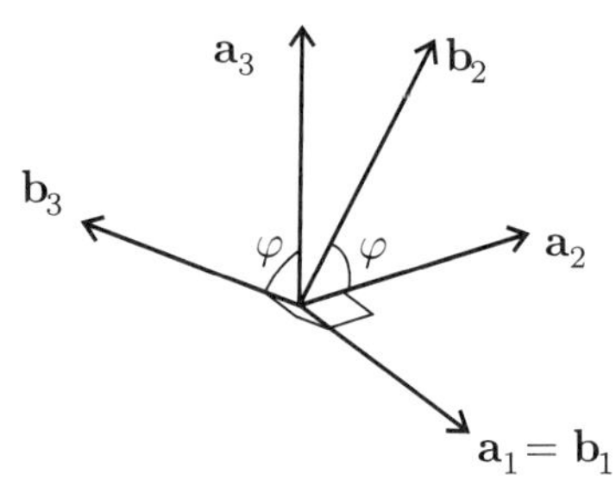


From the definition of the linear and angular velocity, it follows that

$${}^{\mathcal{A}}\mathbf{v}^{\mathcal{B}} = \frac{{}^{a}d\mathbf{r}^{AB}}{dt} = \mathbf{0}$$

and

$${}^{a}\boldsymbol{\omega}^{b} = b\begin{pmatrix} \dot{\varphi} \\ 0 \\ 0 \end{pmatrix},$$

since

$$R_{ba}\frac{dR_{ab}}{dt} = \begin{pmatrix} 0 & 0 & 0 \\ 0 & 0 & -\dot{\varphi} \\ 0 & \dot{\varphi} & 0 \end{pmatrix}.$$

If $P$ is an arbitrary point that is stationary relative to $\mathcal{B}$, i.e., such that

$$\frac{{}^{\mathcal{B}}dP}{dt} = \mathbf{0},$$

then

$$\begin{aligned}
{}^{\mathcal{A}}\frac{dP}{dt} &= {}^{\mathcal{B}}\frac{dP}{dt} + {}^{\mathcal{A}}\mathbf{v}^{\mathcal{B}} + {}^{a}\boldsymbol{\omega}^{b} \times \mathbf{r}^{BP} \\
&= \mathbf{0} + \mathbf{0} + \begin{vmatrix} \mathbf{b}_1 & \mathbf{b}_2 & \mathbf{b}_3 \\ \dot{\varphi} & 0 & 0 \\ {}^{b}\left(\mathbf{r}^{BP}\right)_1 & {}^{b}\left(\mathbf{r}^{BP}\right)_2 & {}^{b}\left(\mathbf{r}^{BP}\right)_3 \end{vmatrix} \\
&= b\begin{pmatrix} 0 \\ -\dot{\varphi}\,{}^{b}\left(\mathbf{r}^{BP}\right)_3 \\ \dot{\varphi}\,{}^{b}\left(\mathbf{r}^{BP}\right)_2 \end{pmatrix}.
\end{aligned}$$

The speed of the point $P$ relative to the $\mathcal{A}$ observer is then

$$\left\| {}^{\mathcal{A}}\frac{dP}{dt} \right\| = \dot{\varphi}\sqrt{{}^{b}\left(\mathbf{r}^{BP}\right)_2^2 + {}^{b}\left(\mathbf{r}^{BP}\right)_3^2},$$

i.e., proportional to the angular speed and the perpendicular distance from the axis through $B$ that is parallel to the $\mathbf{a}_1 = \mathbf{b}_1$ basis vector.

**Illustration 8.8**

If two triads $a$ and $b$ are related through the rotation matrix

$$R_{ab} = R(\varphi, 0, 1, 0) = \begin{pmatrix} \cos\varphi & 0 & \sin\varphi \\ 0 & 1 & 0 \\ -\sin\varphi & 0 & \cos\varphi \end{pmatrix},$$

then the angular velocity becomes

$${}^{a}\boldsymbol{\omega}^{b} = b\begin{pmatrix} 0 \\ \dot{\varphi} \\ 0 \end{pmatrix},$$

since

$$R_{ba}\frac{dR_{ab}}{dt} = \begin{pmatrix} 0 & 0 & \dot{\varphi} \\ 0 & 0 & 0 \\ -\dot{\varphi} & 0 & 0 \end{pmatrix}.$$

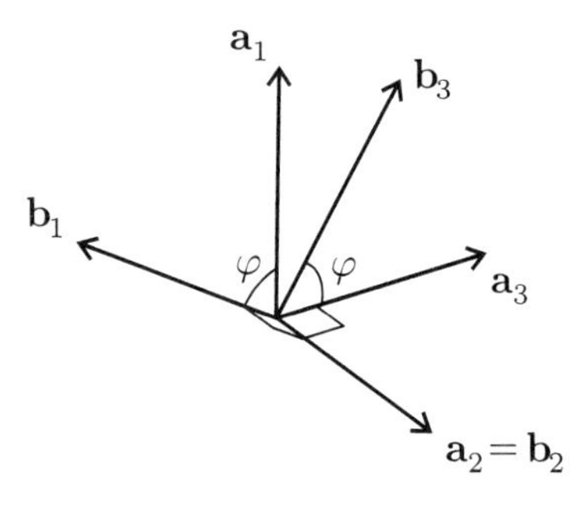


Similarly, if two triads $a$ and $b$ are related through the rotation matrix

$$R_{ab} = R(\varphi, 0, 0, 1) = \begin{pmatrix} \cos\varphi & -\sin\varphi & 0 \\ \sin\varphi & \cos\varphi & 0 \\ 0 & 0 & 1 \end{pmatrix},$$

then the angular velocity becomes

$$^{a}\boldsymbol{\omega}^{b} = b \begin{pmatrix} 0 \\ 0 \\ \dot{\varphi} \end{pmatrix},$$

since

$$R_{ba}\frac{dR_{ab}}{dt} = \begin{pmatrix} 0 & -\dot{\varphi} & 0 \\ \dot{\varphi} & 0 & 0 \\ 0 & 0 & 0 \end{pmatrix}.$$

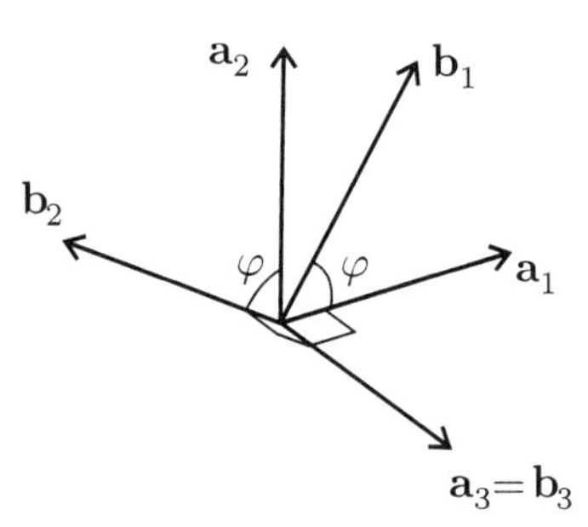


### 8.4.2 Euler Angles

Suppose you want to describe the motion of a free rigid body relative to a background. Introduce an observer $\mathcal{A}$, relative to which the background is stationary, with reference point $A$ and reference triad $a$. Denote by $B$ and $b$ the reference point and reference triad, respectively, of the rigid body.

Since the rigid body's motion is unconstrained, the position of the reference point $B$ relative to the observer $\mathcal{A}$ is given by the coordinate representation

$$^{\mathcal{A}}B = \begin{pmatrix} q_1 \\ q_2 \\ q_3 \end{pmatrix},$$

where $q_1$, $q_2$, and $q_3$ are the corresponding Cartesian coordinates. Similarly, the orientation of the triad $b$ relative to the triad $a$ is given by the rotation matrix

$$R_{ab} = R(q_4, 0, 0, 1)\, R(q_5, 1, 0, 0)\, R(q_6, 0, 0, 1),$$

where $q_4$, $q_5$, and $q_6$ are the Euler angles corresponding to a $3-1-3$ sequence of rotations. From the definitions of the linear and angular velocities of the rigid body, it follows that

$$\frac{^{\mathcal{A}}dB}{dt} = a\frac{d\,^{\mathcal{A}}B}{dt} = a\begin{pmatrix} \dot{q}_1 \\ \dot{q}_2 \\ \dot{q}_3 \end{pmatrix}$$

and

$$^{a}\boldsymbol{\omega}^{b} = b\begin{pmatrix} \dot{q}_4 s_5 s_6 + \dot{q}_5 c_6 \\ \dot{q}_4 s_5 c_6 - \dot{q}_5 s_6 \\ \dot{q}_4 c_5 + \dot{q}_6 \end{pmatrix},$$

since

$$R_{ba}\frac{dR_{ab}}{dt} = \begin{pmatrix} 0 & -\dot{q}_4 c_5 - \dot{q}_6 & \dot{q}_4 s_5 c_6 - \dot{q}_5 s_6 \\ \dot{q}_4 c_5 + \dot{q}_6 & 0 & -\dot{q}_4 s_5 s_6 - \dot{q}_5 c_6 \\ -\dot{q}_4 s_5 c_6 + \dot{q}_5 s_6 & \dot{q}_4 s_5 s_6 + \dot{q}_5 c_6 & 0 \end{pmatrix},$$

where $c_i = \cos q_i$ and $s_i = \sin q_i$ for $i = 4, 5, 6$.

**Illustration 8.9**

Alternatively, introduce two additional triads $t^{(1)}$ and $t^{(2)}$, such that

$$R_{at^{(1)}} = R(q_4, 0, 0, 1),$$
$$R_{t^{(1)}t^{(2)}} = R(q_5, 1, 0, 0),$$

and

$$R_{t^{(2)}b} = R(q_6, 0, 0, 1).$$

As before, it follows that

$$R_{ab} = R_{at^{(1)}} R_{t^{(1)}t^{(2)}} R_{t^{(2)}b}.$$

From the previous section, we find

$$
\begin{aligned}
{}^{a}\boldsymbol{\omega}^{t^{(1)}} &= \dot{q}_4 \mathbf{a}_3, \\
{}^{t^{(1)}}\boldsymbol{\omega}^{t^{(2)}} &= \dot{q}_5 \mathbf{t}_1^{(1)},
\end{aligned}
$$

and

$$
{}^{t^{(2)}}\boldsymbol{\omega}^{b} = \dot{q}_6 \mathbf{t}_3^{(2)}.
$$

Thus,

$$
\begin{aligned}
{}^{a}\boldsymbol{\omega}^{b} &= {}^{a}\boldsymbol{\omega}^{t^{(1)}} + {}^{t^{(1)}}\boldsymbol{\omega}^{t^{(2)}} + {}^{t^{(2)}}\boldsymbol{\omega}^{b} \\
&= bR_{ba}\begin{pmatrix} 0 \\ 0 \\ \dot{q}_4 \end{pmatrix} + bR_{bt^{(1)}}\begin{pmatrix} \dot{q}_5 \\ 0 \\ 0 \end{pmatrix} + bR_{bt^{(2)}}\begin{pmatrix} 0 \\ 0 \\ \dot{q}_6 \end{pmatrix} \\
&= b\begin{pmatrix} \dot{q}_4 \sin q_5 \sin q_6 + \dot{q}_5 \cos q_6 \\ \dot{q}_4 \sin q_5 \cos q_6 - \dot{q}_5 \sin q_6 \\ \dot{q}_4 \cos q_5 + \dot{q}_6 \end{pmatrix}
\end{aligned}
$$

as expected.

### 8.4.3 Multibody Mechanisms

We repeat for reference the discussion from Chapter 6 of the bench-based radial arm saw as shown in the figure below.

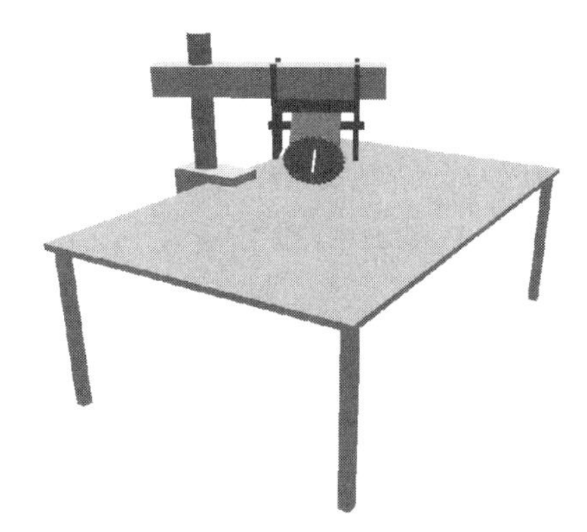

There are four parts that move relative to the stationary workbench, namely, the tool arm, the tool trolley, the blade support, and the blade. A vertical cylinder attached to the workbench provides the support for the mechanism. Specifically:

- The tool arm is free to slide up and down along this cylinder as well as rotate about the cylinder;

- The tool trolley is free to slide along the arm;
- The blade support is free to rotate relative to the tool trolley about an axis parallel to the arm;
- The blade is free to rotate about an arm perpendicular to the axis of rotation of the blade support.

Introduce a main observer $\mathcal{W}$, relative to which the workbench remains stationary. Let its reference point $W$ be located at the center of the vertical cylinder and level with the table top. Let its reference triad $w$ be oriented such that the table top is parallel to the $\mathbf{w}_1$ and $\mathbf{w}_2$ basis vectors and $\mathbf{w}_3$ points away from the table top in the direction of the tool arm.

Introduce an auxiliary observer $\mathcal{A}$, relative to which the tool arm remains stationary. Let its reference point $A$ be located at the center of the vertical cylinder and level with the tool arm. Let its reference triad $a$ be oriented such that $\mathbf{a}_3$ equals $\mathbf{w}_3$ and $\mathbf{a}_2$ is parallel to the tool arm. Then,

$$\mathbf{r}^{WA} = q_1 \mathbf{w}_3 \text{ and } R_{wa} = R\left(q_2, 0, 0, 1\right),$$

where $q_1$ and $q_2$ are configuration coordinates. It follows that

$$^{\mathcal{W}}\mathbf{v}^{\mathcal{A}} = \dot{q}_1 \mathbf{w}_3 \text{ and } ^{w}\boldsymbol{\omega}^{a} = \dot{q}_2 \mathbf{a}_3.$$

Introduce an auxiliary observer $\mathcal{B}$, relative to which the tool trolley remains stationary. Let its reference point $B$ be located on the center line of the tool arm and symmetric relative to the ends of the trolley. Let its reference triad $b$ equal $a$. Then,

$$\mathbf{r}^{AB} = q_3 \mathbf{a}_2 \text{ and } R_{ab} = Id,$$

where $q_3$ is a configuration coordinate. It follows that

$$^{\mathcal{A}}\mathbf{v}^{\mathcal{B}} = \dot{q}_3 \mathbf{a}_2 \text{ and } ^{a}\boldsymbol{\omega}^{b} = \mathbf{0}.$$

Introduce an auxiliary observer $\mathcal{C}$, relative to which the blade support remains stationary. Let its reference point $C$ be located at the center of the axis about which the blade support rotates. Let its reference triad $c$ be oriented such that $\mathbf{c}_2$ equals $\mathbf{a}_2$ and $\mathbf{c}_3$ points toward the center of the blade. Then,

$$\mathbf{r}^{BC} = -p_1 \mathbf{a}_3 \text{ and } R_{bc} = R\left(q_4, 0, 1, 0\right),$$

where $p_1$ is some time-independent parameter and $q_4$ is a configuration coordinate. It follows that

$$^{\mathcal{B}}\mathbf{v}^{\mathcal{C}} = \mathbf{0} \text{ and } ^{b}\boldsymbol{\omega}^{c} = \dot{q}_4 \mathbf{c}_2.$$

Finally, introduce an auxiliary observer $\mathcal{D}$, relative to which the blade remains stationary. Let its reference point $D$ be at the center of the blade. Let its reference triad $d$ be such that $\mathbf{d}_3$ equals $\mathbf{c}_3$. Then,

$$\mathbf{r}^{CD} = p_2\mathbf{c}_3 \text{ and } R_{cd} = R(q_5, 0, 0, 1),$$

where $p_2$ is some time-independent parameter and $q_5$ is a configuration coordinate. It follows that

$${}^{\mathcal{C}}\mathbf{v}^{\mathcal{D}} = \mathbf{0} \text{ and } {}^{c}\boldsymbol{\omega}^{d} = \dot{q}_5\mathbf{d}_3.$$

From the relations derived in Section 8.2, we find

$${}^{w}\boldsymbol{\omega}^{d} = {}^{w}\boldsymbol{\omega}^{a} + {}^{a}\boldsymbol{\omega}^{b} + {}^{b}\boldsymbol{\omega}^{c} + {}^{c}\boldsymbol{\omega}^{d}$$

and

$$\begin{aligned}
{}^{\mathcal{W}}\mathbf{v}^{\mathcal{D}} &= {}^{\mathcal{W}}\mathbf{v}^{\mathcal{A}} + {}^{\mathcal{A}}\mathbf{v}^{\mathcal{D}} + {}^{w}\boldsymbol{\omega}^{a} \times \mathbf{r}^{AD} \\
&= {}^{\mathcal{W}}\mathbf{v}^{\mathcal{A}} + {}^{\mathcal{A}}\mathbf{v}^{\mathcal{B}} + {}^{\mathcal{B}}\mathbf{v}^{\mathcal{D}} \\
&\quad + {}^{a}\boldsymbol{\omega}^{b} \times \mathbf{r}^{BD} + {}^{w}\boldsymbol{\omega}^{a} \times \left(\mathbf{r}^{AB} + \mathbf{r}^{BC} + \mathbf{r}^{CD}\right) \\
&= {}^{\mathcal{W}}\mathbf{v}^{\mathcal{A}} + {}^{\mathcal{A}}\mathbf{v}^{\mathcal{B}} + {}^{\mathcal{B}}\mathbf{v}^{\mathcal{C}} + {}^{\mathcal{C}}\mathbf{v}^{\mathcal{D}} \\
&\quad + {}^{b}\boldsymbol{\omega}^{c} \times \mathbf{r}^{CD} + {}^{a}\boldsymbol{\omega}^{b} \times \left(\mathbf{r}^{BC} + \mathbf{r}^{CD}\right) \\
&\quad + {}^{w}\boldsymbol{\omega}^{a} \times \left(\mathbf{r}^{AB} + \mathbf{r}^{BC} + \mathbf{r}^{CD}\right).
\end{aligned}$$

Alternatively, we may choose to compute ${}^{w}\boldsymbol{\omega}^{d}$ and ${}^{\mathcal{W}}\mathbf{v}^{\mathcal{D}}$ directly from their definitions.

(Ex. 8.16 – Ex. 8.25)

## 8.5 The MAMBO Toolbox

The MAMBO toolbox includes functions for computing the time derivative of an expression (`DiffTime`), the velocity of a point relative to an observer (`LinearVelocity`), and the angular velocity between two triads (`AngularVelocity`). Before invoking either of these procedures, it is necessary to declare which variables are configuration coordinates, and consequently, implicitly dependent on time. This is achieved with the `DeclareStates` procedure[1], which stores the necessary information in the global variables `GlobalExplicit` and `GlobalImplicit`.

In the following MAMBO toolbox session, the `DeclareStates` procedure is invoked to declare the variables $q_1$, $q_2$, and $q_3$ implicitly dependent on time $t$. The subsequent calls to `DiffTime` illustrate its use.

```
> Restart():
> DeclareStates(q1,q2,q3):
```

[1]The declaration of a configuration coordinate as implicitly time-dependent through the `DeclareStates` procedure has no effect on whether to treat the coordinate as a MAMBO state variable or a MAMBO animated variable.

```
> DiffTime(q1-q2*sin(q3));
```

$$q1t - q2t\sin(q3) - q2\cos(q3)\,q3t$$

```
> DiffTime([q1,q2-q3*ln(q1)]);
```

$$[q1t,\ q2t - q3t\ln(q1) - \frac{q3\ q1t}{q1}]$$

```
> DiffTime(q1=q2*q3);
```

$$q1t = q2t\ q3 + q2\ q3t$$

```
> DiffTime(MakeRotations(q1,3));
```

$$\begin{bmatrix} -\sin(q1)\,q1t & -\cos(q1)\,q1t & 0 \\ \cos(q1)\,q1t & -\sin(q1)\,q1t & 0 \\ 0 & 0 & 0 \end{bmatrix}$$

```
> DeclareTriads(a):
> DiffTime(MakeTranslations(a,q1,q2,q3),a);
```

table([
"Type" = "Vector"
"Size" = 1
1 = table([
"Coordinates" = $[q1t,\ q2t,\ q3t]$
"Triad" = $a$
])
])

Note, in particular, the need for a second argument to `DiffTime` when computing the time derivative of a vector relative to some triad.

The `AngularVelocity` and `LinearVelocity` commands compute the angular velocity between two triads and the linear velocity of a point relative to an observer following the rules developed in this chapter. For example,

```
> Restart():
> DeclareTriads(a,b,c):
> DefineTriads([a,b,q1,1],[b,c,q2,cos(t),sin(t),0]):
> DeclareStates(q1,q2):
```

```
> AngularVelocity(a,b);
```

table([
"Type" = "Vector"
"Size" = 1
1 = table([
"Coordinates" = [$q1t$, 0, 0]
"Triad" = $b$
])
])

```
> AngularVelocity(a,c);
```

table([
"Type" = "Vector"
"Size" = 2
1 = table([
"Coordinates" = [$q1t$, 0, 0]
"Triad" = $b$
])
2 = table([
"Coordinates" =
$[\cos(t)\, q2t - \sin(q2)\sin(t),\ \sin(q2)\cos(t) + \sin(t)\, q2t,\ -1 + \cos(q2)]$
"Triad" = $c$
])
])

Similarly,

```
> DeclareObservers(A):
> DeclarePoints(A,B,C):
> DefineObservers([A,A,a]):
> DefinePoints([A,B,a,q3,q4,q5],[B,C,b,0,q6,0]):
> DeclareStates(q3,q4,q5,q6):

> LinearVelocity(A,B);
```

table([
"Type" = "Vector"
"Size" = 1
1 = table([
"Coordinates" = $[q3t,\ q4t,\ q5t]$
"Triad" = $a$
])
])

```
> LinearVelocity(A,C);
```

table([
"Type" = "Vector"
"Size" = 2
1 = table([
"Coordinates" = $[q3t,\ q4t,\ q5t]$
"Triad" = $a$
])
2 = table([
"Coordinates" = $[0,\ q6t,\ q1t\ q6]$
"Triad" = $b$
])
])

where the first argument corresponds to the observer label and the second argument corresponds to the point label.

## 8.6 Exercises

**Exercise 8.1** Let $a$ and $b$ be two arbitrary triads. Show that

$$R_{ba}(t)\,\frac{dR_{ab}(t)}{dt} = \begin{pmatrix} 0 & -\omega_3(t) & \omega_2(t) \\ \omega_3(t) & 0 & -\omega_1(t) \\ -\omega_2(t) & \omega_1(t) & 0 \end{pmatrix}$$

for some functions $\omega_1(t)$, $\omega_2(t)$, and $\omega_3(t)$.

**Solution.** From Chapter 5, we recall that

$$R_{ba}(t)\, R_{ab}(t) = \begin{pmatrix} 1 & 0 & 0 \\ 0 & 1 & 0 \\ 0 & 0 & 1 \end{pmatrix}.$$

Differentiating with respect to $t$ on both sides of this equality yields

$$\frac{dR_{ba}(t)}{dt} R_{ab}(t) + R_{ba}(t) \frac{dR_{ab}(t)}{dt} = \begin{pmatrix} 0 & 0 & 0 \\ 0 & 0 & 0 \\ 0 & 0 & 0 \end{pmatrix},$$

i.e.,

$$\frac{dR_{ba}(t)}{dt} R_{ab}(t) = -R_{ba}(t) \frac{dR_{ab}(t)}{dt}.$$

From matrix algebra, we recall that

$$(AB)^T = B^T A^T$$

for arbitrary matrices $A$ and $B$. Since $(R_{ab})^T = R_{ba}$, it follows that

$$\begin{aligned} \left[R_{ba}(t) \frac{dR_{ab}(t)}{dt}\right]^T &= \frac{d\,(R_{ab}(t))^T}{dt} (R_{ba}(t))^T \\ &= \frac{dR_{ba}(t)}{dt} R_{ab}(t) \\ &= -R_{ba}(t) \frac{dR_{ab}(t)}{dt}. \end{aligned}$$

In terms of indices, this implies that

$$\left[R_{ba}(t) \frac{dR_{ab}(t)}{dt}\right]_{ij} = -\left[R_{ba}(t) \frac{dR_{ab}(t)}{dt}\right]_{ji},$$

which is only possible if

$$R_{ba}(t) \frac{dR_{ab}(t)}{dt} = \begin{pmatrix} 0 & -\omega_3(t) & \omega_2(t) \\ \omega_3(t) & 0 & -\omega_1(t) \\ -\omega_2(t) & \omega_1(t) & 0 \end{pmatrix}$$

as claimed.

**Exercise 8.2** Let $\mathcal{A}$ and $\mathcal{B}$ be two observers with reference points $A$ and $B$ and reference triads $a$ and $b$, such that

$$\mathbf{r}^{AB} = a \begin{pmatrix} t \\ \cos t \\ \sin t \end{pmatrix}$$

and

$$R_{ab} = R\left(\frac{\pi}{2}, \cos t, \sin t, 0\right).$$

Find the linear and angular velocities of $\mathcal{B}$ relative to $\mathcal{A}$.

**Solution.** The linear velocity is given by

$$\begin{aligned} {}^{\mathcal{A}}\mathbf{v}^{\mathcal{B}} &= \frac{{}^a d\mathbf{r}^{AB}}{dt} \\ &= a \frac{d\,{}^a\left(\mathbf{r}^{AB}\right)}{dt} \\ &= a \begin{pmatrix} 1 \\ -\sin t \\ \cos t \end{pmatrix}. \end{aligned}$$

Similarly, since

$$R_{ba} \frac{dR_{ab}}{dt} = \begin{pmatrix} 0 & 1 & \cos t \\ -1 & 0 & \sin t \\ -\cos t & -\sin t & 0 \end{pmatrix},$$

we conclude that the angular velocity equals

$${}^{\mathcal{A}}\boldsymbol{\omega}^{\mathcal{B}} = b \begin{pmatrix} -\sin t \\ \cos t \\ -1 \end{pmatrix}.$$

**Exercise 8.3** Let $\mathcal{A}$ and $\mathcal{B}$ be two observers with reference points $A$ and $B$ and reference triads $a$ and $b$. Find the linear velocity ${}^{\mathcal{A}}\mathbf{v}^{\mathcal{B}}$ when $\mathbf{r}^{AB} =$

a) $a\begin{pmatrix} t^2 \\ \sin t \\ e^{5t} \end{pmatrix}$ b) $a\begin{pmatrix} \cos t \\ t\sin t \\ t^3 e^{5t} \end{pmatrix}$

c) $a\begin{pmatrix} 2 \\ 0 \\ 3 \end{pmatrix}$ d) $a\begin{pmatrix} 2t \\ 0 \\ 3-2t \end{pmatrix}$

e) $a\begin{pmatrix} t-t^2 \\ 1-2t \\ -2 \end{pmatrix}$ f) $a\begin{pmatrix} \cos t^2 \\ \sin t^2 \\ 1 \end{pmatrix}$

**Exercise 8.4** Let $a$ and $b$ be two arbitrary triads. Find the angular velocity ${}^a\boldsymbol{\omega}^b$ when $R_{ab} =$

a) $R(t,1,0,0)$ b) $R(t,1,0,0)\cdot R(2t,0,0,1)$
c) $R(2t,t,0,t)$ d) $R(\cos t,0,0,1)$
e) $R(\cos t,\cos t,0,1)$ f) $R(t,t,1,0)$

**Exercise 8.5** Let $a$ and $b$ be two arbitrary triads and let $\mathbf{v}$ be an arbitrary vector. Show that

$$bR_{ba}\frac{dR_{ab}}{dt}\,{}^b v = {}^a\boldsymbol{\omega}^b \times \mathbf{v}.$$

**Solution.** From Exercise 8.1, we find that

$$\begin{aligned}
&bR_{ba}\frac{dR_{ab}}{dt}{}^b v \\
&= b\begin{pmatrix} 0 & -{}^b\omega_3 & {}^b\omega_2 \\ {}^b\omega_3 & 0 & -{}^b\omega_1 \\ -{}^b\omega_2 & {}^b\omega_1 & 0 \end{pmatrix}\begin{pmatrix} {}^b v_1 \\ {}^b v_2 \\ {}^b v_3 \end{pmatrix} \\
&= b\begin{pmatrix} {}^b\omega_2{}^b v_3 - {}^b\omega_3{}^b v_2 \\ {}^b\omega_3{}^b v_1 - {}^b\omega_1{}^b v_3 \\ {}^b\omega_1{}^b v_2 - {}^b\omega_2{}^b v_1 \end{pmatrix}.
\end{aligned}$$

Similarly,

$$\begin{aligned}
{}^a\boldsymbol{\omega}^b \times \mathbf{v} &= \begin{vmatrix} \mathbf{b}_1 & \mathbf{b}_2 & \mathbf{b}_3 \\ {}^b\omega_1 & {}^b\omega_2 & {}^b\omega_3 \\ {}^b v_1 & {}^b v_2 & {}^b v_3 \end{vmatrix} \\
&= b\begin{pmatrix} {}^b\omega_2{}^b v_3 - {}^b\omega_3{}^b v_2 \\ {}^b\omega_3{}^b v_1 - {}^b\omega_1{}^b v_3 \\ {}^b\omega_1{}^b v_2 - {}^b\omega_2{}^b v_1 \end{pmatrix}
\end{aligned}$$

and the claim follows.

**Exercise 8.6** Let $a$ and $b$ be two arbitrary triads. Show that

$$\begin{aligned}
{}^a\boldsymbol{\omega}^b &= \mathbf{b}_1\left({}^a\frac{d\mathbf{b}_2}{dt}\bullet\mathbf{b}_3\right) \\
&\quad+\mathbf{b}_2\left({}^a\frac{d\mathbf{b}_3}{dt}\bullet\mathbf{b}_1\right) \\
&\quad+\mathbf{b}_3\left({}^a\frac{d\mathbf{b}_1}{dt}\bullet\mathbf{b}_2\right).
\end{aligned}$$

**Solution.** Recall the relation between the derivative of a vector relative to two different triads:

$${}^a\frac{d\mathbf{r}}{dt} = {}^b\frac{d\mathbf{r}}{dt} + {}^a\boldsymbol{\omega}^b \times \mathbf{r}.$$

In particular,

$$\begin{aligned}
{}^a\frac{d\mathbf{b}_i}{dt} &= {}^b\frac{d\mathbf{b}_i}{dt} + {}^a\boldsymbol{\omega}^b \times \mathbf{b}_i \\
&= {}^a\boldsymbol{\omega}^b \times \mathbf{b}_i,
\end{aligned}$$

since the basis vector $\mathbf{b}_i$ has a time-independent matrix representation relative to the $b$ triad. It follows that

$$\begin{aligned}
{}^a\frac{d\mathbf{b}_1}{dt} &= \begin{vmatrix} \mathbf{b}_1 & \mathbf{b}_2 & \mathbf{b}_3 \\ {}^b\omega_1 & {}^b\omega_2 & {}^b\omega_3 \\ 1 & 0 & 0 \end{vmatrix} \\
&= {}^b\omega_3\mathbf{b}_2 - {}^b\omega_2\mathbf{b}_3,
\end{aligned}$$

$$\begin{aligned}
{}^a\frac{d\mathbf{b}_2}{dt} &= \begin{vmatrix} \mathbf{b}_1 & \mathbf{b}_2 & \mathbf{b}_3 \\ {}^b\omega_1 & {}^b\omega_2 & {}^b\omega_3 \\ 0 & 1 & 0 \end{vmatrix} \\
&= -{}^b\omega_3\mathbf{b}_1 + {}^b\omega_1\mathbf{b}_3,
\end{aligned}$$

and

$$\begin{aligned}
{}^a\frac{d\mathbf{b}_3}{dt} &= \begin{vmatrix} \mathbf{b}_1 & \mathbf{b}_2 & \mathbf{b}_3 \\ {}^b\omega_1 & {}^b\omega_2 & {}^b\omega_3 \\ 0 & 0 & 1 \end{vmatrix} \\
&= {}^b\omega_2\mathbf{b}_1 - {}^b\omega_1\mathbf{b}_2.
\end{aligned}$$

Consequently,

$$\mathbf{b}_1\left({}^a\frac{d\mathbf{b}_2}{dt}\bullet\mathbf{b}_3\right)+\mathbf{b}_2\left({}^a\frac{d\mathbf{b}_3}{dt}\bullet\mathbf{b}_1\right)+\mathbf{b}_3\left({}^a\frac{d\mathbf{b}_1}{dt}\bullet\mathbf{b}_2\right)=\mathbf{b}_1{}^b\omega_1+\mathbf{b}_2{}^b\omega_2+\mathbf{b}_3{}^b\omega_3$$

as claimed.

**Exercise 8.7** Let $a$ and $b$ be two arbitrary triads. Compute the time derivatives of the basis vectors of the $b$ triad relative to the $a$ triad, if $R_{ab}=$

a) $R(t,1,0,0)$  b) $R(t,1,0,0)\cdot R(2t,0,0,1)$
c) $R(2t,t,0,t)$  d) $R(\cos t,0,0,1)$
e) $R(\cos t,\cos t,0,1)$  f) $R(t,t,1,0)$

**Exercise 8.8** Let $a$ and $b$ be two arbitrary triads. Show that

$$\frac{{}^a d\,{}^a\boldsymbol{\omega}^b}{dt}=\frac{{}^b d\,{}^a\boldsymbol{\omega}^b}{dt}.$$

**Solution.** For an arbitrary vector $\mathbf{r}$, we have

$$\frac{{}^a d\mathbf{r}}{dt}=\frac{{}^b d\mathbf{r}}{dt}+{}^a\boldsymbol{\omega}^b\times\mathbf{r}.$$

In particular, it follows that

$$\frac{{}^a d\,{}^a\boldsymbol{\omega}^b}{dt}=\frac{{}^b d\,{}^a\boldsymbol{\omega}^b}{dt}+{}^a\boldsymbol{\omega}^b\times{}^a\boldsymbol{\omega}^b=\frac{{}^b d\,{}^a\boldsymbol{\omega}^b}{dt},$$

since

$${}^a\boldsymbol{\omega}^b\times{}^a\boldsymbol{\omega}^b=\mathbf{0}.$$

**Exercise 8.9** Let $a$ and $b$ be two arbitrary triads. Show that

$$\frac{{}^a d\mathbf{r}}{dt}=\frac{{}^b d\mathbf{r}}{dt}$$

for an arbitrary vector $\mathbf{r}$ if and only if $R_{ab}$ is time-independent.

**Solution.** If $R_{ab}$ is time-independent, then

$$R_{ba}\frac{dR_{ab}}{dt}=\begin{pmatrix}0&0&0\\0&0&0\\0&0&0\end{pmatrix},$$

i.e.,

$${}^a\boldsymbol{\omega}^b=\mathbf{0}.$$

It follows that

$$\frac{{}^a d\mathbf{r}}{dt}=\frac{{}^b d\mathbf{r}}{dt}+{}^a\boldsymbol{\omega}^b\times\mathbf{r}=\frac{{}^b d\mathbf{r}}{dt}$$

for any vector $\mathbf{r}$.

Conversely, if

$$\frac{{}^a d\mathbf{r}}{dt}=\frac{{}^b d\mathbf{r}}{dt}$$

for an arbitrary vector $\mathbf{r}$, then

$$\frac{{}^a d\mathbf{b}_i}{dt}=\frac{{}^b d\mathbf{b}_i}{dt}=\mathbf{0}$$

for any basis vector of the triad $b$. This implies that the matrix representations of the basis vectors of $b$ relative to the triad $a$ are time-independent. But

$$b=aR_{ab},$$

i.e., the columns of $R_{ab}$ are time-independent as claimed.

**Exercise 8.10** Show that the angular velocity of a rigid body relative to an observer $\mathcal{A}$ with reference triad $a$ is independent of the choice of reference triad for the rigid body.

**Exercise 8.11** Two bodies in contact are said to be in relative slip relative to an observer $\mathcal{A}$ if the velocities of the points in contact relative to $\mathcal{A}$ are not equal. Show that if two bodies in contact are in relative slip relative to an observer $\mathcal{A}$, then they are in relative slip relative to any other observer $\mathcal{B}$.

**Solution.** Denote the contact points on the two bodies by $P_1$ and $P_2$. It follows that

$$\mathbf{r}^{BP_1} = \mathbf{r}^{BP_2}.$$

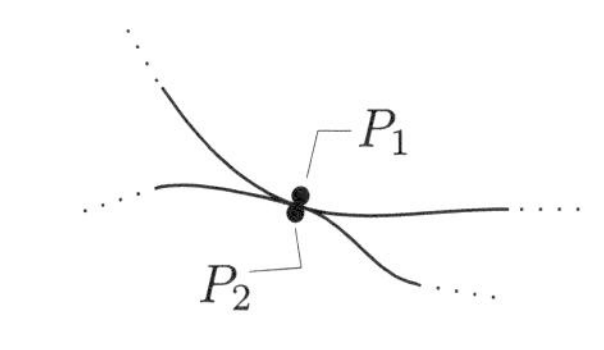


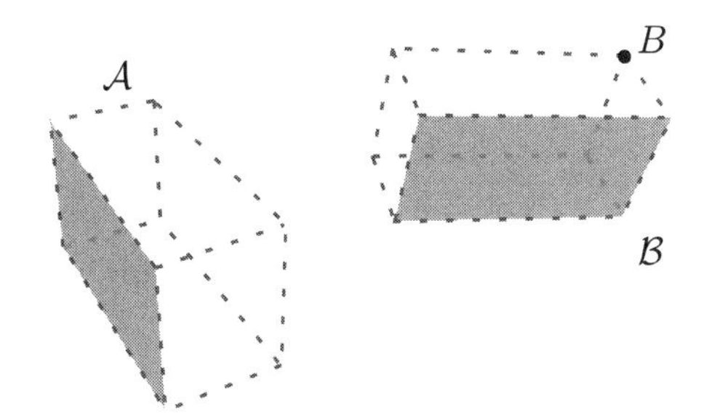


The bodies are in relative slip relative to $\mathcal{A}$ if

$$\frac{{}^{\mathcal{A}}dP_1}{dt} - \frac{{}^{\mathcal{A}}dP_2}{dt} \neq \mathbf{0}.$$

Since

$$\frac{{}^{\mathcal{A}}dP}{dt} = \frac{{}^{\mathcal{B}}dP}{dt} + {}^{\mathcal{A}}\mathbf{v}^{\mathcal{B}} + {}^{a}\boldsymbol{\omega}^{b} \times \mathbf{r}^{BP},$$

it follows that

$$\begin{aligned}\frac{{}^{\mathcal{B}}dP_1}{dt} - \frac{{}^{\mathcal{B}}dP_2}{dt} &= \frac{{}^{\mathcal{A}}dP_1}{dt} - {}^{\mathcal{A}}\mathbf{v}^{\mathcal{B}} - {}^{a}\boldsymbol{\omega}^{b} \times \mathbf{r}^{BP_1} \\ &\quad - \frac{{}^{\mathcal{A}}dP_2}{dt} + {}^{\mathcal{A}}\mathbf{v}^{\mathcal{B}} + {}^{a}\boldsymbol{\omega}^{b} \times \mathbf{r}^{BP_2} \\ &= \frac{{}^{\mathcal{A}}dP_1}{dt} - \frac{{}^{\mathcal{A}}dP_2}{dt} \\ &\neq \mathbf{0}\end{aligned}$$

and the claim follows.

**Exercise 8.12** Let $B_1$ and $B_2$ be two arbitrary points on a rigid body with reference triad $b$. Show that

$$\frac{{}^{\mathcal{A}}dB_2}{dt} - \frac{{}^{\mathcal{A}}dB_1}{dt} = {}^{a}\boldsymbol{\omega}^{b} \times \mathbf{r}^{B_1B_2},$$

where $\mathcal{A}$ is an arbitrary observer with reference point $A$ and reference triad $a$.

[Hint: Let $\mathcal{B}$ be an observer with reference point $B$ and reference triad $b$, relative to which the rigid body is stationary. Use the formula

$$\frac{{}^{\mathcal{A}}dP}{dt} = \frac{{}^{\mathcal{B}}dP}{dt} + {}^{\mathcal{A}}\mathbf{v}^{\mathcal{B}} + {}^{a}\boldsymbol{\omega}^{b} \times \mathbf{r}^{BP},$$

where the first term on the right-hand side vanishes for any point $P$ on the rigid body.]

**Exercise 8.13** Let $a$ and $b$ be two arbitrary triads. Suppose that

$${}^{a}\boldsymbol{\omega}^{b} = b\begin{pmatrix} u_1 \\ u_2 \\ u_3 \end{pmatrix}$$

and find the relationship between the Euler angles and their time derivatives and the quantities $u_1$, $u_2$, and $u_3$ if the rotation matrix $R_{ab}$ corresponds to a

a) $1-3-1$ b) $2-3-1$
c) $2-1-3$ d) $2-1-2$
e) $3-2-1$ f) $1-2-1$

sequence of rotations.

**Exercise 8.14** Let $\mathcal{A}$ be an arbitrary observer with reference point $A$ and reference triad $a$. Introduce a triad $b$, such that $\mathbf{b}_1$ is parallel to the velocity

$$\frac{{}^{\mathcal{A}}dB}{dt}$$

of a point $B$ on some rigid body and $\mathbf{b}_2$ is parallel to the time derivative of the vector $\mathbf{b}_1$ relative to the $a$ triad. Find the angular velocity ${}^{a}\boldsymbol{\omega}^{b}$.

**Solution.** Following the problem statement, we introduce

$$\mathbf{b}_1 = \frac{{}^{\mathcal{A}}dB}{dt} \Big/ \left\| \frac{{}^{\mathcal{A}}dB}{dt} \right\|,$$
$$\mathbf{b}_2 = \frac{1}{\kappa(t)} \frac{{}^{a}d\mathbf{b}_1}{dt},$$

where $\kappa(t) = \left\| \frac{{}^{a}d\mathbf{b}_1}{dt} \right\|$ is known as the *curvature* of the path in space followed by $B$ relative to the observer $\mathcal{A}$, and

$$\mathbf{b}_3 = \mathbf{b}_1 \times \mathbf{b}_2.$$

From a previous exercise, we recall that

$$\begin{aligned} {}^{a}\boldsymbol{\omega}^{b} &= \mathbf{b}_1 \left( \frac{{}^{a}d\mathbf{b}_2}{dt} \bullet \mathbf{b}_3 \right) \\ &\quad + \mathbf{b}_2 \left( \frac{{}^{a}d\mathbf{b}_3}{dt} \bullet \mathbf{b}_1 \right) \\ &\quad + \mathbf{b}_3 \left( \frac{{}^{a}d\mathbf{b}_1}{dt} \bullet \mathbf{b}_2 \right). \end{aligned}$$

Here,

$$\frac{{}^{a}d\mathbf{b}_1}{dt} \bullet \mathbf{b}_2 = \kappa(t)\, \mathbf{b}_2 \bullet \mathbf{b}_2 = \kappa(t)$$

Moreover, since

$$0 = \mathbf{b}_3 \bullet \mathbf{b}_1$$

for all $t$, it follows that

$$\begin{aligned} 0 &= \frac{{}^{a}d}{dt}(\mathbf{b}_3 \bullet \mathbf{b}_1) \\ &= \frac{{}^{a}d\mathbf{b}_3}{dt} \bullet \mathbf{b}_1 + \mathbf{b}_3 \bullet \frac{{}^{a}d\mathbf{b}_1}{dt} \\ &= \frac{{}^{a}d\mathbf{b}_3}{dt} \bullet \mathbf{b}_1 + \kappa(t)\, \mathbf{b}_3 \bullet \mathbf{b}_2 \\ &= \frac{{}^{a}d\mathbf{b}_3}{dt} \bullet \mathbf{b}_1. \end{aligned}$$

Similarly,

$$1 = \mathbf{b}_3 \bullet \mathbf{b}_3$$

for all $t$ implies that

$$\begin{aligned} 0 &= \frac{{}^{a}d}{dt}(\mathbf{b}_3 \bullet \mathbf{b}_3) \\ &= 2\, \frac{{}^{a}d\mathbf{b}_3}{dt} \bullet \mathbf{b}_3. \end{aligned}$$

Since $b$ is a basis, we must have

$$\frac{{}^{a}d\mathbf{b}_3}{dt} = -\tau(t)\, \mathbf{b}_2,$$

where $\tau(t)$ is known as the *torsion* of the path in space followed by $B$ relative to the observer $\mathcal{A}$.

Finally, since

$$0 = \mathbf{b}_2 \bullet \mathbf{b}_3$$

for all $t$, it follows that

$$\begin{aligned} 0 &= \frac{{}^{a}d}{dt}(\mathbf{b}_2 \bullet \mathbf{b}_3) \\ &= \frac{{}^{a}d\mathbf{b}_2}{dt} \bullet \mathbf{b}_3 + \mathbf{b}_2 \bullet \frac{{}^{a}d\mathbf{b}_3}{dt} \\ &= \frac{{}^{a}d\mathbf{b}_2}{dt} \bullet \mathbf{b}_3 - \tau(t)\, \mathbf{b}_2 \bullet \mathbf{b}_2, \end{aligned}$$

i.e.,

$$\frac{{}^{a}d\mathbf{b}_2}{dt} \bullet \mathbf{b}_3 = \tau(t)$$

and, consequently,

$${}^{a}\boldsymbol{\omega}^{b} = \tau(t)\, \mathbf{b}_1 + \kappa(t)\, \mathbf{b}_3.$$

**Exercise 8.15** Let $\mathcal{A}$ be an observer with reference point $A$ and reference triad $a$. Let $B_1$, $B_2$, and $B_3$ be three, non-colinear points on a rigid body with reference triad $b$, such that the differences

$$\mathbf{v}_{2-1} \stackrel{def}{=} \frac{{}^{\mathcal{A}}dB_2}{dt} - \frac{{}^{\mathcal{A}}dB_1}{dt}$$

and

$$\mathbf{v}_{3-1} \stackrel{def}{=} \frac{{}^{\mathcal{A}}dB_3}{dt} - \frac{{}^{\mathcal{A}}dB_1}{dt}$$

are not parallel, and such that the vectors $\mathbf{v}_{2-1} \times \mathbf{v}_{3-1}$ and $\mathbf{r}^{B_1B_2} \times \mathbf{r}^{B_1B_3}$ are not perpendicular. Find expressions for the direction and magnitude of the angular velocity ${}^{a}\boldsymbol{\omega}^{b}$ of the rigid body relative to $\mathcal{A}$ in terms of $\mathbf{v}_{2-1} \times \mathbf{v}_{3-1}$ and $\mathbf{r}^{B_1B_2} \times \mathbf{r}^{B_1B_3}$.

**Solution.** From a previous exercise, it follows that

$$\mathbf{v}_{2-1} = {}^{a}\boldsymbol{\omega}^{b} \times \mathbf{r}^{B_1B_2}$$

and

$$\mathbf{v}_{3-1} = {}^{a}\boldsymbol{\omega}^{b} \times \mathbf{r}^{B_1B_3}.$$

If we form the cross products of the two left-hand sides and the two right-hand sides, we find

$$\begin{aligned}
&\mathbf{v}_{2-1} \times \mathbf{v}_{3-1}\\
&= \left[{}^{a}\boldsymbol{\omega}^{b} \times \mathbf{r}^{B_1B_2}\right] \times \left[{}^{a}\boldsymbol{\omega}^{b} \times \mathbf{r}^{B_1B_3}\right]\\
&= {}^{a}\boldsymbol{\omega}^{b}\left[\left({}^{a}\boldsymbol{\omega}^{b} \times \mathbf{r}^{B_1B_2}\right) \bullet \mathbf{r}^{B_1B_3}\right]\\
&\quad -\mathbf{r}^{B_1B_3}\left[\left({}^{a}\boldsymbol{\omega}^{b} \times \mathbf{r}^{B_1B_2}\right) \bullet {}^{a}\boldsymbol{\omega}^{b}\right]\\
&= {}^{a}\boldsymbol{\omega}^{b}\left[{}^{a}\boldsymbol{\omega}^{b} \bullet \left(\mathbf{r}^{B_1B_2} \times \mathbf{r}^{B_1B_3}\right)\right].
\end{aligned}$$

Here, we used the facts that

$$\begin{aligned}
\mathbf{w}_1 \times (\mathbf{w}_2 \times \mathbf{w}_3) &= \mathbf{w}_2 (\mathbf{w}_1 \bullet \mathbf{w}_3)\\
&\quad -\mathbf{w}_3 (\mathbf{w}_1 \bullet \mathbf{w}_2)
\end{aligned}$$

(cf. result of Exercise 3.74) and

$$(\mathbf{w}_1 \times \mathbf{w}_2) \bullet \mathbf{w}_3 = \mathbf{w}_1 \bullet (\mathbf{w}_2 \times \mathbf{w}_3)$$

for any vectors $\mathbf{w}_1$, $\mathbf{w}_2$, and $\mathbf{w}_3$, and the observation that the vector

$${}^{a}\boldsymbol{\omega}^{b} \times \mathbf{r}^{B_1B_2}$$

is perpendicular to ${}^{a}\boldsymbol{\omega}^{b}$. It follows that the angular velocity vector ${}^{a}\boldsymbol{\omega}^{b}$ is parallel to the cross product $\mathbf{v}_{2-1} \times \mathbf{v}_{3-1}$, i.e., a unit vector parallel to the angular velocity vector is given by

$$\frac{\mathbf{v}_{2-1} \times \mathbf{v}_{3-1}}{\left\|\mathbf{v}_{2-1} \times \mathbf{v}_{3-1}\right\|}.$$

Moreover,

$$\begin{aligned}
&\left(\mathbf{v}_{2-1} \times \mathbf{v}_{3-1}\right) \bullet \left(\mathbf{r}^{B_1B_2} \times \mathbf{r}^{B_1B_3}\right)\\
&= \left\|\mathbf{v}_{2-1} \times \mathbf{v}_{3-1}\right\| \left\|\mathbf{r}^{B_1B_2} \times \mathbf{r}^{B_1B_3}\right\| \cos\theta,
\end{aligned}$$

where $\theta$ is the angle between the vectors $\mathbf{v}_{2-1} \times \mathbf{v}_{3-1}$ and $\mathbf{r}^{B_1B_2} \times \mathbf{r}^{B_1B_3}$, i.e., between ${}^{a}\boldsymbol{\omega}^{b}$ (or $-{}^{a}\boldsymbol{\omega}^{b}$) and $\mathbf{r}^{B_1B_2} \times \mathbf{r}^{B_1B_3}$. On the other hand,

$$\begin{aligned}
&\left(\mathbf{v}_{2-1} \times \mathbf{v}_{3-1}\right) \bullet \left(\mathbf{r}^{B_1B_2} \times \mathbf{r}^{B_1B_3}\right)\\
&= \left[{}^{a}\boldsymbol{\omega}^{b} \bullet \left(\mathbf{r}^{B_1B_2} \times \mathbf{r}^{B_1B_3}\right)\right]^2\\
&= \left\|{}^{a}\boldsymbol{\omega}^{b}\right\|^2 \left\|\mathbf{r}^{B_1B_2} \times \mathbf{r}^{B_1B_3}\right\|^2 \cos^2\theta.
\end{aligned}$$

Substituting for $\cos\theta$ from the first expression then yields

$$\left\|{}^{a}\boldsymbol{\omega}^{b}\right\| = \frac{\left\|\mathbf{v}_{2-1} \times \mathbf{v}_{3-1}\right\|}{\sqrt{\left(\mathbf{v}_{2-1} \times \mathbf{v}_{3-1}\right) \bullet \left(\mathbf{r}^{B_1B_2} \times \mathbf{r}^{B_1B_3}\right)}}.$$

**Exercise 8.16** Use the MAMBO toolbox to compute the angular velocity vector of a rigid body relative to some observer $\mathcal{A}$, where the orientation of the rigid body relative to $\mathcal{A}$ is described in terms of Euler parameters.

[Hint: Recall the Euler parameter representation of the rotation matrix between the reference triad $a$ of the observer $\mathcal{A}$ and the reference triad $b$ of the rigid body in Illustration 6.4.]

**Exercise 8.17** Use the MAMBO toolbox to compute the velocity relative to an observer $\mathcal{A}$ of the end point of a rigid rod connected at the other end to a stationary hinge joint.

**Solution.** Let the joint axis be given by the vector

$$\mathbf{v} = a \begin{pmatrix} v_1 \\ v_2 \\ v_3 \end{pmatrix},$$

where $a$ is the reference triad of the observer $\mathcal{A}$. Introduce an auxiliary observer $\mathcal{B}$, such that the reference point $B$ coincides with the hinge joint and the basis vector $\mathbf{b}_1$ of the corresponding reference triad is parallel to the separation $\overrightarrow{BP}$, where $P$ is the end point of the rod. In particular, let

$$\mathbf{r}^{AB} = a \begin{pmatrix} r_1 \\ r_2 \\ r_3 \end{pmatrix}$$

describe the position of the reference point $B$ relative to $\mathcal{A}$.

The following MAMBO toolbox statements establish the basic geometry.

```
> Restart():
> DeclareObservers(A,B):
> DeclarePoints(A,B,P):
> DeclareTriads(a,b):
> DefineObservers([A,A,a],[B,B,b]):
> DefinePoints([A,B,a,r1,r2,r3],
> [B,P,b,l,0,0]):
> DefineTriads([a,b,q1,v1,v2,v3]):
> DeclareStates(q1):
```

Here, $l$ represents the length of the rod and $q_1$ is the configuration coordinate representing the rotation angle. The velocity of $P$ relative to $\mathcal{A}$ is now obtained from the statement

```
> LinearVelocity(A,P);
```

$$\frac{{}^{\mathcal{A}}dP}{dt} = b \begin{pmatrix} 0 \\ lv_3\dot{q}_1/\sqrt{v_1^2+v_2^2+v_3^2} \\ lv_2\dot{q}_1/\sqrt{v_1^2+v_2^2+v_3^2} \end{pmatrix}.$$

**Exercise 8.18** Use the MAMBO toolbox to compute the velocity relative to an observer $\mathcal{A}$ of the end point of a rigid rod connected at the other end to a stationary spherical joint.

[Hint: Decompose the rotation into a sequence of rotations about basis vectors.]

**Exercise 8.19** Use the MAMBO toolbox to compute the velocity relative to an observer $\mathcal{A}$ of the center point of a rigid square plate connected at one corner to a stationary ball joint.

[Hint:

```
> Restart():
> DeclareObservers(A,B):
> DeclarePoints(A,B,P):
> DeclareTriads(a,b):
> DefineObservers(
> [A,A,a],[B,B,b]):
> DefinePoints([A,B,a,r1,r2,r3],
> [B,P,b,s/2,s/2,0]):
> DefineTriads([a,b,[theta,3],
> [phi,1],[psi,3]]):
> DeclareStates(theta,phi,psi):
> LinearVelocity(A,P);
```

]

**Exercise 8.20** Suppose the coordinate representation of a point $P$ on a rigid body relative to an observer $\mathcal{A}$ is expressed using spherical coordinates, i.e.,

$${}^{\mathcal{A}}P = \begin{pmatrix} q_1 \sin q_2 \cos q_3 \\ q_1 \sin q_2 \sin q_3 \\ q_1 \cos q_2 \end{pmatrix}.$$

Use the MAMBO toolbox to compute the velocity of the point $P$ relative to an observer $\mathcal{B}$, where

$$\mathbf{r}^{BA} = \mathbf{0} \text{ and } R_{ba} = R(\omega t, 0, 0, 1).$$

Here, $A$, $B$, $a$ and $b$ are the reference points and reference triads of the observers $\mathcal{A}$ and $\mathcal{B}$, and $\omega$ is a constant.

**Solution.** The following MAMBO toolbox statements establish the basic geometry:

```
> Restart():
> DeclareObservers(A,B):
> DeclarePoints(A,B,P):
> DeclareTriads(a,b):
> DefineObservers(
> [A,A,a],[B,B,b]):
> DefinePoints([B,A,b,0,0,0],
> [A,P,a,q1*sin(q2)*cos(q3),
> q1*sin(q2)*sin(q3),q1*cos(q2)]):
> DefineTriads([b,a,omega*t,3]):
> DeclareStates(q1,q2,q3):
```

The linear velocity is then obtained from

```
> LinearVelocity(B,P);
```

**Exercise 8.21** Suppose the coordinate representation of a point $P$ on a rigid body relative to an observer $\mathcal{A}$ is expressed using spherical coordinates, i.e.,

$$^{\mathcal{A}}P = \begin{pmatrix} q_1 \sin q_2 \cos q_3 \\ q_1 \sin q_2 \sin q_3 \\ q_1 \cos q_2 \end{pmatrix}.$$

Use the MAMBO toolbox to compute the velocity of the point $P$ relative to an observer $\mathcal{B}$, where

$$\mathbf{r}^{BA} = b \begin{pmatrix} R \sin(\theta + \omega t) \cos \phi \\ R \sin(\theta + \omega t) \sin \phi \\ R \cos(\theta + \omega t) \end{pmatrix},$$

$\mathbf{a}_3$ is parallel to $\mathbf{r}^{BA}$, and $\mathbf{a}_1 = \mathbf{b}_3 \times \mathbf{a}_3$.

**Exercise 8.22** A sphere of radius $R$ is constrained to be in tangential contact with a plane that contains the reference point $A$ of an observer $\mathcal{A}$ and is parallel to the basis vectors $\mathbf{a}_1$ and $\mathbf{a}_2$ of the reference triad $a$ of $\mathcal{A}$. Use the MAMBO toolbox to find the velocity of the contact point on the sphere relative to $\mathcal{A}$.

**Solution.** Let the reference point $B$ of the sphere be located at the center of the sphere and denote by $b$ the reference triad of the sphere. Then,

$$\mathbf{r}^{AB} = a \begin{pmatrix} q_1 \\ q_2 \\ R \end{pmatrix}$$

and

$$R_{ab} = R(q_3, 0, 0, 1)\, R(q_4, 1, 0, 0)\, R(q_5, 0, 0, 1).$$

If $P$ is some arbitrary point on the rigid body, then

$$\begin{aligned} \frac{^{\mathcal{A}}dP}{dt} &= \frac{^{\mathcal{B}}dP}{dt} + {}^{\mathcal{A}}\mathbf{v}^{\mathcal{B}} + {}^{a}\boldsymbol{\omega}^{b} \times \mathbf{r}^{BP} \\ &= {}^{\mathcal{A}}\mathbf{v}^{\mathcal{B}} + {}^{a}\boldsymbol{\omega}^{b} \times \mathbf{r}^{BP}, \end{aligned}$$

since

$$\frac{^{\mathcal{B}}dP}{dt} = \mathbf{0}.$$

If $P$ corresponds to the point currently in contact with the plane, then

$$\mathbf{r}^{BP} = a \begin{pmatrix} 0 \\ 0 \\ -R \end{pmatrix}.$$

The MAMBO toolbox statements

```
> Restart():
> DeclareObservers(A):
> DeclarePoints(A,B):
> DeclareTriads(a,b):
> DefineObservers([A,A,a]):
> DefinePoints([A,B,a,q1,q2,R]):
> DefineTriads([a,b,[q3,3],[q4,1],
> [q5,3]]):
> DeclareStates(q1,q2,q3,q4,q5):
> vel:=LinearVelocity(A,B) &++
> (AngularVelocity(a,b) &xx
> MakeTranslations(a,0,0,-R)):
```

show that

$$\frac{^{\mathcal{A}}dP}{dt}$$

$$= a \begin{pmatrix} \dot{q}_1 + R\dot{q}_5 \cos q_3 \sin q_4 - R\dot{q}_4 \sin q_3 \\ \dot{q}_2 + R\dot{q}_5 \sin q_3 \sin q_4 + R\dot{q}_4 \cos q_3 \\ 0 \end{pmatrix}.$$

**Exercise 8.23** Consider the sphere in the previous exercise. Then,

$$\frac{{}^{\mathcal{A}}dP}{dt} = \frac{{}^{a}d\mathbf{r}^{AP}}{dt} = \frac{{}^{a}d\mathbf{r}^{AB}}{dt} + \frac{{}^{a}d\mathbf{r}^{BP}}{dt}.$$

Is it correct to substitute

$$\mathbf{r}^{BP} = a \begin{pmatrix} 0 \\ 0 \\ -R \end{pmatrix}$$

into this expression?

**Solution.** $P$ is the fixed point on the surface of the sphere that coincides with the current contact point with the plane. Thus, while

$$\mathbf{r}^{BP} = a \begin{pmatrix} 0 \\ 0 \\ -R \end{pmatrix}$$

is true at the moment that $P$ is in contact with the plane, it is not generally true before and after this moment. Since the expression for the velocity of $P$ above involves a derivative with respect to time, we cannot simply use the expression for $\mathbf{r}^{BP}$ at the present time. Clearly,

$$\frac{{}^{a}d}{dt} a \begin{pmatrix} 0 \\ 0 \\ -R \end{pmatrix} = a \frac{d}{dt} \begin{pmatrix} 0 \\ 0 \\ -R \end{pmatrix} = \mathbf{0},$$

while

$$\frac{{}^{a}d\mathbf{r}^{BP}}{dt} \neq \mathbf{0}$$

in general.

Since $P$ is fixed on the surface of the sphere, it follows that

$$\frac{{}^{b}d\mathbf{r}^{BP}}{dt} = \mathbf{0}$$

and thus that

$$\frac{{}^{a}d\mathbf{r}^{BP}}{dt} = \frac{{}^{b}d\mathbf{r}^{BP}}{dt} + {}^{a}\boldsymbol{\omega}^{b} \times \mathbf{r}^{BP} = {}^{a}\boldsymbol{\omega}^{b} \times \mathbf{r}^{BP}.$$

While the derivative on the left requires knowledge of the matrix representation of $\mathbf{r}^{BP}$ relative to $a$ over some interval containing the present time, the expression on the right involves the value of $\mathbf{r}^{BP}$ only at the present time. While we cannot substitute the expression

$$\mathbf{r}^{BP} = a \begin{pmatrix} 0 \\ 0 \\ -R \end{pmatrix}$$

on the left-hand side, it is correct to substitute it on the right-hand side, as done in the previous exercise.

**Exercise 8.24** A cylinder of radius $R$ is constrained to be in tangential contact along a straight line with a plane that contains the reference point $A$ of an observer $\mathcal{A}$ and is parallel to the basis vectors $\mathbf{a}_1$ and $\mathbf{a}_2$ of the reference triad $a$ of $\mathcal{A}$. Use the MAMBO toolbox to find the velocity of any point on the line of contact relative to $\mathcal{A}$.

**Exercise 8.25** A circular disk of radius $R$ is constrained to be in tangential contact with a plane that contains the reference point $A$ of an observer $\mathcal{A}$ and is parallel to the basis vectors $\mathbf{a}_1$ and $\mathbf{a}_2$ of the reference triad $a$ of $\mathcal{A}$. Use the MAMBO toolbox to find the velocity of the contact point on the disk relative to $\mathcal{A}$.

## Summary of notation

A differentiation symbol with a left superscript, such as ${}^{a}\frac{d}{dt}$ or ${}^{\mathcal{A}}\frac{d}{dt}$, was used in this chapter to denote differentiation with respect to time relative to a triad $a$ or an observer $\mathcal{A}$.

A time derivative of a point $B$ with respect to an observer $\mathcal{A}$ was defined in this chapter as the time derivative of the position vector from the observer's reference point to the point $B$ relative to the observer's reference triad and was denoted by ${}^{\mathcal{A}}\frac{dB}{dt}$.

A bold-faced lower-case $\mathbf{v}$ with a left and right superscript, such as ${}^{\mathcal{A}}\mathbf{v}^{\mathcal{B}}$, was used in this chapter to denote the linear velocity of an observer $\mathcal{B}$ relative to an observer $\mathcal{A}$.

A bold-faced lower-case $\boldsymbol{\omega}$ (*omega*) with a left and right superscript, such as ${}^{\mathcal{A}}\boldsymbol{\omega}^{\mathcal{B}}$ or ${}^{a}\boldsymbol{\omega}^{b}$, was used in this chapter to denote the angular velocity of an observer $\mathcal{B}$ relative to an observer $\mathcal{A}$ or the angular velocity between two triads $a$ and $b$.

## Summary of terminology

The limiting direction relative to an observer of the translation between the reference point of some rigid body at times $t$ and $t+h$ as $h \to 0$ is called the *instantaneous direction of translation* of the rigid body relative to the observer. (Page 327)

The limiting value relative to an observer of the ratio between the magnitude of the translation between the reference point of some rigid body at times $t$ and $t+h$ and the time step $h$ as $h \to 0$ is called the *linear speed* of the rigid body relative to the observer. (Page 327)

The limiting direction relative to an observer of the rotation between the reference triad of some rigid body at times $t$ and $t+h$ as $h \to 0$ is called the *instantaneous direction of rotation* of the rigid body relative to the observer. (Page 327)

The limiting value relative to an observer of the ratio between the magnitude of the rotation between the reference triad of some rigid body at times $t$ and $t+h$ and the time step $h$ as $h \to 0$ is called the *angular speed* of the rigid body relative to the observer. (Page 327)

The *linear velocity* of an observer $\mathcal{B}$ relative to an observer $\mathcal{A}$ is the time derivative of the position vector $\mathbf{r}^{AB}$ relative to the reference triad $a$. (Page 328)

The *angular velocity* of an observer $\mathcal{B}$ relative to an observer $\mathcal{A}$ is the vector whose matrix representation relative to the reference triad $b$ is given by the entries of the anti-symmetric matrix $R_{ba}\left(dR_{ab}/dt\right)$. (Page 330)

(Page 331) The *time derivative of a vector* $\mathbf{v}$ relative to a triad $a$ equals the vector whose matrix representation relative to $a$ is the time derivative of the matrix representation of $\mathbf{v}$ relative to $a$.

(Page 348) In the MAMBO toolbox, the procedure `DiffTime` computes derivatives of scalar and vector quantities with respect to time $t$.

(Page 348) In the MAMBO toolbox, the procedure `LinearVelocity` computes the velocity of a point relative to an observer.

(Page 348) In the MAMBO toolbox, the procedure `AngularVelocity` computes the angular velocity between two triads.

(Page 348) In the MAMBO toolbox, the procedure `DeclareStates` declares all variables that are understood to be dependent on time $t$.

# Chapter 9

# Constraints

*wherein the reader learns of:*

- *Formulating configuration and motion constraints to model the limitations imposed by the environment on the configuration and motion of a mechanism;*
- *Introducing independent velocity coordinates to capture the allowable motions that satisfy all imposed constraints;*
- *Formulating kinematic differential equations whose solutions generate allowable motions that satisfy all imposed constraints;*
- *Singularities in the kinematic differential equations.*

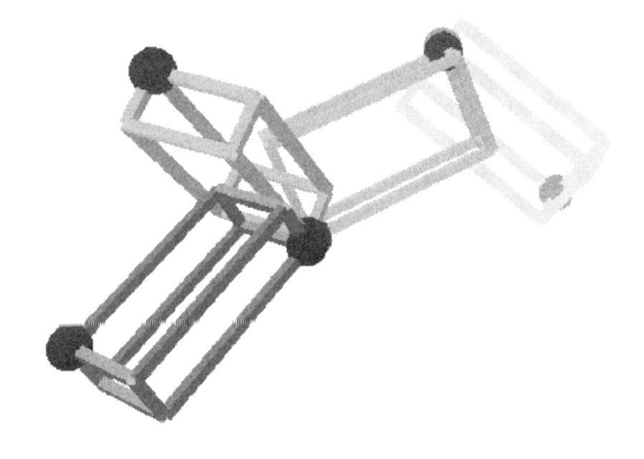

## Practicum

This is it! With the tools of this chapter, you are all set to produce visually satisfying animations of all the model mechanisms in this text. In particular, you are ready to implement configuration and motion constraints on the project mechanisms listed in Appendix C.

I recommend that you consider adopting the methodology in the last section of this chapter, whereby constraints are imposed a few at a time and MAMBO is used to verify the correctness of the constraint formulation. You should also consider experimenting with different choices of independent velocity coordinates to better grasp the notion of singularities. Have fun with it!

# 9.1 Kinematic Differential Equations

(Ex. 9.1 – Ex. 9.8)

## 9.1.1 Complementary Representations

The configuration of a rigid body relative to an observer $\mathcal{W}$ is given by a pure translation $\mathbf{T}_{\mathcal{W}}$ corresponding to the position vector

$$\mathbf{r}^{WB} = w\ {}^{\mathcal{W}}B$$

from the reference point $W$ of the observer to the reference point $B$ of the rigid body; and a pure rotation $\mathbf{R}_{\mathcal{W}}$ corresponding to the rotation matrix

$$R_{wb}$$

from the reference triad $w$ of the observer to the reference triad $b$ of the rigid body.

Every choice of coordinate representation ${}^{\mathcal{W}}B$ and every choice of rotation matrix $R_{wb}$ corresponds to some **configuration** of the rigid body. Similarly, every choice of time-dependence ${}^{\mathcal{W}}B(t)$ and $R_{wb}(t)$ corresponds to some **motion** of the rigid body.

Given a time-dependence ${}^{\mathcal{W}}B(t)$ for the coordinate representation of the point $B$ relative to the observer $\mathcal{W}$, you can compute the velocity

$${}^{\mathcal{W}}\frac{dB}{dt}(t)$$

of the point $B$ relative to $\mathcal{W}$. Similarly, given a time-dependence $R_{wb}(t)$, you can compute the angular velocity

$${}^{w}\boldsymbol{\omega}^{b}(t)$$

between the reference triad $w$ of $\mathcal{W}$ and the reference triad $b$ of the rigid body.

But what about the converse? Is it possible to compute ${}^{\mathcal{W}}B(t)$ and $R_{wb}(t)$ from knowledge of ${}^{\mathcal{W}}\frac{dB}{dt}(t)$ and ${}^{w}\boldsymbol{\omega}^{b}(t)$? Put differently, if I know at what linear speed and in what direction I am traveling at any moment in time, can I deduce my position as a function of time? Similarly, if I know at what angular speed and about what direction I am rotating at any moment in time, can I deduce my orientation as a function of time?

Clearly, two rigid bodies at different positions relative to $\mathcal{W}$ may have the same linear velocity relative to $\mathcal{W}$. Simply knowing the linear velocity is not going to suffice to prescribe the positions of the two bodies as functions of time. So, let me modify the question: if I know my position at some time $t_0$ and I know at what linear speed and in what direction I am traveling at any arbitrary time, can I deduce my position as a function of time? Similarly, if I know my orientation at some time $t_0$ and I know at what angular speed and about what direction I am rotating at any moment in time, can I deduce my orientation as a function of time?

That the answer to both questions is yes follows from the following argument, parts of which are a review from the previous chapter.

Let $q_1$, $q_2$, and $q_3$ represent the Cartesian coordinates of the point $B$ relative to $\mathcal{W}$ and $q_4$, $q_5$, and $q_6$ represent the Euler angles corresponding to a $3-1-3$ sequence of rotations, such that

$$^{\mathcal{W}}B = \begin{pmatrix} q_1 \\ q_2 \\ q_3 \end{pmatrix}$$

and

$$\begin{aligned} R_{wb} &= R(q_4, 0, 0, 1)\, R(q_5, 1, 0, 0)\, R(q_6, 0, 0, 1) \\ &= \begin{pmatrix} c_4 c_6 - s_4 c_5 s_6 & -c_4 s_6 - s_4 c_5 c_6 & s_4 s_5 \\ s_4 c_6 + c_4 c_5 s_6 & s_4 s_6 + c_4 c_5 c_6 & -c_4 s_5 \\ s_5 s_6 & s_5 c_6 & c_5 \end{pmatrix}, \end{aligned}$$

where $c_i = \cos q_i$ and $s_i = \sin q_i$ for $i = 4, 5, 6$. Every configuration of the rigid body corresponds to some set of values of the configuration coordinates $q_1$ through $q_6$. Similarly, every motion of the rigid body corresponds to some set of time-dependencies for the configuration coordinates $q_1(t)$ through $q_6(t)$. In particular, if I know my position and orientation at some time $t_0$, then I can find the corresponding values $q_1(t_0), \ldots, q_6(t_0)$.

The linear and angular velocities of the rigid body relative to $\mathcal{W}$ are

$$\frac{^{\mathcal{W}}dB}{dt} = \frac{^{w}d\mathbf{r}^{WB}}{dt} = w \begin{pmatrix} \dot{q}_1 \\ \dot{q}_2 \\ \dot{q}_3 \end{pmatrix},$$

where the dot superscript is shorthand for a time derivative; and

$$^{w}\boldsymbol{\omega}^{b} = b \begin{pmatrix} \dot{q}_4 s_5 s_6 + \dot{q}_5 c_6 \\ \dot{q}_4 s_5 c_6 - \dot{q}_5 s_6 \\ \dot{q}_4 c_5 + \dot{q}_6 \end{pmatrix},$$

since

$$R_{bw} \frac{dR_{wb}}{dt} = \begin{pmatrix} 0 & -\dot{q}_4 c_5 - \dot{q}_6 & \dot{q}_4 s_5 c_6 - \dot{q}_5 s_6 \\ \dot{q}_4 c_5 + \dot{q}_6 & 0 & -\dot{q}_4 s_5 s_6 - \dot{q}_5 c_6 \\ -\dot{q}_4 s_5 c_6 + \dot{q}_5 s_6 & \dot{q}_4 s_5 s_6 + \dot{q}_5 c_6 & 0 \end{pmatrix}.$$

Now, assume that

$$\frac{^{\mathcal{W}}dB}{dt} = w \begin{pmatrix} u_1 \\ u_2 \\ u_3 \end{pmatrix} \text{ and } {}^{w}\boldsymbol{\omega}^{b} = b \begin{pmatrix} u_4 \\ u_5 \\ u_6 \end{pmatrix},$$

where $u_1$ through $u_6$ are known functions of time. It follows that

$$\begin{aligned}
\dot{q}_1 &= u_1, \\
\dot{q}_2 &= u_2, \\
\dot{q}_3 &= u_3, \\
\dot{q}_4 \sin q_5 \sin q_6 + \dot{q}_5 \cos q_6 &= u_4, \\
\dot{q}_4 \sin q_5 \cos q_6 - \dot{q}_5 \sin q_6 &= u_5, \\
\dot{q}_4 \cos q_5 + \dot{q}_6 &= u_6,
\end{aligned}$$

or, in matrix form,

$$\begin{pmatrix} 1 & 0 & 0 & 0 & 0 & 0 \\ 0 & 1 & 0 & 0 & 0 & 0 \\ 0 & 0 & 1 & 0 & 0 & 0 \\ 0 & 0 & 0 & \sin q_5 \sin q_6 & \cos q_6 & 0 \\ 0 & 0 & 0 & \sin q_5 \cos q_6 & -\sin q_6 & 0 \\ 0 & 0 & 0 & \cos q_5 & 0 & 1 \end{pmatrix} \begin{pmatrix} \dot{q}_1 \\ \dot{q}_2 \\ \dot{q}_3 \\ \dot{q}_4 \\ \dot{q}_5 \\ \dot{q}_6 \end{pmatrix} = \begin{pmatrix} u_1 \\ u_2 \\ u_3 \\ u_4 \\ u_5 \\ u_6 \end{pmatrix}.$$

Since the variables $u_1, \ldots, u_6$ are known functions of time, this is a system of first-order differential equations in the unknowns $q_1$ through $q_6$. Given a set of values for the configuration coordinates at some time $t_0$, the fundamental theory of differential equations guarantees the existence of a unique solution $q_1(t), \ldots, q_6(t)$ over some interval of time around $t = t_0$. The existence and uniqueness theory breaks down when the coefficient matrix becomes singular, i.e., when

$$\begin{vmatrix} 1 & 0 & 0 & 0 & 0 & 0 \\ 0 & 1 & 0 & 0 & 0 & 0 \\ 0 & 0 & 1 & 0 & 0 & 0 \\ 0 & 0 & 0 & \sin q_5 \sin q_6 & \cos q_6 & 0 \\ 0 & 0 & 0 & \sin q_5 \cos q_6 & -\sin q_6 & 0 \\ 0 & 0 & 0 & \cos q_5 & 0 & 1 \end{vmatrix} = -\sin q_5 = 0.$$

Thus, as long as $q_5 \neq n\pi$ for all integers $n$, I may solve the differential equations for the configuration coordinates as functions of time. Substituting the results back into the expressions for ${}^{\mathcal{W}}B$ and $R_{wb}$ then yields the desired position and orientation as functions of time.

The differential equations derived above are known as the *kinematic differential equations*. The discussion shows that a motion of the rigid body can be generated by specifying initial values for the configuration coordinates and explicit functions of time for the *independent velocity coordinates* $u_1, \ldots, u_6$. This representation of the motion is **complementary** to that given by explicit functions of time for the variables $q_1, \ldots, q_6$.

The kinematic differential equations may be hard or even impossible to solve analytically, but a unique solution is guaranteed to exist by the fundamental theory of differential equations. We expect that a numerical algorithm should be able to generate a reasonable approximation.

### 9.1.2 The Free Rigid Body

If the configuration of the rigid body is unconstrained, all six configuration coordinates are **necessary** to describe arbitrary positions and orientations of the rigid body relative to $\mathcal{W}$. These configuration coordinates are **independent**. The free rigid body has six *geometric degrees of freedom*.

Similarly, if the motion of the rigid body is unconstrained, all six components of the linear and angular velocity vectors are **independent**. It takes six independent velocity coordinates to describe arbitrary linear and angular velocities of the rigid body relative to $\mathcal{W}$. The free rigid body has six *dynamic degrees of freedom*.

### 9.1.3 Motion Along a Plane

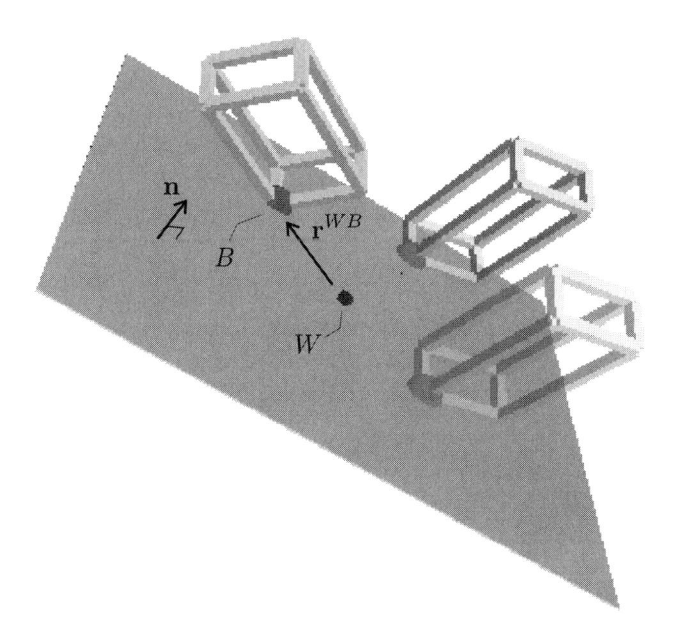


Suppose that the rigid body is constrained in such a way that its reference point $B$ is confined to a plane through the reference point $W$ that is stationary relative to the observer $\mathcal{W}$. If

$$\mathbf{n} = w \begin{pmatrix} n_1 \\ n_2 \\ n_3 \end{pmatrix}$$

is a vector perpendicular to the plane, then

$$\mathbf{r}^{WB} \bullet \mathbf{n} = 0,$$

since $\mathbf{r}^{WB}$ is parallel to the plane. It follows that the **configuration** of the rigid body is constrained. This condition is an example of a *configuration constraint.*

Equivalently,

$$\frac{{}^{\mathcal{W}}dB}{dt} \bullet \mathbf{n} = 0,$$

since the velocity of the point $B$ relative to $\mathcal{W}$ must be parallel to the plane. It follows that the **motion** of the rigid body is constrained. This condition is an example of a *motion constraint.*

If we differentiate the left-hand side of the configuration constraint with respect to time relative to the $w$ triad, we obtain

$$\frac{{}^{w}d}{dt}\left(\mathbf{r}^{WB} \bullet \mathbf{n}\right) = \frac{{}^{w}d\mathbf{r}^{WB}}{dt} \bullet \mathbf{n} = \frac{{}^{\mathcal{W}}dB}{dt} \bullet \mathbf{n},$$

since $\mathbf{n}$ is constant in the $w$ triad. It follows that a motion that satisfies the configuration constraint automatically satisfies the motion constraint. But how about the converse? Specifically, if a motion satisfies the motion constraint, does it automatically satisfy the configuration constraint?

Clearly, the motion constraint requires only that the velocity be perpendicular to $\mathbf{n}$. It follows that the rigid body will move in such a way that $B$ is confined to some plane perpendicular to $\mathbf{n}$, but not necessarily the plane through $W$. However, if $B$ is known to lie on the chosen plane at some time $t_0$, then a motion that satisfies the motion constraint will automatically satisfy the configuration constraint.

This observation may be expressed in terms of a new set of kinematic differential equations. In terms of the configuration coordinates introduced in the initial section of this chapter, the motion constraint becomes

$$0 = \frac{{}^{\mathcal{W}}dB}{dt} \bullet \mathbf{n} = \dot{q}_1 n_1 + \dot{q}_2 n_2 + \dot{q}_3 n_3.$$

The components of the linear velocity of the rigid body relative to $\mathcal{W}$ are no longer independent, since

$$\left(\frac{{}^{\mathcal{W}}dB}{dt} \bullet \mathbf{w}_1\right) n_1 + \left(\frac{{}^{\mathcal{W}}dB}{dt} \bullet \mathbf{w}_2\right) n_2 + \left(\frac{{}^{\mathcal{W}}dB}{dt} \bullet \mathbf{w}_3\right) n_3 = 0.$$

It is not possible to assign arbitrary functions of time to all three components of the linear velocity, since one of these components can always be expressed in terms of the other two.

In light of the motion constraint on the components of the linear velocity, consider introducing known functions of time $u_1(t), \ldots, u_5(t)$, such that

$$^{\mathcal{W}}\frac{dB}{dt} = w\begin{pmatrix} u_1 \\ u_2 \\ \cdot \end{pmatrix} \text{ and } {}^{w}\boldsymbol{\omega}^{b} = b\begin{pmatrix} u_3 \\ u_4 \\ u_5 \end{pmatrix},$$

where the $\cdot$ denotes a quantity that depends on the $u$'s. Using the expressions for $^{\mathcal{W}}\frac{dB}{dt}$ and $^{w}\boldsymbol{\omega}^{b}$ that we derived previously, we must have

$$\begin{aligned} \dot{q}_1 &= u_1, \\ \dot{q}_2 &= u_2, \\ \dot{q}_4 \sin q_5 \sin q_6 + \dot{q}_5 \cos q_6 &= u_3, \\ \dot{q}_4 \sin q_5 \cos q_6 - \dot{q}_5 \sin q_6 &= u_4, \\ \dot{q}_4 \cos q_5 + \dot{q}_6 &= u_5, \end{aligned}$$

and

$$\dot{q}_1 n_1 + \dot{q}_2 n_2 + \dot{q}_3 n_3 = 0,$$

where the last equation restates the motion constraint. In matrix form, these equations become

$$\begin{pmatrix} 1 & 0 & 0 & 0 & 0 & 0 \\ 0 & 1 & 0 & 0 & 0 & 0 \\ 0 & 0 & 0 & \sin q_5 \sin q_6 & \cos q_6 & 0 \\ 0 & 0 & 0 & \sin q_5 \cos q_6 & -\sin q_6 & 0 \\ 0 & 0 & 0 & \cos q_5 & 0 & 1 \\ n_1 & n_2 & n_3 & 0 & 0 & 0 \end{pmatrix} \begin{pmatrix} \dot{q}_1 \\ \dot{q}_2 \\ \dot{q}_3 \\ \dot{q}_4 \\ \dot{q}_5 \\ \dot{q}_6 \end{pmatrix} = \begin{pmatrix} u_1 \\ u_2 \\ u_3 \\ u_4 \\ u_5 \\ 0 \end{pmatrix}.$$

These kinematic differential equations are non-singular as long as the determinant of the coefficient matrix is non-zero, i.e.,

$$\begin{vmatrix} 1 & 0 & 0 & 0 & 0 & 0 \\ 0 & 1 & 0 & 0 & 0 & 0 \\ 0 & 0 & 0 & \sin q_5 \sin q_6 & \cos q_6 & 0 \\ 0 & 0 & 0 & \sin q_5 \cos q_6 & -\sin q_6 & 0 \\ 0 & 0 & 0 & \cos q_5 & 0 & 1 \\ n_1 & n_2 & n_3 & 0 & 0 & 0 \end{vmatrix} = n_3 \sin q_5 \neq 0.$$

If $n_3 = 0$, these equations are everywhere singular, and cannot be solved. This reflects a bad definition of the independent velocity coordinates $u_1$, $\ldots$, $u_5$. If $n_3 \neq 0$, then the kinematic differential equations can be solved away from $q_5 = n\pi$ for any integer $n$.

Alternatively, the choice

$$^{\mathcal{W}}\frac{dB}{dt} = w\begin{pmatrix} u_1 \\ \cdot \\ u_2 \end{pmatrix} \text{ and } {}^{w}\boldsymbol{\omega}^{b} = b\begin{pmatrix} u_3 \\ u_4 \\ u_5 \end{pmatrix}$$

yields

$$\begin{pmatrix} 1 & 0 & 0 & 0 & 0 & 0 \\ 0 & 0 & 1 & 0 & 0 & 0 \\ 0 & 0 & 0 & \sin q_5 \sin q_6 & \cos q_6 & 0 \\ 0 & 0 & 0 & \sin q_5 \cos q_6 & -\sin q_6 & 0 \\ 0 & 0 & 0 & \cos q_5 & 0 & 1 \\ n_1 & n_2 & n_3 & 0 & 0 & 0 \end{pmatrix} \begin{pmatrix} \dot{q}_1 \\ \dot{q}_2 \\ \dot{q}_3 \\ \dot{q}_4 \\ \dot{q}_5 \\ \dot{q}_6 \end{pmatrix} = \begin{pmatrix} u_1 \\ u_2 \\ u_3 \\ u_4 \\ u_5 \\ 0 \end{pmatrix}.$$

which are non-singular as long as

$$\begin{vmatrix} 1 & 0 & 0 & 0 & 0 & 0 \\ 0 & 0 & 1 & 0 & 0 & 0 \\ 0 & 0 & 0 & \sin q_5 \sin q_6 & \cos q_6 & 0 \\ 0 & 0 & 0 & \sin q_5 \cos q_6 & -\sin q_6 & 0 \\ 0 & 0 & 0 & \cos q_5 & 0 & 1 \\ n_1 & n_2 & n_3 & 0 & 0 & 0 \end{vmatrix} = -n_2 \sin q_5 \neq 0.$$

Thus, if the previous choice of independent velocity coordinates yields everywhere singular kinematic differential equations (as would happen if $n_3 = 0$), the present choice need not be everywhere singular (unless $n_2 = 0$ as well).

Finally,

$$\frac{{}^{\mathcal{W}}dB}{dt} = w \begin{pmatrix} \cdot \\ u_1 \\ u_2 \end{pmatrix} \text{ and } {}^{w}\boldsymbol{\omega}^{b} = b \begin{pmatrix} u_3 \\ u_4 \\ u_5 \end{pmatrix}$$

yields

$$\begin{pmatrix} 0 & 1 & 0 & 0 & 0 & 0 \\ 0 & 0 & 1 & 0 & 0 & 0 \\ 0 & 0 & 0 & \sin q_5 \sin q_6 & \cos q_6 & 0 \\ 0 & 0 & 0 & \sin q_5 \cos q_6 & -\sin q_6 & 0 \\ 0 & 0 & 0 & \cos q_5 & 0 & 1 \\ n_1 & n_2 & n_3 & 0 & 0 & 0 \end{pmatrix} \begin{pmatrix} \dot{q}_1 \\ \dot{q}_2 \\ \dot{q}_3 \\ \dot{q}_4 \\ \dot{q}_5 \\ \dot{q}_6 \end{pmatrix} = \begin{pmatrix} u_1 \\ u_2 \\ u_3 \\ u_4 \\ u_5 \\ 0 \end{pmatrix},$$

which are non-singular as long as

$$\begin{vmatrix} 0 & 1 & 0 & 0 & 0 & 0 \\ 0 & 0 & 1 & 0 & 0 & 0 \\ 0 & 0 & 0 & \sin q_5 \sin q_6 & \cos q_6 & 0 \\ 0 & 0 & 0 & \sin q_5 \cos q_6 & -\sin q_6 & 0 \\ 0 & 0 & 0 & \cos q_5 & 0 & 1 \\ n_1 & n_2 & n_3 & 0 & 0 & 0 \end{vmatrix} = n_1 \sin q_5 \neq 0.$$

Thus, if the previous choices of independent velocity coordinates yield everywhere singular kinematic differential equations (as would happen if

$n_2 = n_3 = 0$), the present choice will not be everywhere singular, since $\mathbf{n} \neq \mathbf{0}$ and therefore at least one of the $n_i$'s must be non-zero.

Assuming that the kinematic differential equations are not everywhere singular, we may now solve for the configuration coordinates $q_1(t), \ldots, q_6(t)$ as functions of time. Since the motion constraint is automatically satisfied by such a solution, it follows from the above discussion that the configuration constraint is automatically satisfied by this solution, provided that the initial values $q_1(t_0), \ldots, q_6(t_0)$ satisfy the configuration constraint.

As shown above, only **five** of the components of the linear and angular velocity vectors are independent. It takes **five** independent velocity coordinates to describe arbitrary linear and angular velocities of the rigid body relative to $\mathcal{W}$. This rigid body therefore has **five** dynamic degrees of freedom.

### 9.1.4 Motion Along a Line

Suppose that the rigid body is constrained in such a way that its reference point $B$ is confined to a straight line through the reference point $W$ that is stationary relative to the observer $\mathcal{W}$. If

$$\mathbf{p} = w \begin{pmatrix} p_1 \\ p_2 \\ p_3 \end{pmatrix}$$

is a vector parallel to the line, then

$$\mathbf{r}^{WB} \times \mathbf{p} = \mathbf{0},$$

since $\mathbf{r}^{WB}$ is parallel to the line. It follows that the **configuration** of the rigid body is constrained. This condition is a configuration constraint.

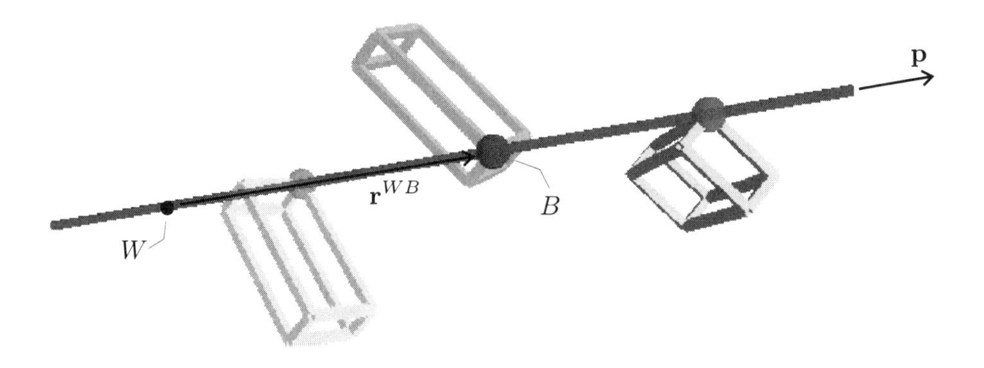


Equivalently,

$$\frac{{}^{\mathcal{W}}dB}{dt} \times \mathbf{p} = \mathbf{0},$$

since the velocity of the point $B$ relative to $\mathcal{W}$ must be parallel to the line. It follows that the **motion** of the rigid body is constrained. This condition is a motion constraint.

If we differentiate the left-hand side of the configuration constraint with respect to time relative to the $w$ triad, we obtain

$$ {}^{w}\frac{d}{dt}\left(\mathbf{r}^{WB} \times \mathbf{p}\right) = {}^{w}\frac{d\mathbf{r}^{WB}}{dt} \times \mathbf{p} = {}^{\mathcal{W}}\frac{dB}{dt} \times \mathbf{p}, $$

since $\mathbf{p}$ is constant in the $w$ triad. It follows that a motion that satisfies the configuration constraint automatically satisfies the motion constraint. As discussed above, the converse is true, provided that $B$ is known to lie on the straight line at some time $t_0$.

In terms of the configuration coordinates introduced previously, the motion constraint implies that

$$ \mathbf{0} = {}^{\mathcal{W}}\frac{dB}{dt} \times \mathbf{p} = w \begin{pmatrix} p_3\dot{q}_2 - p_2\dot{q}_3 \\ -p_3\dot{q}_1 + p_1\dot{q}_3 \\ p_2\dot{q}_1 - p_1\dot{q}_2 \end{pmatrix}. $$

It follows that the components of the linear velocity of the rigid body relative to $\mathcal{W}$ are no longer independent, since

$$ p_3\left({}^{\mathcal{W}}\frac{dB}{dt} \bullet \mathbf{w}_2\right) - p_2\left({}^{\mathcal{W}}\frac{dB}{dt} \bullet \mathbf{w}_3\right) = 0, $$

$$ -p_3\left({}^{\mathcal{W}}\frac{dB}{dt} \bullet \mathbf{w}_1\right) + p_1\left({}^{\mathcal{W}}\frac{dB}{dt} \bullet \mathbf{w}_3\right) = 0, $$

$$ p_2\left({}^{\mathcal{W}}\frac{dB}{dt} \bullet \mathbf{w}_1\right) - p_1\left({}^{\mathcal{W}}\frac{dB}{dt} \bullet \mathbf{w}_2\right) = 0. $$

These are three equations in three unknowns. It would appear that all the components of the linear velocity are determined by these equations. It would appear that we have no freedom at all to assign functions of time to the components of the linear velocity. But that is counterintuitive. The motion constraint says nothing about the linear velocity of the rigid body along the straight line. We must have some freedom left. How do we resolve this?

Let's express the motion constraint in matrix form, such that

$$ \begin{pmatrix} 0 & p_3 & -p_2 \\ -p_3 & 0 & p_1 \\ p_2 & -p_1 & 0 \end{pmatrix} \begin{pmatrix} \dot{q}_1 \\ \dot{q}_2 \\ \dot{q}_3 \end{pmatrix} = \begin{pmatrix} 0 \\ 0 \\ 0 \end{pmatrix}. $$

The determinant of the coefficient matrix

$$ \begin{vmatrix} 0 & p_3 & -p_2 \\ -p_3 & 0 & p_1 \\ p_2 & -p_1 & 0 \end{vmatrix} = 0, $$

i.e., these equations are everywhere singular. For example, if $p_3 \neq 0$, then the third constraint equation can be expressed in terms of the other two:

$$p_2\dot{q}_1 - p_1\dot{q}_2 = -\frac{p_2}{p_3}\left(-p_3\dot{q}_1 + p_1\dot{q}_3\right) - \frac{p_1}{p_3}\left(p_3\dot{q}_2 - p_2\dot{q}_3\right).$$

It follows that a motion that satisfies the first two constraint equations automatically satisfies the third constraint equation. Alternatively, if $p_3 = 0$, but $p_2 \neq 0$, then the second constraint equation can be expressed in terms of the first:

$$p_1\dot{q}_3 = -\frac{p_1}{p_2}\left(-p_2\dot{q}_3\right).$$

It follows that a motion that satisfies the first constraint equation automatically satisfies the second constraint. Finally, if $p_2 = p_3 = 0$, then the first constraint equation is automatically satisfied.

Assume that $p_3 \neq 0$ and let

$$^{\mathcal{W}}\frac{dB}{dt} = w\begin{pmatrix} \cdot \\ \cdot \\ u_1 \end{pmatrix} \text{ and } {}^{w}\boldsymbol{\omega}^{b} = b\begin{pmatrix} u_2 \\ u_3 \\ u_4 \end{pmatrix}.$$

The kinematic differential equations then become

$$\begin{pmatrix} 0 & 0 & 1 & 0 & 0 & 0 \\ 0 & 0 & 0 & \sin q_5 \sin q_6 & \cos q_6 & 0 \\ 0 & 0 & 0 & \sin q_5 \cos q_6 & -\sin q_6 & 0 \\ 0 & 0 & 0 & \cos q_5 & 0 & 1 \\ 0 & p_3 & -p_2 & 0 & 0 & 0 \\ -p_3 & 0 & p_1 & 0 & 0 & 0 \end{pmatrix}\begin{pmatrix} \dot{q}_1 \\ \dot{q}_2 \\ \dot{q}_3 \\ \dot{q}_4 \\ \dot{q}_5 \\ \dot{q}_6 \end{pmatrix} = \begin{pmatrix} u_1 \\ u_2 \\ u_3 \\ u_4 \\ 0 \\ 0 \end{pmatrix},$$

where the last two rows of the coefficient matrix corresponds to the first two motion constraints (recall that the third constraint can be expressed in terms of the first two constraints). These equations are non-singular, as long as

$$\begin{vmatrix} 0 & 0 & 1 & 0 & 0 & 0 \\ 0 & 0 & 0 & \sin q_5 \sin q_6 & \cos q_6 & 0 \\ 0 & 0 & 0 & \sin q_5 \cos q_6 & -\sin q_6 & 0 \\ 0 & 0 & 0 & \cos q_5 & 0 & 1 \\ 0 & p_3 & -p_2 & 0 & 0 & 0 \\ -p_3 & 0 & p_1 & 0 & 0 & 0 \end{vmatrix} = -p_3^2 \sin q_5 \neq 0.$$

Since $p_3 \neq 0$, the kinematic differential equations are not everywhere singular.

If, instead, $p_3 = 0$, but $p_2 \neq 0$, let

$$^{\mathcal{W}}\frac{dB}{dt} = w\begin{pmatrix} \cdot \\ u_1 \\ \cdot \end{pmatrix} \text{ and } {}^{w}\boldsymbol{\omega}^{b} = b\begin{pmatrix} u_2 \\ u_3 \\ u_4 \end{pmatrix}.$$

The kinematic differential equations then become

$$\begin{pmatrix} 0 & 1 & 0 & 0 & 0 & 0 \\ 0 & 0 & 0 & \sin q_5 \sin q_6 & \cos q_6 & 0 \\ 0 & 0 & 0 & \sin q_5 \cos q_6 & -\sin q_6 & 0 \\ 0 & 0 & 0 & \cos q_5 & 0 & 1 \\ 0 & 0 & -p_2 & 0 & 0 & 0 \\ p_2 & -p_1 & 0 & 0 & 0 & 0 \end{pmatrix} \begin{pmatrix} \dot{q}_1 \\ \dot{q}_2 \\ \dot{q}_3 \\ \dot{q}_4 \\ \dot{q}_5 \\ \dot{q}_6 \end{pmatrix} = \begin{pmatrix} u_1 \\ u_2 \\ u_3 \\ u_4 \\ 0 \\ 0 \end{pmatrix},$$

where the last two rows of the coefficient matrix corresponds to the first and last motion constraints (recall that the second constraint can be expressed in terms of the first constraint). These equations are non-singular, as long as

$$\begin{vmatrix} 0 & 1 & 0 & 0 & 0 & 0 \\ 0 & 0 & 0 & \sin q_5 \sin q_6 & \cos q_6 & 0 \\ 0 & 0 & 0 & \sin q_5 \cos q_6 & -\sin q_6 & 0 \\ 0 & 0 & 0 & \cos q_5 & 0 & 1 \\ 0 & 0 & -p_2 & 0 & 0 & 0 \\ p_2 & -p_1 & 0 & 0 & 0 & 0 \end{vmatrix} = p_2^2 \sin q_5 \neq 0.$$

Since $p_2 \neq 0$, the kinematic differential equations are not everywhere singular.

Finally, if $p_2 = p_3 = 0$, but $p_1 \neq 0$, let

$$\frac{{}^{\mathcal{W}}dB}{dt} = w \begin{pmatrix} u_1 \\ \cdot \\ \cdot \end{pmatrix} \text{ and } {}^{w}\boldsymbol{\omega}^{b} = b \begin{pmatrix} u_2 \\ u_3 \\ u_4 \end{pmatrix}.$$

The kinematic differential equations then become

$$\begin{pmatrix} 1 & 0 & 0 & 0 & 0 & 0 \\ 0 & 0 & 0 & \sin q_5 \sin q_6 & \cos q_6 & 0 \\ 0 & 0 & 0 & \sin q_5 \cos q_6 & -\sin q_6 & 0 \\ 0 & 0 & 0 & \cos q_5 & 0 & 1 \\ 0 & 0 & p_1 & 0 & 0 & 0 \\ 0 & -p_1 & 0 & 0 & 0 & 0 \end{pmatrix} \begin{pmatrix} \dot{q}_1 \\ \dot{q}_2 \\ \dot{q}_3 \\ \dot{q}_4 \\ \dot{q}_5 \\ \dot{q}_6 \end{pmatrix} = \begin{pmatrix} u_1 \\ u_2 \\ u_3 \\ u_4 \\ 0 \\ 0 \end{pmatrix},$$

where the last two rows of the coefficient matrix corresponds to the last two motion constraints (recall that the first constraint is automatically satisfied). These equations are non-singular, as long as

$$\begin{vmatrix} 1 & 0 & 0 & 0 & 0 & 0 \\ 0 & 0 & 0 & \sin q_5 \sin q_6 & \cos q_6 & 0 \\ 0 & 0 & 0 & \sin q_5 \cos q_6 & -\sin q_6 & 0 \\ 0 & 0 & 0 & \cos q_5 & 0 & 1 \\ 0 & 0 & p_1 & 0 & 0 & 0 \\ 0 & -p_1 & 0 & 0 & 0 & 0 \end{vmatrix} = -p_1^2 \sin q_5 \neq 0.$$

Since $p_1 \neq 0$, the kinematic differential equations are not everywhere singular.

In any of the three cases, we may now assign arbitrary functions $u_1(t), \ldots, u_4(t)$ and solve for the corresponding $q_1(t), \ldots, q_6(t)$. Since the motion constraint is automatically satisfied by such a solution, it follows from the above discussion that the configuration constraint is automatically satisfied by this solution, provided that the initial values $q_1(t_0), \ldots, q_6(t_0)$ satisfy the configuration constraint.

As shown above, only **four** of the components of the linear and angular velocity vectors are independent. It takes **four** independent velocity coordinates to describe arbitrary linear and angular velocities of the rigid body relative to $\mathcal{W}$. This rigid body therefore has **four** dynamic degrees of freedom.

### 9.1.5 No Translation

Suppose that the rigid body is constrained in such a way that its reference point $B$ coincides with the reference point $W$ of the observer $\mathcal{W}$. Then,

$$\mathbf{r}^{WB} = \mathbf{0}.$$

It follows that the **configuration** of the rigid body is constrained. This condition is a configuration constraint.

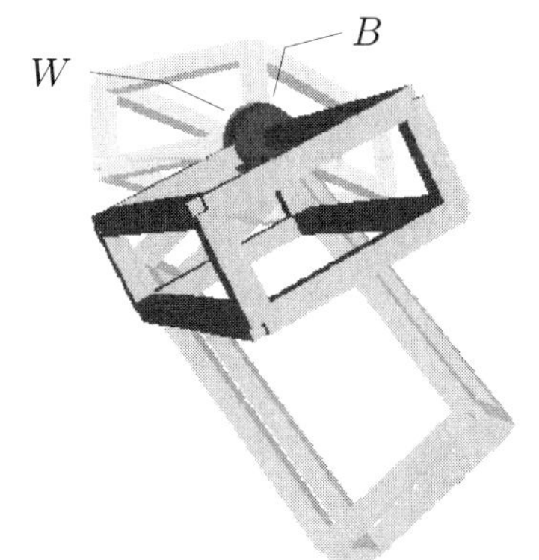


Equivalently, we conclude that

$$\frac{{}^{\mathcal{W}}dB}{dt} = \mathbf{0},$$

since $B$ is stationary relative to $\mathcal{W}$. It follows that the **motion** of the rigid body is constrained. This condition is a motion constraint.

As in previous sections, a motion that satisfies the configuration constraint automatically satisfies the motion constraint. Similarly, a motion that satisfies the motion constraint automatically satisfies the configuration constraint provided that $B$ coincides with $W$ at some time $t_0$.

In terms of the configuration coordinates introduced in previous sections, the motion constraint implies that

$$\mathbf{0} = \frac{{}^{\mathcal{W}}dB}{dt} = w \begin{pmatrix} \dot{q}_1 \\ \dot{q}_2 \\ \dot{q}_3 \end{pmatrix}.$$

Now, let

$${}^{w}\boldsymbol{\omega}^{b} = b \begin{pmatrix} u_1 \\ u_2 \\ u_3 \end{pmatrix}.$$

The kinematic differential equations become

$$\begin{pmatrix} 0 & 0 & 0 & \sin q_5 \sin q_6 & \cos q_6 & 0 \\ 0 & 0 & 0 & \sin q_5 \cos q_6 & -\sin q_6 & 0 \\ 0 & 0 & 0 & \cos q_5 & 0 & 1 \\ 1 & 0 & 0 & 0 & 0 & 0 \\ 0 & 1 & 0 & 0 & 0 & 0 \\ 0 & 0 & 1 & 0 & 0 & 0 \end{pmatrix} \begin{pmatrix} \dot{q}_1 \\ \dot{q}_2 \\ \dot{q}_3 \\ \dot{q}_4 \\ \dot{q}_5 \\ \dot{q}_6 \end{pmatrix} = \begin{pmatrix} u_1 \\ u_2 \\ u_3 \\ 0 \\ 0 \\ 0 \end{pmatrix},$$

where the last three rows correspond to the motion constraints. The equations are non-singular, as long as the determinant of the coefficient matrix is non-zero, i.e.,

$$\begin{vmatrix} 0 & 0 & 0 & \sin q_5 \sin q_6 & \cos q_6 & 0 \\ 0 & 0 & 0 & \sin q_5 \cos q_6 & -\sin q_6 & 0 \\ 0 & 0 & 0 & \cos q_5 & 0 & 1 \\ 1 & 0 & 0 & 0 & 0 & 0 \\ 0 & 1 & 0 & 0 & 0 & 0 \\ 0 & 0 & 1 & 0 & 0 & 0 \end{vmatrix} = \sin q_5 \neq 0.$$

The kinematic differential equations can be solved away from $q_5 = n\pi$ for any integer $n$.

Since the kinematic differential equations are not everywhere singular, we may now assign arbitrary functions $u_1(t), \ldots, u_3(t)$ and solve for the corresponding $q_1(t), \ldots, q_6(t)$. Since the motion constraint is automatically satisfied by such a solution, it follows from the above discussion that the configuration constraint is automatically satisfied by this solution, provided that the initial values $q_1(t_0), \ldots, q_6(t_0)$ satisfy the configuration constraint.

As shown above, only **three** of the components of the linear and angular velocity vectors are independent. It takes **three** independent velocity coordinates to describe arbitrary linear and angular velocities of the rigid body relative to $\mathcal{W}$. This rigid body therefore has **three** dynamic degrees of freedom.

(Ex. 9.9 – Ex. 9.15)

## 9.2 General Formalism

### 9.2.1 Eliminating Redundant Coordinates

In the examples above, constraints on the configuration of a mechanism were formulated as configuration constraints and associated motion constraints. In terms of the configuration coordinates, the configuration constraints were

$$q_1 n_1 + q_2 n_2 + q_3 n_3 = 0$$

in the case of confinement to a plane,

$$\begin{aligned} p_3 q_2 - p_2 q_3 &= 0, \\ -p_3 q_1 + p_1 q_3 &= 0, \\ p_2 q_1 - p_1 q_2 &= 0 \end{aligned}$$

in the case of confinement to a line, and

$$q_1 = q_2 = q_3 = 0$$

in the case of no translation. The corresponding motion constraints were the time derivatives of these equations.

In each of the examples, all six configuration coordinates were retained to describe the configuration of the rigid body and the configuration constraints were only addressed through the solution of an associated system of kinematic differential equations. But, in each of the examples, we could also have eliminated redundant configuration coordinates. For example, if $n_3 \neq 0$ in the case of confinement to a plane, then we could have specified the coordinate representation of $B$ relative to $\mathcal{W}$ as

$$^{\mathcal{W}}B = \begin{pmatrix} q_1 \\ q_2 \\ -\frac{1}{n_3}\left(q_1 n_1 + q_2 n_2\right) \end{pmatrix},$$

such that

$$\mathbf{r}^{WB} \bullet \mathbf{n} = q_1 n_1 + q_2 n_2 + \left(-\frac{1}{n_3}\right)\left(q_1 n_1 + q_2 n_2\right) n_3 = 0,$$

i.e., the configuration constraint would be automatically satisfied.

If we now let

$$\frac{^{\mathcal{W}}dB}{dt} = w \begin{pmatrix} u_1 \\ u_2 \\ \cdot \end{pmatrix} \text{ and } {}^{w}\boldsymbol{\omega}^{b} = b \begin{pmatrix} u_3 \\ u_4 \\ u_5 \end{pmatrix},$$

the kinematic differential equations would be

$$\begin{pmatrix} 1 & 0 & 0 & 0 & 0 \\ 0 & 1 & 0 & 0 & 0 \\ 0 & 0 & \sin q_5 \sin q_6 & \cos q_6 & 0 \\ 0 & 0 & \sin q_5 \cos q_6 & -\sin q_6 & 0 \\ 0 & 0 & \cos q_5 & 0 & 1 \end{pmatrix} \begin{pmatrix} \dot{q}_1 \\ \dot{q}_2 \\ \dot{q}_4 \\ \dot{q}_5 \\ \dot{q}_6 \end{pmatrix} = \begin{pmatrix} u_1 \\ u_2 \\ u_3 \\ u_4 \\ u_5 \end{pmatrix},$$

i.e., a system of five differential equations in five unknowns $q_1$, $q_2$, $q_4$, $q_5$, and $q_6$. Since these are not everywhere singular, we may assign arbitrary functions of time for $u_1(t), \ldots, u_5(t)$ and solve for the configuration coordinates as functions of time. In contrast with the analysis in the previous section, there is no need to be concerned about initial values for the configuration coordinates, since the configuration constraint is automatically satisfied.

Similarly, in the case of confinement to a straight line or the case of no translation, we could eliminate two and three of the configuration coordinates, respectively. In each case, the configuration constraint would be automatically satisfied. The resulting kinematic differential equations would have dimension four and three, respectively. So why did I not follow this course of action? What was the rationale for retaining all configuration coordinates? After all, by retaining all six configuration coordinates, it was necessary to solve a system of six kinematic differential equations instead of five, four, or three, respectively. Moreover, it was necessary to ensure that the initial values of the configuration coordinates satisfy the configuration constraint.

The alternatives are as follows:

**Alternative 1.** Find a solution $q_1(t_0), \ldots, q_n(t_0)$ of the configuration constraints at some initial time and subsequently solve the kinematic differential equations with inputs $u_1(t), \ldots, u_m(t)$ to find $q_1(t), \ldots, q_n(t)$ as functions of time.

**Alternative 2.** Solve the configuration constraints for $n-m$ of the configuration coordinates in terms of the remaining configuration coordinates and subsequently choose between assigning arbitrary functions of time to the remaining configuration coordinates or solving the kinematic differential equations with inputs $u_1(t), \ldots, u_m(t)$.

In the examples above, both alternatives were equally tractable. **In general, this is not the case.** In fact, Alternative 2 suffers from a significant drawback, namely the need to solve the configuration constraints ahead of time for $n-m$ of the configuration coordinates in terms of the remaining coordinates. This was easy in the examples above. In general, this is **impossible**[1]!

**Illustration 9.1**

Assume that the kinematic differential equations of a multibody mechanism are

$$\begin{aligned} \dot{q}_1 &= u_1, \\ \dot{q}_1 \sin q_2 + \dot{q}_2 q_1 &= u_2, \end{aligned}$$

[1] In terms of elementary functions and operations.

which are non-singular, as long as $q_1 \neq 0$. Consider the imposition of the configuration constraint

$$q_1 q_2^2 + \sin(q_1 + q_2) = 0.$$

It is easy to find values for $q_1$ and $q_2$ that satisfy this equation, e.g., $q_1 = q_2 = 0$. It is impossible, however, to solve the equation for $q_1$ in terms of $q_2$, or vice versa. We simply **cannot** pursue Alternative 2.

If, instead, we differentiate the constraint with respect to time, we find

$$\dot{q}_1 q_2^2 + 2\dot{q}_2 q_1 q_2 + (\dot{q}_1 + \dot{q}_2)\cos(q_1 + q_2) = 0,$$

i.e., a motion constraint. It is no longer possible to assign arbitrary functions $u_1(t)$ and $u_2(t)$ in the kinematic differential equations above. The motion constraint implies that $u_1$ and $u_2$ are **dependent**. If, instead, we replace the equation defining $u_2$ with the motion constraint, we obtain the kinematic differential equations

$$\begin{aligned} \dot{q}_1 &= u_1, \\ \dot{q}_1 q_2^2 + 2\dot{q}_2 q_1 q_2 + (\dot{q}_1 + \dot{q}_2)\cos(q_1 + q_2) &= 0, \end{aligned}$$

which are non-singular, as long as

$$2q_1 q_2 + \cos(q_1 + q_2) \neq 0.$$

These equations can be solved for arbitrary functions $u_1(t)$ to generate a motion that automatically satisfies the motion constraint. If $q_1(t_0)$ and $q_2(t_0)$ satisfy the configuration constraint, then the generated motion will automatically satisfy the configuration constraint for all time.

More generally, suppose that

$$f(q_1, \ldots, q_n, t) = 0$$

is a configuration constraint expressed in terms of the configuration coordinates $q_1, \ldots, q_n$. While this equation may be practically impossible to solve for one of the configuration coordinates in terms of the others, it is generally possible to find numerical values for the configuration coordinates that do satisfy the constraint. These can be used as initial values for solving the kinematic differential equations.

Using the chain rule of differentiation, the time derivative of this equation becomes

$$\dot{q}_1 \frac{\partial f}{\partial q_1}(q_1, \ldots, q_n, t) + \cdots + \dot{q}_n \frac{\partial f}{\partial q_n}(q_1, \ldots, q_n, t) + \frac{\partial f}{\partial t}(q_1, \ldots, q_n, t) = 0,$$

i.e., a motion constraint. We can now replace one of the original kinematic differential equations with this motion constraint. If the resulting equations are not everywhere singular, their solution will generate a motion that automatically satisfies the motion constraint. Given the initial values from above, the solution will automatically satisfy the configuration constraint.

### 9.2.2 Integrability

As shown in the previous section, every configuration constraint corresponds to a motion constraint by differentiation with respect to time. Indeed, while the configuration constraint may be very complicated and non-linear in the configuration coordinates, the corresponding motion constraint will always be linear in the derivatives of the configuration coordinates.

But what about the converse? Is every motion constraint that is linear in the derivatives of the configuration coordinates the time derivative of a configuration constraint?

---

**Definition 9.1** A motion constraint is said to be *integrable* or *holonomic* if it can be obtained by differentiation with respect to time of a configuration constraint. In this case, any motion that satisfies the motion constraint, and for which the initial values satisfy the configuration constraint, will automatically satisfy the configuration constraint for all time.

A motion constraint that does not correspond to a configuration constraint as described above is said to be *non-integrable* or *non-holonomic.*

---

**Illustration 9.2**

Suppose that $\phi(t)$ is a solution to the differential equation

$$\frac{dy}{dt} = -\frac{\alpha_2(y,t)}{\alpha_1(y,t)},$$

where $\alpha_1$ and $\alpha_2$ are two differentiable functions of $y$ and $t$ and $\alpha_1$ is not identically equal to zero. In other words,

$$\frac{d}{dt}\phi(t) = -\frac{\alpha_2(\phi(t),t)}{\alpha_1(\phi(t),t)}.$$

Then, the configuration constraint

$$q_1 - \phi(q_2) = 0$$

2

corresponds to the motion constraint

$$\begin{aligned}\dot{q}_1 - \dot{q}_2 \frac{d}{dq_2}\phi(q_2) &= \dot{q}_1 + \frac{\alpha_2(\phi(q_2), q_2)}{\alpha_1(\phi(q_2), q_2)}\dot{q}_2 \\ &= \dot{q}_1 + \frac{\alpha_2(q_1, q_2)}{\alpha_1(q_1, q_2)}\dot{q}_2 \\ &= 0,\end{aligned}$$

where the first equality follows by replacing $t$ by $q_2$ in the differential equation above and the second equality follows from replacing $\phi(q_2)$ by $q_1$ using the configuration constraint.

The result of the illustration shows that any motion constraint of the form

$$\dot{q}_1 + \frac{\alpha_2(q_1, q_2)}{\alpha_1(q_1, q_2)}\dot{q}_2 = 0$$

corresponds to a configuration constraint

$$q_1 - \phi(q_2) = 0,$$

provided that $\phi(t)$ is a solution to the differential equation

$$\frac{dy}{dt} = -\frac{\alpha_2(y, t)}{\alpha_1(y, t)}.$$

The fundamental theory of differential equations guarantees that such a solution can be found as long as $\alpha_1 \neq 0$. It follows that motion constraints that involve only two configuration coordinates are holonomic.

**Illustration 9.3**

Consider the motion constraint

$$\alpha_1\dot{q}_1 + \alpha_2\dot{q}_2 + \alpha_3\dot{q}_3 = 0,$$

where the $\alpha_i$'s are non-zero, differentiable functions of $q_1$, $q_2$, and $q_3$. Suppose that this motion constraint can be obtained from a configuration constraint

$$f(q_1, q_2, q_3) = 0.$$

Differentiation with respect to time then yields the motion constraint

$$\frac{\partial f}{\partial q_1}\dot{q}_1 + \frac{\partial f}{\partial q_2}\dot{q}_2 + \frac{\partial f}{\partial q_3}\dot{q}_3 = 0.$$

It follows that

$$\frac{\partial f}{\partial q_1}/\alpha_1 = \frac{\partial f}{\partial q_2}/\alpha_2 = \frac{\partial f}{\partial q_3}/\alpha_3 = g$$

for some non-zero function[2] $g(q_1, q_2, q_3)$, or

$$\frac{\partial f}{\partial q_1} = g\alpha_1, \frac{\partial f}{\partial q_2} = g\alpha_2, \text{ and } \frac{\partial f}{\partial q_3} = g\alpha_3.$$

Since the second partial derivatives of a multivariable, continuously differentiable function are independent of the order of differentiation, we have

$$\frac{\partial^2 f}{\partial q_i \partial q_j} = \frac{\partial^2 f}{\partial q_j \partial q_i}.$$

Applying this to the expressions above yields

$$\begin{aligned} \frac{\partial g}{\partial q_2}\alpha_1 + g\frac{\partial \alpha_1}{\partial q_2} &= \frac{\partial g}{\partial q_1}\alpha_2 + g\frac{\partial \alpha_2}{\partial q_1}, \\ \frac{\partial g}{\partial q_3}\alpha_1 + g\frac{\partial \alpha_1}{\partial q_3} &= \frac{\partial g}{\partial q_1}\alpha_3 + g\frac{\partial \alpha_3}{\partial q_1}, \end{aligned}$$

and

$$\frac{\partial g}{\partial q_3}\alpha_2 + g\frac{\partial \alpha_2}{\partial q_3} = \frac{\partial g}{\partial q_2}\alpha_3 + g\frac{\partial \alpha_3}{\partial q_2},$$

which can be rewritten as

$$\begin{aligned} g\left(\frac{\partial \alpha_1}{\partial q_2} - \frac{\partial \alpha_2}{\partial q_1}\right) &= \frac{\partial g}{\partial q_1}\alpha_2 - \frac{\partial g}{\partial q_2}\alpha_1, \\ g\left(\frac{\partial \alpha_3}{\partial q_1} - \frac{\partial \alpha_1}{\partial q_3}\right) &= \frac{\partial g}{\partial q_3}\alpha_1 - \frac{\partial g}{\partial q_1}\alpha_3, \\ g\left(\frac{\partial \alpha_2}{\partial q_3} - \frac{\partial \alpha_3}{\partial q_2}\right) &= \frac{\partial g}{\partial q_2}\alpha_3 - \frac{\partial g}{\partial q_3}\alpha_2. \end{aligned}$$

Finally, multiplying the first equation by $\alpha_3$, the second equation by $\alpha_2$, and the third equation by $\alpha_1$, and adding up the results yields

$$\alpha_3\left(\frac{\partial \alpha_1}{\partial q_2} - \frac{\partial \alpha_2}{\partial q_1}\right) + \alpha_2\left(\frac{\partial \alpha_3}{\partial q_1} - \frac{\partial \alpha_1}{\partial q_3}\right) + \alpha_1\left(\frac{\partial \alpha_2}{\partial q_3} - \frac{\partial \alpha_3}{\partial q_2}\right) = 0.$$

The result of the illustration shows that a motion constraint in three configuration coordinates

$$\alpha_1(q_1, q_2, q_3)\dot{q}_1 + \alpha_2(q_1, q_2, q_3)\dot{q}_2 + \alpha_3(q_1, q_2, q_3)\dot{q}_3 = 0$$

[2] $g$ is typically called an *integrating factor*.

2

is holonomic, only if[3] the functions $\alpha_1$, $\alpha_2$, and $\alpha_3$ satisfy the condition

$$\alpha_3\left(\frac{\partial \alpha_1}{\partial q_2}-\frac{\partial \alpha_2}{\partial q_1}\right)+\alpha_2\left(\frac{\partial \alpha_3}{\partial q_1}-\frac{\partial \alpha_1}{\partial q_3}\right)+\alpha_1\left(\frac{\partial \alpha_2}{\partial q_3}-\frac{\partial \alpha_3}{\partial q_2}\right)=0.$$

For example, the velocity constraint

$$\dot{q}_1 q_3+\dot{q}_2=0$$

is non-integrable, since here $\alpha_1 = q_3$, $\alpha_2 = 1$, and $\alpha_3 = 0$, which implies

$$\alpha_3\left(\frac{\partial \alpha_1}{\partial q_2}-\frac{\partial \alpha_2}{\partial q_1}\right)+\alpha_2\left(\frac{\partial \alpha_3}{\partial q_1}-\frac{\partial \alpha_1}{\partial q_3}\right)+\alpha_1\left(\frac{\partial \alpha_2}{\partial q_3}-\frac{\partial \alpha_3}{\partial q_2}\right)=-1\neq 0.$$

### 9.2.3 Physical Modeling

From the previous section, we conclude that it is possible to invent motion constraints that are linear in the derivatives of the configuration coordinates that do not correspond to some configuration constraints. But do such motion constraints occur in practice?

Suppose you want to model the motion of a paddle blade through water. Experience shows you that there is significant resistance from the water to motions of the blade in the direction perpendicular to the blade surface. In contrast, there is very little resistance from the water to motions of the blade in a direction parallel to the blade surface. To *approximate* this behavior, we may impose the motion constraint that the component of the blade's velocity relative to the water that is perpendicular to the blade must be zero.

Introduce a main observer $\mathcal{W}$, relative to which the water is stationary, with reference point $W$ and reference triad $w$. Let $\mathcal{B}$ be an auxiliary observer, relative to which the paddle remains stationary, with reference point $B$ at the center of the paddle blade and reference triad $b$, such that $\mathbf{b}_3$ is along the paddle handle and $\mathbf{b}_1$ is perpendicular to the paddle blade. The constraint may then be formulated as

$$^{\mathcal{W}}\frac{dB}{dt}\bullet \mathbf{b}_1=0.$$

[3]It is, in fact, possible to show that this condition is sufficient for integrability.

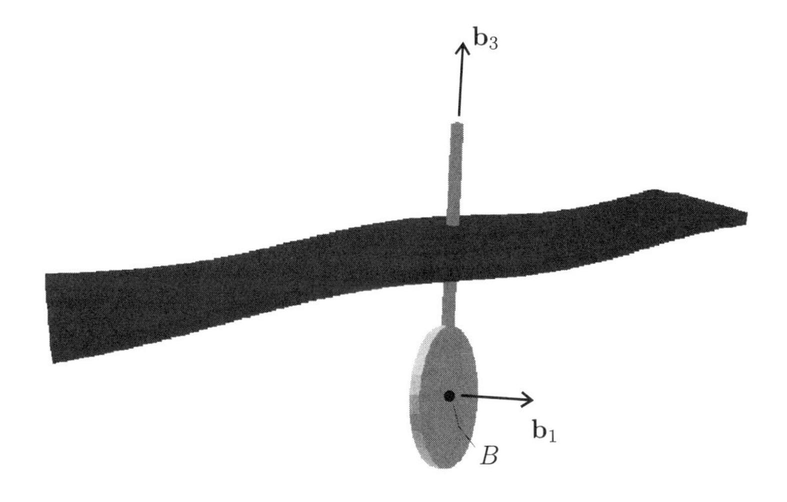


**Illustration 9.4**

Let the coordinate representation of $B$ relative to $\mathcal{W}$ be given by

$$^{\mathcal{W}}B = \begin{pmatrix} q_1 \\ q_2 \\ q_3 \end{pmatrix}.$$

Furthermore, let the orientation of the paddle relative to $\mathcal{W}$ be given by the rotation matrix

$$R_{wb} = R(q_4, 0, 0, 1)\, R(q_5, 1, 0, 0)\, R(q_6, 0, 0, 1).$$

Then, the motion constraint on the velocity of the paddle becomes

$$\dot{q}_1 (\cos q_4 \cos q_6 - \sin q_4 \cos q_5 \sin q_6) \\ + \dot{q}_2 (\sin q_4 \cos q_6 + \cos q_4 \cos q_5 \sin q_6) + \dot{q}_3 \sin q_5 \sin q_6 = 0.$$

This motion constraint is holonomic only if there exists a function

$$f(q_1, q_2, q_3, q_4, q_5, q_6) = 0,$$

such that

$$\frac{\partial f}{\partial q_1} = (\cos q_4 \cos q_6 - \sin q_4 \cos q_5 \sin q_6)\, g(q_1, q_2, q_3, q_4, q_5, q_6),$$
$$\frac{\partial f}{\partial q_2} = (\sin q_4 \cos q_6 + \cos q_4 \cos q_5 \sin q_6)\, g(q_1, q_2, q_3, q_4, q_5, q_6),$$
$$\frac{\partial f}{\partial q_3} = \sin q_5 \sin q_6 g(q_1, q_2, q_3, q_4, q_5, q_6),$$
$$\frac{\partial f}{\partial q_4} = \frac{\partial f}{\partial q_5} = \frac{\partial f}{\partial q_6} = 0$$

2

for some function $g$. The last three conditions imply that $f$ is independent of $q_4$, $q_5$, and $q_6$. But this is only possible if the three expressions

$$\begin{aligned}&(\cos q_4 \cos q_6 - \sin q_4 \cos q_5 \sin q_6)\, g\,(q_1, q_2, q_3, q_4, q_5, q_6)\,,\\&(\sin q_4 \cos q_6 + \cos q_4 \cos q_5 \sin q_6)\, g\,(q_1, q_2, q_3, q_4, q_5, q_6)\,,\end{aligned}$$

and

$$\sin q_5 \sin q_6 g\,(q_1, q_2, q_3, q_4, q_5, q_6)$$

are independent of $q_4$, $q_5$, and $q_6$. The last expression is independent of $q_4$ only if $g$ is independent of $q_4$. But if $g$ is independent of $q_4$, then the first two expressions **will** depend on $q_4$. In conclusion, it is impossible to find a function $f$, such that the motion constraint is equivalent to the configuration constraint

$$f\,(q_1, q_2, q_3, q_4, q_5, q_6) = 0.$$

The motion constraint on the paddle is non-holonomic!

Suppose that you want to model the motion of a skate on an icy surface. Experience shows you that there is significant resistance from the ice to motions of the blade edge of the skate in the direction perpendicular to the blade edge and parallel to the ice. In contrast, there is very little resistance from the ice to motions of the blade edge in a direction parallel to the blade edge and to the ice. To *approximate* this behavior, we may impose the motion constraint that the component of the blade edge's velocity relative to the ice that is perpendicular to the blade edge and parallel to the ice must be zero.

Introduce a main observer $\mathcal{W}$, relative to which the ice is stationary, with reference point $W$ and reference triad $w$, such that $\mathbf{w}_3$ is perpendicular to the ice. Let $\mathcal{B}$ be an auxiliary observer, relative to which the skate remains stationary, with reference point $B$ at the center of the skate's blade edge and reference triad $b$, such that $\mathbf{b}_3$ is perpendicular to the skate blade and $\mathbf{b}_1$ is parallel to the ice. The constraint may then be formulated as

$${}^{\mathcal{W}}\frac{dB}{dt} \bullet (\mathbf{b}_1 \times \mathbf{w}_3) = 0,$$

since

$$\mathbf{b}_1 \times \mathbf{w}_3$$

is a vector that is parallel to the ice and perpendicular to the blade edge.

**Illustration 9.5**

Let the coordinate representation of $B$ relative to $\mathcal{W}$ be given by

$$^{\mathcal{W}}B = \begin{pmatrix} q_1 \\ q_2 \\ q_3 \end{pmatrix}.$$

Furthermore, let the orientation of the skate relative to $\mathcal{W}$ be given by the rotation matrix

$$R_{wb} = R(q_4, 0, 0, 1)\, R(q_5, 1, 0, 0)\, R(q_6, 0, 0, 1).$$

Then, the motion constraint on the velocity of the skate becomes

$$\dot{q}_1 (\cos q_4 \cos q_5 \sin q_6 + \sin q_4 \cos q_6) \\ + \dot{q}_2 (\sin q_4 \cos q_5 \sin q_6 - \cos q_4 \cos q_6) = 0.$$

The result of Exercise 9.10 shows that this constraint is non-holonomic.

In both examples above, the motion constraint was imposed to model your experience of the resistance to motion by the water or the ice. The configuration constraints introduced in previous sections also model your experience of the resistance to motion; away from a plane, away from a straight line, or away from a point, respectively. Ultimately, constraints are models of our experience. They represent our interpretation of the limitations on configurations or motions of a rigid body.

As a final example, suppose you want to model the rolling motion of a rigid body along some stationary surface. Experience shows you that friction between the body and the surface results in a great resistance to sliding motion of the rigid body relative to the stationary surface at the point of contact. To *approximate* this observation, we may impose the motion constraint that the velocity relative to the surface of the contact point parallel to the surface must be zero. In fact, if the rigid body remains in contact with the surface (although the point of contact may change), then the component of the velocity perpendicular to the surface must also be zero. A motion that satisfies this motion constraint is called *rolling without slipping*.

Introduce a main observer $\mathcal{W}$, relative to which the surface is stationary, with reference point $W$ and reference triad $w$. Let $\mathcal{B}$ be an auxiliary observer, relative to which the rigid body remains stationary, with reference point $B$ and reference triad $b$. Denote by $P$ the current contact point between the rigid body and the surface. As shown in Exercises 8.11 and 8.22, the constraint may then be formulated as

$$^{\mathcal{W}}\frac{dP}{dt} = \mathbf{0},$$

i.e.,

$$^{\mathcal{W}}\frac{dB}{dt} + {}^{w}\boldsymbol{\omega}^{b} \times \mathbf{r}^{BP} = \mathbf{0}.$$

**Illustration 9.6**

Specialize to the case of a sphere rolling without slipping on a plane, such that $\mathbf{w}_1$ and $\mathbf{w}_2$ are parallel to the plane and $\mathbf{w}_3$ points away from the plane toward the center of the sphere. Let the coordinate representation of the point $B$ relative to $\mathcal{W}$ be given by

$$^{\mathcal{W}}B = \begin{pmatrix} q_1 \\ q_2 \\ R \end{pmatrix},$$

where $R$ is the radius of the sphere. Let the orientation of the sphere relative to $\mathcal{W}$ be given by

$$R_{wb} = R\left(q_3, 0, 0, 1\right) R\left(q_4, 1, 0, 0\right) R\left(q_5, 0, 0, 1\right).$$

The position of the contact point $P$ relative to $\mathcal{B}$ is given by the position vector

$$\mathbf{r}^{BP} = w \begin{pmatrix} 0 \\ 0 \\ -R \end{pmatrix}.$$

The motion constraint now becomes

$$\begin{aligned} \mathbf{0} &= {}^{\mathcal{W}}\frac{dB}{dt} + {}^{w}\boldsymbol{\omega}^{b} \times \mathbf{r}^{BP} \\ &= w \begin{pmatrix} \dot{q}_1 - R\dot{q}_4 \sin q_3 + R\dot{q}_5 \cos q_3 \sin q_4 \\ \dot{q}_2 + R\dot{q}_4 \cos q_3 + R\dot{q}_5 \sin q_3 \sin q_4 \\ 0 \end{pmatrix}, \end{aligned}$$

i.e.,

$$\dot{q}_1 - R\dot{q}_4 \sin q_3 + R\dot{q}_5 \cos q_3 \sin q_4 = 0,$$
$$\dot{q}_2 + R\dot{q}_4 \cos q_3 + R\dot{q}_5 \sin q_3 \sin q_4 = 0.$$

The result of Exercise 9.11 shows that these motion constraints are nonholonomic.

2

# 9.3 MAMBO

## 9.3.1 Basic Methodology

Given a geometry description in a .geo file, there are three different ways of generating a time-dependent motion in a MAMBO animation, namely:

**Method 1.** Declare all configuration coordinates as MAMBO animated variables and assign explicit functions of time to each variable.

**Method 2.** Declare all configuration coordinates as MAMBO state variables and import a MAMBO .sds file with the values of the state variables listed in tabular format.

**Method 3.** Declare all configuration coordinates as MAMBO state variables, define the kinematic differential equations, and assign explicit functions of time to the independent velocity coordinates.

Methods 1 and 2 only work if all constraints can be formulated as configuration constraints. Moreover, even if that is the case, they require that a sequence of values has been found for the configuration coordinates that satisfy all configuration constraints and generate a motion that appears visually continuous.

Method 3, on the other hand, applies to systems with arbitrary constraints. It does not require one to find values that satisfy the configuration constraints **except for the initial values**. The versatility of Method 3 makes it the method of choice.

As discussed in Chapter 4, the kinematic differential equations are provided to the MAMBO application in a MAMBO motion description, i.e., a .dyn file. For example, suppose that $q_1$, $q_2$, and $q_3$ are configuration coordinates that appear in the MAMBO geometry description. Impose the motion constraints

$$\dot{q}_1 q_2 + \dot{q}_3 = 0,$$
$$\dot{q}_2 q_1^2 = \cos t.$$

Then, it follows that the derivatives of the configuration coordinates are not independent. A possible set of kinematic differential equations is given by

$$\begin{pmatrix} 1 & 0 & 0 \\ q_2 & 0 & 1 \\ 0 & q_1^2 & 0 \end{pmatrix} \begin{pmatrix} \dot{q}_1 \\ \dot{q}_2 \\ \dot{q}_3 \end{pmatrix} = \begin{pmatrix} u_1 \\ 0 \\ \cos t \end{pmatrix},$$

which are non-singular, as long as

$$\begin{vmatrix} 1 & 0 & 0 \\ q_2 & 0 & 1 \\ 0 & q_1^2 & 0 \end{vmatrix} = -q_1^2 \neq 0.$$

The corresponding MAMBO motion description would then include the statements

```
states q1,q2,q3;
time t;
insignals {
    u1 = 1;
}
ode {
    rhs[q1] = u1;
    rhs[q2] = 0;
    rhs[q3] = cos(t);
    mass[q1][q1] = 1;
    mass[q1][q2] = 0;
    mass[q1][q3] = 0;
    mass[q2][q1] = q2;
    mass[q2][q2] = 0;
    mass[q2][q3] = 1;
    mass[q3][q1] = 0;
    mass[q3][q2] = q1^2;
    mass[q3][q3] = 0;
}
```

Here, $u_1(t) = 1$ as specified in the `insignals` block. The `ode` block contains information about the coefficient matrix (`mass`) and the right-hand side (`rhs`) of the kinematic differential equations. The MAMBO state variables are used to label the rows and columns of these matrices. Since the order of the equations is irrelevant, the row indices can be permuted arbitrarily. Moreover, by default, there is no need to include matrix entries that equal zero. Thus, an equivalent `ode` block could read:

```
ode {
    rhs[q2] = u1;
```

```
        rhs[q3] = cos(t);
        mass[q2][q1] = 1;
        mass[q1][q1] = q2;
        mass[q1][q3] = 1;
        mass[q3][q2] = q1^2;
    }
```

The MAMBO toolbox procedure MotionOutput can be invoked to generate a MAMBO motion description including the kinematic differential equations (which can be very complicated). For example, the following statements result in a motion description equivalent to that shown above.

```
> kde:={q1t*q2+q3t=0,q2t*q1^2=cos(t),q1t=u1}:
> MotionOutput(ode=kde,states=[q1,q2,q3],
> insignals=[u1=1]);
```

```
states q1,q2,q3;
time t;
insignals {
    u1 = 1;
    rhs_q1 = cos(t);
    mass_q1_q2 = q1^2;
}
ode {
    rhs[q1] = rhs_q1;
    rhs[q3] = u1;
    mass[q1][q2] = mass_q1_q2;
    mass[q2][q1] = q2;
    mass[q3][q1] = 1;
    mass[q2][q3] = 1;
}
```

The MotionOutput procedure attempts to generate an optimized computation sequence that reduces the number of function calls. This is reflected in the introduction of the rhs_q1 and mass_q1_q2 variables. Their definitions are placed within the **insignals** block, since they effectively provide input signals to the kinematic differential equations.

We can use MAPLE's genmatrix and det procedures to find the determinant of the coefficient matrix and thereby detect any singularities:

```
> det(genmatrix(kde,[q1t,q2t,q3t]));
```

$$q1^2$$

The difference in sign from the determinant computed above is attributable to a switch of two of the rows of the coefficient matrix. The det function works fine for relatively sparse matrices of relatively low dimension. For large matrices that have many non-zero entries, the memory requirements for computing the determinant are beyond what most installations of MAPLE are set up to handle. As an alternative, MAMBO will detect singularities and notify the user as a .dyn file is loaded (if the kinematic differential equations are everywhere singular) or during a simulation (if a singularity is approached). Finally, we may include the option `checksings` when invoking the `MotionOutput` procedure to check whether the initial values for the configuration coordinates correspond to a singularity of the kinematic differential equations as in the following MAMBO toolbox statement:

```
> MotionOutput(ode=kde,states=[q1,q2,q3],
> checksings,insignals=[u1=1]);

Error, (in MotionOutput) Mass matrix initially singular!
```

where the `states=[q1,q2,q3]` argument establishes the default initial values for $q_1$, $q_2$, and $q_3$, i.e.,

$$q_1 = q_2 = q_3 = 0.$$

Since the kinematic differential equations are singular whenever $q_1 = 0$, the `MotionOutput` procedure returns an appropriate error message.

### 9.3.2 Switching Between Constraints

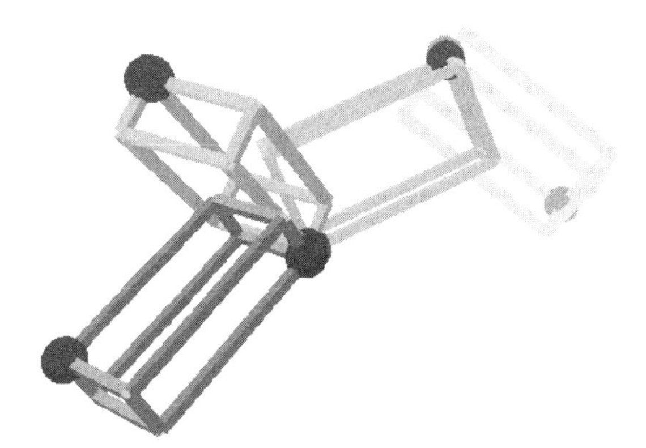

It is possible to switch between different sets of constraints and different sets of kinematic differential equations within the same MAMBO .dyn file, provided that the number of configuration coordinates stays the same. Suppose, for example, that you want to model the motion of a rigid body, such that the point $P_1$ on the rigid body is stationary relative to the main observer when $\sin\left(2\pi\frac{t}{T}\right) \geq 0$ and the point $P_2$ on the rigid body is stationary relative to the main observer when $\sin\left(2\pi\frac{t}{T}\right) < 0$.

We can achieve this by switching between the kinematic differential equations

$$\begin{aligned}
{}^{\mathcal{W}}\frac{dP_1}{dt} \bullet \mathbf{w}_1 &= 0,\\
{}^{\mathcal{W}}\frac{dP_1}{dt} \bullet \mathbf{w}_2 &= 0,\\
{}^{\mathcal{W}}\frac{dP_1}{dt} \bullet \mathbf{w}_3 &= 0,\\
{}^{w}\boldsymbol{\omega}^{b} \bullet \mathbf{b}_1 &= u_1,\\
{}^{w}\boldsymbol{\omega}^{b} \bullet \mathbf{b}_2 &= u_2,\\
{}^{w}\boldsymbol{\omega}^{b} \bullet \mathbf{b}_3 &= u_3,
\end{aligned}$$

which are valid when $\sin\left(2\pi\frac{t}{T}\right) \geq 0$ and

$$\begin{aligned}
{}^{\mathcal{W}}\frac{dP_2}{dt} \bullet \mathbf{w}_1 &= 0,\\
{}^{\mathcal{W}}\frac{dP_2}{dt} \bullet \mathbf{w}_2 &= 0,\\
{}^{\mathcal{W}}\frac{dP_2}{dt} \bullet \mathbf{w}_3 &= 0,\\
{}^{w}\boldsymbol{\omega}^{b} \bullet \mathbf{b}_1 &= u_1,\\
{}^{w}\boldsymbol{\omega}^{b} \bullet \mathbf{b}_2 &= u_2,\\
{}^{w}\boldsymbol{\omega}^{b} \bullet \mathbf{b}_3 &= u_3,
\end{aligned}$$

which are valid when $\sin\left(2\pi\frac{t}{T}\right) < 0$.

Alternatively, we may consider the combined kinematic differential equations

$$\begin{aligned}
\left(k\,{}^{\mathcal{W}}\frac{dP_1}{dt} + (1-k)\,{}^{\mathcal{W}}\frac{dP_2}{dt}\right) \bullet \mathbf{w}_1 &= 0,\\
\left(k\,{}^{\mathcal{W}}\frac{dP_1}{dt} + (1-k)\,{}^{\mathcal{W}}\frac{dP_2}{dt}\right) \bullet \mathbf{w}_2 &= 0,\\
\left(k\,{}^{\mathcal{W}}\frac{dP_1}{dt} + (1-k)\,{}^{\mathcal{W}}\frac{dP_2}{dt}\right) \bullet \mathbf{w}_3 &= 0,\\
{}^{w}\boldsymbol{\omega}^{b} \bullet \mathbf{b}_1 &= u_1,\\
{}^{w}\boldsymbol{\omega}^{b} \bullet \mathbf{b}_2 &= u_2,\\
{}^{w}\boldsymbol{\omega}^{b} \bullet \mathbf{b}_3 &= u_3,
\end{aligned}$$

and set $k = 1$ when $\sin\left(2\pi\frac{t}{T}\right) \geq 0$ and 0 otherwise.

The following MAMBO session generates the necessary kinematic differential equations:

```
> Restart():
> DeclareObservers(W,B):
> DeclarePoints(W,B,P1,P2):
> DeclareTriads(w,b):
> DefineObservers([W,W,w],[B,B,b]):
> DefinePoints([W,B,w,q1,q2,q3],[B,P1,b,p11,p12,p13],
> [B,P2,b,p21,p22,p23]):
> DefineTriads([w,b,[q4,3],[q5,1],[q6,3]]):
> DeclareStates(q1,q2,q3,q4,q5,q6):
> kde:={seq(((k &** LinearVelocity(W,P1)) &++
> ((1-k) &** LinearVelocity(W,P2))) &oo
> MakeTranslations(w,i)=0,i=1..3),
> seq(AngularVelocity(w,b) &oo
> MakeTranslations(b,i)=cat(u,i),i=1..3)}:
> simplify(det(genmatrix(kde,
> [q1t,q2t,q3t,q4t,q5t,q6t])));
```

$$\sin(q5)$$

```
> MotionOutput(ode=kde,states=[q1,q2,q3,q4,q5,q6]);

states q1,q2,q3,q4,q5,q6;
insignals {
    t1 = sin(q5);
    t2 = sin(q6);
    t3 = t1*t2;
    t5 = cos(q6);
    t6 = t5*t1;
    t10 = 1-k;
    t15 = k*(t3*p12-t6*p11)+t10*(t3*p22-t6*p21);
    t16 = cos(q4);
    t20 = cos(q5);
    t28 = k*(t6*p13-t20*p12)+t10*(t6*p23-t20*p22);
    t29 = sin(q4);
    t31 = t16*t20;
    t33 = t29*t5+t31*t2;
    t43 = k*(t20*p11-t3*p13)+t10*(t20*p21-t3*p23);
    t46 = -t29*t2+t31*t5;
    mass_q5_q4 = t20;
    t56 = t29*mass_q5_q4;
    t58 = t16*t5-t56*t2;
    t62 = -t16*t2-t56*t5;
    t72 = k*(t5*p12+t2*p11)+t10*(t5*p22+t2*p21);
    t79 = -k*t2*p13-t10*t2*p23;
    t85 = -k*t5*p13-t10*t5*p23;
    mass_q3_q5 = t5;
    t98 = -k*p12-t10*p22;
    t102 = k*p11+t10*p21;
}
ode {
    rhs[q3] = u1;
    rhs[q4] = u2;
    rhs[q5] = u3;
    mass[q2][q2] = 1;
    mass[q6][q1] = 1;
```

```
        mass[q1][q3] = 1;
        mass[q2][q4] = -t15*t16*t1+t28*t33+t43*t46;
        mass[q1][q4] = t15*t20+t28*t1*t2+t43*t1*t5;
        mass[q3][q4] = t1*t2;
        mass[q4][q4] = t5*t1;
        mass[q5][q4] = mass_q5_q4;
        mass[q6][q4] = t15*t29*t1+t28*t58+t43*t62;
        mass[q2][q5] = -t72*t16*t1+t79*t33+t85*t46;
        mass[q1][q5] = t72*mass_q5_q4+t79*t1*t2+t85*t1*t5;
        mass[q3][q5] = mass_q3_q5;
        mass[q4][q5] = -t2;
        mass[q6][q5] = t72*t29*t1+t79*t58+t85*t62;
        mass[q2][q6] = t98*t33+t102*t46;
        mass[q1][q6] = t98*t1*t2+t102*t1*mass_q3_q5;
        mass[q5][q6] = 1;
        mass[q6][q6] = t98*t58+t102*t62;
    }
```

In addition to specifying $u_1$, $u_2$, and $u_3$ as functions of time in the `insignals` block, we also need to specify $k$. For example,

```
    insignals {
        u1 = cos(t);
        u2 = -1;
        u3 = 0;
        k = sin(2*pi*t/T)>=0;
        ⋮
```

will do the trick. Naturally, $p_{11}, \ldots, p_{23}$, and $T$ all have to be declared as parameters and can subsequently be changed interactively by the user. A syntactically correct MAMBO motion description would thus be generated by the MAMBO toolbox statement

```
> MotionOutput(ode=kde,states=[q1,q2,q3,q4,q5,q6],
> insignals=[u1=cos(t),u2=-1,u3=0,k=(sin(2*Pi*t/T)&>=0)],
> parameters=[p11,p12,p13,p21,p22,p23,T],
> filename="switching.dyn");
```

where the output has been spooled directly to the file `switching.dyn`.

## 9.4 Additional Examples

(Ex. 9.16 – Ex. 9.17)

The ultimate goal of the discussion in the next few subsections is to model the motion of a sphere rolling without slipping on a plane. In a previous section, we accomplished this with a minimal set of five configuration coordinates, two motion constraints, and three independent velocity coordinates. In the present section, we shall retain all six configuration coordinates for the sphere and arrive at the final mechanism only after introducing a sequence of successive motion constraints.

### 9.4.1 Point Contact

Suppose that the configuration of a sphere of radius $\rho$ is constrained in such a way that some point on the surface of sphere always coincides with some point on a plane.

Introduce a main observer $\mathcal{W}$, relative to which the plane is stationary, with reference point $W$ somewhere on the plane and reference triad $w$, such that $\mathbf{w}_1$ and $\mathbf{w}_2$ are parallel to the plane and $\mathbf{w}_3$ points away from the plane toward the center of the sphere. Let $\mathcal{A}$ be an auxiliary observer, relative to which the sphere is stationary, such that the reference point $A$ coincides with the center of the sphere. Let $a$ denote the reference triad of $\mathcal{A}$.

Introduce configuration coordinates, such that

$$\mathbf{r}^{WA} = w \begin{pmatrix} q_1 \\ q_2 \\ q_3 \end{pmatrix}$$

and

$$R_{wa} = R(q_4, 0, 0, 1)\, R(q_5, 1, 0, 0)\, R(q_6, 0, 0, 1).$$

Some point on the sphere's surface will coincide with some point on the plane provided that the distance from the sphere's center to the plane is less than or equal to the radius of the sphere. In other words, the condition on the sphere's configuration is satisfied as long as

$$\left|\mathbf{r}^{WA} \bullet \mathbf{w}_3\right| \le \rho.$$

Since this is an inequality, it **does not** correspond to a configuration constraint. As long as $|q_3| < \rho$, there are no constraints on the values of the configuration coordinates.

**Illustration 9.7**

The following MAMBO toolbox statements establish the basic geometry of the sphere:

```
> Restart():
> DeclareObservers(W,A):
> DeclarePoints(W,A):
> DeclareTriads(w,a):
> DefineObservers([W,W,w],[A,A,a]):
> DefinePoints([W,A,w,q1,q2,q3]):
> DefineTriads([w,a,[q4,3],[q5,1],[q6,3]]):
> DeclareStates(q1,q2,q3,q4,q5,q6):
```

In particular,

```
> FindTranslation(W,A) &oo
> MakeTranslations(w,3)<=rho;
```

$$q3 \leq \rho$$

As long as $|q_3| < \rho$, there is actually an entire circle of points on the sphere that coincide with a circle of points on the plane. As the configuration of the sphere changes, the circle changes position on the surface of the sphere and on the plane.

Suppose now that we want to locate some point $P$ on this circle and trace its motion as the configuration of the sphere changes. To this end, let

$$\mathbf{r}^{AP} = a \begin{pmatrix} \rho \sin q_7 \cos q_8 \\ \rho \sin q_7 \sin q_8 \\ \rho \cos q_7 \end{pmatrix},$$

where $q_7$ and $q_8$ are two additional configuration coordinates that specify the coordinate representation of $P$ relative to $\mathcal{A}$. Since the point $P$ is assumed to coincide with some point on the plane, we must have

$$\mathbf{r}^{WP} \bullet \mathbf{w}_3 = 0.$$

This is a configuration constraint. Equivalently,

$${}^{\mathcal{W}}\frac{dP}{dt} \bullet \mathbf{w}_3 = 0,$$

since the velocity of the point $P$ must be parallel to the plane. This is a motion constraint.

As in previous sections, a motion that satisfies the motion constraint will automatically satisfy the configuration constraint, provided that the configuration constraint is satisfied by the initial values for the configuration coordinates.

**Illustration 9.8**

We continue with the previous MAMBO toolbox session.

```
> DeclarePoints(P):
> DefinePoints([A,P,a,rho*sin(q7)*cos(q8),
> rho*sin(q7)*sin(q8),rho*cos(q7)]):
> DeclareStates(q7,q8):
```

The configuration constraint and corresponding motion constraint are then obtained from the statement

```
> confconst1:=FindTranslation(W,P) &oo
> MakeTranslations(w,3)=0:
> motionconst1:=DiffTime(confconst1):
```

Since there are no constraints on the configuration or motion of the sphere, the components of the linear and angular velocities of $\mathcal{A}$ relative to $\mathcal{W}$ are all independent. We are free to assign arbitrary functions of time to these components, say

$$^{\mathcal{W}}\mathbf{v}^{\mathcal{A}} = w \begin{pmatrix} u_1 \\ u_2 \\ u_3 \end{pmatrix} \text{ and } ^{w}\boldsymbol{\omega}^{a} = a \begin{pmatrix} u_4 \\ u_5 \\ u_6 \end{pmatrix}.$$

Since the position of $P$ relative to $\mathcal{A}$ is described using only two configuration coordinates, at most two of the components of $^{\mathcal{A}}\frac{dP}{dt}$ can be independent from each other. Similarly, it follows that at most two of the components of $^{\mathcal{W}}\frac{dP}{dt}$ can be independent of the components of $^{\mathcal{W}}\mathbf{v}^{\mathcal{A}}$ and $^{w}\boldsymbol{\omega}^{a}$. The additional condition imposed by the motion constraint reduces the number of independent components of $^{\mathcal{W}}\frac{dP}{dt}$ to one, e.g.,

$$^{\mathcal{W}}\frac{dP}{dt} = w \begin{pmatrix} u_7 \\ \cdot \\ \cdot \end{pmatrix}.$$

**Illustration 9.9**

We continue with the previous MAMBO toolbox session. The following statement defines the kinematic differential equations.

```
> kde:={
> seq(LinearVelocity(W,A) &oo
> MakeTranslations(w,i)=cat(u,i),i=1..3),
> seq(AngularVelocity(w,a) &oo
> MakeTranslations(a,i)=cat(u,i+3),i=1..3),
> LinearVelocity(W,P) &oo MakeTranslations(w,1)=u7,
> motionconst1}:
```

From the result of the statement

```
> simplify(det(genmatrix(kde,[seq(cat(q,i,t),i=1..8)]))):
```

we conclude that the kinematic differential equations are not everywhere singular.

Since the kinematic differential equations are not everywhere singular, we may now assign arbitrary functions $u_1(t), \ldots, u_7(t)$ and solve for the corresponding $q_1(t), \ldots, q_8(t)$ until we reach a singularity of the kinematic differential equations. Since the motion constraint is automatically satisfied by such a solution, it follows from the above discussion that the configuration constraint is automatically satisfied by this solution.

### 9.4.2 Tangential Contact

Now, suppose that the sphere in the previous section makes tangential contact with the plane at the point $P$. This implies that all tangent directions to the sphere's surface at the point $P$ are parallel to the plane.

From

$$\mathbf{r}^{AP} = a \begin{pmatrix} \rho \sin q_7 \cos q_8 \\ \rho \sin q_7 \sin q_8 \\ \rho \cos q_7 \end{pmatrix},$$

we conclude that any vector tangent to the sphere's surface at the point $P$ is a linear combination of the vectors

$$\mathbf{t}_1 = a \frac{\partial}{\partial q_7} \begin{pmatrix} \rho \sin q_7 \cos q_8 \\ \rho \sin q_7 \sin q_8 \\ \rho \cos q_7 \end{pmatrix} = a \begin{pmatrix} \rho \cos q_7 \cos q_8 \\ \rho \cos q_7 \sin q_8 \\ -\rho \sin q_7 \end{pmatrix}$$

and

$$\mathbf{t}_2 = a \frac{\partial}{\partial q_8} \begin{pmatrix} \rho \sin q_7 \cos q_8 \\ \rho \sin q_7 \sin q_8 \\ \rho \cos q_7 \end{pmatrix} = a \begin{pmatrix} -\rho \sin q_7 \sin q_8 \\ \rho \sin q_7 \cos q_8 \\ 0 \end{pmatrix}.$$

**Illustration 9.10**

We continue with the previous MAMBO toolbox session. The following statements define the configuration constraints

$$\mathbf{t}_1 \bullet \mathbf{w}_3 = \mathbf{t}_2 \bullet \mathbf{w}_3 = 0.$$

```
> confconst2:=MakeTranslations(a,rho*cos(q7)*cos(q8),
> rho*cos(q7)*sin(q8),-rho*sin(q7)) &oo
> MakeTranslations(w,3)=0:
> confconst3:=MakeTranslations(a,-rho*sin(q7)*sin(q8),
> rho*sin(q7)*cos(q8),0) &oo
> MakeTranslations(w,3)=0:
```

The corresponding motion constraints are obtained by differentiating with respect to time.

```
> motionconst2:=DiffTime(confconst2):
> motionconst3:=DiffTime(confconst3):
```

The imposition of the two additional motion constraints reduces the number of independent components of the velocities discussed above by another two. We expect to only be able to assign arbitrary functions of time to five components of the velocities, e.g.,

$${}^{\mathcal{W}}\mathbf{v}^{\mathcal{A}} = w \begin{pmatrix} u_1 \\ u_2 \\ \cdot \end{pmatrix} \text{ and } {}^{w}\boldsymbol{\omega}^{a} = a \begin{pmatrix} u_4 \\ u_5 \\ u_6 \end{pmatrix}.$$

**Illustration 9.11**

The MAMBO toolbox statement

```
> kde:={seq(LinearVelocity(W,A) &oo
> MakeTranslations(w,i)=cat(u,i),i=1..2),
> seq(AngularVelocity(w,a) &oo MakeTranslations(a,i)
> =cat(u,i+3),i=1..3),motionconst1,
> motionconst2,motionconst3}:
```

defines the corresponding kinematic differential equations. Again, the result of the statement

```
> simplify(det(genmatrix(kde,[seq(cat(q,i,t),i=1..8)]))):
```

shows that the kinematic differential equations are not everywhere singular.

Since the kinematic differential equations are not everywhere singular, we may now assign arbitrary functions $u_1(t)$, $u_2(t)$, $u_4(t)$, $u_5(t)$, and $u_6(t)$ and solve for the corresponding $q_1(t)$, ..., $q_8(t)$ until we reach a singularity of the kinematic differential equations. Since the motion constraints are automatically satisfied by such a solution, it follows from the above discussion that the configuration constraints are automatically satisfied by this solution.

### 9.4.3 Rolling Without Slipping

Now, suppose that the point on the sphere that currently coincides with $P$ has zero velocity relative to $\mathcal{W}$. This is the rolling-without-slipping motion constraint, since the plane is stationary relative to $\mathcal{W}$. Specifically,

$$\frac{{}^{\mathcal{W}}dA}{dt} + {}^{w}\boldsymbol{\omega}^{a} \times \mathbf{r}^{AP} = \mathbf{0}.$$

The configuration constraints imposed in the previous two subsections already guarantee that the velocity component away from the plane, i.e., in the $\mathbf{w}_3$ direction, is zero. The rolling-without-slipping constraint thus implies that

$$\left(\frac{{}^{\mathcal{W}}dA}{dt} + {}^{w}\boldsymbol{\omega}^{a} \times \mathbf{r}^{AP}\right) \bullet \mathbf{w}_1 = 0,$$
$$\left(\frac{{}^{\mathcal{W}}dA}{dt} + {}^{w}\boldsymbol{\omega}^{a} \times \mathbf{r}^{AP}\right) \bullet \mathbf{w}_2 = 0.$$

The number of dynamic degrees of freedom are reduced by another two. We conclude that only three components of the velocities introduced above may be assigned arbitrarily, e.g.,

$${}^{w}\boldsymbol{\omega}^{a} = a \begin{pmatrix} u_4 \\ u_5 \\ u_6 \end{pmatrix}.$$

**Illustration 9.12**

We continue with the same MAMBO toolbox session.

```
> motionconst4:=(LinearVelocity(W,A) &++
> (AngularVelocity(w,a) &xx FindTranslation(A,P))) &oo
> MakeTranslations(w,1)=0:
> motionconst5:=(LinearVelocity(W,A) &++
> (AngularVelocity(w,a) &xx FindTranslation(A,P))) &oo
> MakeTranslations(w,2)=0:
> kde:={seq(AngularVelocity(w,a) &oo
> MakeTranslations(a,i)=cat(u,i+3),i=1..3),motionconst1,
> motionconst2,motionconst3,motionconst4,motionconst5}:
```

The result of the statement

```
> simplify(det(genmatrix(kde,[seq(cat(q,i,t),i=1..8)]))):
```

shows that the resulting kinematic differential equations are not everywhere singular.

Since the kinematic differential equations are not everywhere singular, we may now assign arbitrary functions $u_4(t), \ldots, u_6(t)$ and solve for the corresponding $q_1(t), \ldots, q_8(t)$ until we reach a singularity of the kinematic differential equations. Since the motion constraints are automatically satisfied by such a solution, it follows from the above discussion that the configuration constraints are automatically satisfied by this solution, provided that they are satisfied by the initial values.

## 9.5 Exercises

**Exercise 9.1** Consider the free rigid body in Section 9.1.1. Find the kinematic differential equations corresponding to the independent velocity coordinates $u_1, \ldots, u_6$, where

$$^{\mathcal{W}}\frac{dB}{dt} = w\begin{pmatrix} u_1 \\ u_2 \\ u_3 \end{pmatrix} \text{ and } {}^w\boldsymbol{\omega}^b = w\begin{pmatrix} u_4 \\ u_5 \\ u_6 \end{pmatrix}$$

and determine where they are non-singular.

**Exercise 9.2** Consider the free rigid body in Section 9.1.1. Suppose that the orientation of the rigid body relative to $\mathcal{W}$ is described in terms of Euler parameters (cf. Illustration 6.4). Find the kinematic differential equations corresponding to the independent velocity coordinates $u_1, \ldots, u_6$, where

$$^{\mathcal{W}}\frac{dB}{dt} = w\begin{pmatrix} u_1 \\ u_2 \\ u_3 \end{pmatrix} \text{ and } {}^w\boldsymbol{\omega}^b = b\begin{pmatrix} u_4 \\ u_5 \\ u_6 \end{pmatrix}$$

and determine where they are non-singular.

[Hint: Recall that the Euler parameters $\tilde{q}_i$ satisfy the configuration constraint

$$\tilde{q}_1^2 + \tilde{q}_2^2 + \tilde{q}_3^2 + \tilde{q}_4^2 = 1.$$

]

**Exercise 9.3** Consider the free rigid body in Section 9.1.1. Find the kinematic differential equations corresponding to the independent velocity coordinates $u_1, \ldots, u_6$, where

$$^{\mathcal{W}}\frac{dB}{dt} = w\begin{pmatrix} u_1 \\ u_2 \\ u_3 \end{pmatrix} \text{ and } {}^w\boldsymbol{\omega}^b = w\begin{pmatrix} u_4 \\ u_5 \\ u_6 \end{pmatrix}$$

when the orientation of the rigid body relative to $\mathcal{W}$ is described using a $1-2-3$ sequence of Euler angles.

**Exercise 9.4** Consider the constrained rigid body in Section 9.1.4. Find the kinematic differential equations corresponding to the independent velocity coordinates $u_1, \ldots, u_4$, where

$$^{\mathcal{W}}\frac{dB}{dt} = w\begin{pmatrix} \cdot \\ \cdot \\ u_1 \end{pmatrix} \text{ and } {}^w\boldsymbol{\omega}^b = b\begin{pmatrix} u_2 \\ u_3 \\ u_4 \end{pmatrix}$$

and the motion constraints are

a) $$p_1\dot{q}_3 - p_3\dot{q}_1 = 0$$
$$p_2\dot{q}_1 - p_1\dot{q}_2 = 0$$

b) $$p_3\dot{q}_2 - p_2\dot{q}_3 = 0$$
$$p_2\dot{q}_1 - p_1\dot{q}_2 = 0$$

Determine where the kinematic differential equations are non-singular.

**Exercise 9.5** Consider the constrained rigid body in Section 9.1.4. Find the kinematic differential equations corresponding to the independent velocity coordinates $u_1, \ldots, u_4$, where

$$\frac{{}^{\mathcal{W}}dB}{dt} = w \begin{pmatrix} \cdot \\ u_1 \\ \cdot \end{pmatrix} \text{ and } {}^{w}\boldsymbol{\omega}^{b} = b \begin{pmatrix} u_2 \\ u_3 \\ u_4 \end{pmatrix}$$

and the motion constraints are

$$\text{a)} \begin{array}{l} p_3\dot{q}_2 - p_2\dot{q}_3 = 0 \\ p_1\dot{q}_3 - p_3\dot{q}_1 = 0 \end{array}$$
$$\text{b)} \begin{array}{l} p_1\dot{q}_3 - p_3\dot{q}_1 = 0 \\ p_2\dot{q}_1 - p_1\dot{q}_2 = 0 \end{array}$$

Determine where the kinematic differential equations are non-singular.

**Exercise 9.6** Consider the constrained rigid body in Section 9.1.4. Find the kinematic differential equations corresponding to the independent velocity coordinates $u_1, \ldots, u_4$, where

$$\frac{{}^{\mathcal{W}}dB}{dt} = w \begin{pmatrix} u_1 \\ \cdot \\ \cdot \end{pmatrix} \text{ and } {}^{w}\boldsymbol{\omega}^{b} = b \begin{pmatrix} u_2 \\ u_3 \\ u_4 \end{pmatrix}$$

and the motion constraints are

$$\text{a)} \begin{array}{l} p_3\dot{q}_2 - p_2\dot{q}_3 = 0 \\ p_2\dot{q}_1 - p_1\dot{q}_2 = 0 \end{array}$$
$$\text{b)} \begin{array}{l} p_1\dot{q}_3 - p_3\dot{q}_1 = 0 \\ p_2\dot{q}_1 - p_1\dot{q}_2 = 0 \end{array}$$

Determine where the kinematic differential equations are non-singular.

**Exercise 9.7** Consider a rigid body whose motion is constrained, such that its angular velocity relative to some observer is perpendicular to a plane that is stationary relative to the observer. Introduce configuration coordinates, formulate the corresponding motion constraint, introduce independent velocity coordinates, formulate the kinematic differential equations, and determine where these are non-singular.

**Exercise 9.8** Consider a rigid body whose motion is constrained, such that its angular velocity relative to some observer is parallel to a plane that is stationary relative to the observer. Introduce configuration coordinates, formulate the corresponding motion constraint, introduce independent velocity coordinates, formulate the kinematic differential equations, and determine where these are non-singular.

**Exercise 9.9** For each of the following configuration constraints, find the corresponding motion constraint.

$$\begin{array}{l} \text{a) } q_1^2 + q_2^2 + q_3^2 = 1 \\ \text{b) } q_1 q_2^2 + q_3 \sin q_1 = 0 \\ \text{c) } q_1 \cos(q_2 + q_3) = q_2 \\ \text{d) } q_1 e^{q_2} - q_2 e^{q_1} = 0 \end{array}$$

**Exercise 9.10** Show that the following motion constraint is non-holonomic:

$$\dot{q}_1 (\cos q_4 \cos q_5 \sin q_6 + \sin q_4 \cos q_6) \\ + \dot{q}_2 (\sin q_4 \cos q_5 \sin q_6 - \cos q_4 \cos q_6) = 0.$$

**Exercise 9.11** Show that the following motion constraints are non-holonomic:

$$\dot{q}_1 - R\dot{q}_4 \sin q_3 + R\dot{q}_5 \cos q_3 \sin q_4 = 0,$$
$$\dot{q}_2 + R\dot{q}_4 \cos q_3 + R\dot{q}_5 \sin q_3 \sin q_4 = 0.$$

**Exercise 9.12** For each of the following holonomic motion constraints, find a corresponding configuration constraint.

$$\begin{array}{l} \text{a) } \dot{q}_1 + q_2\dot{q}_2 = 0 \\ \text{b) } f(q_2)\,\dot{q}_1 + q_1\dot{q}_2 = 0 \end{array}$$

**Exercise 9.13** Consider a mechanism whose configuration can be described by three configuration coordinates $q_1$, $q_2$, and $q_3$. Impose the configuration constraint

$$q_1^2 + q_2^2 + q_3^2 = 1$$

and find a set of kinematic differential equations that are nowhere singular.

**Exercise 9.14** Consider a mechanism whose configuration can be described by three configuration coordinates $q_1$, $q_2$, and $q_3$. Impose the configuration constraint

$$q_1^2 + q_2^2 = 1.$$

Find a set of kinematic differential equations that will guarantee that the configuration constraint is approximately satisfied after a sufficiently long time even if the initial conditions do not satisfy the configuration constraint.

**Solution.** The configuration constraint above corresponds to the motion constraint

$$2q_1\dot{q}_1 + 2q_2\dot{q}_2 = 0.$$

As we have three configuration coordinates, it follows that the mechanism has two dynamic degrees of freedom, i.e., that two independent velocity coordinates are necessary to describe all allowable motions of the mechanism, e.g.,

$$u_1 = \dot{q}_1,$$
$$u_2 = \dot{q}_3,$$

which lead to the kinematic differential equations

$$\begin{pmatrix} 1 & 0 & 0 \\ 0 & 0 & 1 \\ 2q_1 & 2q_2 & 0 \end{pmatrix} \begin{pmatrix} \dot{q}_1 \\ \dot{q}_2 \\ \dot{q}_3 \end{pmatrix} = \begin{pmatrix} u_1 \\ u_2 \\ 0 \end{pmatrix},$$

which are non-singular as long as

$$q_2 \neq 0.$$

From the discussion in the text, it follows that the solution to the kinematic differential equations will automatically satisfy the configuration constraints as long as the initial conditions satisfy the configuration constraint. If this is not the case, e.g., if initially

$$q_1^2(0) + q_2^2(0) = 3,$$

then

$$q_1^2(t) + q_2^2(t) = 3$$

for all time.

Alternatively, consider the kinematic differential equations

$$\begin{pmatrix} 1 & 0 & 0 \\ 0 & 0 & 1 \\ 2q_1 & 2q_2 & 0 \end{pmatrix} \begin{pmatrix} \dot{q}_1 \\ \dot{q}_2 \\ \dot{q}_3 \end{pmatrix} = \begin{pmatrix} u_1 \\ u_2 \\ -\left(q_1^2 + q_2^2 - 1\right) \end{pmatrix},$$

which are non-singular as long as $q_2 \neq 0$. If the initial conditions satisfy the configuration constraint, then it follows that the matrix on the right-hand side reduces to the matrix on the right-hand side of the kinematic differential equations proposed above. Since the configuration coordinates will continue to satisfy the configuration constraint, the new set of kinematic differential equations reduces to the previous set for all time.

But what happens if the initial conditions do not satisfy the configuration constraint? Consider the function

$$f(t) = q_1^2(t) + q_2^2(t) - 1.$$

Then, the last of the new set of kinematic differential equations reads

$$\frac{df}{dt} = -f,$$

the solution to which is

$$f(t) = Ce^{-t},$$

where

$$C = f(0) = q_1^2(0) + q_2^2(0) - 1.$$

Clearly, $f \to 0$ independently of $C$ as $t \to \infty$. In fact, the rate of convergence can be increased by increasing the magnitude of the coefficient on the right-hand side. It follows that the solution to the new set of kinematic differential equations will approximately satisfy the configuration constraint after a sufficiently long time independently of the choice of initial conditions for the configuration coordinates.

**Exercise 9.15** Consider the mechanism in the previous exercise and suppose that an approximate solution to the kinematic differential equations will be obtained using a numerical integration routine. Discuss the benefits of using the second set of kinematic differential equations even if the initial conditions for the configuration coordinates do satisfy the configuration constraint.

**Exercise 9.16** Repeat the construction in Section 9.4 for a thin disk rolling without slipping on a plane that is stationary relative to some observer. Determine whether the motion constraints are holonomic or nonholonomic. Find the number of dynamic degrees of freedom. Use the MAMBO toolbox to derive the corresponding kinematic differential equations.

**Exercise 9.17** Repeat the construction in Section 9.4 for a cylinder rolling without slipping on a plane that is stationary relative to some observer. Determine whether the motion constraints are holonomic or nonholonomic. Find the number of dynamic degrees of freedom. Use the MAMBO toolbox to derive the corresponding kinematic differential equations.

### Summary of notation

A lower-case $u$ with suitable subscripts was used in this chapter to denote independent velocity coordinates.

### Summary of terminology

(Page 368) The smallest number of configuration coordinates necessary to describe the configuration of a rigid body is called its number of *geometric degrees of freedom.*

(Page 368) The number of *independent velocity coordinates* necessary to describe the linear and angular velocities of a rigid body is called its number of *dynamic degrees of freedom.*

(Page 367) A complete set of differential equations in the configuration coordinates as functions of time parametrized by the independent velocity coordinates is called a set of *kinematic differential equations.*

(Page 369) A condition on the configuration coordinates is called a *configuration constraint.*

(Page 369) A condition on the configuration coordinates and their derivatives with respect to time is called a *motion constraint.*

(Page 388) Two bodies in contact are said to be *rolling without slipping* on each other if the velocities of the contact points on the two bodies are equal relative to some observer.

(Page 390) The kinematic differential equations are declared in the `ode` block in a MAMBO *motion description*, a MAMBO .dyn file.

(Page 390) The time-dependence of the independent velocity coordinates is declared in the `insignals` block in a MAMBO .dyn file.

(Page 391) Subexpressions generated in optimizing the evaluation of the kinematic differential equations are declared in the `insignals` block in a MAMBO .dyn file.

(Page 391) In the MAMBO toolbox, the procedure `MotionOutput` generates a MAMBO motion description.

# Chapter 10

# Review

*wherein the reader learns of:*

- *Combining the elements developed in previous chapters into a general methodology for describing the geometry and motion of a multibody mechanism.*

Practicum

Having gotten this far, you should feel a great sense of pride. Although at the end of this journey the core thoughts can be summarized in a few key concepts, these are very powerful. They fully provide you with a global strategy for analyzing multibody mechanisms that can reach almost arbitrarily far in complexity.

After completing this chapter, you are encouraged to return to the list of sample projects in Appendix C and throughout the text, and to attempt to implement these in MAMBO. But this is only the beginning. Creative ideas beyond those presented here are bound to spark in you after further experimentation. I wish I could be there to see where they take you.

## 10.1 Rationale

When we left off at the end of Chapter 7, we had formulated a comprehensive methodology for quantitatively describing the instantaneous configuration of a multibody mechanism. With configuration coordinates at least as numerous as the number of geometric degrees of freedom, we were able to identify those configurations that satisfied all configuration constraints, i.e., all *allowable configurations*. Sequences of sets of numerical values for the configuration coordinates that satisfied the configuration constraints could be generated to produce visually appealing animations of a moving multibody mechanism.

But all was not well. In treating the model mechanisms in Chapter 7, two problems of distinct natures were identified, at least one of which was more than a matter of convenience.

### 10.1.1 Animations of Allowable Configurations

To find allowable configurations, we were led to employ numerical equation-solving routines. Given initial guesses for the solutions to a system of configuration constraints, these routines would arrive at an approximate set of values for the configuration coordinates through some iterative and, hopefully, convergent, scheme. Indeed, the rate of convergence could be increased by improved choices for the initial guess given to the numerical scheme. In generating a single frame as was done in Chapter 7, we indicated the possibility of using MAMBO to visually organize the mechanism in an (almost) allowable configuration.

To pass through a pre-processing stage within MAMBO, while adequate for a single frame, turns overwhelmingly cumbersome when generating a sequence of frames for an animation of allowable configurations. It would be helpful with a more automated methodology for optimizing the initial guess given to the numerical scheme. One might imagine some approach of extrapolating from previously found allowable configurations and the associated sets of numerical values for the configuration coordinates. Indeed, as a zeroth-order initial guess, we might simply assume that, after changing the values of a number of configuration coordinates equal to the number of geometric degrees of freedom, all other configuration coordinates remain constant.

**Illustration 10.1**

Suppose, for example, that the allowable configurations of a one-geometric-degree-of-freedom mechanism are those that satisfy the configuration constraint

$$q_1 q_2^2 + \sin(q_1 + q_2) = 0$$

expressed here in terms of two configuration coordinates, $q_1$ and $q_2$. Simple inspection then establishes one allowable configuration corresponding to the choice

$$q_1 = q_2 = 0.$$

Given a small change in $q_1$, say to 0.1, we employ a numerical equation-solving scheme, such as the MAPLE `fsolve` routine to arrive at a corresponding value for $q_2$. Using the zeroth-order guess, $q_2 \approx 0$, we find $q_2 \approx -0.1010205146$ as seen below.

```
> fsolve(subs(q1=0.1,q1*q2^2+sin(q1+q2)=0),{q2=0});
```

$$\{q2 = -.1010205146\}$$

A further change in $q_1$ to 0.2 yields

```
> fsolve(subs(q1=0.2,q1*q2^2+sin(q1+q2)=0),
>{q2=-0.1010205146});
```

$$\{q2 = -.2087122728\}$$

where, again, the previous value for $q_2$ is used as an initial guess.

But we can do better than that. Increased precision in the initial guess for $q_2$ in the above illustration is obtained by a first-order approximation to the change in $q_2$ that results from a change in $q_1$.

**Illustration 10.2**

Return to the mechanism in the previous illustration and suppose that one allowable configuration is given by the choice

$$q_1 = q_1^{\mathrm{ref}},\ q_2 = q_2^{\mathrm{ref}}$$

for the configuration coordinates. Then, if $q_1$ is close to $q_1^{\mathrm{ref}}$, an allowable configuration is obtained from $q_2 = q_2^{\mathrm{ref}} + \Delta q_2$, where the change $\Delta q_2$ is similarly small. In fact, substituting $q_1$ and $q_2 = q_2^{\mathrm{ref}} + \Delta q_2$ into the configuration constraint and subsequently Taylor expanding in $\Delta q_2$, we find

$$\begin{aligned} 0 &= q_1 \left(q_2^{\mathrm{ref}} + \Delta q_2\right)^2 + \sin\left(q_1 + q_2^{\mathrm{ref}} + \Delta q_2\right) \\ &= q_1 \left(q_2^{\mathrm{ref}}\right)^2 + \sin\left(q_1 + q_2^{\mathrm{ref}}\right) \\ &\quad + \left(2 q_1 q_2^{\mathrm{ref}} + \cos\left(q_1 + q_2^{\mathrm{ref}}\right)\right) \Delta q_2 + \text{higher-order terms}, \end{aligned}$$

i.e.,

$$q_2 = q_2^{\mathrm{ref}} + \Delta q_2 \approx q_2^{\mathrm{ref}} - \frac{q_1 \left(q_2^{\mathrm{ref}}\right)^2 + \sin\left(q_1 + q_2^{\mathrm{ref}}\right)}{2 q_1 q_2^{\mathrm{ref}} + \cos\left(q_1 + q_2^{\mathrm{ref}}\right)}$$

With $q_1 = 0.1$ and $q_2^{\mathrm{ref}} = 0$, we find

$$\Delta q_2 \approx -0.1003346721 \Rightarrow q_2 \approx -0.1003346721,$$

which is clearly a superior guess for the value of $q_2$ corresponding to $q_1 = 0.1$ than that found in the previous illustration.

Similarly, with $q_1 = 0.2$ and $q_2^{\text{ref}} \approx -0.1010205146$, we find

$$\Delta q_2 \approx -0.1056449777 \Rightarrow q_2 \approx -0.2066654923,$$

again illustrating the improved accuracy in the initial guess to the numerical equation-solving scheme.

In the previous illustrations, the changes in $q_1$ were small, but distinct. A smoother appearance to an animation sequence of allowable configurations is obtained by reducing the size of the change in $q_1$ between subsequent frames. Repeating the calculation above for decreasing steps in $q_1$, a pattern quickly emerges as to the relationship between the change in $q_1$ and the corresponding change in $q_2$.

**Illustration 10.3**

Return to the mechanism in the previous illustrations and suppose, again, that one allowable configuration is given by the choice

$$q_1 = q_1^{\text{ref}},\ q_2 = q_2^{\text{ref}}$$

for the configuration coordinates. This time, substitute

$$q_1 = q_1^{\text{ref}} + \Delta q_1,\ q_2 = q_2^{\text{ref}} + \Delta q_2$$

into the configuration constraint and perform a two-variable Taylor expansion in $\Delta q_1$ and $\Delta q_2$:

$$\begin{aligned} 0 &= \left(q_1^{\text{ref}} + \Delta q_1\right)\left(q_2^{\text{ref}} + \Delta q_2\right)^2 + \sin\left(q_1^{\text{ref}} + \Delta q_1 + q_2^{\text{ref}} + \Delta q_2\right) \\ &= q_1^{\text{ref}}\left(q_2^{\text{ref}}\right)^2 + \sin\left(q_1^{\text{ref}} + q_2^{\text{ref}}\right) + \left(\left(q_2^{\text{ref}}\right)^2 + \cos\left(q_1^{\text{ref}} + q_2^{\text{ref}}\right)\right)\Delta q_1 \\ &\quad + \left(2q_1^{\text{ref}}q_2^{\text{ref}} + \cos\left(q_1^{\text{ref}} + q_2^{\text{ref}}\right)\right)\Delta q_2 + \text{higher-order terms.} \end{aligned}$$

Since

$$q_1^{\text{ref}}\left(q_2^{\text{ref}}\right)^2 + \sin\left(q_1^{\text{ref}} + q_2^{\text{ref}}\right) = 0,$$

it follows that

$$\Delta q_2 \approx -\frac{\left(q_2^{\text{ref}}\right)^2 + \cos\left(q_1^{\text{ref}} + q_2^{\text{ref}}\right)}{2q_1^{\text{ref}}q_2^{\text{ref}} + \cos\left(q_1^{\text{ref}} + q_2^{\text{ref}}\right)}\Delta q_1.$$

With $q_1^{\text{ref}} = 0$ and $q_2^{\text{ref}} = 0$, we conclude that

$$\Delta q_2 \approx -\Delta q_1$$

with increasing accuracy for decreasing $\Delta q_1$. Similarly, with $q_1^{\text{ref}} = 0.1$ and $q_2^{\text{ref}} \approx -0.1010205146$, we conclude that

$$\Delta q_2 \approx -1.031036325 \Delta q_1$$

with increasing accuracy for decreasing $\Delta q_1$.

Now, consider $\Delta q_1$ and $\Delta q_2$ as the average changes in the configuration coordinates over some interval $\Delta t$ of time and divide the relationship formulated in the previous illustration between $\Delta q_1$ and $\Delta q_2$ by $\Delta t$:

$$\frac{\Delta q_2}{\Delta t} \approx -\frac{\left(q_2^{\text{ref}}\right)^2 + \cos\left(q_1^{\text{ref}} + q_2^{\text{ref}}\right)}{2q_1^{\text{ref}} q_2^{\text{ref}} + \cos\left(q_1^{\text{ref}} + q_2^{\text{ref}}\right)} \frac{\Delta q_1}{\Delta t},$$

thus rephrasing it as a relationship between the average rates of change of the configuration coordinates over the interval $\Delta t$.

The process of successively reducing the length of the time interval $\Delta t$ and, consequently, the size of $\Delta q_1$ and $\Delta q_2$ suggests taking limits as $\Delta t \to 0$ on both sides of the equation. In this limit, the influence of higher-order terms that were disregarded in the derivation of the above equation becomes negligible, thus turning $\approx$ into $=$, i.e.,

$$\lim_{\Delta t \to 0} \frac{\Delta q_2}{\Delta t} = -\frac{\left(q_2^{\text{ref}}\right)^2 + \cos\left(q_1^{\text{ref}} + q_2^{\text{ref}}\right)}{2q_1^{\text{ref}} q_2^{\text{ref}} + \cos\left(q_1^{\text{ref}} + q_2^{\text{ref}}\right)} \lim_{\Delta t \to 0} \frac{\Delta q_1}{\Delta t}$$

provided that these limits both exist. When they do exist, they represent the instantaneous rates of change of the configuration coordinates with respect to time, i.e.,

$$\dot{q}_2 = -\frac{\left(q_2^{\text{ref}}\right)^2 + \cos\left(q_1^{\text{ref}} + q_2^{\text{ref}}\right)}{2q_1^{\text{ref}} q_2^{\text{ref}} + \cos\left(q_1^{\text{ref}} + q_2^{\text{ref}}\right)} \dot{q}_1.$$

Thus, by the desire to generate increasingly smooth animation sequences of allowable configurations, we are naturally led to formulate a motion constraint. It is a straightforward exercise to show that this is the motion constraint obtained by differentiation of the original configuration constraint.

### 10.1.2 Animations of Allowable Motions

While sequences of frames of allowable configurations often suffice to generate visually appealing animations, they fail to address motion constraints that have no counterpart in a configuration constraint, i.e., *nonholonomic constraints*. In contrast to the constraints on the allowable configurations that we considered above, such motion constraints constrain the *allowable changes in configuration*, i.e., the *allowable motions* of the mechanism.

**Illustration 10.4**

Suppose, for example, that the allowable configurations of a two-geometric-degrees-of-freedom mechanism are those that satisfy the configuration constraint

$$q_1 q_2^2 + \sin(q_1 + q_2) = 0$$

and that the allowable motions of this one-dynamic-degree-of-freedom mechanism are those that additionally satisfy the motion constraint

$$\dot{q}_1 q_2 + \dot{q}_3 = 0$$

expressed here in terms of three configuration coordinates, $q_1$, $q_2$, and $q_3$.

From the motion constraint, it follows that small, but finite, changes $\Delta q_1$ over some interval $\Delta t$ in time would correspond to small changes in $q_3$ according to

$$\Delta q_3 \approx -q_2^{\text{ref}} \Delta q_1$$

with increasing accuracy for decreasing $\Delta q_1$. Furthermore, from the previous illustration, we recall the correspondence between $\Delta q_2$ and $\Delta q_1$:

$$\Delta q_2 \approx \frac{\left(q_2^{\text{ref}}\right)^2 + \cos\left(q_1^{\text{ref}} + q_2^{\text{ref}}\right)}{2 q_1^{\text{ref}} q_2^{\text{ref}} + \cos\left(q_1^{\text{ref}} + q_2^{\text{ref}}\right)} \Delta q_1$$

with increasing accuracy for decreasing $\Delta q_1$.

A smooth animation consisting of individual frames of allowable configurations and sequences of frames of allowable motions can be generated as in the illustration. Here, we introduce small changes in the values of some set of configuration coordinates, as numerous as the number of dynamic degrees of freedom, and subsequently compute the associated changes in the remaining configuration coordinates using the motion and configuration constraints. As long as the changes in the configuration coordinates remain small, we might even forego a call to a numerical equation-solving routine in the hope that the configuration constraints remain approximately satisfied throughout the motion.

### 10.1.3 The Kinematic Differential Equations

The first-order relations between small changes in the configuration coordinates that were derived in the above discussion correspond to the iterative Euler method for finding approximate solutions to systems of differential equations. In the last illustration, $\Delta q_1$ was assumed given by some external input (in the language of Chapter 9, this could correspond to the product of an independent velocity coordinate $u_1$ with the length of the time interval $\Delta t$), and all derivatives in the corresponding motion constraints were replaced by finite-difference ratios, i.e.

$$\dot{q}_1 \to \frac{\Delta q_1}{\Delta t}, \; \dot{q}_2 \to \frac{\Delta q_2}{\Delta t}, \text{ and } \dot{q}_3 \to \frac{\Delta q_3}{\Delta t}.$$

Although the methodology offered in the previous section is capable of generating animations of approximately allowable configurations and motions, it might be necessary to reduce the growth of errors introduced in considering finite steps in time $\Delta t$ and space $\Delta q_1$, $\Delta q_2$, and so on. A variety of higher-order methods for finding approximate solutions to systems of differential equations can be employed for this purpose. Indeed, the MAMBO application allows the user to select from a number of routines that are part of the MATLAB suite of differential equation integrators.

For most practical applications of the kinematic differential equations in the realm of visualization, it should be possible to obtain highly satisfactory animations. In applications more geared toward simulating physical behavior, it might still be necessary to augment the kinematic differential equations to control the deviation away from allowable configurations at each instant.

**Illustration 10.5**

Consider again the mechanism in previous illustrations, whose allowable configurations are those that satisfy the configuration constraint

$$f(q_1, q_2) = q_1 q_2^2 + \sin(q_1 + q_2) = 0.$$

The corresponding motion constraint

$$\begin{aligned} \frac{df}{dt}(q_1, q_2) &= \frac{\partial f}{\partial q_1}\dot{q}_1 + \frac{\partial f}{\partial q_2}\dot{q}_2 \\ &= \dot{q}_1 q_2^2 + 2 q_1 q_2 \dot{q}_2 + (\dot{q}_1 + \dot{q}_2)\cos(q_1 + q_2) \\ &= 0 \end{aligned}$$

would then be included in the kinematic differential equations.

Now, suppose that the numerical integration routine used to solve the kinematic differential equations results in a local error, such that

$$-\varepsilon < \frac{df}{dt} < \varepsilon$$

for some small number $\varepsilon > 0$. It follows that

$$-\varepsilon t + f_0 < f < \varepsilon t + f_0,$$

i.e., errors grow at worst linearly in time.

If, instead, we considered the motion constraint

$$\frac{df}{dt} + f = \dot{q}_1 q_2^2 + 2q_1 q_2 \dot{q}_2 + (\dot{q}_1 + \dot{q}_2)\cos(q_1 + q_2) + q_1 q_2^2 + \sin(q_1 + q_2) = 0,$$

then the same local error

$$-\varepsilon < \frac{df}{dt} + f < \varepsilon$$

implies that

$$-\varepsilon + (\varepsilon + f_0)\, e^{-t} < f < \varepsilon + (f_0 - \varepsilon)\, e^{-t},$$

i.e., the error remains bounded.

## 10.2 Modeling Algorithm

Throughout the first six chapters, I advocated the following algorithm for arriving at a complete description of the geometry of a multibody mechanism:

**Step 1.** Identify all constituent rigid bodies. In doing this, I recognize that a rigid body may consist of multiple parts, each of which is a separate rigid body. However, the multiple parts of a rigid body are assumed to be stationary relative to each other. They move as a union relative to all other constituent rigid bodies.

**Step 2.** Introduce a reference point and a reference triad for each constituent rigid body. I usually pick some point that has particular significance for the geometry, say a symmetry point of the rigid body. Similarly, I will pick a triad for which at least one basis vector is parallel to some symmetry line of the rigid body.

**Step 3.** Introduce a main observer, relative to which all configurations are ultimately described. As suggested in Chapter 2, the choice of main observer is motivated by the purpose of the modeling, whether primarily graphics- or physics-oriented. I often pick a reference point and a reference triad of the main observer, such that it is related to the geometry of some object that is stationary relative to the main observer.

**Step 4.** Introduce a separate auxiliary observer for each rigid body whose configuration may change relative to the main observer. I pick the reference point and the reference triad of the auxiliary observer, such that the rigid body remains stationary relative to the auxiliary observer. It is not necessary that the reference point of the auxiliary observer coincides with any point on the corresponding rigid body.

**Step 5.** Arrange the observers and rigid bodies in a tree structure with the main observer as the top node, the auxiliary observers as internal nodes, and the rigid bodies as leaf nodes. I often organize the auxiliary observers to reflect the presence of mechanical joints that restrict the relative motions between different auxiliary observers. This is analogous to the discussion of describing the configurations of the digits on the hand relative to you by describing the digits' configurations relative to the palm and the configuration of the palm relative to you.

**Step 6.** Introduce configuration coordinates to quantify the position vectors and rotation matrices that relate the positions and orientations of successive nodes in the tree structure. I recommend simplicity over cleverness. Often the simplest solution is quite sufficient and will enhance the understanding over a particularly clever solution that may be detrimental to the understanding. I expect that you will have experienced both possibilities when looking at the various examples throughout the text.

**Step 7.** Identify any configuration constraints that restrict the allowable values for the configuration coordinates to actually correspond to geometrically correct configurations of the mechanism.

The insights gained in the previous two chapters allow us to add to this algorithm to arrive at a complete description of the allowable motions of the multibody mechanism. In particular,

**Step 8.** Identify any motion constraints that restrict the allowable changes of the configuration coordinates to actually correspond to kinematically correct motions of the mechanism. This will include differentiating the configuration constraints found in Step 7 with respect to time.

**Step 9.** Introduce a sufficient set of independent velocity coordinates that are linear in the rates of change of the configuration coordinates and derive the corresponding set of kinematic differential equations. In doing this, I am guided by two considerations, namely the desire to avoid everywhere singular kinematic differential equations and the desire to use independent velocity coordinates that have a straightforward physical interpretation. I am not concerned with the complexity of the kinematic differential equations, as these will ultimately be solved numerically in any case. In the development phase, it is often helpful to iterate Steps 8 and 9 one motion constraint at a time. Often this allows for a better understanding of the interpretation of the independent velocity coordinates and enables one to avoid choices that lead to everywhere singular kinematic differential equations.

**Step 10.** Identify initial values of the configuration coordinates that satisfy the configuration constraints while being away from singularities of the kinematic differential equations derived in Step 9. As suggested in Chapter 7, initial values of the configuration coordinates that approximately satisfy the configuration constraints can be obtained through trial and error using MAMBO. Improved accuracy can then be achieved through the use of a numerical equation-solving algorithm, e.g., MAPLE's `fsolve` procedure.

Different choices of time-dependence for the final set of independent velocity coordinates will result in different motions of the mechanism with initial configuration determined in Step 10. It will be possible to (numerically) solve the kinematic differential equations as long as a singularity is not encountered. If the singularity is of coordinate origin, a different choice of configuration coordinates and independent velocity coordinates may eliminate the singularity (but possibly create one for a different configuration). In contrast, if the singularity is physical, it cannot be eliminated through a change in the choice of coordinates. Instead, you must carefully select the time-dependence for the independent velocity coordinates to avoid the singularity.

## 10.3 A Bicycle

The algorithm in the previous section establishes the complete description of the allowable configurations and motions of a multibody mechanism. In this section, we return to the bicycle discussed in Chapter 7 and implement the last steps of the modeling algorithm. For reference, we summarize the development of the corresponding geometry hierarchy detailed in Chapter 7.

### 10.3.1 Geometry Hierarchy

Introduce a main observer $\mathcal{W}$ with reference point $W$ and reference triad $w$, such that the stationary plane with which the bicycle makes contact is the plane through $W$ spanned by $\mathbf{w}_1$ and $\mathbf{w}_2$.

Introduce four auxiliary observers $\mathcal{A}_{\text{rear wheel}}$, $\mathcal{A}_{\text{front wheel}}$, $\mathcal{A}_{\text{frame}}$, and $\mathcal{A}_{\text{steering}}$, relative to which the rear wheel, front wheel, frame, and steering column, respectively, are stationary. In particular, let the reference point $A_{\text{rear wheel}}$ of the rear wheel observer $\mathcal{A}_{\text{rear wheel}}$ be located at the center of the rear wheel. Choose the reference triad $a^{(\text{rear wheel})}$ of $\mathcal{A}_{\text{rear wheel}}$, such that the wheel axis is parallel to the vector $\mathbf{a}_3^{(\text{rear wheel})}$. Similarly, let the reference point $A_{\text{front wheel}}$ of the front wheel observer $\mathcal{A}_{\text{front wheel}}$ be located at the center of the front wheel. Choose the reference triad $a^{(\text{front wheel})}$ of $\mathcal{A}_{\text{front wheel}}$, such that the wheel axis is parallel to the vector $\mathbf{a}_3^{(\text{front wheel})}$.

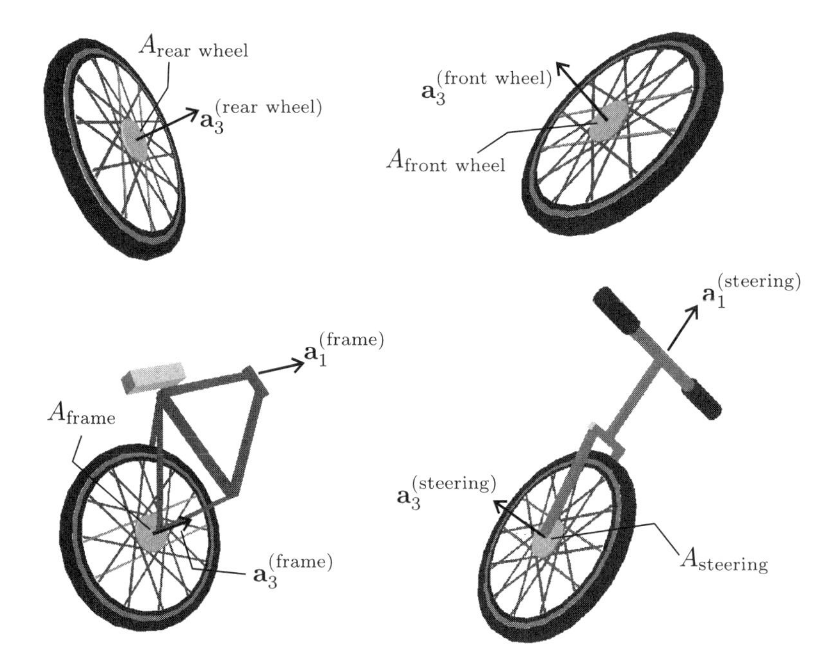


Let the reference point $A_{\text{frame}}$ of the frame observer $\mathcal{A}_{\text{frame}}$ coincide with $A_{\text{rear wheel}}$. Choose the reference triad $a^{(\text{frame})}$ of $\mathcal{A}_{\text{frame}}$, such that $\mathbf{a}_3^{(\text{frame})}$ equals $\mathbf{a}_3^{(\text{rear wheel})}$ and $\mathbf{a}_1^{(\text{frame})}$ is parallel to the forward direction of the bicycle saddle. Finally, let the reference point $A_{\text{steering}}$ of the steering column observer $\mathcal{A}_{\text{steering}}$ coincide with $A_{\text{front wheel}}$. Choose the reference triad $a^{(\text{steering})}$ of $\mathcal{A}_{\text{steering}}$, such that $\mathbf{a}_3^{(\text{steering})}$ equals $\mathbf{a}_3^{(\text{front wheel})}$ and $\mathbf{a}_1^{(\text{steering})}$ is parallel to the axis of rotation of the steering column.

The configuration of the observer $\mathcal{A}_{\text{frame}}$ relative to the main observer $\mathcal{W}$ is described by a pure translation $\mathbf{T}_{\mathcal{W}\to\mathcal{A}_{\text{frame}}}$ corresponding to the position vector

$$\mathbf{r}^{WA_{\text{frame}}} = w \begin{pmatrix} q_1 \\ q_2 \\ q_3 \end{pmatrix}$$

and a pure rotation $\mathbf{R}_{\mathcal{W}\to\mathcal{A}_{\text{frame}}}$ corresponding to the rotation matrix

$$R_{wa^{(\text{frame})}} = R(q_4, 0, 0, 1)\, R(q_5, 1, 0, 0)\, R(q_6, 0, 0, 1)\,.$$

The configuration of the observer $\mathcal{A}_{\text{rear wheel}}$ relative to the observer $\mathcal{A}_{\text{frame}}$ is described by a pure translation $\mathbf{T}_{\mathcal{A}_{\text{frame}}\to\mathcal{A}_{\text{rear wheel}}}$ corresponding to the position vector

$$\mathbf{r}^{A_{\text{frame}}A_{\text{rear wheel}}} = \mathbf{0}$$

and a pure rotation $\mathbf{R}_{\mathcal{A}_{\text{frame}}\to\mathcal{A}_{\text{rear wheel}}}$ corresponding to the rotation matrix

$$R_{a^{(\text{frame})}a^{(\text{rear wheel})}} = R(q_7, 0, 0, 1)\,.$$

The configuration of the observer $\mathcal{A}_{\text{steering}}$ relative to the observer $\mathcal{A}_{\text{frame}}$ is described by a pure translation $\mathbf{T}_{\mathcal{A}_{\text{frame}}\to\mathcal{A}_{\text{steering}}}$ corresponding to the position vector

$$\mathbf{r}^{A_{\text{frame}}A_{\text{steering}}} = a^{(\text{frame})} \begin{pmatrix} p_3 + \frac{p_1 p_5}{\sqrt{p_1^2+p_2^2}} - \frac{p_2 p_6}{\sqrt{p_1^2+p_2^2}} \cos q_8 \\ p_4 + \frac{p_2 p_5}{\sqrt{p_1^2+p_2^2}} + \frac{p_1 p_6}{\sqrt{p_1^2+p_2^2}} \cos q_8 \\ p_6 \sin q_8 \end{pmatrix}$$

and a pure rotation $\mathbf{R}_{\mathcal{A}_{\text{frame}}\to\mathcal{A}_{\text{steering}}}$ corresponding to the rotation matrix

$$R_{a^{(\text{frame})}a^{(\text{steering})}} = R(\theta, 0, 0, 1)\, R(q_8, 1, 0, 0)\,,$$

where

$$\cos\theta = \frac{p_1}{\sqrt{p_1^2 + p_2^2}} \text{ and } \sin\theta = \frac{p_2}{\sqrt{p_1^2 + p_2^2}}.$$

Finally, the configuration of the observer $\mathcal{A}_{\text{front wheel}}$ relative to the observer $\mathcal{A}_{\text{steering}}$ is described by a pure translation $\mathbf{T}_{\mathcal{A}_{\text{steering}}\to\mathcal{A}_{\text{front wheel}}}$ corresponding to the position vector

$$\mathbf{r}^{A_{\text{steering}}A_{\text{front wheel}}} = \mathbf{0}$$

and a pure rotation $\mathbf{R}_{\mathcal{A}_{\text{steering}}\to\mathcal{A}_{\text{front wheel}}}$ corresponding to the rotation matrix

$$R_{a^{(\text{steering})}a^{(\text{front wheel})}} = R(q_9, 0, 0, 1)\,.$$

### 10.3.2 Configuration Constraints

The only configuration constraints restricting allowable configurations of the bicycle are those requiring that the rear and front wheels make tangential contact with the stationary plane. To simplify the formulation of these constraints, we introduced the points $P_{\text{rear contact}}$ and $P_{\text{front contact}}$ to represent the points on the rear wheel and front wheel, respectively, that make contact with the plane and at which points the tangent direction to the corresponding wheel is parallel to the plane. In particular, we let

$$\mathcal{A}_{\text{rear wheel}} P_{\text{rear contact}} = \begin{pmatrix} R\cos q_{10} \\ R\sin q_{10} \\ 0 \end{pmatrix}$$

and

$$\mathcal{A}_{\text{front wheel}} P_{\text{front contact}} = \begin{pmatrix} R\cos q_{11} \\ R\sin q_{11} \\ 0 \end{pmatrix},$$

where $R$ is the wheel radius. Then, tangential contact with the plane is ensured if

$$\mathbf{w}_3 \bullet \mathbf{r}^{WP_{\text{rear contact}}} = 0,$$
$$\mathbf{w}_3 \bullet \mathbf{r}^{WP_{\text{front contact}}} = 0,$$

$$\mathbf{r}^{A_{\text{rear wheel}}P_{\text{rear contact}}} \bullet \left(\mathbf{w}_3 \times \mathbf{a}_3^{(\text{rear wheel})}\right)$$
$$= \mathbf{w}_3 \bullet \left(\mathbf{a}_3^{(\text{rear wheel})} \times \mathbf{r}^{A_{\text{rear wheel}}P_{\text{rear contact}}}\right) = 0,$$

and

$$\mathbf{r}^{A_{\text{front wheel}}P_{\text{front contact}}} \bullet \left(\mathbf{w}_3 \times \mathbf{a}_3^{(\text{front wheel})}\right)$$
$$= \mathbf{w}_3 \bullet \left(\mathbf{a}_3^{(\text{front wheel})} \times \mathbf{r}^{A_{\text{front wheel}}P_{\text{front contact}}}\right) = 0.$$

### 10.3.3 Motion Constraints

We begin by deriving the motion constraints corresponding to the configuration constraints formulated above. Specifically,

$$\mathbf{w}_3 \bullet \frac{{}^{w}d\mathbf{r}^{WP_{\text{rear contact}}}}{dt} = 0,$$
$$\mathbf{w}_3 \bullet \frac{{}^{w}d\mathbf{r}^{WP_{\text{front contact}}}}{dt} = 0,$$
$$\mathbf{w}_3 \bullet \frac{{}^{w}d}{dt}\left(\mathbf{a}_3^{(\text{rear wheel})} \times \mathbf{r}^{A_{\text{rear wheel}}P_{\text{rear contact}}}\right) = 0,$$

and

$$\mathbf{w}_3 \bullet \frac{{}^{w}d}{dt}\left(\mathbf{a}_3^{(\text{front wheel})} \times \mathbf{r}^{A_{\text{front wheel}} P_{\text{front contact}}}\right) = 0.$$

Now, impose the condition that the bicycle's wheels are rolling without slipping on the stationary plane. In Chapters 8 and 9, this was found to imply that the velocity of the point on the rim of each of the wheels that is currently in contact with the plane has zero velocity relative to the plane, i.e., relative to $\mathcal{W}$. Following Chapter 9, we thus find that the rolling-without-slipping constraints can be written as

$$\frac{{}^{\mathcal{W}}dA_{\text{rear wheel}}}{dt} + {}^{w}\boldsymbol{\omega}^{a^{(\text{rear wheel})}} \times \mathbf{r}^{A_{\text{rear wheel}} P_{\text{rear contact}}} = \mathbf{0}$$

and

$$\frac{{}^{\mathcal{W}}dA_{\text{front wheel}}}{dt} + {}^{w}\boldsymbol{\omega}^{a^{(\text{front wheel})}} \times \mathbf{r}^{A_{\text{front wheel}} P_{\text{front contact}}} = \mathbf{0}.$$

Since each of these equations corresponds to three motion constraints, the rolling-without-slipping constraint coupled with the previously formulated configuration constraints add up to a total of 10 motion constraints. If all of these constraints were independent, it would follow that the number of dynamic degrees of freedom of the bicycle would be one. Equivalently, it would follow that, at most, one independent velocity coordinate would be necessary to describe all allowable motions of the bicycle. But this is counterintuitive. After all, a given angular speed of the rear wheel does not uniquely define the angular speed of the front wheel, since the latter also depends on the rate of change of the steering angle. As was the case with the motion constraints in Section 9.1.4, the 10 motion constraints derived here cannot be independent.

In fact, in Section 9.4.3, we similarly concluded that the velocity component in the $\mathbf{w}_3$ direction already had to be zero as a consequence of the previously imposed configuration constraints. Here,

$$\begin{aligned}
&\frac{{}^{\mathcal{W}}dA_{\text{rear wheel}}}{dt} + {}^{w}\boldsymbol{\omega}^{a^{(\text{rear wheel})}} \times \mathbf{r}^{A_{\text{rear wheel}} P_{\text{rear contact}}} \\
&= \frac{{}^{w}d\mathbf{r}^{WA_{\text{rear wheel}}}}{dt} + {}^{w}\boldsymbol{\omega}^{a^{(\text{rear wheel})}} \times \mathbf{r}^{A_{\text{rear wheel}} P_{\text{rear contact}}} \\
&= \frac{{}^{w}d\left(\mathbf{r}^{WP_{\text{rear contact}}} - \mathbf{r}^{A_{\text{rear wheel}} P_{\text{rear contact}}}\right)}{dt} \\
&\quad + {}^{w}\boldsymbol{\omega}^{a^{(\text{rear wheel})}} \times \mathbf{r}^{A_{\text{rear wheel}} P_{\text{rear contact}}}
\end{aligned}$$

$$= \frac{{}^{w}d\mathbf{r}^{WP_{\text{rear contact}}}}{dt} - \frac{{}^{w}d\mathbf{r}^{A_{\text{rear wheel}}P_{\text{rear contact}}}}{dt} + {}^{w}\boldsymbol{\omega}^{a^{(\text{rear wheel})}} \times \mathbf{r}^{A_{\text{rear wheel}}P_{\text{rear contact}}}$$

$$= \frac{{}^{w}d\mathbf{r}^{WP_{\text{rear contact}}}}{dt} - \frac{{}^{a^{(\text{rear wheel})}}d\mathbf{r}^{A_{\text{rear wheel}}P_{\text{rear contact}}}}{dt}$$

$$= \frac{{}^{w}d\mathbf{r}^{WP_{\text{rear contact}}}}{dt} - \dot{q}_{10}a^{(\text{rear wheel})}\begin{pmatrix} -R\sin q_{10} \\ R\cos q_{10} \\ 0 \end{pmatrix}$$

$$= \frac{{}^{w}d\mathbf{r}^{WP_{\text{rear contact}}}}{dt} - \dot{q}_{10}\left(\mathbf{a}_3^{(\text{rear wheel})} \times \mathbf{r}^{A_{\text{rear wheel}}P_{\text{rear contact}}}\right).$$

It follows that

$$\mathbf{w}_3 \bullet \left(\frac{{}^{\mathcal{W}}dA_{\text{rear wheel}}}{dt} + {}^{w}\boldsymbol{\omega}^{a^{(\text{rear wheel})}} \times \mathbf{r}^{A_{\text{rear wheel}}P_{\text{rear contact}}}\right)$$
$$= \mathbf{w}_3 \bullet \frac{{}^{w}d\mathbf{r}^{WP_{\text{rear contact}}}}{dt} - \dot{q}_{10}\mathbf{w}_3 \bullet \left(\mathbf{a}_3^{(\text{rear wheel})} \times \mathbf{r}^{A_{\text{rear wheel}}P_{\text{rear contact}}}\right),$$

which automatically equals zero if the configuration constraints are satisfied throughout the motion. Following an identical argument for the front wheel, we conclude that the 10 motion constraints are not independent. Indeed, we are free to eliminate the first two motion constraints listed at the top of this section, leaving a total of eight motion constraints. Although it is not immediately evident that these are independent, this would follow if it were possible to choose a set of independent velocity coordinates that results in a not-everywhere-singular set of kinematic differential equations.

### 10.3.4 Independent Velocity Coordinates and Kinematic Differential Equations

We turn to the selection of independent velocity coordinates to uniquely specify all allowable motions of the bicycle. Under the assumption that the eight motion constraints derived in the previous section are actually independent, it follows that the bicycle has three dynamic degrees of freedom, i.e., that three independent velocity coordinates suffice to specify all allowable motions of the bicycle.

While there are infinitely many choices for the independent velocity coordinates, we are guided by our experience with bicycles to consider the three different ways in which the rider controls the bicycle motion, namely:

- Controlling the rate of change of the orientation of the real wheel relative to the frame, $\dot{q}_7$;

- Controlling the rate of change of the tilt of the frame relative to the stationary plane, $\dot{q}_5$;
- Controlling the rate of change of the orientation of the steering column relative to the frame, $\dot{q}_8$.

Consequently, we propose the following choice of independent velocity coordinates:

$$
\begin{aligned}
u_1 &= \dot{q}_5, \\
u_2 &= \dot{q}_7, \\
u_3 &= \dot{q}_8.
\end{aligned}
$$

We thus arrive at a complete set of kinematic differential equations:

$$
\begin{aligned}
\dot{q}_5 &= u_1, \\
\dot{q}_7 &= u_2, \\
\dot{q}_8 &= u_3, \\
\mathbf{w}_3 \bullet \frac{{}^{\mathcal{W}}d}{dt}\left(\mathbf{a}_3^{(\text{rear wheel})} \times \mathbf{r}^{A_{\text{rear wheel}} P_{\text{rear contact}}}\right) &= 0, \\
\mathbf{w}_3 \bullet \frac{{}^{\mathcal{W}}d}{dt}\left(\mathbf{a}_3^{(\text{front wheel})} \times \mathbf{r}^{A_{\text{front wheel}} P_{\text{front contact}}}\right) &= 0, \\
\mathbf{w}_1 \bullet \left(\frac{{}^{\mathcal{W}}dA_{\text{rear wheel}}}{dt} + {}^{\mathcal{W}}\boldsymbol{\omega}^{a^{(\text{rear wheel})}} \times \mathbf{r}^{A_{\text{rear wheel}} P_{\text{rear contact}}}\right) &= 0, \\
\mathbf{w}_2 \bullet \left(\frac{{}^{\mathcal{W}}dA_{\text{rear wheel}}}{dt} + {}^{\mathcal{W}}\boldsymbol{\omega}^{a^{(\text{rear wheel})}} \times \mathbf{r}^{A_{\text{rear wheel}} P_{\text{rear contact}}}\right) &= 0, \\
\mathbf{w}_3 \bullet \left(\frac{{}^{\mathcal{W}}dA_{\text{rear wheel}}}{dt} + {}^{\mathcal{W}}\boldsymbol{\omega}^{a^{(\text{rear wheel})}} \times \mathbf{r}^{A_{\text{rear wheel}} P_{\text{rear contact}}}\right) &= 0, \\
\mathbf{w}_1 \bullet \left(\frac{{}^{\mathcal{W}}dA_{\text{front wheel}}}{dt} + {}^{\mathcal{W}}\boldsymbol{\omega}^{a^{(\text{front wheel})}} \times \mathbf{r}^{A_{\text{front wheel}} P_{\text{front contact}}}\right) &= 0, \\
\mathbf{w}_2 \bullet \left(\frac{{}^{\mathcal{W}}dA_{\text{front wheel}}}{dt} + {}^{\mathcal{W}}\boldsymbol{\omega}^{a^{(\text{front wheel})}} \times \mathbf{r}^{A_{\text{front wheel}} P_{\text{front contact}}}\right) &= 0, \\
\mathbf{w}_3 \bullet \left(\frac{{}^{\mathcal{W}}dA_{\text{front wheel}}}{dt} + {}^{\mathcal{W}}\boldsymbol{\omega}^{a^{(\text{front wheel})}} \times \mathbf{r}^{A_{\text{front wheel}} P_{\text{front contact}}}\right) &= 0.
\end{aligned}
$$

where the last eight equations correspond to the motion constraints retained at the end of the previous section. As quickly becomes evident, the above set of kinematic differential equations is algebraically very complicated. Any effort to locate singularities by computing the determinant of the coefficient matrix as in Chapter 9 is thus doomed to fail. Instead, we will rely on the ability of MAMBO and the MAMBO toolbox to detect configurations that correspond to singularities.

### 10.3.5 An Initial Configuration

Initial values for the configuration coordinates that would correspond to an allowable configuration were already found at the end of Section 7.3. There it was suggested that a collection of such sets of values for the configuration coordinates could be generated to give the appearance of a smooth allowable motion. As suggested there, in addition to being very cumbersome, this methodology would fail to address the rolling-without-slipping constraints on the bicycle wheels.

In contrast, the formulation derived above in terms of the kinematic differential equations guarantees that the configuration constraints **and** the motion constraints are satisfied for all time as long as the initial configuration is an allowable configuration and as long as singularities are not encountered. Now, every choice of time-dependence of the three independent velocity coordinates, $u_1(t)$, $u_2(t)$, and $u_3(t)$ will generate an allowable motion.

### 10.3.6 MAMBO

We recall the formulation of the geometric hierarchy of the bicycle in the MAMBO toolbox from Chapter 7:

```
> Restart():
> DeclareObservers(W,Arear,Afront,Aframe,Asteer):
> DeclarePoints(W,Arear,Afront,Aframe,Asteer):
> DeclareTriads(w,arear,afront,aframe,asteer):
> DefineObservers([W,W,w],[Arear,Arear,arear],
> [Afront,Afront,afront],[Aframe,Aframe,aframe],
> [Asteer,Asteer,asteer]):
> DefinePoints([W,Aframe,w,q1,q2,q3]):
> DefineTriads([w,aframe,[q4,3],[q5,1],[q6,3]]):
> DefinePoints([Aframe,Arear,NullVector()]):
> DefineTriads([aframe,arear,[q7,3]]):
> v:=MakeTranslations(aframe,p1,p2,0):
> aframe1:=MakeTranslations(aframe,1):
> aframe2:=MakeTranslations(aframe,2):
> aframe3:=MakeTranslations(aframe,3):
> b1:=(1/VectorLength(v)) &** v:
> b2:=(1/VectorLength(aframe3 &xx b1)) &**
> (aframe3 &xx b1):
> b3:=b1 &xx b2:
> DefineTriads([aframe,asteer,
> [matrix(3,3,(i,j)->cat(aframe,i) &oo cat(b,j))],
> [q8,1]]):
> DefinePoints([Aframe,Asteer,[aframe,p3,p4,0],
> [asteer,p5,p6,0]]):
> DefinePoints([Asteer,Afront,NullVector()]):
> DefineTriads([asteer,afront,q9,3]):
> DefineNeighbors([W,Aframe],[Aframe,Arear],
> [Aframe,Asteer],[Asteer,Afront]):
```

and the implementation of the configuration constraints using the following MAMBO toolbox statements:

```
> DeclarePoints(Prear,Pfront):
> DefinePoints([Arear,Prear,arear,R*cos(q10),
> R*sin(q10),0],[Afront,Pfront,afront,R*cos(q11),
> R*sin(q11),0]):
> f1:=simplify(FindTranslation(Arear,Prear) &oo
> (MakeTranslations(w,3) &xx
> MakeTranslations(arear,3)))=0:
> f2:=FindTranslation(W,Prear) &oo
> MakeTranslations(w,3)=0:
> f3:=simplify(FindTranslation(Afront,Pfront) &oo
> (MakeTranslations(w,3) &xx
> MakeTranslations(afront,3)))=0:
> f4:=FindTranslation(W,Pfront) &oo
> MakeTranslations(w,3)=0:
```

The kinematic differential equations derived above are obtained from the MAMBO toolbox statements

```
> DeclareStates(seq(cat(q.i),i=1..11)):
> kde:= {q5t=u1,q7t=u2,q8t=u3} union DiffTime({f1,f3})
> union {seq((LinearVelocity(W,Arear) &++
> (AngularVelocity(w,arear) &xx
> FindTranslation(Arear,Prear)))
> &oo MakeTranslations(w,i)=0,i=1..3),
> seq((LinearVelocity(W,Afront) &++
> (AngularVelocity(w,afront)
> &xx FindTranslation(Afront,Pfront)))
> &oo MakeTranslations(w,i)=0,i=1..3)}:
```

Finally, we use `MotionOutput` to check whether the initial configuration found in Chapter 7 corresponds to a singularity of the kinematic differential equations and, if not, to export the motion description to the file `bike.dyn`.

```
> MotionOutput(ode=kde,states=[q1=0,q2=0,q3=.4207354924,
> q4=0,q5=1,q6=.1280053711e-1,q7=0,q8=.8,q9=0,
> q10=-1.583596864,q11=3.288069682],parameters=
> [p1=-.348968837,p2 =.937134317,p3=1.296859217,
> p4=.6774200135,p5=-.7601008180,p6=-.10,R=.5],
> insignals=[u1=0,u2=0,u3-0],checksingo,
> filename="bike.dyn");
```

## 10.4 A Desk Lamp

As a final example, we return to the desk lamp discussed in Chapter 7 and implement the last steps of the modeling algorithm as suggested. For reference, we summarize the development of the corresponding geometry hierarchy detailed in Chapter 7.

### 10.4.1 Geometry Hierarchy

Introduce a main observer $\mathcal{W}$ with reference point $W$ and reference triad $w$. Introduce six auxiliary observers $\mathcal{A}_{\text{base}}$, $\mathcal{A}_{\text{lower beam}}$, $\mathcal{A}_{\text{middle beam}}$, $\mathcal{A}_{\text{upper beam}}$, $\mathcal{A}_{\text{bracket}}$, and $\mathcal{A}_{\text{lamp shade}}$, relative to which the base, the lower, middle, and upper beams, the bracket, and the lamp shade, respectively, are stationary.

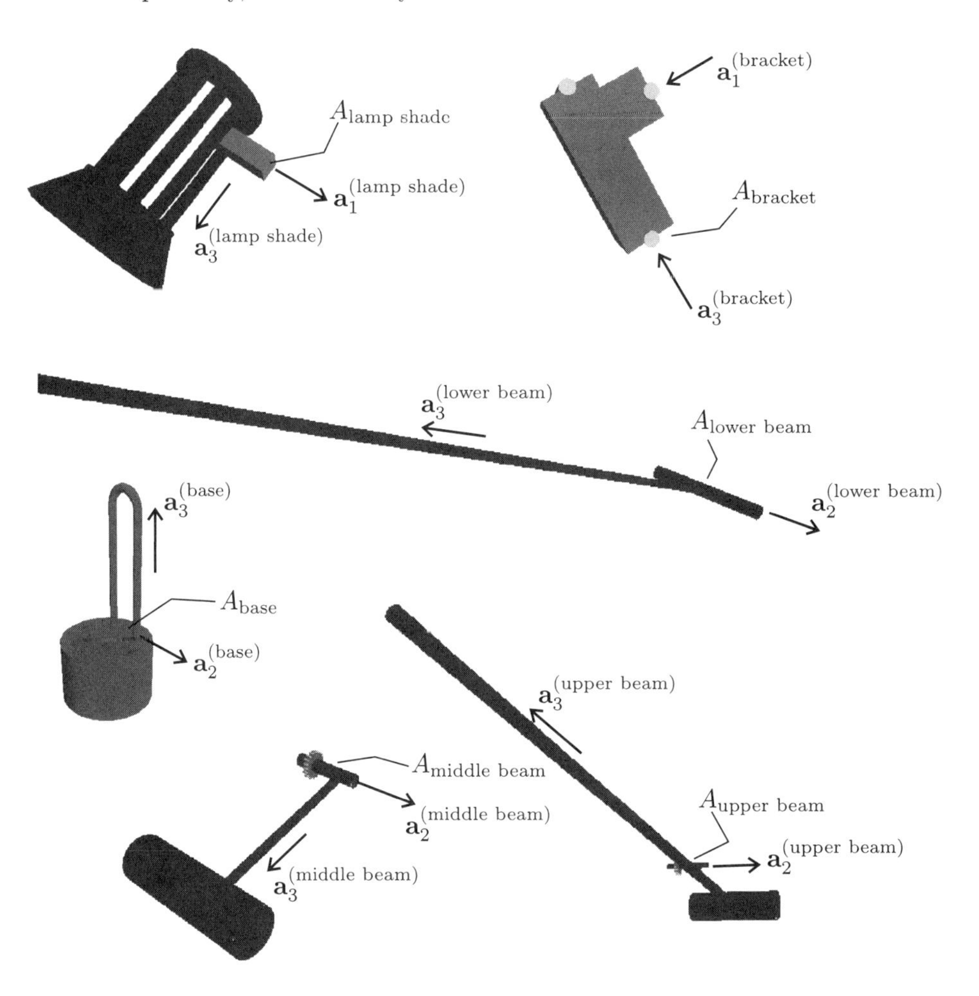


In particular, let the reference point $A_{\text{base}}$ of the base observer $\mathcal{A}_{\text{base}}$ be located at the top of the base centered between the vertical posts. Choose the reference triad $a^{(\text{base})}$ of $\mathcal{A}_{\text{base}}$, such that the vertical posts are parallel to the vector $\mathbf{a}_3^{(\text{base})}$ and are separated in a direction parallel to the vector $\mathbf{a}_2^{(\text{base})}$. Let the reference point $A_{\text{lower beam}}$ of the lower beam observer $\mathcal{A}_{\text{lower beam}}$ be located at the center of the horizontal bar connecting the beam to the base. Choose the reference triad $a^{(\text{lower beam})}$

of $\mathcal{A}_{\text{lower beam}}$, such that the beam is parallel to the vector $\mathbf{a}_3^{(\text{lower beam})}$ and the horizontal bar is parallel to $\mathbf{a}_2^{(\text{lower beam})}$. Similarly, let the reference point $A_{\text{middle beam}}$ of the middle beam observer $\mathcal{A}_{\text{middle beam}}$ be located at the center of the horizontal bar connecting the beam to the base. Choose the reference triad $a^{(\text{middle beam})}$ of $\mathcal{A}_{\text{middle beam}}$, such that the beam is parallel to the vector $\mathbf{a}_3^{(\text{middle beam})}$ and the horizontal bar is parallel to $\mathbf{a}_2^{(\text{middle beam})}$. Moreover, let the reference point $A_{\text{upper beam}}$ of the upper beam observer $\mathcal{A}_{\text{upper beam}}$ be located at the center of the horizontal bar connecting the beam to the base. Choose the reference triad $a^{(\text{upper beam})}$ of $\mathcal{A}_{\text{upper beam}}$, such that the beam is parallel to the vector $\mathbf{a}_3^{(\text{upper beam})}$ and the horizontal bar is parallel to $\mathbf{a}_2^{(\text{upper beam})}$.

Let the reference point $A_{\text{bracket}}$ of the bracket observer $\mathcal{A}_{\text{bracket}}$ coincide with the hinge joint connecting the bracket to the lower beam. Choose the reference triad $a^{(\text{bracket})}$ of $\mathcal{A}_{\text{bracket}}$, such that the line between the hinge joints connecting the bracket to the lower beam and the lamp shade, respectively, is parallel to $\mathbf{a}_3^{(\text{bracket})}$ and the hinge axes are parallel to $\mathbf{a}_2^{(\text{bracket})}$. Finally, let the reference point $A_{\text{lamp shade}}$ of the lamp shade observer $\mathcal{A}_{\text{lamp shade}}$ coincide with the hinge joint connecting the lamp shade to the bracket. Choose the reference triad $a^{(\text{lamp shade})}$ of $\mathcal{A}_{\text{lamp shade}}$, such that the symmetry axis of the lamp shade is parallel to $\mathbf{a}_3^{(\text{lamp shade})}$ and the hinge axis is parallel to $\mathbf{a}_2^{(\text{lamp shade})}$.

The configuration of the observer $\mathcal{A}_{\text{base}}$ relative to the main observer $\mathcal{W}$ is described by a pure translation $\mathbf{T}_{\mathcal{W}\to\mathcal{A}_{\text{base}}}$ corresponding to the position vector

$$\mathbf{r}^{WA_{\text{base}}} = w \begin{pmatrix} q_1 \\ q_2 \\ q_3 \end{pmatrix}$$

and a pure rotation $\mathbf{R}_{\mathcal{W}\to\mathcal{A}_{\text{base}}}$ corresponding to the rotation matrix

$$R_{wa^{(\text{base})}} = R(q_4, 0, 0, 1)\, R(q_5, 1, 0, 0)\, R(q_6, 0, 0, 1).$$

The configuration of the observer $\mathcal{A}_{\text{lower beam}}$ relative to the observer $\mathcal{A}_{\text{base}}$ is described by a pure translation $\mathbf{T}_{\mathcal{A}_{\text{base}}\to\mathcal{A}_{\text{lower beam}}}$ corresponding to the position vector

$$\mathbf{r}^{A_{\text{base}}A_{\text{lower beam}}} = a^{(\text{base})} \begin{pmatrix} 0 \\ 0 \\ p_1 \end{pmatrix}$$

and a pure rotation $\mathbf{R}_{\mathcal{A}_{\text{base}}\to\mathcal{A}_{\text{lower beam}}}$ corresponding to the rotation matrix

$$R_{a^{(\text{base})}a^{(\text{lower beam})}} = R(q_7, 0, 1, 0).$$

The configuration of the observer $\mathcal{A}_{\text{middle beam}}$ relative to the observer $\mathcal{A}_{\text{base}}$ is described by a pure translation $\mathbf{T}_{\mathcal{A}_{\text{base}} \to \mathcal{A}_{\text{middle beam}}}$ corresponding to the position vector

$$\mathbf{r}^{A_{\text{base}} A_{\text{middle beam}}} = a^{(\text{base})} \begin{pmatrix} 0 \\ 0 \\ p_2 \end{pmatrix}$$

and a pure rotation $\mathbf{R}_{\mathcal{A}_{\text{base}} \to \mathcal{A}_{\text{middle beam}}}$ corresponding to the rotation matrix

$$R_{a^{(\text{base})} a^{(\text{middle beam})}} = R\left(q_8, 0, 1, 0\right).$$

The configuration of the observer $\mathcal{A}_{\text{upper beam}}$ relative to the observer $\mathcal{A}_{\text{base}}$ is described by a pure translation $\mathbf{T}_{\mathcal{A}_{\text{base}} \to \mathcal{A}_{\text{upper beam}}}$ corresponding to the position vector

$$\mathbf{r}^{A_{\text{base}} A_{\text{upper beam}}} = a^{(\text{base})} \begin{pmatrix} 0 \\ 0 \\ p_3 \end{pmatrix}$$

and a pure rotation $\mathbf{R}_{\mathcal{A}_{\text{base}} \to \mathcal{A}_{\text{upper beam}}}$ corresponding to the rotation matrix

$$R_{a^{(\text{base})} a^{(\text{upper beam})}} = R\left(q_9, 0, 1, 0\right).$$

The configuration of the observer $\mathcal{A}_{\text{bracket}}$ relative to the observer $\mathcal{A}_{\text{lower beam}}$ is described by a pure translation $\mathbf{T}_{\mathcal{A}_{\text{lower beam}} \to \mathcal{A}_{\text{bracket}}}$ corresponding to the position vector

$$\mathbf{r}^{A_{\text{lower beam}} A_{\text{bracket}}} = a^{(\text{lower beam})} \begin{pmatrix} 0 \\ 0 \\ p_4 \end{pmatrix}$$

and a pure rotation $\mathbf{R}_{\mathcal{A}_{\text{lower beam}} \to \mathcal{A}_{\text{bracket}}}$ corresponding to the rotation matrix

$$R_{a^{(\text{lower beam})} a^{(\text{bracket})}} = R\left(q_{10}, 0, 1, 0\right).$$

Finally, the configuration of the observer $\mathcal{A}_{\text{lamp shade}}$ relative to the observer $\mathcal{A}_{\text{bracket}}$ is described by a pure translation $\mathbf{T}_{\mathcal{A}_{\text{bracket}} \to \mathcal{A}_{\text{lamp shade}}}$ corresponding to the position vector

$$\mathbf{r}^{A_{\text{bracket}} A_{\text{lamp shade}}} = a^{(\text{bracket})} \begin{pmatrix} 0 \\ 0 \\ p_5 \end{pmatrix}$$

and a pure rotation $\mathbf{R}_{\mathcal{A}_{\text{bracket}} \to \mathcal{A}_{\text{lamp shade}}}$ corresponding to the rotation matrix

$$R_{a^{(\text{bracket})} a^{(\text{lamp shade})}} = R\left(q_{11}, 0, 1, 0\right).$$

### 10.4.2 Configuration Constraints

The only configuration constraints restricting allowable configurations of the desk lamp are those requiring that the upper beam and the bracket connect at the appropriate hinge joint. To simplify the formulation of these constraints, we introduced the points $H_{\text{bracket}}$ and $H_{\text{upper beam}}$ to represent the points on the bracket and upper beam, respectively, that coincide with the corresponding hinge joint. In particular, let

$$^{\mathcal{A}_{\text{upper beam}}}H_{\text{upper beam}} = \begin{pmatrix} 0 \\ 0 \\ p_6 \end{pmatrix}$$

and

$$^{\mathcal{A}_{\text{bracket}}}H_{\text{bracket}} = \begin{pmatrix} -p_7 \\ 0 \\ p_8 \end{pmatrix}.$$

Then, the points $H_{\text{bracket}}$ and $H_{\text{upper beam}}$ will coincide with the corresponding hinge joint, provided that

$$\begin{aligned} \mathbf{a}_1^{(\text{base})} \bullet \mathbf{r}^{H_{\text{bracket}}H_{\text{upper beam}}} &= 0, \\ \mathbf{a}_3^{(\text{base})} \bullet \mathbf{r}^{H_{\text{bracket}}H_{\text{upper beam}}} &= 0. \end{aligned}$$

### 10.4.3 Motion Constraints

We begin by deriving the motion constraints corresponding to the configuration constraints formulated above. Specifically,

$$\begin{aligned} \mathbf{a}_1^{(\text{base})} \bullet \,^{a^{(\text{base})}}\frac{d\mathbf{r}^{H_{\text{bracket}}H_{\text{upper beam}}}}{dt} &= 0, \\ \mathbf{a}_3^{(\text{base})} \bullet \,^{a^{(\text{base})}}\frac{d\mathbf{r}^{H_{\text{bracket}}H_{\text{upper beam}}}}{dt} &= 0. \end{aligned}$$

Now, impose the condition that changes in the orientations of the upper and middle beams are constrained by the presence of the spur gears. Here, we will model this constraint as a rolling-without-slipping constraint on the two gears, i.e., that the contact points on the two gears have identical velocities.

Let $P_1$ and $P_2$ denote the contact points on the upper and lower gears, respectively. Similarly, let $C_1$ and $C_2$ denote the points at the center of the upper and lower gears, respectively. Then, it follows that

$$^{a^{(\text{base})}}\frac{dP_1}{dt} = \,^{a^{(\text{base})}}\frac{dC_1}{dt} + \,^{a^{(\text{base})}}\boldsymbol{\omega}^{a^{(\text{upper beam})}} \times \mathbf{r}^{C_1P_1}$$

and

$$^{a^{(\text{base})}}\frac{dP_2}{dt} = {}^{a^{(\text{base})}}\frac{dC_2}{dt} + {}^{a^{(\text{base})}}\boldsymbol{\omega}^{a^{(\text{middle beam})}} \times \mathbf{r}^{C_2 P_2}.$$

But, the center points are stationary relative to the base, i.e.,

$$^{a^{(\text{base})}}\frac{dC_1}{dt} = {}^{a^{(\text{base})}}\frac{dC_2}{dt} = \mathbf{0}.$$

Also,

$$\begin{aligned}
{}^{a^{(\text{base})}}\boldsymbol{\omega}^{a^{(\text{upper beam})}} &= \dot{q}_9 \mathbf{a}_2^{(\text{base})}, \\
{}^{a^{(\text{base})}}\boldsymbol{\omega}^{a^{(\text{middle beam})}} &= \dot{q}_8 \mathbf{a}_2^{(\text{base})}, \\
\mathbf{r}^{C_1 P_1} &= -R_1 \mathbf{a}_3^{(\text{base})}, \\
\mathbf{r}^{C_2 P_2} &= R_2 \mathbf{a}_3^{(\text{base})},
\end{aligned}$$

from which we conclude that

$$^{a^{(\text{base})}}\frac{dP_1}{dt} = -R_1 \dot{q}_9 \mathbf{a}_1^{(\text{base})}$$

and

$$^{a^{(\text{base})}}\frac{dP_2}{dt} = R_2 \dot{q}_8 \mathbf{a}_1^{(\text{base})},$$

where $R_1$ and $R_2$ are the radii of the upper and lower gears, respectively. The rolling-without-slipping constraint implies that

$$^{a^{(\text{base})}}\frac{dP_1}{dt} = {}^{a^{(\text{base})}}\frac{dP_2}{dt},$$

i.e., the motion constraint becomes

$$R_2 \dot{q}_8 = -R_1 \dot{q}_9.$$

It is interesting to note that the motion constraint derived here is actually holonomic in the sense that it can be obtained by differentiation with respect to time of the configuration constraint

$$R_2 q_8 = -R_1 q_9 + \text{arbitrary constant}.$$

We might be led to conclude that the desk lamp only has eight geometric degrees of freedom, instead of the nine that were previously found. Indeed, this would be true if there was a way by which a unique value could be selected for the arbitrary constant in the constraint equation above. That there is no unique choice of value for the arbitrary constant corresponds to the absence of a unique choice of initial relative configuration of the upper and middle beams. We conclude that, although holonomic, the motion constraint is only a constraint on *changes* to the relative orientation.

As in the case of the bicycle, it is not immediately evident that the three motion constraints derived here are independent. This would follow, however, if it were possible to choose a set of independent velocity coordinates that results in a not-everywhere-singular set of kinematic differential equations.

### 10.4.4 Independent Velocity Coordinates and Kinematic Differential Equations

We turn to the selection of independent velocity coordinates to uniquely specify all allowable motions of the desk lamp. Under the assumption that the three motion constraints derived in the previous section are actually independent, it follows that the desk lamp has eight dynamic degrees of freedom, i.e., that eight independent velocity coordinates suffice to specify all allowable motions of the desk lamp.

While there are infinitely many choices for the independent velocity coordinates, we are guided by our experience with desk lamps to consider the eight different ways in which a person controls the desk lamp motion, namely:

- Controlling the components of the linear velocity of the base relative to $\mathcal{W}$, ${}^{\mathcal{W}}\mathbf{v}^{\mathcal{A}_{(\text{base})}}$;
- Controlling the components of the angular velocity of the base relative to $\mathcal{W}$, ${}^{w}\boldsymbol{\omega}^{a^{(\text{base})}}$;
- Controlling the angular velocity of the upper beam relative to the base, ${}^{a^{(\text{base})}}\boldsymbol{\omega}^{a^{(\text{upper beam})}}$;
- Controlling the angular velocity of the lamp shade relative to the bracket, ${}^{a^{(\text{bracket})}}\boldsymbol{\omega}^{a^{(\text{lamp shade})}}$.

Consequently, we propose the following choice of independent velocity coordinates:

$$
\begin{aligned}
u_1 &= \mathbf{w}_1 \bullet {}^{\mathcal{W}}\mathbf{v}^{\mathcal{A}_{(\text{base})}},\\
u_2 &= \mathbf{w}_2 \bullet {}^{\mathcal{W}}\mathbf{v}^{\mathcal{A}_{(\text{base})}},\\
u_3 &= \mathbf{w}_3 \bullet {}^{\mathcal{W}}\mathbf{v}^{\mathcal{A}_{(\text{base})}},\\
u_4 &= \mathbf{a}_1^{(\text{base})} \bullet {}^{w}\boldsymbol{\omega}^{a^{(\text{base})}},\\
u_5 &= \mathbf{a}_2^{(\text{base})} \bullet {}^{w}\boldsymbol{\omega}^{a^{(\text{base})}},\\
u_6 &= \mathbf{a}_3^{(\text{base})} \bullet {}^{w}\boldsymbol{\omega}^{a^{(\text{base})}},
\end{aligned}
$$

$$u_7 = \mathbf{a}_2^{(\text{base})} \bullet {}^{a^{(\text{base})}}\boldsymbol{\omega}^{a^{(\text{upper beam})}},$$
$$u_8 = \mathbf{a}_2^{(\text{base})} \bullet {}^{a^{(\text{bracket})}}\boldsymbol{\omega}^{a^{(\text{lamp shade})}}.$$

We thus arrive at a complete set of kinematic differential equations:

$$\begin{aligned}
\dot{q}_1 &= u_1\\
\dot{q}_2 &= u_2\\
\dot{q}_3 &= u_3\\
\dot{q}_4 \sin q_5 \sin q_6 + \dot{q}_5 \cos q_6 &= u_4\\
\dot{q}_4 \sin q_5 \cos q_6 - \dot{q}_5 \sin q_6 &= u_5\\
\dot{q}_4 \cos q_5 + \dot{q}_6 &= u_6\\
\dot{q}_9 &= u_7\\
\dot{q}_{11} &= u_8
\end{aligned}$$

$$\begin{aligned}
&- p_4\dot{q}_7 \cos q_7 + p_6\dot{q}_9 \cos q_9\\
&- (\dot{q}_7 + \dot{q}_{10})(p_7 \sin(q_7 + q_{10}) + p_8 \cos(q_7 + q_{10})) = 0\\
&p_4\dot{q}_7 \sin q_7 - p_7\dot{q}_9 \sin q_9\\
&- (\dot{q}_7 + \dot{q}_{10})(p_7 \cos(q_7 + q_{10}) - p_8 \sin(q_7 + q_{10})) = 0\\
&\dot{q}_8 + \dot{q}_9 = 0,
\end{aligned}$$

where the last three equations correspond to the motion constraints derived in the previous section, and we have assumed that the spur gears have identical radii. These differential equations are non-singular, as long as

$$q_5 \neq n\pi$$

for any integer $n$ and

$$p_8 \sin q_{10} - p_7 \cos q_{10} \neq 0.$$

Of these two singularities, we have encountered the first one on numerous previous occasions whenever the orientation of a rigid body was described in terms of $3-1-3$ Euler angles and the components of the angular velocity vector in the body-fixed triad were chosen as independent velocity coordinates. A different choice of Euler angles or a different choice of independent velocity coordinates would move this singularity to a different configuration. In fact, the singularity could be entirely eliminated by retaining the definition of the independent velocity coordinates but describing the orientation using Euler parameters (cf. Exercise 9.2).

To understand the significance of the second singularity, consider the cross product

$$\mathbf{r}^{A_{\text{bracket}}H_{\text{bracket}}} \times \mathbf{r}^{A_{\text{lower beam}}A_{\text{bracket}}} = -p_4 (p_8 \sin q_{10} - p_7 \cos q_{10})\, \mathbf{a}_2^{(\text{base})}.$$

The singular configuration evidently corresponds to the case when this cross product equals zero, i.e., when the vectors

$$\mathbf{r}^{A_{\text{bracket}}H_{\text{bracket}}} \text{ and } \mathbf{r}^{A_{\text{lower beam}}A_{\text{bracket}}}$$

are parallel as shown in the figure below.

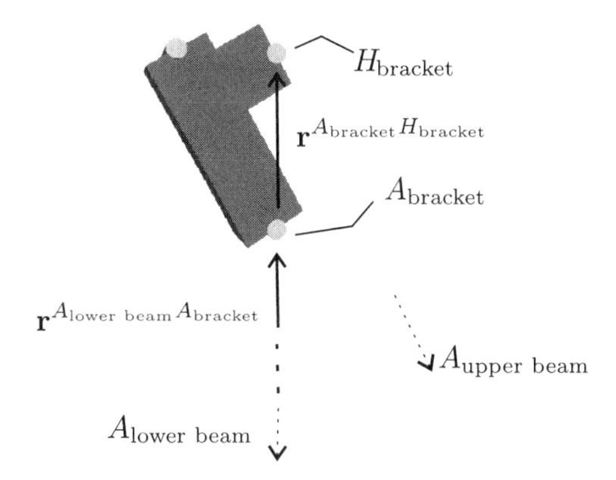


Clearly, it is no longer possible to increase the inclination of the upper beam any further without breaking the constraint at the hinge joint. The origin of the singularity must therefore lie in our choice to control the rate of change of the orientation of the upper beam relative to the base through the independent velocity coordinate $u_7$. If, instead, we had chosen to control the rate of change of the orientation of the lower beam relative to the base, the above configuration would not have corresponded to a singularity of the kinematic differential equations. Instead, a singularity would have occurred when the vectors $\mathbf{r}^{A_{\text{bracket}}H_{\text{bracket}}}$ and $\mathbf{r}^{A_{\text{upper beam}}H_{\text{upper beam}}}$ became parallel.

Retaining the present definition of $u_7$, the singularity could also be avoided by appropriately designing the geometry of the desk lamp. In fact, it is possible to select values for the parameters $p_1$, $p_2$, $p_3$, $p_4$, $p_5$, $p_6$, $p_7$, and $p_8$, such that the singular configuration cannot occur for any values of the configuration coordinates that satisfy the configuration constraints. It is an interesting exercise to show that the choice

$$p_1 = 11, p_2 = 12.5, p_3 = 14, p_4 = 45,$$
$$p_5 = 3.5, p_6 = 45, p_7 = 1.5, p_8 = 2.6$$

that was used in Chapter 7 accomplishes this.

### 10.4.5 An Initial Configuration

Initial values for the configuration coordinates that would correspond to an allowable configuration were already found at the end of Section 7.4. There it was suggested that a collection of such sets of values for the configuration coordinates could be generated to give the appearance of a smooth allowable motion. As suggested there, in addition to being very cumbersome, this methodology would fail to address the rolling-without-slipping constraint on the spur gears.

In contrast, the formulation derived above in terms of the kinematic differential equations guarantees that the configuration constraints **and** the motion constraints are satisfied for all time as long as the initial configuration is an allowable configuration and as long as singularities are not encountered. Now, every choice of time-dependence of the eight independent velocity coordinates, $u_1(t)$, $u_2(t)$, $u_3(t)$, $u_4(t)$, $u_5(t)$, $u_6(t)$, $u_7(t)$, and $u_8(t)$ will generate an allowable motion.

### 10.4.6 MAMBO

We recall the formulation of the geometric hierarchy of the desk lamp in the MAMBO toolbox from Chapter 7:

```
> Restart():
> DeclareObservers(W,Base,Lamp,Bracket,UpperBeam,
> LowerBeam,MiddleBeam):
> DeclarePoints(W,Base,Lamp,Bracket,UpperBeam,
> LowerBeam,MiddleBeam):
> DeclareTriads(w,base,lamp,bracket,upperbeam,
> lowerbeam,middlebeam):
> DefineNeighbors([W,Base],[Base,LowerBeam],
> [LowerBeam,Bracket],[Bracket,Lamp],[Base,MiddleBeam],
> [Base,UpperBeam]):
> DefineObservers([W,W,w],[Base,Base,base]
> [LowerBeam,LowerBeam,lowerbeam],
> [MiddleBeam,MiddleBeam,middlebeam],
> [UpperBeam,UpperBeam,upperbeam],
> [Bracket,Bracket,bracket],[Lamp,Lamp,lamp]):
> DefinePoints([W,Base,w,q1,q2,q3],
> [Base,LowerBeam,base,0,0,p1],
> [Base,MiddleBeam,base,0,0,p2],
> [Base,UpperBeam,base,0,0,p3],
> [LowerBeam,Bracket,lowerbeam,0,0,p4],
> [Bracket,Lamp,bracket,0,0,p5]):
> DefineTriads([w,base,[q4,3],[q5,1],[q6,3]],
> [base,lowerbeam,q7,2],[lowerbeam,bracket,q10,2],
> [bracket,lamp,q11,2],[base,middlebeam,q8,2],
> [base,upperbeam,q9,2]):
```

and the implementation of the configuration constraints using the following MAMBO toolbox statements:

```
> DeclarePoints(HUpperBeam,HBracket):
> DefinePoints(
> [UpperBeam,HUpperBeam,upperbeam,0,0,p6],
> [Bracket,HBracket,bracket,-p7,0,p8]):
```

```
> f1:=FindTranslation(HBracket,HUpperBeam) &oo
> MakeTranslations(base,1)=0:
> f2:=FindTranslation(HBracket,HUpperBeam) &oo
> MakeTranslations(base,3)=0:
```

The kinematic differential equations derived above are obtained from the MAMBO toolbox statements

```
> kde:={seq(LinearVelocity(W,Base) &oo
> MakeTranslations(w,i)=cat(u,i),i=1..3),
> seq(AngularVelocity(w,base) &oo
> MakeTranslations(base,i)=cat(u,i+3),i=1..3),
> AngularVelocity(base,upperbeam) &oo
> MakeTranslations(base,2)=u7,
> AngularVelocity(bracket,lamp) &oo
> MakeTranslations(base,2)=u8}
> union {q8t+q9t=0} union DiffTime({f1,f2}):
```

Finally, we use `MotionOutput` to export the motion description to the file `lamp.dyn`.

```
> MotionOutput(ode=kde,parameters=[p1=11,p2=12.5,p3=14,
> p4=45,p5=3.5,p6=45,p7=1.5,p8=2.6],states=[q1,q2,q3,q4,
> q5=.01,q6,q7=1.200039725,q8=1.8,q9=1.2,
> q10=-.6769771893,q11=2],insignals=[u1=0,u2=0,u3=0,
> u4=0,u5=0,u6=0,u7=-.2*cos(t),u8=cos(t)],
> filename="lamp.dyn"):
```

Here, the choice $q_5 = 0.01$ avoids the singularity at $q_5 = n\pi$ for some integer $n$.

# Chapter 11

# A Look Ahead

*wherein the reader learns of:*

- *Fundamental means of describing mass distributions;*
- *The influence of the distribution of mass on the motion of real-world objects;*
- *The motion of isolated rigid bodies relative to inertial observers;*
- *Formulating dynamic differential equations for the motion of rigid bodies relative to inertial observers;*
- *The physical modeling of constraints.*

Practicum

Use MAMBO to explore the implications of the discussion in this chapter. Investigate the visual characteristics of the motion of isolated rigid bodies relative to inertial observers. Investigate the visual characteristics of the motion of isolated rigid bodies relative to non-inertial observers.

Implement the dynamic and kinematic differential equations for the example systems in MAMBO and study the resulting motion. Discuss the process of modeling physical phenomena and the method for validating or discarding a particular model. Consider, in particular, the notions of inertial observers, forces, and torques.

## 11.1 Principles of Newtonian Mechanics

(Ex. 11.1 – Ex. 11.18)

### 11.1.1 Mass Distributions

The motion of real-world objects is determined by their interactions with other real-world objects. Experience shows you that the effects of these interactions on the motion of a given real-world object depend on the amount of matter contained within the object and the way in which that matter is distributed throughout the object. It is dramatically more difficult to affect the motion of a train engine traveling even at slow speeds than it is to affect the motion of a baseball traveling at the same speed. It is significantly easier to slow the turning of a pencil about its symmetry axis than it is to slow the rotation of the wings of a windmill.

The physical quantity representing the total amount of matter in a real-world object is called its *mass*. It is far from obvious how to measure the mass of an object. All means of measurement, e.g., using a balance scale or looking at collisions between objects, rely on scientific theory of the outcome of such experiments as a function of the mass. The validity of such theory can only be ascertained through experiments.

The distribution of matter throughout an object may differ significantly between different objects of the same shape. Let $P$ denote a point within an object and consider the ratio between the amount of matter contained within a volume element centered on $P$ and the volume of the element. Experience shows you that for small (but not microscopic) values of the radius, this ratio is more or less independent of radius. The value for the ratio in this range of radii is called the *density at the point* $P$ and is denoted by $\rho(P)$. Variations in $\rho$ as a function of location throughout an object quantify the way in which matter is distributed throughout the object.

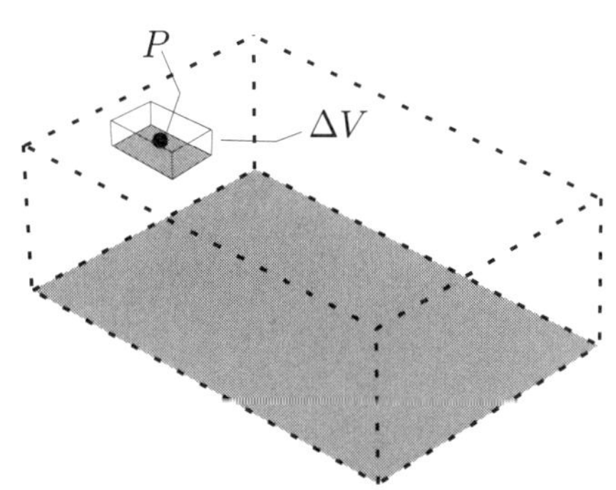


**Illustration 11.1**

If $\Delta V$ is the volume of a small element of matter centered on a point $P$ of an object, then

$$\Delta M \approx \rho(P)\,\Delta V$$

is the mass of the matter contained within the volume. Experience shows you that the mass of a composite object equals the sum of the masses of all its constituent parts. In other words,

$$M = \sum \Delta M \approx \sum \rho(P)\,\Delta V.$$

In the limit that the volume $\Delta V$ goes to zero, the sum becomes infinite and turns into an integral over all points in the object, i.e.,

$$M = \int_V \rho(P)\,dV.$$

If $\rho$ is constant throughout the object, i.e., $\rho(P) = \rho_0$, then

$$M = \int_V \rho(P)\,dV = \rho_0 \int_V dV = \rho_0 V,$$

i.e., the mass equals the density multiplied by the volume occupied by the object. When $\rho$ is constant, the object is said to be *homogeneous*.

Suppose $\mathbf{n}$ is an arbitrary vector of unit length. The *first moment of inertia* of a mass distribution about the straight line through a point $A$ that is spanned by $\mathbf{n}$ is given by the integral

$$\int_V \mathbf{n} \times \mathbf{r}^{AP} \rho(P)\,dV.$$

The contribution to this integral of a small volume of matter at a point $P$ is proportional to its distance to the straight line through $A$. The integral is a measure of how evenly spread out the mass is about the point $A$.

Since $\mathbf{n}$ is independent of the point $P$, we may rewrite the above integral as

$$\mathbf{n} \times \int_V \mathbf{r}^{AP} \rho(P)\,dV,$$

i.e., the first moment of inertia about an arbitrary straight line through $A$ can be computed from the vector

$$\int_V \mathbf{r}^{AP} \rho(P)\,dV.$$

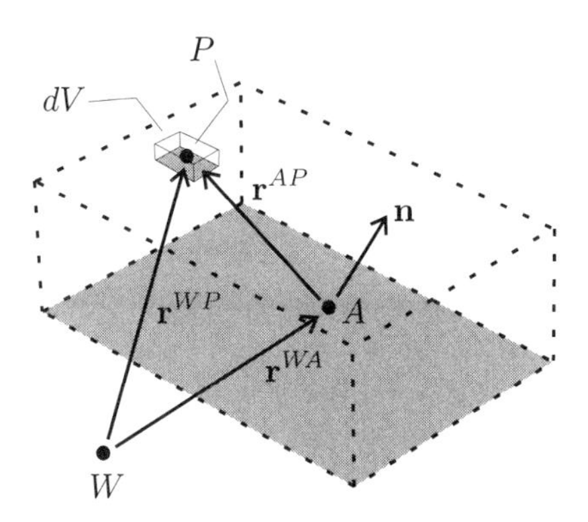


Suppose that $W$ is the reference point of an observer $\mathcal{W}$. Then, since $\mathbf{r}^{AP} = \mathbf{r}^{WP} - \mathbf{r}^{WA}$ and $A$ is independent of $P$, we find

$$\begin{aligned}\int_V \mathbf{r}^{AP} \rho(P)\, dV &= \int_V \mathbf{r}^{WP} \rho(P)\, dV - \mathbf{r}^{WA} \int_V \rho(P)\, dV \\ &= \int_V \mathbf{r}^{WP} \rho(P)\, dV - M\mathbf{r}^{WA}.\end{aligned}$$

It follows that if the point $A$ is defined by

$$\mathbf{r}^{WA} = \frac{1}{M} \int_V \mathbf{r}^{WP} \rho(P)\, dV,$$

then the first moment of inertia of the mass distribution about $A$ is zero for all directions.

The point $A$ whose position relative to the observer $\mathcal{W}$ is defined by the position vector above is called the *center of mass* of the mass distribution. Since the first moment of inertia about the center of mass is zero for arbitrary directions, the mass is, in some sense, evenly distributed about the center of mass.

### Illustration 11.2

The center of mass of a homogeneous body is independent of the density, since

$$\frac{1}{M} \int_V \mathbf{r}^{WP} \rho(P)\, dV = \frac{1}{\rho_o V} \int_V \mathbf{r}^{WP} \rho_0 dV = \frac{1}{V} \int_V \mathbf{r}^{WP} dV.$$

This point is called the *centroid* of the volume occupied by the object.

For example, suppose you want to find the center of mass of a homogeneous sphere of radius $R$. Let $W$ correspond to the center of the sphere and let $w$ be some arbitrary triad. We may express the position vector

$\mathbf{r}^{WP}$ in spherical coordinates:

$$\mathbf{r}^{WP} = w \begin{pmatrix} r\sin\phi\cos\theta \\ r\sin\phi\sin\theta \\ r\cos\phi \end{pmatrix}.$$

It follows that

$$\begin{aligned}\frac{1}{V}\int_V \mathbf{r}^{WP} dV &= \frac{3}{4\pi R^3}\int_0^R \int_0^\pi \int_0^{2\pi} w \begin{pmatrix} r\sin\phi\cos\theta \\ r\sin\phi\sin\theta \\ r\cos\phi \end{pmatrix} r^2 \sin\phi d\theta d\phi dr \\ &= \mathbf{0},\end{aligned}$$

i.e., the center of mass of the sphere coincides with the sphere's center.

Two real-world objects may share the same center of mass, yet have very different mass distributions. To detect such differences, it is necessary to consider higher-order measures of the spread of matter about the center of mass.

Suppose $\mathbf{n}$ and $\mathbf{m}$ are two arbitrary vectors of unit length. The *second moment of inertia* of a mass distribution about the straight lines through a point $A$ that are parallel to $\mathbf{n}$ and $\mathbf{m}$, respectively, is given by the integral

$$\int_V \left(\mathbf{n} \times \mathbf{r}^{AP}\right) \bullet \left(\mathbf{m} \times \mathbf{r}^{AP}\right) \rho(P)\, dV.$$

The contribution to this integral of a small volume of matter at a point $P$ is proportional to the product of the distances from the point to the two straight lines through $A$. As with the first moment of inertia, this integral is a measure of how spread out the mass is about the point $A$.

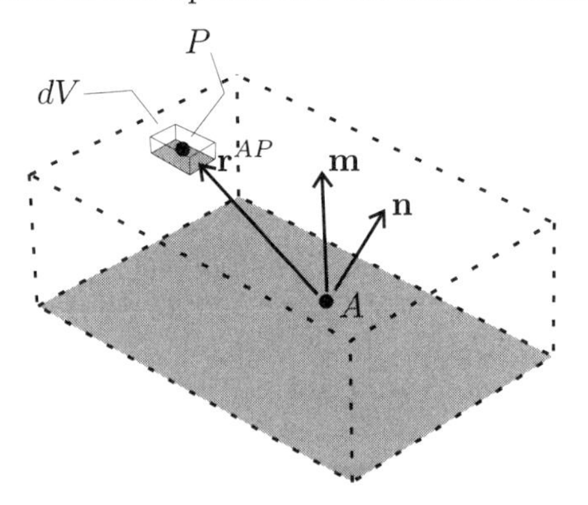


The result of Exercise 11.1 shows that

$$\int_V \left(\mathbf{n} \times \mathbf{r}^{AP}\right) \bullet \left(\mathbf{m} \times \mathbf{r}^{AP}\right) \rho(P)\, dV = \left({}^w n\right)^T \, {}^w I^A \, {}^w m,$$

where ${}^w n$ and ${}^w m$ are the matrix representations of $\mathbf{n}$ and $\mathbf{m}$ relative to a triad $w$ and the $[i,j]$-th entry of the *moment of inertia matrix* ${}^w I^A$ *about the point* $A$ *relative to the triad* $w$ is given by the formula[1]

$$ {}^w I^A_{ij} = \int_V \left( \left\| \mathbf{r}^{AP} \right\|^2 \delta_{ij} - {}^w\left(\mathbf{r}^{AP}\right)_i {}^w\left(\mathbf{r}^{AP}\right)_j \right) \rho\left(P\right) dV. $$

**Illustration 11.3**

Consider the homogeneous sphere in the previous illustration with

$$ \mathbf{r}^{WP} = w \begin{pmatrix} r\sin\phi\cos\theta \\ r\sin\phi\sin\theta \\ r\cos\phi \end{pmatrix}. $$

Then,

$$ \begin{aligned} {}^w I^W_{12} = {}^w I^W_{21} &= -\rho_0 \int_0^R \int_0^\pi \int_0^{2\pi} r^4 \sin^3\phi\cos\theta\sin\theta d\theta d\phi dr = 0, \\ {}^w I^W_{13} = {}^w I^W_{31} &= -\rho_0 \int_0^R \int_0^\pi \int_0^{2\pi} r^4 \sin^2\phi\cos\phi\cos\theta d\theta d\phi dr = 0, \\ {}^w I^W_{23} = {}^w I^W_{32} &= -\rho_0 \int_0^R \int_0^\pi \int_0^{2\pi} r^4 \sin^2\phi\cos\phi\sin\theta d\theta d\phi dr = 0, \\ {}^w I^W_{11} &= \rho_0 \int_0^R \int_0^\pi \int_0^{2\pi} r^4 \sin\phi\left(\sin^2\phi\sin^2\theta + \cos^2\phi\right) d\theta d\phi dr \\ &= \frac{8\pi R^5}{15}\rho_0, \\ {}^w I^W_{22} &= \rho_0 \int_0^R \int_0^\pi \int_0^{2\pi} r^4 \sin\phi\left(\sin^2\phi\cos^2\theta + \cos^2\phi\right) d\theta d\phi dr \\ &= \frac{8\pi R^5}{15}\rho_0, \end{aligned} $$

and

$$ {}^w I^W_{33} = \rho_0 \int_0^R \int_0^\pi \int_0^{2\pi} r^4 \sin^3\phi d\theta d\phi dr = \frac{8\pi R^5}{15}\rho_0, $$

where we have again used the volume element expressed in spherical coordinates:

$$ dV = r^2 \sin\phi d\theta d\phi dr. $$

Since

$$ M = \frac{4\pi R^3}{3}\rho_0, $$

[1]Recall the Kronecker delta notation $\delta_{ij} = \begin{cases} 1 & i=j \\ 0 & i \neq j \end{cases}$.

it follows that the moment of inertia matrix about the center of mass of the sphere relative to the arbitrary triad $w$ is given by

$$ {}^{w}I^{W} = \begin{pmatrix} \frac{2}{5}MR^2 & 0 & 0 \\ 0 & \frac{2}{5}MR^2 & 0 \\ 0 & 0 & \frac{2}{5}MR^2 \end{pmatrix}. $$

The first moment of inertia equals zero about arbitrary straight lines through the center of mass of an arbitrary mass distribution. It is natural to ask whether there might exist a (possibly different) point, such that the second moment of inertia equals zero about arbitrary pairs of straight lines through this second point. But, if $\mathbf{n} = \mathbf{m}$, then the second moment of inertia becomes

$$ \int_V \left\| \mathbf{n} \times \mathbf{r}^{AP} \right\|^2 \rho(P)\, dV. $$

Since the integrand is $\geq 0$, this integral can only equal zero if the integrand is zero everywhere. Since the density cannot equal zero everywhere, we must have

$$ \mathbf{n} \times \mathbf{r}^{AP} = \mathbf{0} $$

for all points in the mass distribution. This is only possible if all points lie on the straight line through $A$ that is spanned by $\mathbf{n}$. But real-world objects always have non-zero volume, and so the second moment of inertia cannot equal zero for arbitrary pairs of straight lines through any point in a mass distribution.

### 11.1.2 Linear and Angular Momentum

Suppose that the reference point $B$ of a rigid body is located at its center of mass and let $b$ denote the rigid body's reference triad. As discussed in the previous chapter, the motion of the rigid body is entirely determined by the velocity of the reference point $B$ relative to an observer $\mathcal{W}$, i.e.,

$$ {}^{\mathcal{W}}\frac{dB}{dt} $$

and the angular velocity between the reference triad $w$ of the observer $\mathcal{W}$ and the reference triad $b$, i.e.,

$$ {}^{w}\boldsymbol{\omega}^{b}. $$

From experience, we conclude that the degree to which the motion of a rigid body can be changed depends not only on the velocity of its reference point and the angular velocity of its reference triad, but also on the amount of matter contained within the rigid body and the way the matter is distributed throughout the body. To account for these observations, we introduce the dynamical quantities ${}^{\mathcal{W}}\mathbf{p}$ and ${}^{\mathcal{W}}\mathbf{h}$, called the *linear momentum* and *angular momentum*, respectively, of the rigid body relative to $\mathcal{W}$.

In particular, let

$$ {}^{\mathcal{W}}\mathbf{p} = M \frac{{}^{\mathcal{W}}dB}{dt} = w\ M \frac{d\ {}^{w}\left(\mathbf{r}^{WB}\right)}{dt}, $$

where $M$ is the total mass of the rigid body, and

$$ {}^{\mathcal{W}}\mathbf{h} = b\ {}^{b}I^{B}\ {}^{b}\left({}^{w}\boldsymbol{\omega}^{b}\right), $$

where ${}^{b}I^{B}$ is the moment of inertia matrix about the center of mass relative to the body-fixed reference triad $b$.

### 11.1.3 Isolated Rigid Bodies and Inertial Observers

The motion of real-world objects is determined by interactions with other real-world objects. An object that does not interact with any other objects is said to be *isolated*. The first law of Newtonian mechanics establishes the existence of a privileged collection of observers, relative to which the linear and angular momenta of any arbitrary, isolated rigid bodies **do not change with time**. Such observers are known as *inertial observers*.

If $\mathcal{N}$ is an inertial observer and $B$ is the center of mass of an isolated rigid body, it follows that

$$ {}^{\mathcal{N}}\mathbf{p} = M \frac{{}^{\mathcal{N}}dB}{dt} $$

is constant relative to $\mathcal{N}$. Since the amount of matter does not change with time, the velocity of the center of mass relative to the inertial observer must also be constant. It follows that the center of mass moves along a straight line relative to the inertial observer. This conclusion is contained in **Newton's first law of motion**:

> The center of mass of an isolated rigid body moves along a straight line with constant speed relative to an inertial observer.

**Illustration 11.4**

Let the reference point $B$ of an isolated rigid body be located at the center of mass of the body. The position of the rigid body relative to an inertial observer $\mathcal{N}$ is then given by the position vector

$$\mathbf{r}^{NB} = n \begin{pmatrix} q_1 \\ q_2 \\ q_3 \end{pmatrix},$$

where $N$ is the reference point of the inertial observer, $n$ is the reference triad of the inertial observer, and $q_1$, $q_2$, and $q_3$ are configuration coordinates.

Since the rigid body is isolated, its motion cannot be constrained. Thus, we can introduce three independent velocity coordinates $u_1$, $u_2$, and $u_3$, such that

$$\frac{{}^{\mathcal{N}}dB}{dt} = n \begin{pmatrix} u_1 \\ u_2 \\ u_3 \end{pmatrix},$$

from which the kinematic differential equations

$$\dot{q}_1 = u_1,$$
$$\dot{q}_2 = u_2,$$
$$\dot{q}_3 = u_3$$

follow.

The linear momentum of the rigid body now becomes

$${}^{\mathcal{N}}\mathbf{p} = M \frac{{}^{\mathcal{N}}dB}{dt} = n \begin{pmatrix} Mu_1 \\ Mu_2 \\ Mu_3 \end{pmatrix}.$$

Since the linear momentum is constant relative to the inertial observer,

$$\frac{{}^{n}d\,{}^{\mathcal{N}}\mathbf{p}}{dt} = \mathbf{0},$$

i.e.,

$$\dot{u}_1 = 0,$$
$$\dot{u}_2 = 0,$$
$$\dot{u}_3 = 0.$$

These differential equations in the independent velocity coordinates are called the *dynamic differential equations*. Since the derivatives with respect to time of the independent velocity coordinates all equal zero, the independent velocity coordinates must also be constant, i.e., $u_1(t) = u_1(0)$, $u_2(t) = u_2(0)$, and $u_3(t) = u_3(0)$. The kinematic differential equations then yield

$$q_1(t) = u_1(0)\,t + q_1(0),$$
$$q_2(t) = u_2(0)\,t + q_2(0),$$
$$q_3(t) = u_3(0)\,t + q_3(0)$$

corresponding to motion along the straight line through the point

$$\mathbf{r}^{NB} = n \begin{pmatrix} q_1(0) \\ q_2(0) \\ q_3(0) \end{pmatrix}$$

with constant tangent vector

$$n \begin{pmatrix} u_1(0) \\ u_2(0) \\ u_3(0) \end{pmatrix}.$$

From Exercise 11.4, we find that the moment of inertia matrices of a rigid body about its center of mass $B$ relative to two different triads $n$ and $b$ are related by the formula

$$R_{nb}\,{}^{b}I^{B} R_{bn} = {}^{n}I^{B}.$$

Thus, if $\mathcal{N}$ is an inertial observer and $b$ is the reference triad of an isolated rigid body, it follows that

$$\begin{aligned} {}^{\mathcal{N}}\mathbf{h} &= b\,{}^{b}I^{B}\,{}^{b}\left({}^{n}\boldsymbol{\omega}^{b}\right) \\ &= nR_{nb}\,{}^{b}I^{B}R_{bn}\,{}^{n}\left({}^{n}\boldsymbol{\omega}^{b}\right) \\ &= n\,{}^{n}I^{B}\,{}^{n}\left({}^{n}\boldsymbol{\omega}^{b}\right) \end{aligned}$$

is constant relative to $\mathcal{N}$. In contrast with the case of the linear velocity, this does not imply that the angular velocity of the rigid body relative to $\mathcal{N}$ is constant. Instead, it is only required that the orientation of the rigid body relative to $\mathcal{N}$ changes in such a way that the matrix product

$${}^{n}I^{B}\,{}^{n}\left({}^{n}\boldsymbol{\omega}^{b}\right)$$

remains constant.

**Illustration 11.5**

Let $b$ denote the reference triad of the rigid body in the previous illustration. The orientation of the rigid body relative to an inertial observer $\mathcal{N}$ is given by the rotation matrix

$$R_{nb} = R(q_4, 0, 0, 1)\, R(q_5, 1, 0, 0)\, R(q_6, 0, 0, 1),$$

where $n$ and $b$ are the reference triads of the inertial observer and the rigid body and $q_4$, $q_5$, and $q_6$ are configuration coordinates.

Since the rigid body is isolated, its motion cannot be constrained. Thus, we can introduce three independent velocity coordinates $u_4$, $u_5$, and $u_6$, such that

$$^n\boldsymbol{\omega}^b = b\begin{pmatrix} u_4 \\ u_5 \\ u_6 \end{pmatrix},$$

from which the kinematic differential equations

$$\begin{aligned} \dot{q}_4 \sin q_5 \sin q_6 + \dot{q}_5 \cos q_6 &= u_4, \\ \dot{q}_4 \sin q_5 \cos q_6 - \dot{q}_5 \sin q_6 &= u_5, \\ \dot{q}_4 \cos q_5 + \dot{q}_6 &= u_6 \end{aligned}$$

follow.

The angular momentum becomes

$$^{\mathcal{N}}\mathbf{h} = b\begin{pmatrix} {}^bI_{11}^B u_4 + {}^bI_{12}^B u_5 + {}^bI_{13}^B u_6 \\ {}^bI_{21}^B u_4 + {}^bI_{22}^B u_5 + {}^bI_{23}^B u_6 \\ {}^bI_{31}^B u_4 + {}^bI_{32}^B u_5 + {}^bI_{33}^B u_6 \end{pmatrix}.$$

Since the angular momentum is constant relative to the inertial observer,

$$\frac{^n d\, ^{\mathcal{N}}\mathbf{h}}{dt} = \frac{^b d\, ^{\mathcal{N}}\mathbf{h}}{dt} + {}^n\boldsymbol{\omega}^b \times {}^{\mathcal{N}}\mathbf{h} = \mathbf{0},$$

i.e.,

$$\begin{aligned} &{}^bI_{11}^B \dot{u}_4 + {}^bI_{12}^B \dot{u}_5 + {}^bI_{13}^B \dot{u}_6 + {}^bI_{31}^B u_4 u_5 + {}^bI_{32}^B u_5^2 + {}^bI_{33}^B u_5 u_6 \\ &\qquad - {}^bI_{21}^B u_4 u_6 - {}^bI_{22}^B u_5 u_6 - {}^bI_{23}^B u_6^2 = 0, \end{aligned}$$

$$\begin{aligned} &{}^bI_{21}^B \dot{u}_4 + {}^bI_{22}^B \dot{u}_5 + {}^bI_{23}^B \dot{u}_6 + {}^bI_{11}^B u_4 u_6 + {}^bI_{12}^B u_5 u_6 + {}^bI_{13}^B u_6^2 \\ &\qquad - {}^bI_{31}^B u_4^2 - {}^bI_{32}^B u_4 u_5 - {}^bI_{33}^B u_4 u_6 = 0, \end{aligned}$$

and

$$\begin{aligned} &{}^bI_{31}^B \dot{u}_4 + {}^bI_{32}^B \dot{u}_5 + {}^bI_{33}^B \dot{u}_6 + {}^bI_{21}^B u_4^2 + {}^bI_{22}^B u_4 u_5 + {}^bI_{23}^B u_4 u_6 \\ &\qquad - {}^bI_{11}^B u_4 u_5 - {}^bI_{12}^B u_5^2 - {}^bI_{13}^B u_5 u_6 = 0. \end{aligned}$$

These dynamic differential equations are called *Euler's equations*. In contrast to the dynamic differential equations corresponding to the time-independence of the linear momentum relative to the inertial observer, these equations do not imply that the independent velocity coordinates $u_4$, $u_5$, and $u_6$ are constant. In fact, their time-dependence can be quite complicated. The resulting changes in the orientation of the rigid body relative to the inertial observer can be quite dramatic.

Now, suppose that $\mathcal{N}_1$ and $\mathcal{N}_2$ are two inertial observers with reference points $N_1$ and $N_2$ and reference triads $n^{(1)}$ and $n^{(2)}$, respectively. If $B$ is the center of mass of an arbitrary isolated rigid body, then

$$\frac{{}^{\mathcal{N}_1}dB}{dt} = \frac{{}^{\mathcal{N}_2}dB}{dt} + {}^{\mathcal{N}_1}\mathbf{v}^{\mathcal{N}_2} + {}^{n^{(1)}}\boldsymbol{\omega}^{n^{(2)}} \times \mathbf{r}^{N_2 B}.$$

Since

$$\frac{{}^{\mathcal{N}_1}dB}{dt} \text{ and } \frac{{}^{\mathcal{N}_2}dB}{dt}$$

must both be constant, it follows that

$${}^{\mathcal{N}_1}\mathbf{v}^{\mathcal{N}_2} + {}^{n^{(1)}}\boldsymbol{\omega}^{n^{(2)}} \times \mathbf{r}^{N_2 B}$$

must also be constant. This must be true of an arbitrary isolated rigid body, including all those for which $\mathbf{r}^{N_2 B}$ changes with time. We conclude that, for any two inertial observers $\mathcal{N}_1$ and $\mathcal{N}_2$,

$${}^{\mathcal{N}_1}\mathbf{v}^{\mathcal{N}_2}$$

is constant and

$${}^{n^{(1)}}\boldsymbol{\omega}^{n^{(2)}} = \mathbf{0}.$$

The converse is also true, namely that if

$${}^{\mathcal{N}_1}\mathbf{v}^{\mathcal{N}_2}$$

is constant and

$${}^{n^{(1)}}\boldsymbol{\omega}^{n^{(2)}} = \mathbf{0}$$

and $\mathcal{N}_1$ is an inertial observer, then $\mathcal{N}_2$ is also an inertial observer.

> Any observer with constant linear velocity and vanishing angular velocity relative to an inertial observer is also an inertial observer.

**Illustration 11.6**

Since the velocity of the center of mass of an isolated rigid body is constant relative to an inertial observer $\mathcal{N}_1$, an observer $\mathcal{N}_2$ with reference point at the center of mass of the rigid body and with constant orientation relative to $\mathcal{N}_1$ is also an inertial observer.

Note, however, that the rigid body need not be stationary relative to $\mathcal{N}_2$, since its orientation may be changing with time.

Suppose the center of mass $P$ of an isolated rigid body moves with constant velocity relative to an inertial observer $\mathcal{N}$ along the straight line spanned by the vector $\mathbf{n}_1 + \mathbf{n}_3$ through the reference point of $\mathcal{N}$. If $\mathcal{A}$ is a second observer, such that

$$\mathbf{r}^{NA} = \mathbf{0} \text{ and } R_{na} = R(\omega t, 1, 0, 0),$$

then the point $P$ moves along a spiral relative to $\mathcal{A}$. Clearly, in this case, the observer $\mathcal{A}$ is not an inertial observer.

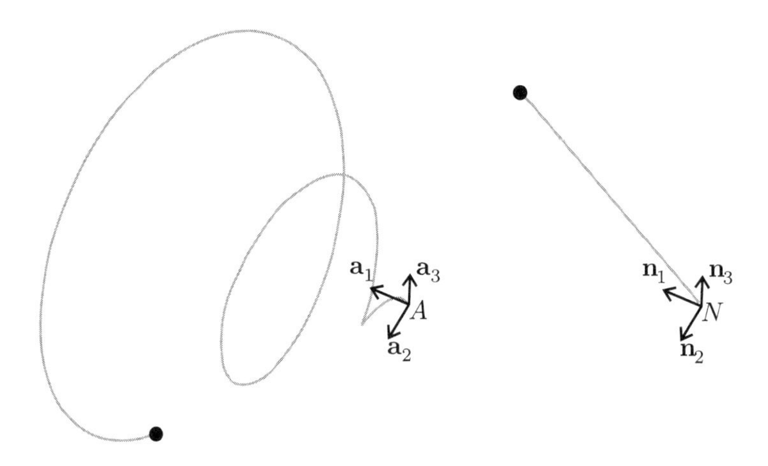


Suppose, more generally, that the center of mass of a rigid body moves along a curved line relative to an observer $\mathcal{A}$. Then, if the rigid body is isolated, $\mathcal{A}$ cannot be an inertial observer. Similarly, if $\mathcal{A}$ is an inertial observer, the rigid body cannot be isolated. In other words, if $\mathcal{A}$ is thought to be an inertial observer, it follows that the rigid body must be interacting with other objects in the world, since otherwise it would move along a straight line with constant velocity relative to $\mathcal{A}$. Similarly, if the rigid body is thought to be isolated, we must conclude that $\mathcal{A}$ is not an inertial observer.

The introduction of inertial observers was based entirely on the notion of isolated rigid bodies and their motion. We can detect whether a rigid body is isolated by observing its motion relative to an inertial observer. Conversely, we can detect whether an observer is an inertial observer by observing the motion of an isolated rigid body relative to the observer. If this sounds like a circular definition, that's because it is! There appears to be no way of separating the notion of an inertial observer from that of an isolated rigid body.

### 11.1.4 Kinetic Energy

Suppose that $\mathcal{N}$ is an inertial observer. Then, the result of Exercise 11.10 shows that

$$\begin{aligned}\frac{d}{dt}\left(\frac{1}{2}{}^{\mathcal{N}}\mathbf{p}\bullet\ {}^{\mathcal{N}}\frac{dB}{dt}+\frac{1}{2}{}^{\mathcal{N}}\mathbf{h}\bullet\ {}^{n}\boldsymbol{\omega}^{b}\right) &= {}^{n}\frac{d\,{}^{\mathcal{N}}\mathbf{p}}{dt}\bullet\ {}^{\mathcal{N}}\frac{dB}{dt} \\ &\quad+{}^{n}\frac{d\,{}^{\mathcal{N}}\mathbf{h}}{dt}\bullet\ {}^{n}\boldsymbol{\omega}^{b} \\ &= 0\end{aligned}$$

for an isolated rigid body, where the reference point $B$ of the rigid body is at the center of mass and $b$ is a body-fixed reference triad. The quantity

$${}^{\mathcal{N}}K=\frac{1}{2}{}^{\mathcal{N}}\mathbf{p}\bullet\ {}^{\mathcal{N}}\frac{dB}{dt}+\frac{1}{2}{}^{\mathcal{N}}\mathbf{h}\bullet\ {}^{n}\boldsymbol{\omega}^{b}$$

is known as the *kinetic energy* of the rigid body relative to the inertial observer. Since the kinetic energy of an isolated rigid body remains constant relative to an inertial observer, the motion of an isolated rigid body is said to be *conservative*.

**Illustration 11.7**

Using the expressions for the linear and angular momenta derived in previous illustrations, we find

$${}^{\mathcal{N}}K=\frac{M}{2}\begin{pmatrix}u_1 & u_2 & u_3\end{pmatrix}\begin{pmatrix}u_1\\u_2\\u_3\end{pmatrix}+\frac{1}{2}\begin{pmatrix}u_4 & u_5 & u_6\end{pmatrix}{}^{b}I^{B}\begin{pmatrix}u_4\\u_5\\u_6\end{pmatrix}.$$

### 11.1.5 MAMBO

The motion of an isolated rigid body may be visualized with MAMBO by appending the independent velocity coordinates to the MAMBO state variables and the corresponding dynamic differential equations to the `ode` block.

The MAMBO toolbox procedures `LinearMomentum` and `AngularMomentum` can be invoked to compute the linear and angular momentum of a rigid body relative to an observer. Their use is illustrated in the following sequence of statements:

```
> Restart():
> DeclareObservers(N):
> DeclarePoints(N,B):
> DeclareTriads(n,b):
> DefineObservers(N,N,n):
> DefinePoints(N,B,n,q1,q2,q3):
```

```
> DefineTriads(n,b,[q4,3],[q5,1],[q6,3]):
> DeclareStates(q1,q2,q3,q4,q5,q6):
> p:=LinearMomentum(N,B);
> h:=AngularMomentum(N,b);
```

$$
\begin{aligned}
&p := \text{table}([\\
&\text{“Size”} = 1\\
&1 = \text{table}([\\
&\text{“Triad”} = n\\
&\text{“Coordinates”} = [M\, q1t,\ M\, q2t,\ M\, q3t]\\
&])\\
&\text{“Type”} = \text{“Vector”}\\
&])
\end{aligned}
$$

$$
\begin{aligned}
&h := \text{table}([\\
&\text{“Size”} = 1\\
&1 = \text{table}([\\
&\text{“Triad”} = b\\
&\text{“Coordinates”} = [\mathit{Inertia11}\,(\sin(q5)\, q4t \sin(q6) + q5t \cos(q6))\\
&\ + \mathit{Inertia12}\,(\cos(q6)\, q4t \sin(q5) - \sin(q6)\, q5t)\\
&\ + \mathit{Inertia13}\,(q6t + q4t \cos(q5)),\\
&\mathit{Inertia21}\,(\sin(q5)\, q4t \sin(q6) + q5t \cos(q6))\\
&\ + \mathit{Inertia22}\,(\cos(q6)\, q4t \sin(q5) - \sin(q6)\, q5t)\\
&\ + \mathit{Inertia23}\,(q6t + q4t \cos(q5)),\\
&\mathit{Inertia31}\,(\sin(q5)\, q4t \sin(q6) + q5t \cos(q6))\\
&\ + \mathit{Inertia32}\,(\cos(q6)\, q4t \sin(q5) - \sin(q6)\, q5t)\\
&\ + \mathit{Inertia33}\,(q6t + q4t \cos(q5))]\\
&])\\
&\text{“Type”} = \text{“Vector”}\\
&])
\end{aligned}
$$

We may replace every occurrence of a derivative of a configuration coordinate by an expression linear in the independent velocity coordinates through a suitable introduction of kinematic differential equations.

```
> kde:={seq(LinearVelocity(N,B) &oo
> MakeTranslations(n,i)=cat(u,i),i=1..3),
> seq(AngularVelocity(n,b) &oo
> MakeTranslations(b,i)=cat(u,i+3),i=1..3)}:
> kde:=solve(kde,{q1t,q2t,q3t,q4t,q5t,q6t});
```

$$
\begin{aligned}
&kde := \{q5t = -u5 \sin(q6) + \cos(q6)\, u4,\ q3t = u3,\\
&q4t = \frac{\cos(q6)\, u5 + u4 \sin(q6)}{\sin(q5)},\ q1t = u1,\ q2t = u2,\\
&q6t = -\frac{\cos(q5)\cos(q6)\, u5 + \cos(q5)\, u4 \sin(q6) - u6 \sin(q5)}{\sin(q5)}\}
\end{aligned}
$$

```
> p:=Simplify(subs(kde,eval(p)));
```

$p :=$ table([
"Size" $= 1$
$1 =$ table([
"Triad" $= n$
"Coordinates" $= [M\, u1,\ M\, u2,\ M\, u3]$
])
"Type" = "Vector"
])

```
> h:=Simplify(subs(kde,eval(h)));
```

$h :=$ table([
"Size" $= 1$
$1 =$ table([
"Triad" $= b$
"Coordinates" $= [\mathit{Inertia11}\ u4 + \mathit{Inertia12}\ u5 + \mathit{Inertia13}\ u6,$
$\mathit{Inertia21}\ u4 + \mathit{Inertia22}\ u5 + \mathit{Inertia23}\ u6,$
$\mathit{Inertia31}\ u4 + \mathit{Inertia32}\ u5 + \mathit{Inertia33}\ u6]$
])
"Type" = "Vector"
])

The dynamic differential equations are obtained by differentiating the linear and angular momenta with respect to time relative to the inertial observer and setting the result equal to zero. In particular, in order to turn these vector equations into scalar equations, we take the dot product with the basis vectors of the inertial observer and those of the rigid body reference triad, respectively.

```
> DeclareStates(u1,u2,u3,u4,u5,u6);
> dde:={
> seq(DiffTime(p,n) &oo MakeTranslations(n,i)=0,i=1..3),
> seq(DiffTime(h,n) &oo MakeTranslations(b,i)=0,i=1..3)}:
```

Finally, the `MotionOutput` procedure may be invoked to formulate a MAMBO-compatible description of the kinematic and dynamic differential equations.

```
> MotionOutput(ode=kde union dde,
> states=[seq(cat(q,i),i=1..6),seq(cat(u,i),i=1..6)],
> parameters=[M,seq(seq(cat(Inertia,i,j),
> i=1..3),j=1..3)]);
```

```
states q1,q2,q3,q4,q5,q6,u1,u2,u3,u4,u5,u6;
parameters M,Inertia11,Inertia21,Inertia31,Inertia12,Inertia22,
                   Inertia32,Inertia13,Inertia23,Inertia33;
insignals {
    mass_u1_q6 = -Inertia21*u4-Inertia22*u5-Inertia23*u6;
```

```
        mass_u6_q6 = Inertia11*u4+Inertia12*u5+Inertia13*u6;
        t7 = cos(q5);
        t8 = -mass_u1_q6;
        t10 = cos(q6);
        t11 = sin(q5);
        t12 = t10*t11;
        t16 = Inertia31*u4+Inertia32*u5+Inertia33*u6;
        t19 = sin(q6);
        t20 = t11*t19;
        mass_q1_q4 = t7;
        mass_u5_q5 = t10;
    }
    ode {
        rhs[u4] = u3;
        rhs[q6] = u1;
        rhs[q4] = u2;
        rhs[u5] = u4;
        rhs[u3] = u5;
        rhs[q1] = u6;
        mass[u1][u4] = Inertia11;
        mass[q5][u4] = Inertia31;
        mass[u6][u4] = Inertia21;
        mass[q2][u1] = M;
        mass[u1][q6] = mass_u1_q6;
        mass[q1][q6] = 1;
        mass[u6][q6] = mass_u6_q6;
        mass[u1][q4] = -t7*t8+t12*t16;
        mass[q5][q4] = -t12*mass_u6_q6+t20*t8;
        mass[u5][q4] = t11*t19;
        mass[u3][q4] = t10*t11;
        mass[q1][q4] = mass_q1_q4;
        mass[u6][q4] = -t20*t16+mass_q1_q4*mass_u6_q6;
        mass[u1][q5] = -t19*t16;
        mass[q5][q5] = t19*mass_u6_q6+t10*t8;
        mass[u5][q5] = mass_u5_q5;
        mass[u3][q5] = -t19;
        mass[u6][q5] = -mass_u5_q5*t16;
        mass[u1][u5] = Inertia12;
        mass[q5][u5] = Inertia32;
        mass[u6][u5] = Inertia22;
        mass[u2][u3] = M;
        mass[q6][q1] = 1;
        mass[q4][q2] = 1;
        mass[u4][q3] = 1;
        mass[q3][u2] = M;
        mass[u1][u6] = Inertia13;
        mass[q5][u6] = Inertia33;
        mass[u6][u6] = Inertia23;
    }
```

### 11.1.6 Forces and Torques

Suppose that $\mathcal{N}$ is an inertial observer with reference point $N$ and reference triad $n$. Then, the linear and angular momentum of an isolated rigid body are constant relative to $\mathcal{N}$, i.e.,

$$\frac{{}^n d\, {}^{\mathcal{N}}\mathbf{p}}{dt} = \mathbf{0}$$

and

$$\frac{{}^n d\, {}^{\mathcal{N}}\mathbf{h}}{dt} = \mathbf{0}.$$

This is not generally true for a rigid body that is interacting with its environment. Instead, the rate of change of the linear momentum relative to the inertial observer is given by the differential equation

$$\frac{{}^n d\, {}^{\mathcal{N}}\mathbf{p}}{dt} = \mathbf{F},$$

where $\mathbf{F}$ is known as the *force acting on the rigid body*. This is the content of **Newton's second law of motion**.

Similarly, the rate of change of the angular momentum relative to the inertial observer is given by the differential equation

$$\frac{{}^n d\, {}^{\mathcal{N}}\mathbf{h}}{dt} = \mathbf{T},$$

where $\mathbf{T}$ is known as the *torque acting on the rigid body.*

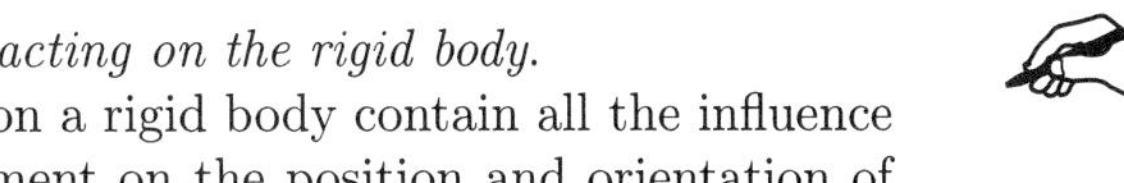

The force and torque acting on a rigid body contain all the influence of interactions with the environment on the position and orientation of the rigid body relative to an inertial observer. Clearly, for the isolated rigid body, $\mathbf{F} = \mathbf{T} = \mathbf{0}$. The converse, however, does not follow, since it is possible to imagine that the interactions with the environment cancel each other.

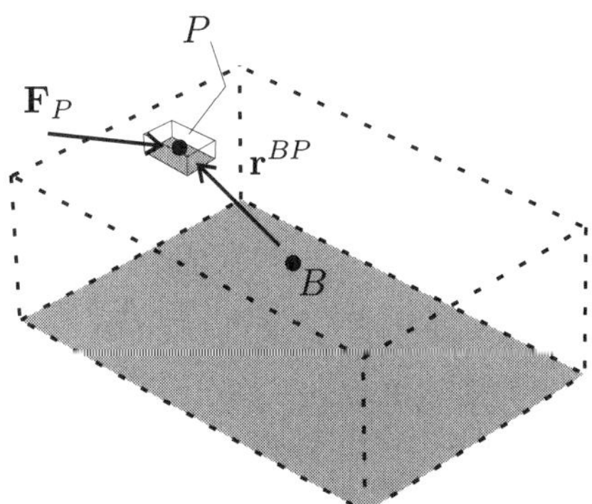


If a small volume of matter centered at a point $P$ of a rigid body experiences a net interaction with its environment, we say that a force $\mathbf{F}_P$ is applied at the point $P$. Experience shows you that a force $\mathbf{F}_P$

applied at a point $P$ results in a net force **and a net torque** on the rigid body. Indeed, we find

$$\mathbf{F} = \mathbf{F}_p \text{ and } \mathbf{T} = \mathbf{r}^{BP} \times \mathbf{F}_p,$$

where $B$ is the center of mass of the rigid body.

Moreover, if forces $\mathbf{F}_P$ and $\mathbf{F}_Q$ are applied at two points $P$ and $Q$ of the rigid body, then experience shows that

$$\mathbf{F} = \mathbf{F}_P + \mathbf{F}_Q \text{ and } \mathbf{T} = \mathbf{r}^{BP} \times \mathbf{F}_P + \mathbf{r}^{BQ} \times \mathbf{F}_Q.$$

As an example, if $\mathbf{F}_Q = -\mathbf{F}_P$ then there is no net force acting on the rigid body. The net torque

$$\begin{aligned} \mathbf{T} &= \mathbf{r}^{BP} \times \mathbf{F}_P - \mathbf{r}^{BQ} \times \mathbf{F}_P \\ &= \left(\mathbf{r}^{BP} - \mathbf{r}^{BQ}\right) \times \mathbf{F}_P \\ &= \mathbf{r}^{QP} \times \mathbf{F}_P, \end{aligned}$$

depends only on the relative position of the points $P$ and $Q$ and not on their absolute position relative to the center of mass. Such a pair of forces is called a *couple*.

**Illustration 11.8**

Suppose that the interactions of a rigid body with its environment are captured by a force

$$\mathbf{F} = -Mg\mathbf{n}_3$$

and torque

$$\mathbf{T} = \mathbf{0},$$

where $g$ is some constant and $\mathbf{n}_3$ is the third basis vector of the reference triad of an inertial reference frame $\mathcal{N}$. It follows that

$$\frac{{}^n d\, {}^{\mathcal{N}}\mathbf{p}}{dt} = \mathbf{F} = -Mg\mathbf{n}_3$$

and

$$\frac{{}^n d\, {}^{\mathcal{N}}\mathbf{h}}{dt} = \mathbf{T} = \mathbf{0}.$$

The dynamic differential equations corresponding to the angular momentum equation are identical to those derived in Illustration 11.5. The dynamic differential equations corresponding to the linear momentum equation become

$$\begin{aligned} \dot{u}_1 &= 0, \\ \dot{u}_2 &= 0, \\ \dot{u}_3 &= -g, \end{aligned}$$

which can be solved to yield

$$u_1(t) = u_1(0), \ u_2(t) = u_2(0), \text{ and } u_3(t) = -gt + u_3(0).$$

From the kinematic differential equations derived previously, we conclude that

$$\begin{aligned} q_1(t) &= u_1(0)\,t + q_1(0), \\ q_2(t) &= u_2(0)\,t + q_2(0), \\ q_3(t) &= -\frac{gt^2}{2} + u_3(0)\,t + q_3(0). \end{aligned}$$

The path followed by the center of mass of the rigid body relative to the inertial observer deviates from a straight line by an amount proportional to the square of the time variable. It follows that the path is given by a *parabola.*

The parabolic path followed by the center of mass of the rigid body in the illustration is independent of the mass distribution of the rigid body. The same path is followed by all objects with the same initial position and initial linear velocity. As an example, all objects with zero initial linear velocity relative to the inertial observer will move in the opposite direction to the $\mathbf{n}_3$ vector in such a way that the distance from the starting position increases by a factor of four as the elapsed time doubles.

Experiments by the Italian renaissance scientist Galileo Galilee showed that this type of motion is exhibited by objects released from rest near the Earth's surface. Thus, if there exists an inertial observer stationary relative to the Earth's surface, this implies that objects near the Earth's surface experience a net interaction in the local downward direction that is proportional to the mass of the objects.

**Illustration 11.9**

We may simulate the motion of the rigid body in the previous illustration using MAMBO. Continuing with the same MAMBO session as in the previous section, we may redefine the dynamic differential equations and re-invoke the `MotionOutput` procedure.

```
> dde:={
> seq((DiffTime(p,n) &-- MakeTranslations(n,0,0,-M*g))
> &oo MakeTranslations(n,i)=0,i=1..3),
> seq(DiffTime(h,n)
> &oo MakeTranslations(b,i)=0,i=1..3)}:
> MotionOutput(ode=kde union dde,
> states=[seq(cat(q,i),i=1..6),seq(cat(u,i),i=1..6)],
> parameters=[M,seq(seq(cat(Inertia,i,j),
> i=1..3),j=1..3),g]);
```

Here, we have included the union of the kinematic and dynamic differential equations in the `ode` block of the MAMBO motion description in order to ensure that there are as many differential equations as there are states.

---

Suppose that the interactions of a rigid body with its environment are captured by a force

$$\mathbf{F} = k\left(l_0 - \left\|\mathbf{r}^{NP}\right\|\right)\frac{\mathbf{r}^{NP}}{\left\|\mathbf{r}^{NP}\right\|}$$

and a torque

$$\mathbf{T} = \mathbf{r}^{BP} \times \mathbf{F},$$

where $k$ and $l_0$ are two constants and $P$ is some point on the rigid body, such that

$$\mathbf{r}^{BP} = b\begin{pmatrix} p_1 \\ p_2 \\ p_3 \end{pmatrix}.$$

This formulation is a close approximation to the interactions between a rigid body and a spring attached at one end to the reference point of the inertial observer and at the other end to the point $P$.

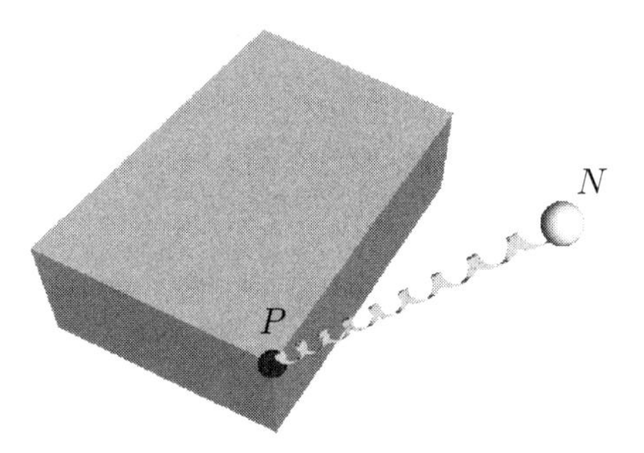


We may simulate the motion of the rigid body under these interactions using MAMBO. Continuing with the same MAMBO session as in the previous section, we may redefine the dynamic differential equations and re-invoke the `MotionOutput` procedure.

```
> DeclarePoints(P):
> DefinePoints([B,P,b,p1,p2,p3]):
> rNP:=FindTranslation(N,P):
> rBP:=FindTranslation(B,P):
```

```
> dde:={
> seq((DiffTime(p,n) &--
> ((k*(l0/VectorLength(rNP)-1)) &** rNP)) &oo
> MakeTranslations(n,i)=0,i=1..3),
> seq((DiffTime(h,n) &-- (rBP &xx
> ((k*(l0/VectorLength(rNP)-1)) &** rNP))) &oo
> MakeTranslations(b,i)=0,i=1..3)}:
> MotionOutput(ode=kde union dde,
> states=[seq(cat(q,i),i=1..6),seq(cat(u,i),i=1..6)],
> parameters=[M,seq(seq(cat(Inertia,i,j),
> i=1..3),j=1..3),g,p1,p2,p3,k,l0]);
```

### 11.1.7 Non-inertial Observers

Suppose that $\mathcal{N}$ is an inertial observer and $\mathcal{A}$ is a non-inertial observer. From a previous chapter, we recall that

$$\frac{{}^{\mathcal{N}}dB}{dt} = \frac{{}^{\mathcal{A}}dB}{dt} + \frac{{}^{\mathcal{N}}dA}{dt} + {}^{n}\boldsymbol{\omega}^{a} \times \mathbf{r}^{AB},$$

where $B$ is a point on a rigid body, $A$ is the reference point of $\mathcal{A}$, and $n$ and $a$ are the reference triads of $\mathcal{N}$ and $\mathcal{A}$, respectively. If we differentiate with respect to time relative to the $n$ triad on both sides of this relation, we obtain

$$\begin{aligned}
\frac{{}^{\mathcal{N}}d^2B}{dt^2} &= \frac{{}^{n}d}{dt}\left(\frac{{}^{\mathcal{A}}dB}{dt} + \frac{{}^{\mathcal{N}}dA}{dt} + {}^{n}\boldsymbol{\omega}^{a} \times \mathbf{r}^{AB}\right) \\
&= \frac{{}^{\mathcal{A}}d^2B}{dt^2} + {}^{n}\boldsymbol{\omega}^{a} \times \frac{{}^{\mathcal{A}}dB}{dt} + \frac{{}^{\mathcal{N}}d^2A}{dt^2} + \frac{{}^{n}d\,{}^{n}\boldsymbol{\omega}^{a}}{dt} \times \mathbf{r}^{AB} \\
&\quad + {}^{n}\boldsymbol{\omega}^{a} \times \left(\frac{{}^{\mathcal{A}}dB}{dt} + {}^{n}\boldsymbol{\omega}^{a} \times \mathbf{r}^{AB}\right).
\end{aligned}$$

But if $B$ is the center of mass of a rigid body, then this implies that

$$\begin{aligned}
\mathbf{F} &= \frac{{}^{n}d\,{}^{\mathcal{N}}\mathbf{p}}{dt} \\
&= M\,\frac{{}^{\mathcal{N}}d^2B}{dt^2} \\
&= M\,\frac{{}^{\mathcal{A}}d^2B}{dt^2} + M\,\frac{{}^{\mathcal{N}}d^2A}{dt^2} + 2M\,{}^{n}\boldsymbol{\omega}^{a} \times \frac{{}^{\mathcal{A}}dB}{dt} \\
&\quad + M\,{}^{n}\boldsymbol{\omega}^{a} \times \left({}^{n}\boldsymbol{\omega}^{a} \times \mathbf{r}^{AB}\right) + M\,\frac{{}^{n}d\,{}^{n}\boldsymbol{\omega}^{a}}{dt} \times \mathbf{r}^{AB},
\end{aligned}$$

i.e.,

$$\begin{aligned}
\frac{{}^{a}d\,{}^{\mathcal{A}}\mathbf{p}}{dt} &= \mathbf{F} - M\,\frac{{}^{\mathcal{N}}d^2A}{dt^2} - 2M\,{}^{n}\boldsymbol{\omega}^{a} \times \frac{{}^{\mathcal{A}}dB}{dt} \\
&\quad - M\,{}^{n}\boldsymbol{\omega}^{a} \times \left({}^{n}\boldsymbol{\omega}^{a} \times \mathbf{r}^{AB}\right) - M\,\frac{{}^{n}d\,{}^{n}\boldsymbol{\omega}^{a}}{dt} \times \mathbf{r}^{AB}.
\end{aligned}$$

If we wish to describe the motion relative to the non-inertial observer using Newton's second law, we will find that there are forces acting on the rigid body that cannot be explained in terms of interactions with the environment. From the above expression, we see that these terms originate in the relative motion of the two observers. The term

$$-M \, {}^{\mathcal{N}}\frac{d^2 A}{dt^2}$$

gives a non-zero contribution if the reference point of $\mathcal{A}$ is accelerated relative to $\mathcal{N}$. If the reference triad of $\mathcal{A}$ is rotating relative to $\mathcal{N}$, the *Coriolis* term

$$-2M \, {}^{n}\boldsymbol{\omega}^{a} \times {}^{\mathcal{A}}\frac{dB}{dt}$$

gives a non-zero contribution that is perpendicular to the velocity of the point $B$ relative to $\mathcal{A}$. Similarly, the *centrifugal* term

$$-M \, {}^{n}\boldsymbol{\omega}^{a} \times \left({}^{n}\boldsymbol{\omega}^{a} \times \mathbf{r}^{AB}\right)$$

gives a non-zero contribution that is proportional to the distance from the point $B$ to the instantaneous axis of rotation of $\mathcal{A}$ relative to $\mathcal{N}$. Finally, the term

$$-M \, {}^{n}\frac{d \, {}^{n}\boldsymbol{\omega}^{a}}{dt} \times \mathbf{r}^{AB}$$

gives a non-zero contribution if the angular velocity of $\mathcal{A}$ relative to $\mathcal{N}$ is time-dependent relative to $\mathcal{N}$.

Thus, if $\mathcal{A}$ knows of the time-dependence of its configuration relative to the inertial observer $\mathcal{N}$, then $\mathcal{A}$ can add the correction terms to Newton's second law and proceed to analyze the motion of the rigid body. If, instead, $\mathcal{A}$ is unaware of it being non-inertial, then it will conclude that its inability to predict the motion of the rigid body without including additional forces is due to yet-undiscovered interactions between the rigid body and its environment. This begs the question that was raised in an earlier section as to whether $\mathcal{A}$ can determine with any certainty whether it is inertial or not. As intricate as the entanglement of the notions of an inertial observer and an isolated rigid body were, as intricate is the connection between an inertial observer and the notion of forces and interactions between a rigid body and its environment.

### 11.1.8 Action – Reaction

Interactions between real-world objects come at a cost. When a force is applied to a rigid body, there results a change in its linear momentum. **Newton's third law of motion** establishes a strict bookkeeping, whereby that change is balanced by an opposite change in the total linear momentum of the objects in the rigid body's environment. In particular,

> If the interaction between two objects results in a force $\mathbf{F}$ on one of the objects, then the other object experiences a force $-\mathbf{F}$.

When two bodies are in contact, the force imposed by one body on the other at the contact point is opposed by a force of equal magnitude but opposite direction imposed by the second body on the first. If $B_1$ and $B_2$ denote the centers of mass of the two bodies and the corresponding contact points are $P_1$ and $P_2$, then a force $\mathbf{F}$ applied to the second body at $P_2$ results in a torque

$$\mathbf{T}_2 = \mathbf{r}^{B_2 P_2} \times \mathbf{F}$$

on the second body and a torque

$$\mathbf{T}_1 = \mathbf{r}^{B_1 P_1} \times (-\mathbf{F}) \neq -\mathbf{T}_2$$

on the first body. Thus, the change in angular momentum of the second body is not generally balanced by an opposite change in angular momentum of the first body.

### 11.1.9 Constraint Forces

The motion of an isolated rigid body is entirely unconstrained. In contrast, constraints on the motion of the rigid body must correspond to the imposition of forces and torques that restrict the allowable motions of the body. Such forces and torques are called *constraint forces* and *constraint torques.*

Suppose that the motion of a rigid body is constrained, such that its center of mass stays at a fixed distance $l$ from the reference point $N$ of an inertial reference frame $\mathcal{N}$. In particular, let

$$\mathbf{r}^{NB} = n \begin{pmatrix} q_1 \\ q_2 \\ q_3 \end{pmatrix},$$

where $B$ denotes the center of mass of the rigid body and $n$ is the reference triad of $\mathcal{N}$. It follows that

$$\frac{{}^{\mathcal{N}}dB}{dt} = n \begin{pmatrix} \dot{q}_1 \\ \dot{q}_2 \\ \dot{q}_3 \end{pmatrix}.$$

The constraint on the position of the center of mass implies that the components of the velocity of $B$ relative to $\mathcal{N}$ are not independent. Indeed, since the distance from $N$ to $B$ is constant, we must have

$$\mathbf{r}^{NB} \bullet \mathbf{r}^{NB} = l^2 \Leftrightarrow q_1^2 + q_2^2 + q_3^2 = l^2.$$

The corresponding motion constraint is

$$2\,\frac{{}^{\mathcal{N}}dB}{dt}\bullet\mathbf{r}^{NB}=0\Leftrightarrow 2q_1\dot{q}_1+2q_2\dot{q}_2+2q_3\dot{q}_3=0.$$

Following the methodology introduced in previous chapters, we could introduce two independent velocity coordinates and derive the corresponding kinematic differential equations, guaranteeing that the motion constraint would be satisfied for all time.

Here, we will take a different approach. Let's, instead, treat the rigid body as if it were free, but acted upon by a constraint force

$$\mathbf{F}_c=n\begin{pmatrix}F_{c1}\\F_{c2}\\F_{c3}\end{pmatrix},$$

whose sole purpose is to ensure that the motion constraint is satisfied. Introduce a full complement of independent velocity coordinates

$$\frac{{}^{\mathcal{N}}dB}{dt}=n\begin{pmatrix}u_1\\u_2\\u_3\end{pmatrix},$$

from which we find the kinematic differential equations

$$\begin{aligned}\dot{q}_1&=u_1,\\\dot{q}_2&=u_2,\\\dot{q}_3&=u_3.\end{aligned}$$

The linear momentum of the rigid body relative to the inertial observer now becomes

$${}^{\mathcal{N}}\mathbf{p}=n\begin{pmatrix}Mu_1\\Mu_2\\Mu_3\end{pmatrix}$$

and Newton's second law of motion

$$\frac{{}^{n}d\,{}^{\mathcal{N}}\mathbf{p}}{dt}=\mathbf{F}_c$$

yields the dynamic differential equations

$$\begin{aligned}M\dot{u}_1&=F_{c1},\\M\dot{u}_2&=F_{c2},\\M\dot{u}_3&=F_{c3}.\end{aligned}$$

The question is what form the coordinates $F_{ci}$ of the constraint force relative to the reference triad $n$ must take in order to ensure that the solution to the kinematic and dynamic differential equations is an allowable motion.

Recall that any set of functions $q_1(t)$, $q_2(t)$, $q_3(t)$ that satisfies the motion constraint

$$2q_1\dot{q}_1 + 2q_2\dot{q}_2 + 2q_3\dot{q}_3 = 0$$

and for which the set of values $q_1(t_0)$, $q_2(t_0)$, and $q_3(t_0)$ satisfies the configuration constraint

$$q_1^2 + q_2^2 + q_3^2 = l^2$$

will satisfy the configuration constraint for all time.

Now, consider the time derivative of the motion constraint[2]

$$2\dot{q}_1^2 + 2q_1\ddot{q}_1 + 2\dot{q}_2^2 + 2q_2\ddot{q}_2 + 2\dot{q}_3^2 + 2q_3\ddot{q}_3 = 0$$

called an *acceleration constraint.* In analogy with the previous statement, any set of functions $q_1(t)$, $q_2(t)$, $q_3(t)$ that satisfies the acceleration constraint and for which the set of values $q_1(t_0)$, $q_2(t_0)$, $q_3(t_0)$, $\dot{q}_1(t_0)$, $\dot{q}_2(t_0)$, and $\dot{q}_3(t_0)$ satisfies the motion constraint will satisfy the motion constraint for all time.

In conclusion, any set of functions $q_1(t)$, $q_2(t)$, $q_3(t)$, $u_1(t)$, $u_2(t)$, and $u_3(t)$ that satisfies the acceleration constraint and the kinematic differential equations, and for which the set of values $q_1(t_0)$, $q_2(t_0)$, $q_3(t_0)$, $u_1(t_0)$, $u_2(t_0)$, and $u_3(t_0)$ satisfies the motion and configuration constraints will satisfy the configuration constraint for all time. In particular, using the kinematic and dynamic differential equation, the acceleration constraint becomes

$$\begin{aligned} 0 &= 2u_1^2 + 2q_1\frac{F_{c1}}{M} + 2u_2^2 + 2q_2\frac{F_{c2}}{M} + 2u_3^2 + 2q_3\frac{F_{c3}}{M} \\ &= \frac{2}{M}\left(M\left\|\frac{{}^{\mathcal{N}}dB}{dt}\right\|^2 + \mathbf{r}^{NB} \bullet \mathbf{F}_c\right). \end{aligned}$$

If this condition is satisfied by the coordinates of the constraint force, then the acceleration constraint is automatically satisfied.

To uniquely determine the constraint force, it is necessary to impose additional conditions on its coordinates. Recall that the kinetic energy relative to an inertial observer is constant for an isolated rigid body. Here, the body is no longer isolated. Nevertheless, it appears reasonable to expect that the constraint force will be unable to change the kinetic energy of the rigid body. Since

$$\begin{aligned} \frac{d\,{}^{\mathcal{N}}K}{dt} &= \frac{{}^{n}d\,{}^{\mathcal{N}}\mathbf{p}}{dt} \bullet \frac{{}^{\mathcal{N}}dB}{dt} + \frac{{}^{n}d\,{}^{\mathcal{N}}\mathbf{h}}{dt} \bullet {}^{n}\boldsymbol{\omega}^{b} \\ &= \mathbf{F}_c \bullet \frac{{}^{\mathcal{N}}dB}{dt}, \end{aligned}$$

[2]The double dots refer to a second derivative with respect to time.

it follows that the kinetic energy remains unchanged only if $\mathbf{F}_c$ is perpendicular to ${}^{\mathcal{N}}\frac{dB}{dt}$ for arbitrary velocities. From the motion constraint, we recall that the velocity is constrained to be perpendicular to $\mathbf{r}^{NB}$. It follows that $\mathbf{F}_c$ must be parallel to $\mathbf{r}^{NB}$, i.e.,

$$\mathbf{F}_c = F_c \frac{\mathbf{r}^{NB}}{\|\mathbf{r}^{NB}\|}.$$

Using the acceleration constraint, it follows that

$$F_c = \frac{\mathbf{r}^{NB} \bullet \mathbf{F}_c}{\|\mathbf{r}^{NB}\|} = -M \left\| {}^{\mathcal{N}}\frac{dB}{dt} \right\|^2 / \left\|\mathbf{r}^{NB}\right\|.$$

## 11.2 Dynamics

The analysis of the possible motions of a multibody mechanism affected by forces and torques is known as *dynamics*. Rather than impose a certain motion on a mechanism as was done in previous chapters, the goal of a dynamical analysis is to understand the motion that results from the imposition of forces. These forces may be due to fundamental physical processes in the universe or a result of mechanical actuation, e.g., using motors, muscles, and so on.

The laws of motion that were presented in the previous section have been found to describe the dynamics of real-world mechanisms to a great degree of accuracy and are therefore universally accepted when studying a majority of everyday mechanisms. Their failure to adequately address physical processes involving extreme length and timescales (for which the theories of quantum mechanics and relativity need to be invoked) in no way reduces their importance to the engineer and physicist.

A successful application of the laws of Newtonian mechanics relies entirely on the ability to formulate expressions for the forces and torques affecting real-world objects. To formulate such expressions is the role of physical modeling. This is no longer an exact science. Instead, experimental observations and theoretical considerations are combined to suggest possible mathematical models for the interactions between objects. These models can be validated only by comparison of the predicted dynamics with actual experimental observations. Improvements in the models typically follow upon such comparisons, and the process is repeated until a desired degree of accuracy has been achieved.

The combined kinematic and dynamic differential equations governing the motion of a multibody mechanism are generally too complicated to be solved analytically. Instead, we resort to any one of a large number of approximate numerical algorithms. While numerical approximations can be computed through hand calculations, it is only with the advent of high-speed computing that it has become possible to implement such numerical algorithms with a large degree of accuracy for simulations over long periods of time.

But with numerical prediction does not necessarily come cognition. As scientists, we strive beyond the ability to simulate multi-degree-of-freedom systems with specific initial values for the configuration coordinates and the independent velocity coordinates. Instead, we seek to make qualitative and quantitative statements regarding the behavior of these systems under a variety of different initial conditions and a variety of different force and torque descriptions.

### 11.2.1 Central Forces

Suppose that $\mathcal{N}$ is an inertial observer with reference point $N$ and reference triad $n$. Denote by $B$ the center of mass of a rigid body of total mass $M$ and suppose that the rigid body's motion is governed by a force

$$\mathbf{F} = f\left(\left\|\mathbf{r}^{NB}\right\|\right)\mathbf{r}^{NB},$$

where $f$ is some arbitrary function, and torque $\mathbf{T} = \mathbf{0}$. The rigid body is said to be affected by a *central force*. Newton's second law of motion yields

$$\frac{{}^n d\,{}^{\mathcal{N}}\mathbf{p}}{dt} = M\frac{{}^n d^2\mathbf{r}^{NB}}{dt^2} = f\left(\left\|\mathbf{r}^{NB}\right\|\right)\mathbf{r}^{NB}.$$

The force acting on the rigid body is always parallel to the position vector from the reference point of the inertial observer to the center of mass of the rigid body. Moreover, its magnitude depends only on the distance between $N$ and $B$.

Now, consider the vector quantity

$$\mathbf{H} = \mathbf{r}^{NB} \times \frac{{}^n d\mathbf{r}^{NB}}{dt}.$$

Then,

$$\begin{aligned}\frac{{}^n d\mathbf{H}}{dt} &= \frac{{}^n d}{dt}\left(\mathbf{r}^{NB} \times \frac{{}^n d\mathbf{r}^{NB}}{dt}\right)\\ &= \frac{{}^n d\mathbf{r}^{NB}}{dt} \times \frac{{}^n d\mathbf{r}^{NB}}{dt} + \frac{f\left(\left\|\mathbf{r}^{NB}\right\|\right)}{M}\mathbf{r}^{NB} \times \mathbf{r}^{NB}\\ &= \mathbf{0}.\end{aligned}$$

It follows that $\mathbf{H}$ is constant relative to the inertial observer. Since

$$\mathbf{r}^{NB} \bullet \mathbf{H} = 0,$$

it follows that $\mathbf{r}^{NB}$ lies in the plane perpendicular to $\mathbf{H}$ for all time, i.e., that the center of mass of the rigid body stays in a plane through the reference point $N$ perpendicular to the vector $\mathbf{H}$. Given an initial position and initial velocity of the point $B$ relative to $\mathcal{N}$, we can compute $\mathbf{H}$ and, consequently, find the plane in which the point $B$ will remain.

**Illustration 11.10**

Suppose that the force acting on the rigid body is given by

$$\mathbf{F} = -k\mathbf{r}^{NB},$$

i.e., that the magnitude of the force is proportional to the distance between $N$ and $B$. Suppose that the reference triad $n$ is chosen such that $\mathbf{H}$ is parallel to $\mathbf{n}_3$, i.e., such that

$$\mathbf{H} = \mathbf{n}\begin{pmatrix} 0 \\ 0 \\ h \end{pmatrix},$$

where $h$ is constant throughout the motion.

Now, consider the quantities

$$E_i = \frac{M}{2}\left({}^n\frac{d\mathbf{r}^{NB}}{dt} \bullet \mathbf{n}_i\right)^2 + \frac{k}{2}\left(\mathbf{r}^{NB} \bullet \mathbf{n}_i\right)^2, \; i = 1, 2.$$

Then,

$$\begin{aligned} \frac{dE_i}{dt} &= M\left({}^n\frac{d^2\mathbf{r}^{NB}}{dt^2} \bullet \mathbf{n}_i\right)\left({}^n\frac{d\mathbf{r}^{NB}}{dt} \bullet \mathbf{n}_i\right) \\ &\quad + k\left({}^n\frac{d\mathbf{r}^{NB}}{dt} \bullet \mathbf{n}_i\right)\left(\mathbf{r}^{NB} \bullet \mathbf{n}_i\right) \\ &= -k\left(\mathbf{r}^{NB} \bullet \mathbf{n}_i\right)\left({}^n\frac{d\mathbf{r}^{NB}}{dt} \bullet \mathbf{n}_i\right) \\ &\quad + k\left({}^n\frac{d\mathbf{r}^{NB}}{dt} \bullet \mathbf{n}_i\right)\left(\mathbf{r}^{NB} \bullet \mathbf{n}_i\right) \\ &= 0, \end{aligned}$$

i.e., $E_1$ and $E_2$ are constant relative to the inertial observer.

Since $\mathbf{r}^{NB}$ is perpendicular to $\mathbf{H}$, we have

$$\mathbf{r}^{NB} = n\begin{pmatrix} q_1 \\ q_2 \\ 0 \end{pmatrix}, \; {}^n\frac{d\mathbf{r}^{NB}}{dt} = n\begin{pmatrix} \dot{q}_1 \\ \dot{q}_2 \\ 0 \end{pmatrix}.$$

It follows that

$$h = q_1\dot{q}_2 - q_2\dot{q}_1$$

and

$$E_1 = \frac{M}{2}\dot{q}_1^2 + \frac{k}{2}q_1^2, \; E_2 = \frac{M}{2}\dot{q}_2^2 + \frac{k}{2}q_2^2.$$

Eliminating $\dot{q}_1$ and $\dot{q}_2$ from these equations yields

$$h = \sqrt{\frac{2}{M}} \left( q_1 s_2 \sqrt{E_2 - \frac{k}{2} q_2^2} - q_2 s_1 \sqrt{E_1 - \frac{k}{2} q_1^2} \right),$$

where $s_1$ and $s_2$ equal plus or minus 1. Solving this equation for $q_1$ and reorganizing the result then yields

$$\frac{\left(4E_2 q_1 \pm 2q_2\sqrt{4E_1E_2 - kMh^2}\right)^2}{8ME_2h^2} + \frac{kq_2^2}{2E_2} = 1,$$

i.e., the equation of an ellipse centered at $N$.

Suppose that the force acting on the rigid body is given by

$$\mathbf{F} = -\frac{k}{\|\mathbf{r}^{NB}\|^3}\mathbf{r}^{NB},$$

i.e., that the magnitude of the force is inversely proportional to the square of the distance between $N$ and $B$. Suppose that the reference triad $n$ is chosen such that $\mathbf{H}$ is parallel to $\mathbf{n}_3$, i.e., such that

$$\mathbf{H} = n \begin{pmatrix} 0 \\ 0 \\ h \end{pmatrix},$$

where $h$ is constant throughout the motion.

Now, consider the quantity

$$E = \frac{M}{2} \frac{{}^n d\mathbf{r}^{NB}}{dt} \bullet \frac{{}^n d\mathbf{r}^{NB}}{dt} - \frac{k}{\sqrt{\mathbf{r}^{NB} \bullet \mathbf{r}^{NB}}}.$$

Then,

$$\begin{aligned} \frac{dE}{dt} &= M \frac{{}^n d^2\mathbf{r}^{NB}}{dt^2} \bullet \frac{{}^n d\mathbf{r}^{NB}}{dt} + k \frac{\frac{{}^n d\mathbf{r}^{NB}}{dt} \bullet \mathbf{r}^{NB}}{\|\mathbf{r}^{NB}\|^3} \\ &= -\frac{k}{\|\mathbf{r}^{NB}\|^3}\mathbf{r}^{NB} \bullet \frac{{}^n d\mathbf{r}^{NB}}{dt} + k \frac{\frac{{}^n d\mathbf{r}^{NB}}{dt} \bullet \mathbf{r}^{NB}}{\|\mathbf{r}^{NB}\|^3} \\ &= 0, \end{aligned}$$

i.e., $E$ is constant throughout the motion.

Similarly, consider the vector quantity

$$\mathbf{R} = M\mathbf{H} \times \frac{{}^n d\mathbf{r}^{NB}}{dt} + k \frac{\mathbf{r}^{NB}}{\sqrt{\mathbf{r}^{NB} \bullet \mathbf{r}^{NB}}}.$$

Since $\mathbf{H}$ is constant, it follows that

$$\begin{aligned}
\frac{{}^{n}d\mathbf{R}}{dt} &= M\mathbf{H}\times\frac{{}^{n}d^{2}\mathbf{r}^{NB}}{dt^{2}}+k\frac{\left[\begin{array}{c}\frac{{}^{n}d\mathbf{r}^{NB}}{dt}\left(\mathbf{r}^{NB}\bullet\mathbf{r}^{NB}\right)\\ -\mathbf{r}^{NB}\left(\frac{{}^{n}d\mathbf{r}^{NB}}{dt}\bullet\mathbf{r}^{NB}\right)\end{array}\right]}{\left\|\mathbf{r}^{NB}\right\|^{3}}\\
&= -k\frac{\left[\begin{array}{c}\left(\mathbf{r}^{NB}\times\frac{{}^{n}d\mathbf{r}^{NB}}{dt}\right)\times\mathbf{r}^{NB}-\frac{{}^{n}d\mathbf{r}^{NB}}{dt}\left\|\mathbf{r}^{NB}\right\|^{2}\\ +\mathbf{r}^{NB}\left(\frac{{}^{n}d\mathbf{r}^{NB}}{dt}\bullet\mathbf{r}^{NB}\right)\end{array}\right]}{\left\|\mathbf{r}^{NB}\right\|^{3}}\\
&= \mathbf{0},
\end{aligned}$$

where we have used the identity

$$(\mathbf{c}\times\mathbf{b})\times\mathbf{a}=\mathbf{a}\times(\mathbf{b}\times\mathbf{c})=\mathbf{b}\,(\mathbf{a}\bullet\mathbf{c})-\mathbf{a}\,(\mathbf{b}\bullet\mathbf{c})$$

which was shown in Exercise 3.74. From the definitions, we see that

$$\mathbf{R}\bullet\mathbf{H}=\mathbf{0},$$

i.e.,

$$\mathbf{R}=n\begin{pmatrix} r_1\\ r_2\\ 0\end{pmatrix},$$

where $r_1$ and $r_2$ are constant throughout the motion.

Since $\mathbf{r}^{NB}$ is perpendicular to $\mathbf{H}$, we have

$$\mathbf{r}^{NB}=n\begin{pmatrix}\rho\cos\theta\\ \rho\sin\theta\\ 0\end{pmatrix},\quad \frac{{}^{n}d\mathbf{r}^{NB}}{dt}=n\begin{pmatrix}\dot{\rho}\cos\theta-\dot{\theta}\rho\sin\theta\\ \dot{\rho}\sin\theta+\dot{\theta}\rho\cos\theta\\ 0\end{pmatrix},$$

where $\rho$ and $\theta$ are polar coordinates in the $\mathbf{n}_1$, $\mathbf{n}_2$ plane. It follows that

$$\begin{aligned}
h &= \rho^{2}\dot{\theta},\\
E &= \frac{M}{2}\left(\rho^{2}\dot{\theta}^{2}+\dot{\rho}^{2}\right)-\frac{k}{\rho},\\
r_1 &= -M\rho^{2}\dot{\theta}\dot{\rho}\sin\theta-M\rho^{3}\dot{\theta}^{2}\cos\theta+k\cos\theta,
\end{aligned}$$

and

$$r_2=M\rho^{2}\dot{\theta}\dot{\rho}\cos\theta-M\rho^{3}\dot{\theta}^{2}\sin\theta+k\sin\theta.$$

Eliminating $\dot{\rho}$ and $\dot{\theta}$ from these equations yields

$$E=\frac{1}{2}\frac{r_1^{2}+r_2^{2}-k^{2}}{Mh^{2}}$$

and

$$\rho = \frac{h^2 M}{k - r_1 \cos\theta - r_2 \sin\theta},$$

i.e., the equation of a *conical section* (i.e., the curve created when slicing a cone with a plane) with focus at $N$. Depending on the values of $E$, $h$, $k$, and $M$, this curve is either an ellipse, a parabola, or a hyperbola.

### 11.2.2 The Two-body Problem

Let $B_1$ and $B_2$ denote the centers of mass of two rigid bodies of mass $M_1$ and $M_2$. *Newton's law of gravity* states that, to the lowest approximation, the two rigid bodies interact by means of a mutually attracting force that is proportional in magnitude to the masses of the rigid bodies and inversely proportional to the square of the distance between $B_1$ and $B_2$. Specifically, denote by $\mathbf{F}_1$ and $\mathbf{F}_2$ the forces on the two bodies. Then,

$$\mathbf{F}_1 = -G\frac{M_1 M_2}{\|\mathbf{r}^{B_1B_2}\|^3}\mathbf{r}^{B_2B_1} \text{ and } \mathbf{F}_2 = -G\frac{M_1 M_2}{\|\mathbf{r}^{B_2B_1}\|^3}\mathbf{r}^{B_1B_2} = -\mathbf{F}_1,$$

where $G$ is some constant of nature. In the absence of other interactions, Newton's second law of motion implies that

$$M_1 \frac{{}^n d^2\mathbf{r}^{NB_1}}{dt^2} = -G\frac{M_1 M_2}{\|\mathbf{r}^{B_2B_1}\|^3}\mathbf{r}^{B_2B_1},$$
$$M_2 \frac{{}^n d^2\mathbf{r}^{NB_2}}{dt^2} = -G\frac{M_1 M_2}{\|\mathbf{r}^{B_1B_2}\|^3}\mathbf{r}^{B_1B_2},$$

where $n$ and $N$ are the reference triad and reference point of an inertial observer $\mathcal{N}$.

Now, consider a point $P$, such that

$$\mathbf{r}^{NP} = \frac{M_1\mathbf{r}^{NB_1} + M_2\mathbf{r}^{NB_2}}{M_1 + M_2}.$$

Using the dynamic differential equations, it follows that

$$\frac{{}^n d^2\mathbf{r}^{NP}}{dt^2} = \mathbf{0},$$

i.e., the point $P$ moves with constant velocity relative to the inertial observer. Let $\mathcal{N}^*$ be an observer with reference point $N^*$ at $P$ and reference triad $n^*$ coinciding with $n$. Then, $\mathcal{N}^*$ is also an inertial observer. Since

$$\mathbf{0} = \mathbf{r}^{N^*P} = \frac{M_1\mathbf{r}^{N^*B_1} + M_2\mathbf{r}^{N^*B_2}}{M_1 + M_2},$$

it follows that

$$\mathbf{r}^{B_1 B_2} = \mathbf{r}^{N^* B_2} - \mathbf{r}^{N^* B_1} = \left(1 + \frac{M_2}{M_1}\right) \mathbf{r}^{N^* B_2}$$

and

$$\mathbf{r}^{B_2 B_1} = \mathbf{r}^{N^* B_1} - \mathbf{r}^{N^* B_2} = \left(1 + \frac{M_1}{M_2}\right) \mathbf{r}^{N^* B_1}.$$

It follows that the dynamic differential equations relative to the $\mathcal{N}^*$ inertial observer are

$$M_1 \frac{{}^{n^*}d^2\mathbf{r}^{N^* B_1}}{dt^2} = -G \frac{M_1 M_2}{\left(1 + \frac{M_1}{M_2}\right)^2} \frac{\mathbf{r}^{N^* B_1}}{\|\mathbf{r}^{N^* B_1}\|^3}$$

and

$$M_1 \frac{{}^{n^*}d^2\mathbf{r}^{N^* B_2}}{dt^2} = -G \frac{M_1 M_2}{\left(1 + \frac{M_2}{M_1}\right)^2} \frac{\mathbf{r}^{N^* B_2}}{\|\mathbf{r}^{N^* B_2}\|^3}.$$

By comparison with the equations in the previous section, we conclude that the centers of mass of the two rigid bodies move on conical sections with focus at the reference point $N^*$. For example, as long as the initial velocities of the points $B_1$ and $B_2$ are sufficiently small, the resulting orbits are ellipses.

## 11.3 Exercises

**Exercise 11.1** Show that there exists a matrix ${}^wI^A$, such that

$$\int_V \left(\mathbf{n} \times \mathbf{r}^{AP}\right) \bullet \left(\mathbf{m} \times \mathbf{r}^{AP}\right) \rho(P)\, dV = \left({}^w n\right)^T \, {}^wI^A \, {}^w m,$$

and find a formula for the $[i, j]$-th entry of ${}^wI^A$.

[Hint: Let

$$\mathbf{r}^{AP} = w \begin{pmatrix} {}^w\left(\mathbf{r}^{AP}\right)_1 \\ {}^w\left(\mathbf{r}^{AP}\right)_2 \\ {}^w\left(\mathbf{r}^{AP}\right)_3 \end{pmatrix}$$

and expand the product $\left(\mathbf{n} \times \mathbf{r}^{AP}\right) \bullet \left(\mathbf{m} \times \mathbf{r}^{AP}\right)$. Then compare the result to that obtained from the product

$$\left({}^w n\right)^T \; B \; {}^w m$$

for an arbitrary matrix $B$.]

**Exercise 11.2** Show that the moment of inertia matrix ${}^bI^B$ is symmetric for any choice of reference triad $b$ and reference point $B$.

**Exercise 11.3** Find the moment of inertia matrix about the center of mass for a homogeneous rectangular block relative to a triad with basis vectors parallel to the edges of the block.

**Exercise 11.4** Show that the moment of inertia matrices ${}^aI^Q$ and ${}^bI^Q$ about a point $Q$ relative to two different triads $a$ and $b$ are related by the formula

$$ {}^bI^Q = R_{ba}\ {}^aI^Q R_{ab}. $$

**Solution.** From Exercise 11.1, we recall that

$$ \int_V \left(\mathbf{n} \times \mathbf{r}^{QP}\right) \bullet \left(\mathbf{m} \times \mathbf{r}^{QP}\right) \rho(P)\, dV $$
$$ = \left({}^an\right)^T\ {}^aI^Q\ {}^am $$
$$ = \left({}^bn\right)^T\ {}^bI^Q\ {}^bm, $$

since the left-hand side is independent of the triad. Since

$$ \begin{aligned} {}^an &= R_{ab}\ {}^bn \\ {}^am &= R_{ab}\ {}^bm \end{aligned} $$

we find

$$ \left({}^an\right)^T\ {}^aI^Q\ {}^am $$
$$ = \left(R_{ab}\ {}^bn\right)^T\ {}^aI^Q\ R_{ab}\ {}^bm $$
$$ = \left({}^bn\right)^T R_{ba}\ {}^aI^Q R_{ab}\ {}^bm $$

and the claim follows, since the statement must be true for arbitrary $\mathbf{n}$ and $\mathbf{m}$.

**Exercise 11.5** Show that the moment of inertia matrix of a homogeneous cube about its centroid (i.e., its geometric center) is independent of the triad.

[Hint: Compute the moment of inertia matrix about the centroid relative to a triad whose basis vectors are parallel to the edges of the cube. Then use the result of the previous exercise to find the moment of inertia matrix relative to some other triad.]

**Exercise 11.6** Show that for every point $A$ there exists a triad, relative to which the moment of inertia matrix about $A$ is diagonal. The directions through $A$ spanned by the basis vectors of this triad are called the *principal inertia directions* at $A$. The corresponding diagonal entries of the moment of inertia matrix are called the *principal moments of inertia* at $A$.

**Solution.** Recall from linear algebra that for every symmetric matrix $S$ there exists an orthogonal matrix $V$, such that the matrix product

$$ V^T S V $$

is diagonal with the eigenvalues of $S$ along the diagonal.

Now, consider the moment of inertia matrix ${}^wI^A$ about a point $A$ relative to a triad $w$. From a previous exercise, we recall that ${}^wI^A$ is symmetric. Let $p$ be a new triad, such that $R_{wp}$ equals the corresponding $V$ matrix. Then,

$$ {}^pI^A = R_{pw}\ {}^wI^A R_{wp}, $$

i.e., ${}^pI^A$ is diagonal.

**Exercise 11.7** Use the result of Exercise 11.4 to show that the angular momentum of a rigid body with reference triad $b$ relative to an inertial observer $\mathcal{N}$ is given by

$$ \mathbf{h} = a\ {}^aI^B\ {}^a\left({}^n\boldsymbol{\omega}^b\right), $$

where $a$ is some arbitrary triad. Comment on the advantage of using ${}^bI^B$ in the original definition of the angular momentum.

[Hint: ${}^bI^B$ is a time-independent matrix.]

**Exercise 11.8** Formulate Euler's equations in the triad that was found in Exercise 11.6. Show that these are satisfied by

$$ \begin{aligned} u_4(t) &= const,\ u_5(t) = u_6(t) = 0, \\ u_5(t) &= const,\ u_4(t) = u_6(t) = 0, \end{aligned} $$

and

$$ u_6(t) = const, u_4(t) = u_5(t) = 0. $$

**Exercise 11.9** Consider initial values in the vicinity of any of the three solutions discussed in the previous exercise and use MAMBO to study the subsequent rotational motion of the rigid body.

**Exercise 11.10** Show that

$$\frac{d}{dt}\left(\frac{1}{2}{}^{\mathcal{N}}\mathbf{p} \bullet {}^{\mathcal{N}}\frac{dB}{dt} + \frac{1}{2}{}^{\mathcal{N}}\mathbf{h} \bullet {}^{n}\boldsymbol{\omega}^{b}\right)$$
$$= \frac{{}^{n}d\,{}^{\mathcal{N}}\mathbf{p}}{dt} \bullet {}^{\mathcal{N}}\frac{dB}{dt} + \frac{{}^{n}d\,{}^{\mathcal{N}}\mathbf{h}}{dt} \bullet {}^{n}\boldsymbol{\omega}^{b}$$

**Exercise 11.11** Suppose that there is a force applied to every element of matter in a rigid body whose magnitude is proportional to the amount of mass in the element and whose direction is independent of position in the rigid body. Show that the net force acting on the rigid body is independent of position and orientation of the rigid body and has a magnitude that is proportional to the total mass $M$ of the rigid body. Show that there is no net torque acting on the rigid body.

**Exercise 11.12** Suppose that the center of mass of a rigid body moves along a helical curve with constant linear speed. Find the net force that must be acting on the rigid body.

**Exercise 11.13** Suppose that the center of mass of a rigid body moves along an elliptical curve with linear speed inversely proportional to the distance from the center of the ellipse. Find the net force that must be acting on the rigid body.

**Exercise 11.14** Use MAMBO to visualize the motion of the rigid body in Section 11.1.9. What happens if the initial conditions on the independent velocity coordinates do not satisfy the motion constraint?

**Exercise 11.15** Repeat the derivation of the constraint force for the rigid body in Section 11.1.9 under the assumption that it is affected by an additional force $\mathbf{F}$.

**Exercise 11.16** Use MAMBO to visualize the motion of the rigid body in the previous exercise when $\mathbf{F}$ is a constant force.

**Exercise 11.17** Use MAMBO to visualize the motion of the rigid body in the previous exercise with the imposition of an additional force opposite in direction and proportional in magnitude to the velocity of the center of mass.

**Exercise 11.18** Use MAMBO to visualize the motion of an otherwise free rigid body that is acted upon by a constraint force $\mathbf{F}_c$ that ensures that the center of mass of the rigid body stays on a given helical curve.

## Summary of notation

The Greek letter $\rho$ (*rho*) was used in this chapter to denote the density of a mass distribution.

An upper-case $I$ with a left and a right superscript, such as ${}^{w}I^{A}$, was used in this chapter to denote the moment of inertia matrix of a mass distribution about a point relative to a triad.

A bold-faced lower-case $\mathbf{p}$ with a left superscript was used in this chapter to denote the linear momentum of a rigid body relative to an observer.

A bold-faced lower-case $\mathbf{h}$ with a left superscript was used in this chapter to denote the angular momentum of a rigid body relative to an observer.

An upper-case $K$ with a left superscript was used in this chapter to denote the kinetic energy of a rigid body relative to an observer.

## Summary of terminology

The *mass* of an object is a measure of the amount of matter in the object. (Page 439)

The *density* of a mass distribution at a point is a measure of the average mass per unit volume contained in a small volume around the point. (Page 439)

A mass distribution is *homogeneous* if the density is the same at all interior points. (Page 440)

The *first moment of inertia* of a mass distribution about a straight line through a point is a measure of how the matter is distributed about the point. (Page 440)

The first moment of inertia is zero about any straight line through the *center of mass* of a mass distribution. (Page 441)

The *second moment of inertia* of a mass distribution about two straight lines through a point is a measure of how the matter is distributed about the point. (Page 442)

The *moment of inertia matrix* of a mass distribution about a point relative to a triad consists of the second moments of inertia of the mass distribution about any pair of straight lines through the point that are parallel to the basis vectors of the triad. (Page 443)

The *linear momentum* of a rigid body relative to an observer is the product of the mass of the body with the velocity of the center of mass of the body relative to the observer. (Page 445)

(Page 445) The *angular momentum* of a rigid body relative to an observer is a vector whose matrix representation relative to a body-fixed triad equals the product of the moment of inertia matrix of the body about the center of mass relative to the body-fixed triad and the matrix representation relative to the body-fixed triad of the angular velocity of the rigid body relative to the observer.

(Page 445) A rigid body is *isolated* if it does not interact with its environment.

(Page 445) The linear and angular momenta of an isolated rigid body are constant relative to an *inertial observer.*

(Page 445) *Newton's first law of motion* states that the center of mass of an isolated rigid body moves along a straight line relative to an inertial observer.

(Page 446) The *dynamic differential equations* govern the rate of change of the independent velocity coordinates.

(Page 451) The *kinetic energy* of an isolated rigid body is constant relative to an inertial observer.

(Page 455) *Newton's second law of motion* states that the rate of change of the linear momentum of a rigid body relative to an inertial observer equals the *force* applied to the body.

(Page 455) The rate of change of the angular momentum of a rigid body relative to an inertial observer equals the *torque* applied to the body.

(Page 461) *Newton's third law of motion* states that if the interaction between two rigid bodies results in a force on the first body, then a force of equal magnitude but opposite direction acts on the second body.

(Page 461) Forces and torques that act to sustain a configuration or motion constraint on a rigid body are called *constraint forces* and *constraint torques.*

(Page 451) In the MAMBO toolbox, the procedure `LinearMomentum` computes the linear momentum of a rigid body relative to an observer.

(Page 451) In the MAMBO toolbox, the procedure `AngularMomentum` computes the angular momentum of a rigid body relative to an observer.

# Appendix A

# MAPLE and MATLAB

As becomes evident when treating complicated mechanisms with long chains of linked observers, the intensity of the algebra grows rapidly. Only the simplest of problems are amenable to back-of-the-envelope calculations, and there is a serious risk for errors at any step in more complex situations. In the past several decades, tools have emerged for performing algebraic manipulations using interactive computer environments. Programs such as MAPLE and MATHEMATICA have grown to encompass much of algebra, analysis, and numerics. In addition to a plethora of commands for prepackaged mathematical operations, these programs generally contain a programming environment reminiscent of high-level programming languages such as Pascal or C, allowing the user to add functionality and to automate repetitive actions.

## A.1 General Syntax

The MAMBO toolbox described in the main body of this text is based on a set of procedures written using the MAPLE programming language and compatible with MAPLE V and later versions as well as with MATLAB's extended symbolic toolbox.

**Illustration A.1**

In the sample MAPLE session below, we highlight some fundamental syntactic conventions and basic operations of the MAPLE system. For example, the result of a computation is only revealed if the command line ends with a semicolon, being suppressed if it ends with a colon. If either is missing, the system will not recognize the end of the line.

```
> factor(a^2-b^2);
```

$$(a-b)(a+b)$$

```
> simplify((x^3-y^3)/(x-y));
```

$$x^2 + y\,x + y^2$$

```
> soln:=solve(x^2+p*x+q,x);
```

$$soln := -\frac{1}{2}\,p + \frac{1}{2}\,\sqrt{p^2 - 4\,q},\ -\frac{1}{2}\,p - \frac{1}{2}\,\sqrt{p^2 - 4\,q}$$

```
> simplify(subs(x=soln[1],x^2+p*x+q));
```

$$0$$

```
> diff(x*ln(x)-sin(x)^2,x);
```

$$\ln(x) + 1 - 2\sin(x)\cos(x)$$

```
> int(1/sqrt(1+x^2),x=0..t);
```

$$\ln(t + \sqrt{t^2 + 1})$$

```
> with(linalg):

Warning, new definition for norm

Warning, now definition for trace
> A:=matrix(3,3,[[1,0,0],[0,cos(t),sin(t)],
[0,-sin(t),cos(t)]]);
```

$$A := \begin{bmatrix} 1 & 0 & 0 \\ 0 & \cos(t) & \sin(t) \\ 0 & -\sin(t) & \cos(t) \end{bmatrix}$$

```
> eigenvals(A);
```

$$1,\ \cos(t) + \sqrt{\cos(t)^2 - 1},\ \cos(t) - \sqrt{\cos(t)^2 - 1}$$

```
> simplify(multiply(A,transpose(A)));
```

$$\begin{bmatrix} 1 & 0 & 0 \\ 0 & 1 & 0 \\ 0 & 0 & 1 \end{bmatrix}$$

Placeholders for complicated algebraic expressions are assigned through the assignment operator `:=` as in the example with the matrix `A` above. The statement

```
> A:='A';
```

$$A := A$$

unassigns the placeholder `A`. To unassign all placeholders and global variables, use the `restart` command.

Note the excellent MAPLE help facility, accessible through the menu system or by typing a question mark followed by the command one requires help with, e.g., `?subs`. For the majority of commands, the help window will contain a number of examples that further highlight the grammar and syntax. Take some time to explore MAPLE and familiarize yourself with its functionality.

All MAPLE statements included in this text will run in a MATLAB system with the extended symbolic toolbox, provided that they are entered as `maple('statement')`, where `statement` is one or several syntactically correct MAPLE statements. For example,

```
>> maple('factor(a^2-b^2);')

ans =

(a-b)*(a+b)

>> maple('simplify((x^3-y^3)/(x-y));')

ans =

x^2+y*x+y^2
```

are equivalent to the first two MAPLE statements above. Here, the MAPLE statements are passed to the `maple` function in MATLAB within single apostrophes corresponding to a MATLAB string. Note that MATLAB uses a semicolon to suppress output, so that the statement

```
>> maple('factor(a^2-b^2);');
```

results in no output.

In a MAPLE session, it is possible to break statements across several lines without any special indication. In MATLAB, a statement that doesn't fit on one line may be continued to the next line, provided that an ellipsis (i.e., ...) is inserted at the line break. Thus, we might write

```
>> 2+3*...
(2-3)

ans =

    -1
```

To break a MATLAB string across several lines requires extra care, since the ellipsis should not be included in the string. Consider the following examples:

```
>> ['tree',...
'stump']

ans =

treestump
```

```
>> maple(['with(linalg):',...
'A:=matrix(3,3,[[1,0,0],[0,0,-1],[0,1,0]])'])

ans =

A := matrix([[1, 0, 0], [0, 0, -1], [0, 1, 0]])
```

where we have enclosed the substrings within square brackets and separated the two substrings with a comma followed by the ellipsis.

MAPLE variable assignments, such as the above definition of the matrix `A`, allocate memory within the MATLAB session and can be accessed in subsequent MAPLE statements. For example,

```
>> maple('eigenvals(A)')

ans =

1, i, -i
```

These variables do not become part of the MATLAB workspace, however.

To allow the inclusion of single apostrophes within the MAPLE statement (as within any MATLAB string), it is necessary to double each apostrophe as in

```
>> 'The dog''s owner'
ans =
The dog's owner

>> maple('A:=''A'';')
ans =
A := 'A'
```

Finally, it is syntactically incorrect to use an upper-case `I` on the left-hand side of an assignment within MAPLE, since `I` represents the imaginary unit and may not be redefined. In contrast, a lower-case `i` or `j` represents the imaginary unit in MAPLE statements within a MATLAB session and must be avoided on the left-hand side of an assignment or within `for` loops and `seq` statements (see below).

## A.2 MAPLE Data Structures

We differentiate between data structures and the actions that can be performed on them. The MAMBO toolbox makes use of four different types of constructs that serve different, and quite intuitive, purposes in the programming paradigm. These are the *set*, the *list*, the *table*, and the *array*.

### Sets

In MAPLE, the set data structure retains the everyday meaning of the terminology. A set is an *unordered* sequence of elements, delimited by curly braces. Thus, the statements

```
> r:={a,b,c}:
> s:={}:
```

assign a set with three elements, `a`, `b`, and `c`, to the variable `r`, and an empty set to `s`. The following statements perform some useful actions on sets:

```
> r[2];
```

$$b$$

```
> r union {f,d};
```

$$\{a, b, f, c, d\}$$

```
> r minus {f,c};
```

$$\{a, b\}$$

```
> member(a,r);
```

$$true$$

The first line selects the second element of `r`, but since `r` is unordered, the answer may change if `r` is assigned the same set again. The second line creates a new set, which is the union of `r` and `{f,d}`, while the set that results from the third line is the difference between `r` and `{f,c}`. Finally, the result of the last line is true if `a` is an element in `r`, and false otherwise.

### Lists

The list data structure differs from the set in that the elements are ordered, and thus selecting the i-th element always returns the same variable. Lists are delimited by square brackets. Except for the set-theoretic statements above, everything else would hold by replacing braces with brackets.

```
> r:=[a,b,c]:
> s:=[]:
> r[2];
```

$$b$$

Instead of the `union` command, we would write

```
> [op(r),f,d];
```

$$[a, b, c, f, d]$$

to create a list that contains the elements of `r` followed by `f` and `d`. Here, the operator `op` extracts the elements of `r`. A related useful command is `nops`, which returns the number of elements.

### Tables

A MAPLE table is essentially a look-up table, such that to each index element there corresponds at most one expression. For example, the statements

```
> r:=table():
> r[a]:=15:
> r[a,b]:=Pi:
```

assign an empty table to `r` after which the correspondences `a`$\leftrightarrow$`15` and `(a,b)`$\leftrightarrow \pi$ are established within `r`. In fact, any expressions are valid as indices. To check whether a table contains a particular index element, one can use the `assigned` command, as in

```
> assigned(r[a]);
```

$$true$$

which returns true since `r[a]` does exist. Similarly, an index together with the corresponding entry may be removed from a table using the `unassign` command, as in

```
> unassign('r[a]');
```

after which the statement

```
> assigned(r[a]);
```

$$false$$

returns false. It should be noted that the statement

```
> r;
```

$$r$$

will return only the name `r`. In order to see the actual table, one could write

```
> print(r);
```

$$\begin{array}{l} \text{table}([ \\ (a,\, b) = \pi \\ ]) \end{array}$$

Similarly, to access the content of a table, we use the `eval` command, as in

```
> subs(Pi=3.14,eval(r));
```

table([
(a, b) = 3.14
])

where every occurrence of $\pi$ in `r` is replaced by `3.14`.

### Arrays

The array construct differs from tables in that the indices are limited to integers. For example, the statement

```
>  r:=array(1..3,1..3):
```

assigns a three-by-three array to `r`. Array entries are accessed by the selection operator, but an error is returned if an index lies outside the prescribed range. Thus, `r[1,2]` is allowed, while `r[0,2]` would return an error. A special type of array is a matrix. This is a two-dimensional array, whose indices always start at 1. MAPLE contains all the standard linear algebra operations on matrices, including determinants, eigenvalues and eigenvectors, and many others. These are accessed by first typing `with(linalg):` at the command prompt. As with tables, one can use the `print` command to see the actual array or the `eval` command to access the actual array.

## A.3 MAPLE Programming

We turn to actual programming in MAPLE. We will describe an implementation of the quicksort algorithm for sorting a list of floats.

**Illustration A.2**

Consider the utility function below that swaps two entries in a list.

```
swap:=proc(v,i,j)
  local local_v,temp;
    local_v:=v;
    temp:=local_v[i]; local_v[i]:=local_v[j]; local_v[j]:=temp;
    RETURN(local_v);
  end;
```

As with tables and arrays, the procedure is assigned to the variable `swap`. In order to view the procedure definition, it is necessary to give the command `print(swap)`. We have chosen to follow MAPLE's internal typographical settings, which display reserved keywords in the MAPLE language in bold-face. The procedure definition begins with the keyword `proc` followed by a, possibly empty, sequence of arguments, separated by commas, within the parentheses. If the arguments are omitted, then they can be accessed through the local variable `args` and their number through the local variable `nargs`.

We remark that procedure calls are "call by value" – the procedure has no access to the variable that was used in a procedure call, only its value. That is the reason for the local variable `local_v` and the statement `local_v:=v`. Note that no semicolon is necessary on the procedure line. Following the procedure head, we declare the local variables that we will use. This list is optional, since any non-global variable that appears within a procedure is assumed to be local. It is also possible to declare global variables at this point. The procedure is exited via a `RETURN` statement. This returns a value to the calling environment that can then be assigned to a variable or as part of an expression. We note that there is no type checking of the inputs in this example (although this is possible, if desired), nor of the output. Such error checking will be omitted in this text.

The following sequence of statements illustrate the function of the procedure `swap`.

```
> names:=["Hanna","Harriet","Haley"];
```

$$names := [\text{"Hanna"}, \text{"Harriet"}, \text{"Haley"}]$$

```
> swap(names,1,3);
```

$$[\text{"Haley"}, \text{"Harriet"}, \text{"Hanna"}]$$

With the help of the `swap` function, we can now show the complete `quicksort` program:

```
quicksort:=proc()
  local v,n,i,temp,last;
    v:=args; n:=nops(v);
    if (n<=1) then
      RETURN(v);
    else
      v:=swap(v,1,trunc((1+n)/2));
      last:=1;
      for i from 2 to n do
        if (v[i]<v[1]) then last:=last+1; v:=swap(v,last,i); fi;
      od;
      v:=swap(v,1,last);
      RETURN([op(quicksort([op(1..(last-1),v)])),
        v[last],op(quicksort([op((last+1)..n,v)]))]);
    fi;
  end;
```

This procedure expects a list of numbers as the argument. It reorganizes the list so that all elements smaller than the middle element (note the `trunc` command) are moved to the left of this element. The procedure then calls itself recursively with the sublists to the left and to the right of the middle element. Sorting ends with lists containing one or fewer elements and control is passed back up to the calling routine. This procedure makes use of some standard programming statements for execution control. These constructs – the `for` loop and the `if` conditional statement – are fairly self-explanatory. Note the `if-fi` and `do-od` pairing to delimit the block. A sample run with `quicksort` would be

```
>  unirand:=rand(1..100):
>  numbers:=[seq(unirand(),i=1..10)];
```

$$numbers := [62, 49, 4, 24, 96, 74, 90, 38, 58, 100]$$

```
>  quicksort(numbers);
```

$$[4, 24, 38, 49, 58, 62, 74, 90, 96, 100]$$

Here, `unirand` is defined as a procedure that returns random integers between 1 and 100. The `seq` statement creates a sequence of 10 integers separated by commas.

In a MATLAB session, MAPLE procedures may be defined at the command prompt as they would within MAPLE. It is also possible to store the procedure definitions (and other MAPLE statements) without the enclosing `maple(' ')` in a separate text file and load this into the MATLAB session using the MATLAB `procread` command. Thus, if the file `quicksort.src` contains the definitions of the `swap` and `quicksort` procedures as shown above, then the statements

```
>> procread('quicksort.src');
>> maple('unirand:=rand(1..100);');
>> maple('numbers:=[seq(unirand(),k=1..10)];')

ans =

numbers := [56, 64, 58, 61, 75, 86, 17, 62, 8, 50]

>> maple('quicksort(numbers);')

ans =

[8, 17, 50, 56, 58, 61, 62, 64, 75, 86]
```

illustrate the use of the `quicksort` routine within a MATLAB session. Note the use of the symbol `k` instead of `i` in the call to `seq` as per the discussion in Section A.1.

# Appendix B

# The MAMBO Toolbox

The MAMBO toolbox contains procedures for establishing a geometry, extracting information about the geometry, and formulating conditions on the geometry. Within a MAPLE session, on-line help with MAMBO toolbox commands is offered through the MAPLE help facility, i.e., by typing a question mark followed by the procedure name at the command prompt, e.g.,

```
>  ?DeclareStates
```

To load the MAMBO toolbox into MAPLE, it is necessary to append the folder containing the MAMBO toolbox source to the MAPLE library path, followed by a `with` command, e.g.

```
>  libname:=libname,"D:\\Mambo":
>  with(Mambo):
Warning, new definition for norm
Warning, new definition for trace
Warning, new definition for fortran
```

"Mambo, version 1.0"

The syntax for specifying a path differs between operating systems and should be found from the MAPLE manual.

To load the MAMBO toolbox into MATLAB, change the current directory to the folder containing the MAMBO toolbox source and use the `procread` command, as described in Appendix A.

## B.1 Establishing a Geometry

A MAMBO toolbox geometry is spanned by a set of observers, a set of points, and a set of triads. Specifically, an observer is defined by its corresponding reference point and reference triad; a pair of points is defined by the position vector from the first point to the second point, and vice versa; and a pair of triads is defined by the rotation matrix between the first triad and the second triad, and vice versa.

If the position vector (rotation matrix) between a pair of points (triads) has been explicitly defined or can be computed from explicitly defined position vectors (rotation matrices) using vector addition (matrix multiplication), the points (triads) are said to be *related.*

An observer hierarchy is established within the MAMBO toolbox by declaring pairs of observers as immediate neighbors within the hierarchy. If two observers have been declared immediate neighbors, they are said to be *directly related.* If two observers belong to the same hierarchy but are not immediate neighbors, they are said to be *indirectly related.*

Three separate steps are involved in establishing a MAMBO toolbox geometry, namely:

- Declaring the observer, point, and triad labels;
- Defining the reference point and reference triads of declared observers, the position vectors between declared points, and the rotation matrices between declared triads;
- Defining the observer hierarchy.

In following these steps, it is entirely acceptable to declare (and define) points and triads that do not correspond to reference points and reference triads of any observers. Such unassociated points and triads may represent geometrical features of objects that are stationary relative to an observer.

### B.1.1 Global Variables

The MAMBO toolbox geometry is represented by three pairs of global variables as listed below.

| | |
|---|---|
| `GlobalObserverDeclarations` | `GlobalObserverDefinitions` |
| `GlobalPointDeclarations` | `GlobalPointDefinitions` |
| `GlobalTriadDeclarations` | `GlobalTriadDefinitions` |

The global variable `GlobalObserverDeclarations` is a MAPLE table with indices given by declared observer labels. If `A` is the label of a declared observer, then the entry `GlobalObserverDeclarations[A]` is a set containing the labels of all directly related observers.

The global variable `GlobalPointDeclarations` is a MAPLE table with indices given by declared point labels. If `A` is the label of a declared point, then the entry `GlobalPointDeclarations[A]` is a set containing the labels of all directly related points.

The global variable `GlobalTriadDeclarations` is a MAPLE table with indices given by declared triad labels. If `a` is the label of a declared triad, then the entry `GlobalTriadDeclarations[a]` is a set containing the labels of all directly related triads.

The global variable `GlobalObserverDefinitions` is a MAPLE table with indices given by declared observer labels. If `A` is the label of a declared observer, then the entry `GlobalObserverDefinitions[A]` is a table with entries given by the corresponding reference point and reference triad.

The global variable `GlobalPointDefinitions` is a MAPLE table with indices given by pairs of declared point labels. If `A` and `B` are the labels of two declared points, then the entry `GlobalPointDefinitions[A,B]` is given by the corresponding position vector from `A` to `B`.

The global variable `GlobalTriadDefinitions` is a MAPLE table with indices given by pairs of declared triad labels. If `a` and `b` are the labels of two declared triads, then the entry `GlobalTriadDefinitions[a,b]` is given by the corresponding rotation matrix between `a` and `b`.

Upon loading the MAMBO toolbox, an empty MAMBO toolbox geometry is established. Subsequent declarations and definitions modify this MAMBO toolbox geometry. The statement

```
Restart();
```

resets the MAMBO toolbox geometry to its initial, empty state. The statement

```
Undo();
```

undoes any changes to the MAMBO toolbox geometry that resulted from the previous procedure call.

### B.1.2 Declaring a Geometry

The procedures `DeclareObservers`, `DeclarePoints`, and `DeclareTriads` can be invoked to declare the labels of an arbitrary sequence of observers, points, and triads, respectively.

```
DeclareObservers(A1,...,An);
DeclarePoints(P1,...,Pn);
DeclareTriads(a1,...,an);
```

Each of the arguments to these procedures **must be an unassigned Maple name**. It is syntactically correct to use the same MAPLE name both as an observer label, a point label, and a triad label. It is not possible to declare a MAPLE name twice for the same object type. These commands are independent of one another and can occur in any order.

### B.1.3 Vectors and Rotation Matrices

The procedure `MakeTranslations` can be invoked with an arbitrary sequence of vector specifications to generate a representation of a composite vector relative to the MAMBO toolbox geometry. When multiple vector specifications are contained within a single procedure call, the parameters governing each separate vector specification are enclosed within brackets. The result of a call to `MakeTranslations` that includes multiple vector specifications is the vector sum of the individual vectors.

Vector specifications can take one of three syntactically correct forms. Specifically:

- `a1,i` – denotes the `i` th basis vector in the triad `a1`;
- `a1,v1,v2,v3` – denotes the vector whose matrix representation in the `a1` triad is given by the algebraic quantities `v1`, `v2`, and `v3`;
- `pv` – denotes a previously computed vector.

All three forms are allowed within a single call to `MakeTranslations`. The statement

```
NullVector();
```

returns the zero vector.

The procedure `MakeRotations` can be invoked with an arbitrary sequence of rotation-matrix specifications to generate a representation of a composite rotation matrix. When multiple rotation-matrix specifications are contained within a single procedure call, the parameters governing each separate rotation-matrix specification are enclosed within brackets. The result of a call to `MakeRotations` that includes multiple rotation-matrix specifications is the matrix product of the individual rotation matrices.

Rotation-matrix specifications can take one of three syntactically correct forms. Specifically:

- `phi1,i` – denotes the rotation matrix corresponding to a rotation by an angle `phi1` about the `i`-th basis vector in the current triad;
- `phi1,v1,v2,v3` – denotes the rotation matrix corresponding to a rotation by an angle `phi1` about the vector whose matrix representation in the current triad is given by the algebraic quantities `v1`, `v2`, and `v3`;

- `rm` – denotes a previously computed orthogonal matrix.

All three forms are allowed within a single call to `MakeRotations`.

### B.1.4 Defining a Geometry

The procedures `DefineObservers`, `DefinePoints`, and `DefineTriads` can be invoked to define an arbitrary sequence of observers, points, and triads, respectively. When defining multiple elements within a single procedure call, the parameters governing each separate definition are enclosed within brackets.

```
DefineObservers(A,P,a);
DefineObservers([A1,P1,a1],...,[An,Pn,an]);
```

Here, the first parameter is the label of a declared observer, the second parameter is the label of a declared point, and the third parameter is the label of a declared triad. An observer may only be defined once.

```
DefinePoints(P,Q,pv);
DefinePoints([P1,Q1,pv1],...,[Pn,Qn,pvn]);
```

Here, the first and second parameters are the labels of two declared points and the third parameter contains a specification of the position vector from the first to the second point. Specifically, the syntax for the third parameter is identical to that for the argument list in a call to the `MakeTranslations` procedure. It is not possible to define the position vector between a pair of already related points.

```
DefineTriads(p,q,rm);
DefineTriads([p1,q1,rm1],...,[pn,qn,rmn]);
```

Here, the first and second parameters are the labels of two declared triads and the third parameter contains a specification of the rotation matrix between the first and the second triad. Specifically, the syntax for the third parameter is identical to that for the argument list in a call to the `MakeRotations` procedure. It is not possible to define the rotation matrix between a pair of already related triads.

### B.1.5 Defining Neighbors

The procedure `DefineNeighbors` can be invoked to define an arbitrary sequence of pairs of observers as directly related. When defining multiple pairs of neighbors within a single procedure call, the parameters governing each separate definition are enclosed within brackets.

```
DefineNeighbors(A,B);
```

```
DefineNeighbors([A1,B1],...,[An,Bn]);
```

Here, the first and second parameters are the labels of two declared observers. It is not possible to define a direct relation between a pair of already related observers.

### B.1.6 Defining Objects

The procedure `DefineObjects` can be invoked to associate an arbitrary sequence of MAMBO objects, such as spheres, cylinders, and blocks, with defined observers. When associating multiple objects within a single procedure call, the parameters governing each separate definition are enclosed within brackets.

```
DefineObjects(A,Type,properties);
DefineObjects([A1,Type1,properties1],...,
              [A2,Type2,properties2]);
```

Here, the first and second parameters are the label of a defined observer and one of the object type keywords `'Sphere'`, `'Cylinder'`, or `'Block'`, including the apostrophes[1]. The specification of the object properties is given by a sequence of optional arguments:

- `point=Point` – where `Point` is the label of a point that is related to the reference point of the parent observer or a position vector describing the position of the reference point of the object relative to the parent observer. In the absence of a `point` specification, the reference point of the object is assumed to coincide with that of the parent observer;

- `orient=Triad` – where `Triad` is the label of a triad that is related to the reference triad of the parent observer or a rotation matrix describing the orientation of the reference triad of the object relative to the parent observer. In the absence of an `orient` specification, the reference triad of the object is assumed to coincide with that of the parent observer;

- `radius=R`, `length=L`, `xlength=XL`, `ylength=YL`, `zlength=ZL` – where `R`, `L`, `XL`, `YL`, and `ZL` are algebraic expressions. Of these, the `radius` specification only applies to spheres and cylinders, the `length` specification only applies to cylinders, and the `xlength`, `ylength`, and `zlength` specifications only apply to blocks. In the absence of any of these specifications, the corresponding dimensions (when appropriate) are assumed to equal 1;

[1] Note the need to double the apostrophes whenever `DefineObjects` is part of a MAPLE statement entered at the command prompt in MATLAB (see Appendix A).

- `color=Color` – where `Color` is the label `white`, `black`, `red`, `green`, `blue`, `yellow`, `magenta`, or `cyan`, or is a MAPLE string of the form "`{r,g,b}`", including the double quotes, where `r`, `g`, and `b` are algebraic expressions. In the absence of a `color` specification, the object color is assumed to be white.

The definition of objects associated with defined observers is reflected in the global variable `GlobalObjectDefinitions`.

### B.1.7 Declaring Coordinates

The procedure `DeclareStates` can be invoked to declare an arbitrary sequence of variables time-dependent.

```
DeclareStates(q1,...,qn);
```

This declaration is reflected in the global variables `GlobalExplicit` and `GlobalImplicit`, which are referenced by any MAMBO toolbox computation involving differentiation with respect to the time variable `t`. This declaration does not imply that the corresponding variables will be declared as MAMBO `states` in an associated MAMBO motion description.

## B.2 Extracting Information

The relative position and orientation of points, triads, and observers within the MAMBO toolbox geometry may be extracted through a number of MAMBO toolbox utilities:

- `FindRotation(a,b)` – returns the rotation matrix between the related triads `a` and `b`;
- `FindOrientation(A,B)` – returns the rotation matrix between the related reference triads of the defined observers `A` and `B`;
- `FindTranslation(P,Q)` – returns the position vector from the point `P` to the related point `Q`;
- `FindPosition(A,B)` – returns the position vector from the reference point of the defined observer `A` to the related reference point of the defined observer `B`;
- `FindCoordinates(A,P)` – returns the coordinate representation of the declared point `P` relative to the defined observer `A`, provided that the reference point of `A` is related to `P`.

The MAMBO toolbox utility GeometryOutput can be invoked to generate a MAMBO geometry description corresponding to a connected component of the MAMBO toolbox geometry based at some declared observer.

```
GeometryOutput(main=observer,filename=string,
          states=statelist,parameters=parlist,
          anims=animlist,time=timevar,
          checkargs,checktree);
```

Here, the main observer is declared to GeometryOutput through the main=observer argument.

The optional argument filename is used to spool the output straight to the file whose path (unless you are saving in the working directory) and filename are specified by the given string string. Furthermore, the optional argument checkargs together with specifications of any MAMBO state variables, parameters, animated variables, and time variable contained in the MAPLE lists statelist, parlist, and/or animlist and the label timevar aborts the creation of a MAMBO geometry description if some algebraic symbols in the geometry description have not been declared. Finally, the optional argument checktree aborts the creation of a MAMBO geometry description if the reference point and/or reference triad of some observer have not been defined. In the absence of the checktree argument, undefined reference points and/or reference triads result in omitted **POINT** and **ORIENT** statements in the geometry description. Similarly, if the reference points and reference triads of an observer have been defined but are not related to the parent reference point and reference triad, respectively, the corresponding **POINT** and/or **ORIENT** statements are empty, resulting in a syntactically incorrect MAMBO geometry description.

Changes in the relative position and orientation of points, triads, and observers within the MAMBO toolbox geometry and the associated linear and angular momenta may be extracted through the LinearVelocity, AngularVelocity, LinearMomentum, and AngularMomentum functions:

- LinearVelocity(A,B) – returns the velocity of the point B relative to the observer A, provided that the point B is related to the reference point of the observer A;

- AngularVelocity(a,b) – returns the angular velocity between the related triads a and b;

- LinearMomentum(A,B) – returns the linear momentum of a rigid body with center of mass at the point B relative to the observer A, provided that the point B is related to the reference point of the observer A;

- `AngularMomentum(A,b)` – returns the angular momentum of a rigid body with reference triad `b` relative to the observer `A`, provided that the triad `b` is related to the reference triad of the observer `A`.

# B.3 Formulating Conditions on a Geometry

## B.3.1 Vector Utilities

The five binary operations `&++`, `&**`, `&--`, `&oo`, and `&xx` implement vector addition, scalar multiplication, vector subtraction, the vector dot product, and the vector cross product. The function `VectorLength` returns the length of a vector as defined by the square root of the vector dot product of the vector with itself.

```
v &++ w;
v &-- w;
v &oo w;
v &xx w;
k &** v;
VectorLength(v);
```

Here, `v` and `w` are two arbitrary vectors and `k` is an algebraic expression. The binary operations all have the same precedence, and parentheses should be employed where necessary to enforce a particular order of evaluation.

The `Express` procedure converts an arbitrary vector `v` to its expression relative to an arbitrary declared triad `a`, provided that all the triads in the description of `v` are related to `a`.

```
Express(v,a);
```

The `Simplify` procedure applies MAPLE's simplify command to the coordinates of an arbitrary vector `v`.

```
Simplify(v);
```

## B.3.2 Constraints

The vector utilities and the MAMBO toolbox query utilities described in previous sections can be invoked to formulate constraints on the MAMBO toolbox geometry. Following the methodology in the main text, all such constraints should be converted to linear constraints in the derivatives of the configuration coordinates.

The function `DiffTime` returns the derivative with respect to time (assumed to be denoted by `t`) of its argument.

```
DiffTime(expr);
DiffTime(v,a);
```

Here, the first form of the function call applies as long as `expr` is an algebraic expression, an equality involving algebraic expressions, or a set, list, table, or array of algebraic expressions. In the second form of the function call, the MAMBO toolbox vector `v` is differentiated with respect to time relative to the declare triad `a`, provided that the triad `a` is related to all the triads in the specification of `v`.

The MAMBO toolbox utility `MotionOutput` exports the kinematic and dynamic differential equations corresponding to the MAMBO toolbox geometry, the motion constraints, and the equations of motion into a MAMBO motion description.

```
MotionOutput(ode=odeset,filename=string,states=statelist,
             parameters=parlist,anims=animlist,
             insignals=insiglist,time=timevar,
             checkargs,checksings);
```

Here, the kinematic and dynamic differential equations are declared to `MotionOutput` through the `ode=odeset` argument, where `odeset` is a MAPLE set of equations. The `MotionOutput` command will fail to generate a motion description if the number of differential equations is different from the number of state variables declared in the MAPLE list `statelist`.

The optional argument `filename` is used to spool the output straight to the file whose path (unless you are saving in the working directory) and filename are specified by the given string `string`. Furthermore, the optional argument `checkargs` together with specifications of any MAMBO state variables, parameters, animated variables, insignals and time variable contained in the MAPLE lists `statelist`, `parlist`, `animlist` and/or `insiglist`, and the label `timevar` aborts the creation of a MAMBO motion description if some algebraic symbols in the motion description have not been declared. Finally, the optional argument `checksings` aborts the creation of a MAMBO motion description if the differential equations are singular for the initial configuration given by the initial values for the state variables and values for the parameters provided in `statelist` and `parlist`.

# Appendix C

# Simulation and Visualization Projects

As indicated in the preface, this text has served as course literature for sophomore-level, senior-level, and beginning graduate-level courses on multibody mechanics and visualization offered at Virginia Polytechnic Institute and State University in Blacksburg, Virginia, USA, and at the Royal Institute of Technology in Stockholm, Sweden. A substantial component of the performance assessment in these courses is based on a team simulation and visualization project, in which the students (in groups of two or three) are challenged to implement the theoretical knowledge imparted to them in the regular lectures to model, simulate, and animate the motion of an actual mechanism.

The following is a list of suggested team projects. Additional projects can be found in the Exercise sections in Chapters 2, 4, and 6.

1. A puppeteer performing for money in a crowded train station with escalators.

2. A couple of trapeze artists performing advanced tricks over a group of acrobats.

3. A six-degree-of-freedom flight simulator at Disney World on which a cricket has landed.

4. A four-degree-of-freedom industrial robot working at a conveyor belt in a plant where the operators move around on unicycles.

5. A rowboat race down the Charles River in Boston.

6. An NHL hockey game with crowds doing the "wave."

7. A mosquito biting a person on the neck.

8. A person demonstrating the functionality of an umbrella stroller.
9. A person rollerblading and opening an umbrella.

There is clearly no unique desirable solution to any of these animation tasks. In all cases, the groups are challenged to make modeling decisions as to the visual complexity of their multibody mechanism, the mathematical complexity of its kinematics, and the physical complexity of any kinematical constraints.

## C.1 Project Presentation

At the conclusion of the semester, the teams present the results of their efforts in oral and written presentations. The evaluation of these presentations is based on the following desirable high-level objectives:

- Content – reasonable modeling, correct mathematics, relevant material, integrating quantitative and qualitative material, clear presentation;
- Form – adequate composition, clear description of basic assumptions, approach and results, logically organized presentation of theory and method;
- Language and style – appropriate use of terminology, high level of readability, correct grammar and spelling.

The following table describes a suggested detailed structure for the final written report.

Introduction
- Overview of challenges in modeling a general problem
  - Choice of configuration coordinates and parameters
  - Formulating constraints
  - Solving constraints
  - Controlling the motion
  - Explaining physical implications of constraints
  - Modeling of dynamics
- Possible applications of skills and knowledge
- Table of contents
- List of figures
- List of tables
- Nomenclature

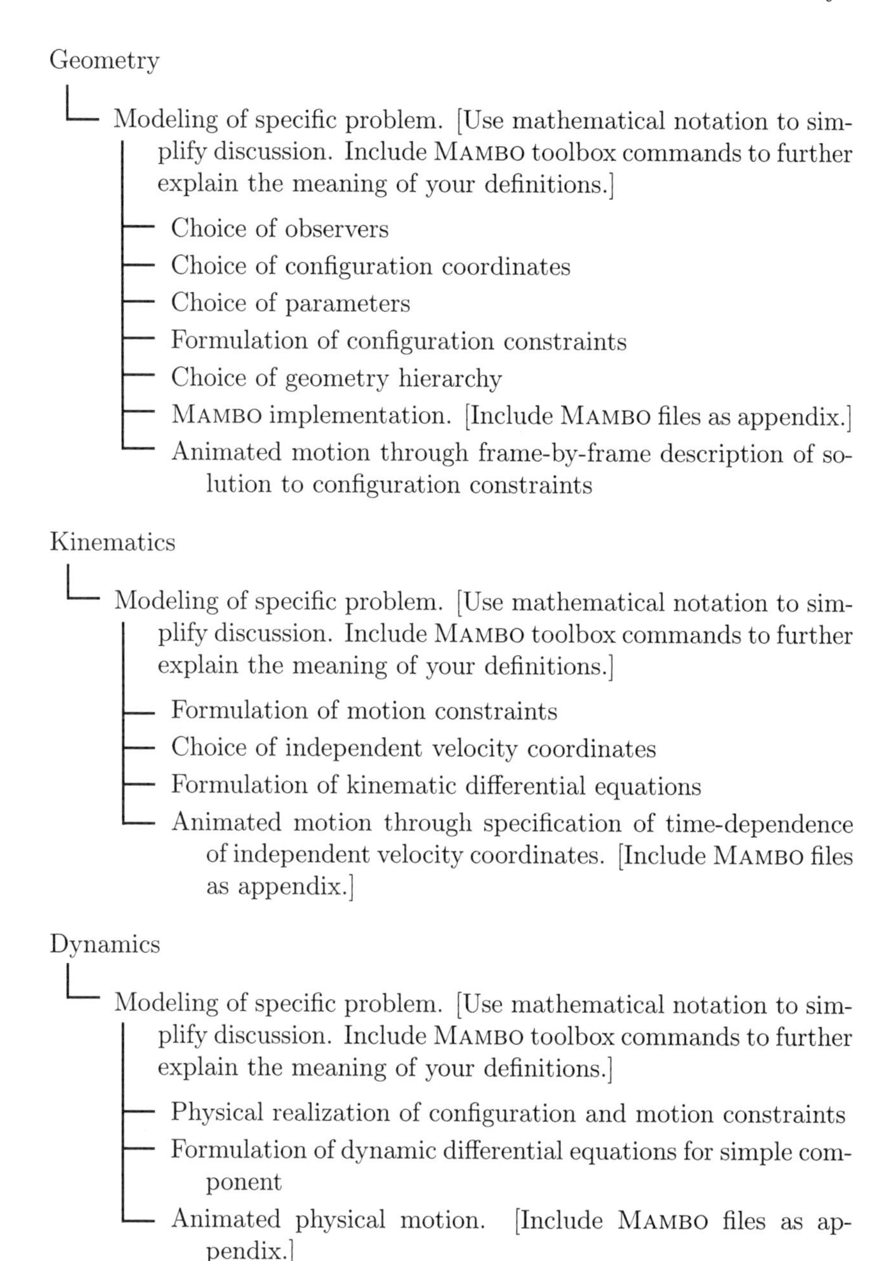
Geometry
Modeling of specific problem. [Use mathematical notation to simplify discussion. Include MAMBO toolbox commands to further explain the meaning of your definitions.]
Choice of observers
Choice of configuration coordinates
Choice of parameters
Formulation of configuration constraints
Choice of geometry hierarchy
MAMBO implementation. [Include MAMBO files as appendix.]
Animated motion through frame-by-frame description of solution to configuration constraints
Kinematics
Modeling of specific problem. [Use mathematical notation to simplify discussion. Include MAMBO toolbox commands to further explain the meaning of your definitions.]
Formulation of motion constraints
Choice of independent velocity coordinates
Formulation of kinematic differential equations
Animated motion through specification of time-dependence of independent velocity coordinates. [Include MAMBO files as appendix.]
Dynamics
Modeling of specific problem. [Use mathematical notation to simplify discussion. Include MAMBO toolbox commands to further explain the meaning of your definitions.]
Physical realization of configuration and motion constraints
Formulation of dynamic differential equations for simple component
Animated physical motion. [Include MAMBO files as appendix.]

# Subject Index